THEORY OF SUPERCONDUCTIVITY

This is Volume 17 in
PURE AND APPLIED PHYSICS

A Series of Monographs and Textbooks

Consulting Editors: H. S. W. MASSEY AND KEITH A. BRUECKNER

A complete list of titles in this series appears at the end of this volume.

THEORY OF SUPERCONDUCTIVITY

John M. Blatt

Applied Mathematics Department
University of New South Wales
Kensington, N.S.W., Australia

1964

ACADEMIC PRESS • New York and London

ACADEMIC PRESS, INC.
111 Fifth Avenue, New York, New York 10003

United Kingdom Edition published by
ACADEMIC PRESS, INC. (LONDON) LTD.
Berkeley Square House, London W1X 6BA

LIBRARY OF CONGRESS CATALOG CARD NUMBER: 63-22333

Second Printing, 1971

PRINTED IN THE UNITED STATES OF AMERICA

Preface

For many decades, superconductivity was the shame and despair of theoretical physics. The phenomenon was discovered by Kamerlingh Onnes in 1911. The quantum theory of normal metals was developed fully by the early 1930's. A successful set of phenomenological equations for superconducting metals was given by F. London in 1935. Yet, in 1950, almost 40 years after the original discovery, there was not even a proper beginning of an adequate microscopic theory of superconductivity!

This abject failure of theory is even worse when considered from our present vantage point. In fact, every single element necessary to a successful theory of superconductivity was known to theorists, in one connection or another, by 1935. The existence of an electron-phonon interaction and its importance in metals emerged from Bloch's work on the theory of electrical resistance. The peculiar condensation of a Bose-Einstein gas had been discovered by Einstein in 1925. The idea that pairs of Fermions can combine to form effective Bosons was well known in chemistry, equally well known in molecular spectroscopy, and had, in 1931, proved of decisive importance in our understanding of the composition of atomic nuclei. Yet, the first published suggestion of a Bose-Einstein condensation of electron pairs as the cause of superconductivity was made in 1946, and then not by a theoretical physicist, but rather by an experimental chemist, R. Ogg!

The state of our theoretical ideas in the year 1950 has been summarized lucidly by F. London in his book "Superfluids," Vol. 1. By accident, this same year also marked the beginning of the modern period in the theory of superconductivity, through Fröhlich's discovery of the role of the electron-phonon interaction. Fröhlich's work proved of decisive importance, in spite of the fact that his detailed theory has not stood the test of time. The ideas which he introduced have been so fruitful that, within a decade of his paper, a broadly satisfactory theory of superconductivity has come into existence.

This book is intended to present and explain the advances of this decade in a connected and readily understandable fashion, and to summarize the present state of our knowledge, as well as ignorance, of the superconducting state. The treatment presupposes a first course in quantum mechanics, and a descriptive course in solid state physics, as exemplified by such books as those of Schiff[1] and Kittel,[2] respectively. Wherever other material is required, it is developed in the text or in Appendices.

In spite of the comparative youth of the microscopic theory of superconductivity, there is already so much of it that an encyclopedic treatment would have more than doubled the length of this book. Any selection is a matter of taste; my principle has been to emphasize the "super"-properties of superconductors, those truly spectacular phenomena which have given the subject its peculiar fascination for so many years; these properties are the well-nigh complete expulsion of magnetic fields from superconducting materials (the Meissner-Ochsenfeld effect), and the persistent currents in superconducting rings. In my opinion, the success or failure of a theory of superconductivity depends on the extent to which these particular properties are understood and explained by the theory, starting from fundamental principles. Thermodynamic properties, especially the nature of the thermodynamic transition between the normal and superconducting states and the exponentially decreasing specific heat at low temperatures, are next in importance. Unfortunately, an adequate treatment of these matters leaves all too little space for the many interesting "normal fluid" properties which are associated with the elementary excitations above the superconducting ground state. In all cases where shortage of space has precluded detailed treatment, summaries are given along with extensive references to the literature. The Bibliography includes the papers of most importance historically, as well as complete references to the recent literature so that the reader interested in a special area can obtain a balanced view.

In accordance with the general theoretical level of this book, I have avoided the use of sophisticated methods wherever possible. The most important such omission is the use of thermodynamic Green's functions. This is a big subject in its own right, and all I can do is to refer the interested reader to the excellent book by Bonch-Bruevich and Tyablikov (Bonch-Bruevich, 62), in particular to Chapter VI of that book.

The manuscript was finished early in 1963, and references to later work are only sporadic. I have tried to maintain a reasonable balance between the effort to make the treatment complete and free from errors, and the effort to clarify the presentation of the subject matter.

The first chapter contains a survey of the situation as of 1950, and states briefly those properties of superconductors which are considered basic, and which any theory of superconductivity must explain. Chapter II

[1] "Quantum Mechanics." McGraw-Hill, New York, 1949.

[2] "Introduction to Solid State Physics." Wiley, New York, 1953.

treats the model of a charged Bose-Einstein gas, the first adequate theoretical model of a superconductor, following the work of Schafroth. The third chapter is devoted to the concept of correlated electron pairs in metals, and the "quasi-chemical equilibrium theory" which is built on that concept. The treatment is qualitative and descriptive; the rather complex mathematical theory is given in Appendix B. In Chapter III, we show that, in spite of the Fermi-Dirac statistics of unpaired electrons, and in spite of the high density of the electron gas in metals, a Bose-Einstein-like condensation of electron pairs is possible, and that this is the basic mechanism producing the superconducting state.

Chapter IV then departs from the historical order, by giving a self-consistent treatment of the pairing wave function for the ground state, and developing the qualitative consequences of the resulting solution. In this discussion, it is not necessary to relax the law of conservation of particles, and thus elementary methods suffice. The, historically prior, work of Bardeen, Cooper, and Schrieffer, and of Bogoliubov, Tolmachov, and Shirkov, is presented in Chapters V and VI. Chapter V starts with a simple introduction to the method of creation and destruction operators which is required for those theories; we then show how the ground state wave functions of the BCS theory and of the Bogoliubov theory are related to each other, and to the pairing wave function of Chapter IV. Chapter VI contains the thermodynamic theory of Bogoliubov, Zubarev, and Tserkovnikov, and the BCS theory of transport properties associated with "Bogolon" excitations. These theories are based on the "reduced Hamiltonian" of BCS; the meaning of this assumption is elucidated in Chapter VII, by means of a comparison with a certain specialization of the quasi-chemical equilibrium theory.

The Meissner-Ochsenfeld magnetic field expulsion, and the persistent electric currents in superconducting rings, form the subject of Chapters VIII and IX, respectively. The theory of the Meissner effect is based directly on the ground state pairing wave function, and is closely analogous to Schafroth's calculation of the Meissner effect of the ideal Bose gas. When we come to persistent currents, however, in Chapter IX, we are at the limit of present theory. Calculations are given only for the ideal Bose gas model, not for the electron pair gas. The Bose gas model accounts nicely for certain properties, in particular for the fact that supercurrents are insensitive to the presence of scattering centers, and for the "flux quantization." But there is no theory for the decay, if any, of these persistent currents. The Bose gas model gives no decay at all, and it is possible that a theory of the lifetime of supercurrents requires an improved understanding of the fundamentals of statistical mechanics, in particular of the complex of questions related to the ergodic problem and the irreversibility of the sense of time.

Chapter X is devoted to other problems, requiring further theoretical and experimental elucidation, such as properties of "hard" superconductors, the mysterious regularities in alloys found by Matthias and collaborators, and the properties of superconducting thin films.

Appendix A is a self-contained introduction to the relevant parts of statistical mechanics. The treatment is unconventional and, I hope, clearer than elsewhere, particularly in the discussion of the basis of ensemble theory, and the relation between statistical mechanics and classical thermodynamics.

Appendix B summarizes the mathematical formulation of the quasi-chemical equilibrium theory, and gives derivations of most of the formulas used in the body of the book.

I am happy to thank Dr. P. W. Anderson and Dr. L. M. Delves for their critical reading of the whole manuscript, and their many valuable suggestions for improvement. The ideas on which this book is based were developed in close collaboration with S. T. Butler and M. R. Schafroth, the prime movers in the conception of the quasi-chemical equilibrium theory, and T. Matsubara, without whom its subsequent mathematical formulation would have been impossible. I have learned much from discussions with F. Bloch, N. Bogoliubov, A. Bohr, M. Buckingham, A. deShalit, H. Fröhlich, M. Girardeau, A. Katz, A. Klein, R. Kubo, R. M. May, J. E. and M. G. Mayer, L. Onsager, W. Pauli, I. Talmi, C. Thompson, G. Wentzel, V. F. Weisskopf, E. P. Wigner, and B. Zumino. They have influenced, directly or indirectly, whatever is good in this book; they are not responsible for its shortcomings.

JOHN M. BLATT

Sydney, Australia
August, 1963

Contents

Chapter VIII

The Meissner Effect

Chapter IX

Persistent Currents

Chapter X

Further Problems

Appendix A

Some Concepts from Statistical Mechanics

Appendix B

Mathematical Formulation of the Quasi-Chemical Equilibrium Theory

I

The Phenomenon of Superconductivity

This chapter contains a brief introduction to the state of our knowledge of the phenomenon of superconductivity in the early 1950's, just before the development of a microscopic theory. The chapter ends with a series of questions, which a microscopic theory of superconductivity should be able to answer.

1. The Thermodynamic Transition

Many metals and alloys undergo a change of state at low temperatures, into a "superconducting state." The transition is an abrupt, *thermodynamic transition* in pure specimens; the transition temperature ranges from below 1°K to about 20°K, in different materials. In the absence of an applied magnetic field, there is no latent heat associated with the transition. Such transitions are often called "second-order," to distinguish them from "first-order" transitions such as the liquid-solid transition (Ehrenfest 33).

Although there is no latent heat, the transition to the superconducting state shows itself clearly in the behavior of the specific heat c_V, which has a discontinuous jump at the transition point. This is illustrated schematically in Fig. 1.1. As far as is known at present (Corak 56, Cochran 60, 61, 62) the change in the specific heat is really sudden; there is no evidence for a logarithmic infinity in the specific heat, of the type which is known to occur in liquid He^4 at the lambda-transition (Buckingham 61). Furthermore, again unlike liquid helium, there is at present no clear evidence for unusual behavior *above* the transition temperature, in superconductors. The metal behaves perfectly normally right down to the transition temperature, giving no indication of anything unusual until, all of a sudden, it changes over into the superconducting state, with a number of truly spectacular properties.

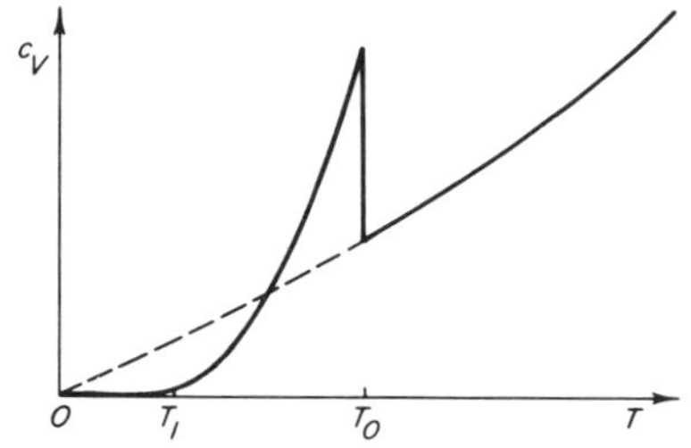

FIG. 1.1. Schematic behavior of the specific heat at constant volume, c_V, as a function of the absolute temperature, T. The specific heat shows a sudden jump at the superconducting transition temperature T_0. Above this temperature, the metal is normal, and its specific heat is given by Eq. (1.3). Below the transition temperature, the metal is superconducting, and the specific heat drops rapidly below the extrapolated normal curve (the dashed curve).

Contrary to general impression, superconductivity is *not* a rare or even infrequent phenomenon. It is observed in many metals and alloys, under widely varying conditions of lattice structure, perfection or otherwise of the lattice, presence of chemical impurities, past history of the specimen, etc. Superconductivity is not a "dirt effect"—it is well-nigh universal, and an even more challenging phenomenon than ordinary metallic conductivity.

From the point of view of the development of a theory of superconductivity (and this is the point of view emphasized throughout this book), the frequent occurrence of superconductivity is a most helpful clue; it excludes, from the start, all sorts of "special" explanations. In particular, since widely different crystal lattices occur among superconductors, the lattice structure cannot be of qualitative importance. Again, impurities and lattice imperfections leave the basic phenomenon of superconductivity unaltered, even though they alter details, such as the transition temperature and the sharpness of the transition. Furthermore, there is every indication that the purer and more perfect a superconducting specimen, the more clearly does it display the basic properties of superconductivity.[1] Thus, again as a starting point, it is permissible to ignore impurities and lattice imperfections altogether, in basic ques-

[1] For example, purification of the sample makes the transition "clean" in tantalum (Budnick 60, Milne 61) and in lead (Franck 61).

tions concerning superconductivity. There is only one exception to this rule: When discussing the difference between electrical conduction in normal conductors and superconductors, impurities and/or imperfections of some kind *must* be introduced into the treatment, since in the absence of them, even the ordinary metallic conductivity would be infinite (Peierls 55).

An important property of superconductors follows directly from inspection of Fig. 1.1. Imagine that we heat a specimen through a small temperature interval dT, at constant volume. Then the increase in entropy S is given by

$$dS = \frac{dQ}{T} = \frac{c_V \, dT}{T} \tag{1.1}$$

where dQ is the heat added, and c_V the specific heat. Integrating from $T = 0$ up to some finite temperature T and using the third law of thermodynamics (in the form $S = 0$ at the absolute zero of temperature), we get the following expression for the entropy S of the specimen:

$$S = \int_0^T \frac{c_V(T')}{T'} \, dT' \tag{1.2}$$

where the integration is to be carried out at constant volume V of the specimen.

Now consider a temperature T_1 well below the transition temperature T_0, as indicated schematically in Fig. 1.1. Also indicated in this figure is the extrapolated specific heat curve for the "normal" state of the metal (the dashed curve). It is clear that the integral (1.2), for $T = T_1$, is much smaller for the superconductor than it would be for a normal metal.

In a normal metal, the low temperature behavior of the specific heat is given by

$$c_V = aT + bT^3 \tag{1.3}$$

The bT^3 term is primarily the specific heat associated with the Debye vibrations of the lattice; at very low temperatures, the linear term aT dominates; this term arises from the kinetic energy of the heat motions of the electron gas. This latter term contributes, according to (1.2), an amount aT to the entropy of the specimen. We have just seen that this term, by itself, already greatly exceeds the total entropy of the metal in the superconducting state.

In statistical mechanics, entropy is identified with the "degree of disorder" of the system. A low entropy indicates low disorder, that is, a highly ordered state. We can therefore conclude, from thermodynamics alone, not only that the superconducting state is more ordered in some sense than the normal state of the metal (that is obvious anyway, since it is the low-temperature preferred state), but also that the new ordering must involve, in some direct fashion, the motion of the "free electrons" in the metal.

2. The Meissner-Ochsenfeld Effect

This thermodynamic conclusion is reinforced strongly by the spectacular electromagnetic properties of superconductors. The first of these is the behavior in a magnetic field.

When a superconductor is inserted between the poles of a magnet, the magnetic field lines do not penetrate the specimen, but rather go around the specimen. The exclusion of the magnetic field from the specimen is nearly perfect; there is a narrow penetration layer, of the order of 10^{-5} to 10^{-6} cm, near the surface.

This phenomenon, though quite spectacular in itself, is *not* the famous Meissner-Ochsenfeld effect (Meissner 33). Rather, Meissner and Ochsenfeld placed a single crystal of tin into a magnetic field at room temperature (so that the field lines penetrated the tin), and *then* cooled the specimen without moving it. When the superconducting transition temperature was reached, they observed that the magnetic field strength outside the metal changed abruptly, consistent with *the specimen expelling the field when cooled through the superconducting transition point.*

Whereas the refusal of a magnetic field to penetrate into a superconductor could have been interpreted in terms of generation of eddy currents on the surface of a specimen of infinite electrical conductivity (and was in fact so interpreted before the Meissner-Ochsenfeld experiment), the expulsion of a magnetic field from a specimen already containing that field allows no such interpretation. The magnetic field expulsion is not a mere consequence of an infinite electrical conductivity, but is a magnetic phenomenon in its own right, a well-nigh *perfect diamagnetism.* Speaking thermodynamically, the state finally reached in *both* experiments (cool first, apply field afterward; and: apply field first, cool afterward) must be the true thermodynamic equilibrium state of the specimen, not merely a long-lived metastable state.

London (35, 37, 50) was the first to draw attention to the fundamental implications of the Meissner-Ochsenfeld experiment, and to develop a set of descriptive, phenomenological equations for superconductors suggested by this experiment. The first observation is that superconductivity is a magnetic property every bit as much as an electrical (conduction) property.

The usual normal magnetic specimen is described by the equation

$$\mathbf{B} = \mu\mathbf{H} \tag{2.1}$$

which relates the magnetic flux density, **B**, to the magnetic field intensity **H**. The quantity μ is a "material constant," which depends on the specimen and its detailed properties (temperature, pressure, etc.). From the fundamental point of view, **B** is a simpler quantity than **H**, since **B** is the average of the microscopic magnetic field throughout a small volume of the specimen (Becker 33). Furthermore, the magnetization density **M** is related to **B** and **H** through

$$\mathbf{H} = \mathbf{B} - 4\pi\mathbf{M} \tag{2.2}$$

and **M** is a direct indication of the response of the medium to the applied field ($\mathbf{M} = 0$ in a vacuum). The relation corresponding to (2.1) is usually written as

$$\mathbf{M} = \chi\mathbf{H} \tag{2.3}$$

where χ is called the "magnetic susceptibility" and is related to μ of (2.1) by [from (2.2)]

$$\mu = 1 + 4\pi\chi \tag{2.4}$$

Since **H** is best considered a derived quantity, neither (2.1) nor (2.3) are really desirable forms for a basic equation of state. We shall write the basic magnetic equation in the form

$$\mathbf{M} = K\mathbf{B} \tag{2.5}$$

where, clearly,

$$K = \frac{\chi}{\mu} = \frac{\mu - 1}{4\pi\mu} = \frac{\chi}{1 + 4\pi\chi} \tag{2.6}$$

Equation (2.5), with K a material constant, provides a good description of the static magnetic properties of most normal materials. However, (2.5) fails badly in some important materials, of which the best-known are the ferromagnets. In ferromagnets, the magnetization **M** is by no means directly proportional to the applied field.

London (35) pointed out that superconductors, as a result of the Meissner-Ochsenfeld experiment, are a second class of materials in which (2.5) fails completely. There are strong and important magnetic effects in superconductors, but they cannot be described by (2.5), no matter what value we use for the "material constant" K. Rather, the equation itself must be changed. London wrote his equation in terms of a current density of "supercurrent," $\mathbf{j}_s$, as follows:

$$\text{curl}\,(\Lambda \mathbf{j}_s) = -\mathbf{B}/c \tag{2.7}$$

where Λ is a new "material constant," and c is the velocity of light.

If the current density $\mathbf{j}_s$ arises from a spatially non-uniform magnetization density $\mathbf{M}(r)$, then $\mathbf{j}_s$ is given by the relation (Becker 33)

$$\mathbf{j}_s = c\,\text{curl}\,\mathbf{M} \tag{2.8}$$

Substitution of (2.8) into (2.7) yields the London equation in a form directly comparable to (2.5), namely,

$$\text{curl}\,(\Lambda\,\text{curl}\,\mathbf{M}) = -\mathbf{B}/c^2 \tag{2.9}$$

or, in the usual case of constant Λ,

$$\text{curl}\,\text{curl}\,\mathbf{M} = -\mathbf{B}/\Lambda c^2 \tag{2.10}$$

Comparison of (2.5) and (2.10) shows a fundamental change in the magnetic response of the material, a change which alters the constitutive equation, not merely a coefficient in an equation.

Before going on, let us first show just how (2.10) leads to a strong field expulsion. Rather than considering the general case, let us take a simple example, namely an infinite slab of superconducting material, filling the region between $x = -a$ and $x = a$. Suppose a magnetic field B_0, parallel to the z-axis, is maintained on both sides of the slab; let us compute $B(x)$ in the interior of the slab.

Since the vectors $\mathbf{M}$ and $\mathbf{B}$ have z-components only, Eq. (2.10) assumes the form

$$\frac{d^2M}{dx^2} = +\frac{B(x)}{\Lambda c^2} \tag{2.11}$$

This must be combined with the Maxwell equation

$$c\,\text{curl}\,\mathbf{H} = 4\pi\mathbf{j}_{\text{ex}} + \frac{\partial \mathbf{D}}{\partial t} \tag{2.12}$$

where $\mathbf{H}$ is given by (2.2), $\mathbf{j}_{\text{ex}}$ is the current density of an externally imposed current [i.e., does *not* include the contribution $\mathbf{j}_s$, Eq. (2.8)], and $\mathbf{D}$ is the electric displacement vector. Under the static conditions envisaged here, both terms on the right-hand side of (2.12) vanish, and we get

$$dH/dx = 0 \tag{2.13}$$

so that

$$H(x) = B(x) - 4\pi M(x) = \text{constant} = B_0 \tag{2.14}$$

Here the value of the constant is determined from the boundary condition on the two sides of the slab. We solve (2.14) for $M(x)$ in terms of $B(x)$, and substitute into (2.11) to obtain

$$\frac{d^2B}{dx^2} = \frac{4\pi B}{\Lambda c^2} \tag{2.15}$$

The denominator on the right-hand side of (2.15) has the dimension of a squared length; we introduce the "London penetration depth" λ by the definition

$$\lambda = c(\Lambda/4\pi)^{1/2} \tag{2.16}$$

The solution of (2.15) with the boundary conditions $B(a) = B(-a) = B_0$ is unique and is given by

$$B(x) = B_0 \frac{\cosh(x/\lambda)}{\cosh(a/\lambda)} \tag{2.17}$$

Since the London depth λ is of the order of 10^{-5} to 10^{-6} cm in most superconductors, the penetration law (2.17) amounts to an almost complete expulsion of the magnetic field from the specimen. The value of the field at the center of the slab, $B(0)$, is exponentially small and utterly negligible for slabs of ordinary size. It is this exponential dependence, rather than the mere smallness of the penetrating field, which is characteristic for the Meissner effect.

Let us compare the penetration law (2.17) for superconductors, with the field penetration into an ordinary magnetic material. If we had used Eq. (2.5) instead of (2.10), the same derivation would have led to the result

$$B(x) = \frac{B_0}{1 - 4\pi K} \quad \text{for} \quad -a < x < a \tag{2.18}$$

In a normal, diamagnetic material, K is negative, and thus $B(x)$ inside the specimen is less than the applied field B_0. To that extent, ordinary diamagnets also expel a magnetic field. But the dependence of $B(x)$ on the depth of penetration, x, is completely different for (2.17) and (2.18). *Superconductivity is not merely a particularly strong diamagnetism, but rather diamagnetism of a quite new kind.*

The magnetization density $M(x)$ in the superconducting slab is given by (2.14) and (2.17) as

$$M(x) = -\frac{B_0 - B(x)}{4\pi} = -\frac{B_0}{4\pi}\left[1 - \frac{\cosh(x/\lambda)}{\cosh(a/\lambda)}\right] \tag{2.19}$$

The distribution of "supercurrent" density j_s is then obtained from (2.8) to be

$$j_s(x) = -\frac{cB_0}{4\pi\lambda}\frac{\sinh(x/\lambda)}{\cosh(a/\lambda)} \tag{2.20}$$

where j_s is a vector in the y-direction, of the above magnitude. These quantities are shown schematically in Fig. 1.2.

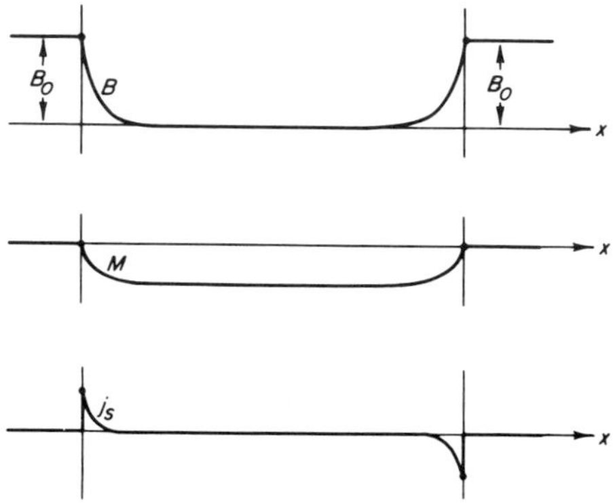

FIG. 1.2. Schematic drawing of the Meissner-Ochsenfeld field expulsion from a uniform slab. (a) The magnetic flux density B is equal to B_0 on the two sides of the slab, but drops exponentially within the slab itself; (b) The density of magnetization M vanishes at the faces of the slab, and is largest in the interior; (c) The magnetization is produced by a supercurrent j_s, which flows only in the London penetration layer near the faces of the slab; the current has opposite sign at the two faces, so that there is no net transport of charge in this situation.

We see that the supercurrent density is strongest near the surface of the slab (at $x = a$ and at $x = -a$), and this strong supercurrent shields the inside of the slab from the applied magnetic field. Although the current density $j_s(x)$ is strongly concentrated near the surface, it is not a mathematical surface current; the latter would be represented by a delta-function term in $j_s(x)$, proportional to $\delta(x - a)$ or $\delta(x + a)$. Mathematical surface currents are unphysical abstractions and cannot be expected to occur in actual physical materials. In terms of the magnetization density $\mathbf{M}$, a mathematical surface current is related to a discontinuity in the parallel component $M_{\parallel}$ of the vector $\mathbf{M}(r)$, on the two sides of the surface:

$$J_{\text{surface}} = c\,\Delta M_{\parallel} \tag{2.21}$$

To exclude such unphysical currents, we must impose the boundary condition

$$M_{\parallel} = \text{continuous on the surface} \tag{2.22}$$

which reduces, for a specimen exposed to vacuum ($\mathbf{M} = 0$) on all sides, to the boundary condition:

$$\mathbf{M}(r) = \text{normal to surface of specimen, at the boundary} \tag{2.23}$$

The magnetization distribution (2.19) obeys this condition, since the parallel (and only) component of the magnetization vector, $M(x) = M_z(x)$ vanishes at $x = a$ and at $x = -a$.

Actually, the penetration law for the normal diamagnetic material, (2.18), violates condition (2.23), and is to that extent unphysical. The reason is quite straightforward: The little Ampere "circular" currents responsible for ordinary diamagnetism are induced currents within each atom, in a gaseous material; in an ordinary metal, the circular currents have diameters of the order of the de Broglie wavelength of an electron at the Fermi surface of the metal. In either case, the ring currents extend over some finite, though small, region, and this finite extension is neglected in the simple point relationship (2.5). The magnetization density $\mathbf{M}(r)$ at a point r depends on the magnetic field $\mathbf{B}(r')$ at neighboring points r', not merely on the magnetic field $\mathbf{B}(r)$ at the same point r. That is, we expect a relationship of type

$$\mathbf{M}(r) = \int K(r, r')\, \mathbf{B}(r')\, d^3r' \tag{2.24}$$

rather than the simple point-relation (2.5). Replacing (2.24) by (2.5) amounts to ignoring the size of the little ring currents responsible for the diamagnetism, and this neglect leads to the appearance of mathematical surface currents. This is of no practical importance in ordinary magnetic materials, but is worth noting as a matter of principle.

Although the magnetization density (2.19) in the slab gives rise to an effective current density (2.20), this does not mean that the slab as a whole carries a net current under these conditions. Rather $j_s(x)$, Eq. (2.20), is an odd function of x, so that the total current vanishes:

$$I = \int_{-a}^{a} j_s(x)\, dx = 0 \tag{2.25}$$

The shielding current near one surface of the slab cancels the shielding current near the opposite surface.

This cancellation is not merely accidental to this particular geometry, but is a general property of "magnetic" currents, i.e., of currents expressible in the form (2.8) (Schafroth 56). *A current density due to a distribution of magnetization always leads to zero net charge transport.* The proof of this theorem is simple: Let S be a cross section of the material (say, of a wire), and let C be the closed circumference of S (the curve in which S intersects the surface of the wire): let $\mathbf{n}$ be a unit normal vector to the surface element of area dS. If the current density $\mathbf{j}$ is given by (2.8), the net current through S is

$$I = \int_S \mathbf{j} \cdot \mathbf{n}\, dS = c \int_S \operatorname{curl} \mathbf{M} \cdot \mathbf{n}\, dS = c \oint_C \mathbf{M} \cdot \mathbf{ds} = 0 \tag{2.26}$$

where the last integrand vanishes because of the boundary condition (2.23).

This theorem is important because it shows that charge transport in superconductors (which will be discussed later on) cannot be reduced immediately to magnetic properties such as the Meissner effect. Charge transport and the Meissner effect are separate phenomena which must not be confused.

The exponential field expulsion property has been proved here only for the special case of the infinite slab. A general proof for arbitrary geometry is given in von Laue's book (Laue 52), in which there are also many explicit solutions of the London equation for various geometries. These are sometimes of considerable practical usefulness (Cioffi 62; Hildebrandt 62, 62a).

3. Further Phenomenology of the Meissner Effect

Although the London equation (2.7) or its equivalent (2.10), was a very good guess and contains the main elements of a descriptive theory of the Meissner effect, it has been found that the newer, more accurate experiments are not in full agreement with (2.10), but require a somewhat modified description. This section is devoted to developing a more general phenomenological framework, to prepare the way for the fundamental theory of the subsequent chapters; we shall pay especial attention to the task of isolating those aspects of the theory which are essential, as opposed to mere modifications of detail.

Since the basic equations (2.10) and (2.12) are *linear*, it is permissible to make a Fourier decomposition of these equations, and this turns out to be rather useful. Let us make the decomposition first and discuss the results afterward.

In a slab occupying the symmetrical region $-a < x < +a$, and with symmetrical boundary conditions ($B = B_0$ on both sides), the natural Fourier expansion of the magnetization is in terms of a cosine series:

$$M(x) = \sum_{n=1}^{\infty} M_n\, u_n(x) \tag{3.1}$$

where

$$u_n(x) = a^{-1/2} \cos\left[(n - \tfrac{1}{2})\pi x/a\right] \tag{3.2}$$

The component $M(x) = M_z(x)$ is parallel to the surface of the slab; thus, according to (2.23), $M(x)$ must vanish at $x = a$ and $x = -a$; the argument of the cosine functions has been chosen so that this is the case, separately for each function $u_n(x)$.

For a volume of arbitrary shape, we need a basic set of vector fields $\mathbf{u}_n(r)$; since the operator *curl curl* enters the London equation (2.10), it is convenient to make the vector functions $\mathbf{u}_n(r)$ eigenfunctions of this operator; they should also satisfy the boundary condition (2.23). Thus, the generalization of (3.1) and (3.2) for an arbitrary volume is

$$\mathbf{M}(r) = \sum_{n=1}^{\infty} M_n \mathbf{u}_n(r) \tag{3.3}$$

with

$$\text{curl curl}\ \mathbf{u}_n = q_n^2 \mathbf{u}_n \tag{3.4a}$$

$$\text{div}\ \mathbf{u}_n = 0 \tag{3.4b}$$

$$\mathbf{u}_n = \text{normal to surface of } V\text{, at the boundary} \tag{3.4c}$$

$$\int (\mathbf{u}_n)^2\, dV = 1 \tag{3.4d}$$

Conditions (3.4a) and (3.4c) are essential; (3.4d) is merely a convenient normalization condition of no special significance. If we take the divergence of both sides of (3.4a), and divide both sides by q_n^2, condition (3.4b) follows immediately. Thus, *if* q_n^2 is non-zero, (3.4b) is not really a separate condition. Conversely, however, (3.4b) together with (3.4c) can be shown to restrict the eigenvalue equation (3.4a) to positive, non-zero eigenvalues q_n^2. This motivates the use of (3.4b) as a condition on the basic vector functions $\mathbf{u}_n$.

For the special case of the slab, the functions (3.2), interpreted as z-components of vector functions $\mathbf{u}_n(r) = [0, 0, u_n(x)]$, are a subset of the complete set $\mathbf{u}_n(r)$ defined by (3.4); by symmetry, they are the only functions $\mathbf{u}_n$ of interest in that problem. The eigenvalues q_n in (3.41) are given by

$$q_n = (n - \tfrac{1}{2})\pi/a \tag{3.5}$$

This relation would of course be different in a volume of different shape.

Since the set of functions $\mathbf{u}_n$ is a complete set for vector fields $\mathbf{B}$ with $\operatorname{div} \mathbf{B} = 0$ (Schafroth 60), we can also expand the flux density vector $\mathbf{B}(r)$ in terms of the functions $\mathbf{u}_n$:

$$\mathbf{B}(r) = \sum_{n=1}^{\infty} B_n \mathbf{u}_n(r) \tag{3.6}$$

For the special case of the slab, this expansion assumes the form

$$B(x) = \sum_{n=1}^{\infty} B_n u_n(x) \tag{3.7}$$

with $u_n(x)$ given by (3.2).

The expansion (3.6) or (3.7) differs from (3.1) or (3.3) in one respect: the parallel component of $\mathbf{B}$, unlike the parallel component of $\mathbf{M}$, need not vanish at the surface of the superconductor, and frequently does not in fact vanish [for example, see (2.17) and (2.19)]. This does not prevent us from expanding $B(x)$ in a Fourier series of functions $u_n(x)$, all of which vanish at the surface; but it does mean that the expansion (3.7) is less rapidly convergent than the expansion (3.1).

Let us now write down the London equation (2.10) in terms of these expansions. Using (3.4a) and assuming that we can differentiate the series (3.3) term by term, we obtain

$$M_n = K(q_n)\, B_n \tag{3.8}$$

where

$$K(q) = -\frac{1}{\Lambda c^2 q^2} \qquad \text{(London kernel)} \tag{3.9}$$

It is apparent that the "kernel" $K(q)$ is just the Fourier transform of the general kernel $K(\mathbf{r}, \mathbf{r}')$ which appears in (2.24); equally, $K(q)$ is a natural generalization of the material constant K which appears in the basic magnetic equation (2.5) of a normal magnetic material.

The London kernel $K(q)$, Eq. (3.9), differs from the kernel for a normal magnetic material, i.e., from $K(q) = K =$ constant, both for large q and for small q. Large q means large wave numbers, and hence small distances. Thus the behavior of $K(q)$ for large q determines the field penetration law in the immediate neighborhood of the surface of the specimen. There, the constant $K(q) = K$ leads to a physically unreasonable true surface current, as we have remarked before [see the discussion of Eq. (2.24)], and should really be modified appropriately. It turns out that a falloff of $K(q)$ for large q, proportional to $1/q^2$, is sufficiently rapid to ensure physically reasonable magnetizations; thus a normal material is better described by a kernel of type

$$K(q) = \frac{K}{1 + (\mu q)^2} \qquad \text{(normal material)} \tag{3.10}$$

in which μ, with the dimension of a length, is of the order of magnitude of the charged particle orbits responsible for the magnetic behavior, and K is the material constant which enters into the approximate equation (2.5), and is related to the susceptibility χ by (2.6).

With this correction to the "normal" kernel, the normal and superconducting kernels (3.10) and (3.9) do not differ qualitatively for large q. The crucial qualitative difference occurs for very *small* q where (3.9) becomes infinite whereas (3.10) stays finite. Small q means a small wave number, hence a long wavelength. The behavior of $K(q)$ for small q governs the field penetration law for *large* distances x into the specimen. The pole of the London kernel $K(q)$ at $q = 0$ is correlated directly with the exponential penetration law (2.17) for large x. Thus, the *exponential field expulsion characteristic of the London-type Meissner effect corresponds to a kernel function* $K(q)$ *with a* $1/q^2$ *singularity at* $q = 0$.

We are now in a position to distinguish essential modifications of the magnetic behavior from minor modifications. The essential aspect of $K(q)$ is its behavior at $q = 0$; this is what distinguishes a superconductor of London type from all other materials. Modifications to $K(q)$ at intermediate values of q have been proposed on experimental grounds by Pippard (53, 53a). By studying the sharpness of the transition in zero field, the size of the "surface energy" (see Section 5), and the variation of the field penetration at microwave frequency with impurity content of the superconducting specimen (Faber 55), Pippard was led to the conclusion that there exists a characteristic length in superconductors in addition to the London penetration depth λ, Eq. (2.16). He thought of this length in terms of a correlated wave function of some kind, and named it the "coherence length." This coherence length ξ depends on impurity content much more strongly than the London depth λ; ξ decreases as we add more impurities. In very pure materials, ξ is of order 10^{-4} cm, some 10 to 100 times as large as the London depth λ. From the physical point of view, the existence of such a characteristic length is of major importance, and any theory of superconductivity must not only allow an understanding of what this coherence length means from the microscopic point of view, but also of its surprisingly large order of magnitude.

However, from the point of view of the descriptive theory of the

Meissner effect, it turns out that the Pippard modification of the kernel function $K(q)$ is only of minor importance. Pippard's kernel can be written in the form (Schafroth 56, 60)

$$K(q) = -\frac{1}{\Lambda c^2 q^2} \frac{\arctan(\xi q)}{\xi_0 q} \qquad \text{(Pippard kernel)} \tag{3.11}$$

where ξ is the Pippard coherence length in the actual material, and ξ_0 is the limit of ξ as the impurity content of the material is reduced to zero. Although (3.11) differs from the London kernel (3.9) for intermediate values of q, the behavior for very small q is unaltered, except for a constant factor ξ/ξ_0 which amounts merely to a redefinition of the material constant Λ. Thus, the Pippard kernel leads to a field penetration law which differs considerably from (2.17) for small x, i.e., in the neighborhood of the surface,[1a] but which has substantially the same exponential falloff for large x as the London penetration law (2.17). As we shall see in Chapter IV, the microscopic theory of superconductivity does lead to a reasonable explanation for the meaning of the coherence length ξ.

There is a general relationship between the behavior of a system of charges in a constant magnetic field, and of the same system in a rotating frame of reference; this general relationship is embodied in Larmor's theorem. It turns out that there exists an analogous relationship for superfluids (Schafroth 55); we may define an "equilibrium superfluid" to be a substance which, when placed in a rotating container, yields less than the classically expected moment of inertia, even after full thermodynamic equilibrium has been obtained. That is, part or all of the superfluid fails to rotate along with the container, no matter how long we wait, and this lack of rotation is the true equilibrium state of such a fluid. For our present purpose, we may define a "perfect superconductor" to be a magnetic substance with a kernel $K(q)$ which has a $1/q^2$ singularity at $q = 0$, such as (3.9) or (3.11) for example.[1b] Thus a "perfect superconductor" is a substance with a perfect (exponential) field expulsion for large distances x into the specimen. Schafroth (55) then proved the connection theorem: If a system of identical spinless particles is an equilibrium superfluid, then the same system, when the particles are given a test charge ϵ, is also a perfect superconductor;

[1a] See Peter (58) for a theoretical calculation of the field penetration, in connection with an experiment by Schawlow (58a).

[1b] The definition of a "perfect superconductor" in Schafroth (55) is phrased differently, but these two definitions can be shown to be equivalent.

and, conversely, any perfect superconductor consisting of identical spinless particles is an equilibrium superfluid. Although the restriction to spinless particles appears awkward, it can be relaxed easily if spin-orbit coupling can be ignored. The relation is then one between the *orbital* part of the magnetic response, and the moment of inertia under rotation.

Thus, to the extent that the electron spins in metals are unimportant for superconductivity (this is a very reasonable assumption), and to the extent that a London-type field expulsion is really present, the gas of "superelectrons" must also show a non-classical moment of inertia under rotation. This behavior was surmised already by London (50), by more qualitative reasoning. Measurements on the gyromagnetic effect (Doll 58) do not contradict this view.

Both perfect superconductivity and equilibrium superfluidity are strictly quantum mechanical effects which cannot occur in purely classical systems (Blatt 55). They are related to correlated wave functions, with effective de Broglie wavelengths of the same order of magnitude as the size of the containing vessel. London in particular stressed the fact that superconductivity and superfluidity are macroscopic manifestations of quantum effects, and his contention can be supported by a more formal argument.[2]

It is not at all easy to accept that correlated quantum mechanical wave functions can exist over "miles of dirty lead wire," to use Casimir's expressive phrase. Attempts have been made to develop alternative phenomenological equations for superconductors (Schafroth 56), equations consistent with a finite range of correlation. The characteristic feature of these alternative equations is a kernel $K(q)$ with a singularity weaker than $1/q^2$, in fact no stronger than $1/q$. However, the modern experimental work is not consistent with such a limited field expulsion (Sturge 58). Thus, reluctant as we may well be to accept quantum correlations over arbitrarily large distances, it seems difficult indeed to escape this conclusion, and it becomes one of the tasks of a microscopic theory of superconductivity to make such quantum correlations understandable.

---

[2] This argument, contained in Blatt (55), has been criticized by Prange (61, 61a), who shows that additional assumptions are required for the proof, and that these additional assumptions may not hold if the system has an energy gap. We shall discuss the energy gap argument for superconductivity in more detail, in Chapter VIII. The rotating superconductor is also discussed by Grin (61), and a general theory of "off-diagonal long-range order" is given by Yang (62).

4. The Critical Field

Although the Meissner-Ochsenfeld effect is described quite accurately by *linear* relations between the magnetization and the magnetic field, this linearity does not persist with arbitrarily high fields. The most striking non-linear effect is the existence of a *critical field* for superconductivity. Suppose a superconducting specimen, in the shape of a cylinder, is exposed to a constant magnetic field $B = H_0$ parallel to its surface, everywhere on its surface. (We choose this particular geometry to avoid having to consider the "demagnetization factor" by which an arbitrary superconducting specimen would alter the field external to itself.) For small values of H_0, the field distribution inside the superconductor is as expected from the Meissner effect, and can be computed from knowledge of the kernel $K(q)$ in (3.8), together with the geometry. But at a certain critical field strength H_0, the situation changes abruptly: the specimen loses the property of superconductivity altogether, there is a thermodynamic transition back to the normal state, with a latent heat of transition [i.e., a first-order transition in the sense of Ehrenfest (33)]. Heat must be supplied to the specimen to enable it to make the transition from the superconducting to the normal state, at constant temperature and constant applied magnetic field, just as heat must be supplied to ice to make it transform into water at the melting point.

Although this phenomenon appears at first sight quite spectacular, the explanation is straightforward thermodynamically (Gorter 33, 34; Rutgers 33, 36; Keesom 24). It is just a matter of working out the free energy balance, taking account of the Meissner effect. In the superconducting state of the material, the magnetic field is expelled, whereas the magnetic field penetrates the material in its normal state. The critical field H_c is reached when the difference in field energy between these two situations just balances the difference in intrinsic free energy between the normal and superconducting states. The only non-trivial thing here is the *sign* of the field energy contribution: energy must be supplied (work must be done) to expel the field from the specimen. The negative sign of this contribution can be seen physically most simply by imagining the specimen, in its superconducting state, in a field-free region; we then transport the specimen, through the fringing field of a magnet, into a final region where there is a magnetic field H_0 everywhere along its surface, parallel to the surface. In the fringing field region, the field is stronger closer to the magnet; the induced magnetization $\mathbf{M}$ is opposite to the direction of the fringing field (diamagnetism!); thus there is a net force on the specimen, pushing it away from the magnet. This means mechanical work is done *on* the specimen in order to place it into the magnetic field region, i.e., to produce the final state in which the flux density **B** is expelled from the specimen. By contrast, if we ignore the very slight magnetism of the metal in its normal state, no such mechanical work need be done on the normal specimen. This accounts for the sign of the field energy contribution to the thermodynamic (free) energy balance.

More formally, the argument proceeds as follows (Schafroth 60). The magnetic work stored per cm³ of material is given by

$$F_m = (4\pi)^{-1} \int_0^B H \, dB \tag{4.1}$$

The form $H\,dB$, rather than $B\,dH$, is deduced from Poynting's equation in electromagnetic theory (Abraham 49). This is the magnetic work under isothermal conditions, and is therefore the magnetic contribution to the free energy per unit volume. The total free energy per cm³ is then

$$F(B, T) = F_0(T) + (4\pi)^{-1} \int_0^B H \, dB \tag{4.2}$$

If the temperature T and the magnetic induction vector B are given, then $F(B, T)$ is the appropriate free energy to use. However, in practice (for example, in the case of the cylinder with given parallel magnetic field H_0 on its surface) the given quantity is H, not B. B adjusts itself, inside the specimen, in accordance with the Meissner effect, whereas $H = B - 4\pi M$ is constant inside the cylindrical specimen. We therefore introduce a thermodynamic potential $G(H, T)$ appropriate to H rather than B as independent variable. $G(H, T)$ is given by

$$G(H, T) = F - BH/4\pi = F_0(T) - (4\pi)^{-1} \int_0^H B(H) \, dH \tag{4.3}$$

The minus sign is apparent in this form.

If we ignore the London field penetration layer altogether, the superconducting magnetic equation of state is simply $B(H) = 0$, giving (the subscript "s" stands for superconducting state):

$$G_s(H, T) = F_0(T), \qquad H \leqslant H_c \tag{4.4}$$

However, as soon as H reaches the critical field value H_c, the Meissner field expulsion ceases, and is replaced by the ordinary magnetic behavior of the normal metal; this normal susceptibility is so small that we may ignore it in this discussion; the normal magnetic equation of state is thus

$$B = H \qquad \text{for} \qquad H > H_c \tag{4.5}$$

leading to

$$G_n(H, T) = F_0(T) - (8\pi)^{-1}(H^2 - H_c^2)\,, \qquad H > H_c \tag{4.6}$$

$G(H, T)$ is one continuous function, with the distinct analytic forms (4.4) and (4.6) for the two regions of H.

The entropy density S is

$$S = -\left(\frac{\partial G}{\partial T}\right)_H = -F_0'(T), \qquad H < H_c \tag{4.7a}$$

$$= -F_0'(T) - (4\pi)^{-1}H_c\,\frac{dH_c}{dT}\,, \qquad H > H_c \tag{4.7b}$$

The latent heat of transition per unit volume is then given by

$$Q = T(S_n - S_s) = -\frac{T}{4\pi}H_c\,\frac{dH_c}{dT} \tag{4.8}$$

Since it takes a positive latent heat to transform the superconducting specimen at $H = H_c$ into a normal specimen at the same H, Eq. (4.8) implies that $dH_c/dT < 0$, i.e., the critical field decreases with increasing temperature. Equation (4.8) checks quantitatively with observation, confirming thereby that thermodynamics is applicable to the problem, i.e., the Meissner field expulsion is truly reversible.

The thermodynamic transition discussed in Section 1 was in the absence of an external magnetic field H. This is the limiting case of (4.8) for $H_c = 0$, $T = T_0$; the latent heat Q vanishes under those conditions, i.e., the transition becomes second-order. At that point, there exists a simple relation for the jump in the specific heat $C = -T(\partial^2 G/\partial T^2)_H$, first derived by Rutgers (33):

$$(C_s - C_n)_{T=T_c} = \frac{T_c}{4\pi}\left(\frac{dH_c}{dT}\right)^2 \tag{4.9}$$

As expected from this equation, the limiting value of the specific heat

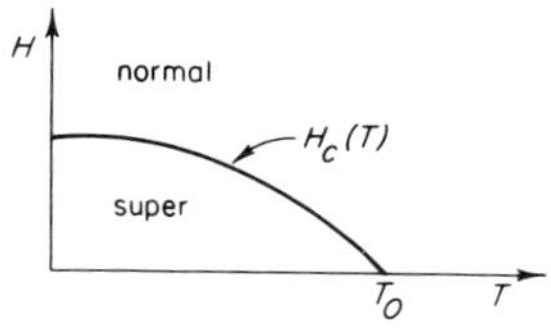

FIG. 1.3. Phase diagram of a superconductor in the H-T plane. The critical field curve, $H = H_c(T)$, divides the phase plane into a normal and a superconducting region. The thermodynamic transition is second order (no latent heat) in zero field, but is first order everywhere else along the critical field curve. $H_c(T)$ approaches zero at $T = T_0$ with a finite slope, given by Eq. (4.9). At other points, the slope dH_c/dT is related to the latent heat through Eq. (4.8). The slope is zero at $T = 0$.

from below (in the superconducting region) exceeds the normal specific heat just above the zero-field transition temperature T_c. Equation (4.9) also checks quantitatively. The critical field curve $H_c = H_c(T)$ divides the H-T plane into a superconducting and a normal region, as shown in Fig. 1.3.

5. Surface Energy and Ginzburg-Landau Theory

So far, we have considered only bulk superconductors. However, if a superconducting specimen is subjected to a strong magnetic field (H of order H_c) under more general conditions than the ones envisaged so far (uniform parallel field applied to the surface of a cylinder), theoretical arguments (Landau 38, 43) lead one to expect breakup into a domain structure, with alternating normal and superconducting regions. This is called the "intermediate state" of superconductors (London 50). Landau's theoretical work stimulated a set of experiments (Shalnikov 45; Meshkovsky 47) in which this domain structure was actually observed. Landau's original theory envisaged the domains becoming infinitely fine near the surface of the specimen; thus the original experiment was designed to detect a domain structure in the deep interior of the specimen, and did so. However, the experimenters also looked at the surface of the specimen (Meshkovsky 47a) and found an easily observable domain structure there. In his second paper (Landau 43), Landau had anticipated this possibility, through the introduction of a positive *surface energy* associated with the creation of an interface between a normal and a superconducting region, in the presence of a magnetic field. The measurements on the intermediate state allow one to determine the order of magnitude of this surface energy term (Makei 58).

The applied magnetic field penetrates into the normal region, and penetrates partially (to distances of the order of a London depth) into the superconducting regions. This partial penetration gives rise to a surface energy contribution, of course. If we retain the *linear* relation between magnetization M and flux density B postulated in the London equation (2.10), or in its generalization (3.8), then the surface free energy can be deduced directly from (4.3) and the field penetration law. The crucial point is the *sign* of this surface free energy: it is negative, thereby favoring the creation of such surfaces.

Let us consider what happens as we decrease the applied field from

$H > H_c$ to lower values. It turns out (Abrikosov 52, 57, 57a) that the lowest free energy is produced by a network of thin "filaments" or "flux lines" imbedded in the bulk superconducting specimen. These flux filaments are spaced apart from each other by less than a London depth; hence effectively the magnetic field still penetrates the whole specimen, and no sharp Meissner effect is observed at $H = H_c$. Rather, there is an entire region of fields, $H_{c1} < H < H_{c2}$, over which the field expulsion gradually takes place, with the flux filaments becoming less important as the field decreases. Below the "lower critical field" H_{c1}, the entire material is superconducting, and the Meissner effect is perfect.[3]

This sort of behavior is not observed in so-called "soft" superconductors (tin, lead, mercury, etc.); therefore, in these materials, at least, the surface energy must be positive.[4]

A positive surface energy cannot be explained on a purely linear theory. On the other hand, the experiments on the field penetration law are quite consistent with purely linear behavior in low fields, and give only small non-linearities even quite close to the critical field (Pippard 50, Sharvin 51, Richards 62). The penetration depth varies by only a few percent all the way up to $H = H_c$ at the surface.

Thus, the non-linear theory must change the sign of the surface energy term, and yet give corrections of only a few percent for the field penetration law. The Ginzburg-Landau theory (Ginzburg 45, 50, 50a, 52, 52b, 53, 55, 56, 56a, 57, 58, 58a, 59) not only achieved this astounding feat, but it actually *predicted* the smallness in the non-linearities in the Meissner effect, *before* the experiment was performed.

Ginzburg and Landau start from a two-fluid model of superconductivity, in which there are "normal electrons" and "superelectrons" at any temperature $T < T_c$. In a bulk superconductor, the density n_s of superelectrons is temperature-dependent but spatially constant. In the neighborhood of a "domain boundary" between a superconducting and a normal region, however, n_s must be a function of position, which vanishes once we are well inside the "normal" region. It is then a

---

[3] The "lower critical field" is not obtained from a formal, thermodynamic approach (Schafroth 60), since such an approach permits the flux filaments to become infinitely thin. However, if we define a length L characteristic of the surface energy σ by $\sigma = LH_c^2/8\pi$, then flux filaments of thickness much less than $|L|$ have a decreased surface energy, and are not favored. Related difficulties of the London theory have been pointed out by Beck (55).

[4] See Chapter X for a discussion of "hard" superconductors. The Abrikosov flux lines have been observed experimentally (Kinsel 62), and so have the upper and lower critical fields (Stromberg 62).

reasonable guess to set n_s equal to the square of a "wave function for superelectrons,"

$$n_s = | \psi_s(x) |^2 \tag{5.1}$$

The assumption (5.1) *means physically that all the "superelectrons" are described by one and the same, coherent, wave function* $\psi_s(x)$.

If n_s, and hence ψ_s, varies in space, there is a kinetic energy associated with such a variation; if in addition there is a magnetic field, also space-dependent, described by a vector potential $\mathbf{A}(x)$, this kinetic energy is presumably given by

$$E_{\text{kin}} = \frac{1}{2m^*} \left(\psi_s, \left(-i\hbar\nabla - \frac{e^*}{c}\mathbf{A}(x)\right)^2 \psi_s\right) \tag{5.2}$$

where m^* is the "effective mass of a superelectron" and e^* is its effective charge.

In the *absence* of a magnetic field, and for small n_s (i.e., close to the transition temperature T_c), we may expand the free energy F_{s0} of the superconducting state in a power series in n_s and in $T_c - T$; the leading terms are

$$F_{s0} = F_{n0} - \kappa(T_c - T)\, n_s + \tfrac{1}{2}\beta n_s^2 + \ldots \tag{5.3}$$

where, to a first approximation, κ and β are temperature-independent positive constants, and where F_{n0} is the free energy density in the normal state, in the absence of a magnetic field.

The total free energy density in the transition region between superconducting and normal domains is the sum of three terms: (i) the zero-field term (5.3); (ii) the magnetic field energy density $H^2/8\pi$; and (iii) the kinetic energy term for the "superelectrons," (5.2):

$$F_{sH} = F_{s0} + H^2/8\pi + E_{\text{kin}} \tag{5.4}$$

From here on, everything is routine: the behavior of the wave function $\psi_s(x)$ and of the vector potential $\mathbf{A}(x)$ are determined by the condition that the free energy should be a minimum. This leads to a set of self-consistent equations similar to the Hartree-Fock equation in ordinary atomic physics (Schiff 49), and non-linear for the same reason. Although (5.4) is clearly gauge-invariant, it is convenient to employ a particular gauge here, defined by

$$\text{div}\, \mathbf{A}_1 = 0 \tag{5.5}$$

and

$$\left(-i\hbar\nabla\psi_s - \frac{e^*}{c}\mathbf{A}_1(x)\,\psi_s\right)_\perp = 0 \quad \text{on the surface of the specimen} \tag{5.6}$$

This gauge is called the "London gauge" since it was first introduced by F. London, and used extensively by him. It has the advantage that a physical current density **j**, which must satisfy the conditions

$$\operatorname{div} \mathbf{j} = 0 \tag{5.7}$$

and

$$j_\perp = 0 \qquad \text{on the surface of the specimen} \tag{5.8}$$

could conceivably turn out to be proportional to $\mathbf{A}_1$. In fact, the London equation (2.7) can be written in the form

$$\operatorname{curl} (\Lambda \mathbf{j}_s) = - c^{-1} \operatorname{curl} \mathbf{A} \tag{5.9}$$

If we fix the gauge of **A** uniquely as the London gauge, we can "cancel the curls" in (5.9) to get the alternative form

$$\mathbf{j}_s = - \frac{1}{\Lambda c} \mathbf{A}_1(x) \tag{5.10}$$

Returning now to the Ginzburg-Landau theory, variation of (5.4) with respect to **A** and ψ_s^* yields the self-consistent equations

$$(2m^*)^{-1} (- i\hbar\nabla - e^*\mathbf{A}_1/c)^2 \psi_s + \frac{\partial F_{s0}}{\partial \psi_s^*} = 0 \tag{5.11}$$

[where of course ψ_s^* enters F_{s0}, Eq. (5.3), through the relation (5.1)], and

$$\nabla^2 \mathbf{A}_1 = - 4\pi \mathbf{j}_s/c = \frac{2\pi i e^* \hbar}{m^* c} (\psi_s^* \nabla \psi_s - \psi_s \nabla \psi_s^*) + \frac{4\pi (e^*)^2}{m^* c^2} | \psi_s |^2 \mathbf{A}_1 \tag{5.12}$$

Equations (5.5), (5.6), (5.11), and (5.12) are a coupled set of non-linear equations for $\psi_s(x)$ and $\mathbf{A}_1(x)$, which require numerical solution in the general case.

For weak fields, however, analytic solution is possible and yields the London theory! In this limit, we may ignore $\nabla \psi_s$ altogether; only the second term of (5.11) contributes and this gives just the ordinary thermal equilibrium in the absence of any field; in (5.12), also, the second term on the right is the only contributor, and this yields

$$\mathbf{j}_s = - \frac{(e^*)^2}{m^* c} | \psi_s |^2 \mathbf{A}_1 = - \frac{(e^*)^2 n_s}{m^* c} \mathbf{A}_1 \tag{5.13}$$

Comparison with (5.10) shows that this is just the London equation, provided that the density n_s of superelectrons is constant in space, as it would be in the weak-field situation. By assuming that m^* and e^* are of the same order of magnitude as the mass and charge of an ordinary electron, and that n_s is of the order of the observed number of conduction electrons per cm^3, we obtain an estimate for the London constant Λ in (2.7), and hence of the London penetration depth λ, Eq. (2.16). These order-of-magnitude estimates are in good agreement with experiment. The temperature dependence [near $T = T_c$, which is the only region in which the expansion (5.3) is valid] is also reasonable: ignoring the first term of (5.11) yields the relation

$$n_s = |\psi_s|^2 \cong \frac{\kappa}{\beta}(T_c - T) \tag{5.14}$$

which means in turn that the London depth λ deduced from this theory increases near $T = T_c$ proportionately to $(T_c - T)^{-1/2}$, becoming infinite for $T = T_c$.

This weak-field solution provides one relation between the (initially arbitrary) parameters κ, β, m^*, and e^*, in terms of an experimentally measurable quantity, $\lambda = \lambda(T)$ near $T = T_c$. A second relationship is provided by the observed critical field $H_c = H_c(T)$ in that same region. Writing the theory in non-dimensional form eliminates one more parameter, so that finally there is only one free parameter remaining. If this free parameter is taken to be the effective charge e^*, and is set equal to one electronic charge, there are no adjustable parameters left at all. The theory is therefore surprisingly definite in its predictions, and the observed qualitative and even semiquantitative agreement with experiment is a considerable success. The predictions are two: (1) The value of the surface energy, which now indeed turns out to have the right sign (positive) as well as the right order of magnitude, and (2) The coefficient γ in the non-linear correction to the London penetration depth:

$$\lambda(H) = \lambda(0)\,[1 + \gamma(H/H_c)^2 + \ldots] \tag{5.15}$$

γ was predicted by Ginzburg and Landau to be of the order of a few percent, in complete agreement with the later measurements (Pippard 50, Sharvin 51). The agreement with experiment is improved slightly by assuming $e^* = 2e$, i.e., a double electronic charge for the "super-electrons."

The modifications to the *linear* field penetration law suggested by Pippard (53, 53a) and discussed in Section 3, do not follow from the Landau-Ginzburg theory which, as we have just seen, leads to the unmodified London kernel (3.9). However, it turns out that the Pippard modifications are not needed, experimentally, in the temperature region

$T \sim T_c$ to which the Landau-Ginzburg theory applies. In that region, the London kernel (3.9) is by itself in good agreement with experiment.

In view of the great success of the Ginzburg-Landau theory, and its very suggestive character, any microscopic theory of superconductivity should allow us to understand the initial assumptions of the Ginzburg-Landau theory from first principles; and we shall see that this is indeed the case.

At this stage, however, it may help to emphasize just what the crucial physical assumption of the theory is: it is the relation (5.1), which assumes that *all* the "superelectrons" can be described by *one and the same* quantum-mechanical wave function $\psi_s(x)$. Such an assumption would be completely wrong for the "normal" electrons in the metal; these electrons occupy a Fermi-Dirac distribution, and obey the Pauli exclusion principle; thus, far from all electrons being described by one and the same wave function, *no two* "normal" electrons can have the same wave function! At first sight, then, the basic assumption of the Ginzburg-Landau theory appears to be in direct conflict with the Pauli exclusion principle, and this situation must be clarified by the microscopic theory, before any more detailed discussion makes sense.

6. Persistent Ring Currents

So far, we have been concerned primarily with the superconducting specimen in thermodynamic equilibrium, perhaps subjected to an applied magnetic field. This, however, is not how superconductivity was discovered by Kammerlingh Onnes (Onnes 11). Onnes observed a sudden decrease of the electrical resistance to unmeasurably small values, and accordingly coined the name "superconductivity." The most spectacular manifestation of vanishing electrical resistance is provided by *persistent currents induced in superconducting rings.*[4a]

The experiment proceeds in three steps, as follows:

(1) A wire of lead or tin, in the form of a closed ring, is placed between the poles of a magnet, at room temperature. Some of the flux lines pass through the hole in the ring, others through the material of the ring, the remainder pass around the outside of the ring.

(2) The system is cooled down below the superconducting transition

[4a] For a nice account of the early experiments, with many references, see Serin (56).

temperature T_c. As a result of the Meissner effect (which was not known to Onnes, however) the flux lines now avoid the material of the ring itself. But a certain amount of flux passes through the hole in the ring.

(3) The ring is now pulled out from the magnet, without allowing it to heat up. The flux through the hole in the ring is trapped (*initially*, this would also be true for a normally conducting ring); this flux is maintained by a current around the ring, which is induced by pulling the ring out of the magnet.

The striking thing about superconducting rings is, not that such a ring current is generated by this process, but rather that *this ring current persists indefinitely*. Ring currents have been kept going for hours, for days, for months, and for years! Not only has no ring current died down, but the experiments give no indication of any decrease at all in the strength of the current, as measured by the magnetic field in the neighborhood of the ring.

"Persistent" though these ring currents certainly are, there is no doubt that they represent metastable states, not the thermodynamic equilibrium state of the ring. When the ring carries a current, the magnetic field created by this current gives rise to a field energy. The energy of the system is easily seen to exceed the energy of the quiescent ring, with no current or field.[5]

London (35) augmented his magnetic response equation (2.7) by a second phenomenological equation, so as to include this phenomenon of persistent currents. The second London equation, unlike (2.7), involves the time. It reads

$$\frac{\partial}{\partial t}(\Lambda \mathbf{j}_s) = \mathbf{E} \tag{6.1}$$

where Λ is the same material constant which appears in (2.7) and $\mathbf{E}$ is the electric field.

This equation is the replacement, in the London phenomenology, of Ohm's law for normal conductors; in its differential form (Abraham 49) Ohm's law reads

$$\mathbf{j}_n = \sigma \mathbf{E} \tag{6.2}$$

where $\mathbf{j}_n$ is the current density of a "normal" current and σ is the

[5] Strictly speaking, we should look at free energies $F = E - TS$, not energies E. However, at low temperatures T, F nearly equals E, and the term $-TS$ makes no qualitative difference in this discussion. The free energy F rather than the G, Eq. (4.3), is relevant here since there is no externally imposed field H in this problem.

electrical conductivity of the material. Actually, the London phenomenology envisages both types of current, normal and supercurrent, co-existing in the same material; thus the total density of electric current is given by

$$\mathbf{j} = \mathbf{j}_n + \mathbf{j}_s \tag{6.3}$$

The main difference between (6.1) and its "normal" counterpart, (6.2), is the appearance of the time derivative of the supercurrent in (6.1). A vanishing electric field, $\mathbf{E} = 0$, leads to vanishing normal current $\mathbf{j}_n$; it leads to an unchanging, but *not* necessarily vanishing, supercurrent $\mathbf{j}_s$. Qualitatively speaking, this is sufficient to allow for infinitely persistent currents in superconducting rings: the normal current contribution (6.2) dies down quickly, and with it the electric field around the ring; but the supercurrent contribution maintains its value indefinitely if $\mathbf{E}$ remains zero thereafter. Equation (6.1) permits no further change in $\mathbf{j}_s$ under those conditions.

More formally, the London theory gives rise to a special conservation law, the "conservation of fluxoid." Let C be a closed curve, drawn within the superconducting material of the ring; let S be the surface spanned by C; this surface does not lie entirely within the superconducting material, rather it covers the hole in the ring. The ordinary magnetic flux through the surface S is given by

$$\Phi = \int_S \mathbf{B} \cdot \mathbf{n} \, dS \tag{6.4}$$

the "fluxoid" of London is defined by

$$\Psi_C = \Phi + c\Lambda \oint_C \mathbf{j}_s \cdot d\mathbf{s} \tag{6.5}$$

If the superconducting wire is of ordinary laboratory dimensions, its thickness is much larger than the London depth λ, Eq. (2.16). We can then choose the curve C to lie well inside the wire; as a result of the Meissner effect, the supercurrent density $\mathbf{j}_s$ is concentrated within a layer of thickness λ near the surface of the wire. The line integral in (6.5) is then nearly zero, so that the fluxoid can be interpreted physically as the total flux through the hole in the ring, including the flux through the London penetration layer within the material of the wire.

We now prove (London 50) that Ψ_C is conserved, and is independent of the details of the curve C, depending only on the *topology* of this curve. We start from Faraday's law of electromagnetic induction:

$$\frac{d\Phi}{dt} + c \oint \mathbf{E} \cdot d\mathbf{s} = 0 \tag{6.6}$$

Using Eq. (6.1) to replace **E** in (6.6), we immediately obtain the conservation law for the fluxoid:

$$d\Psi_C/dt = 0 \tag{6.7}$$

It remains to prove that Ψ_C depends only on the topology of the curve C. We start by proving that Ψ_C is independent of the choice of the mathematical surface S spanned by C. Let S' be a second surface spanned by C; then

$$\Psi_C - \Psi_{C'} = \int_S \mathbf{B} \cdot \mathbf{n}\, dS - \int_{S'} \mathbf{B} \cdot \mathbf{n}\, dS = \int_{V'} \operatorname{div} \mathbf{B}\, dV = 0 \tag{6.8}$$

where V' is the volume enclosed between the surfaces S and S', both of which have the common boundary curve C. Next, we consider another curve C', deformed in continuous fashion from the original curve C, in such a way that during the entire process of deformation the curve remains within the superconducting volume. This determines the topology of the curve. For example, if C follows the wire once around, so does C'; if C winds twice around the wire before closing in on itself, so does C', etc. Let S' be a surface between the curves C and C', S' remaining entirely within the superconducting volume (this latter condition is necessary in order to make the London equations applicable everywhere on S'). Then the difference of the two fluxoids is, from (6.5),

$$\Psi_C - \Psi_{C'} = \int_{S'} \mathbf{B} \cdot \mathbf{n}\, dS + c\Lambda \oint_{C-C'} \mathbf{j}_s \cdot \mathbf{ds} \tag{6.9}$$

where $C - C'$ is the curve spanning S', consisting of C traversed forward, and C' traversed backward. Since S', by assumption, lies completely within the superconducting volume, we may use the first London Eq. (2.7) solved for B, to obtain

$$\Psi_C - \Psi_{C'} = -c\Lambda \int_{S'} (\operatorname{curl} \mathbf{j}_s) \cdot \mathbf{n}\, dS + c\Lambda \oint_{C-C'} \mathbf{j}_s \cdot \mathbf{ds} = 0 \tag{6.10}$$

which vanishes by Stokes' theorem. Thus, we have established that the value of the fluxoid Ψ_C not only is independent of time, but also depends only on the topology of the curve C.

If the curve C can be shrunk continuously to a point, without going outside the superconducting volume, the fluxoid clearly vanishes identically, since it vanishes for the deformed curve (the point). Thus, non-trivial values of the fluxoid are possible only in multiply connected superconductors, e.g., superconducting rings.

A particularly neat experimental proof of this fluxoid conservation law has been given by Rose-Innes (61a), who shows that fluxoid conservation makes it impossible to use a system of circulating supercurrents as a perfect gyroscope.

To obtain a better insight into the physical assumptions underlying these phenomenological equations, let us consider in some detail the relationship between the first London equation, in the form (2.7), and the second London equation, (6.1). Equation (2.7) does not contain the time; if we assume that (2.7) holds not only in full thermodynamic equilibrium, but also holds quite generally, at all times t, in time-dependent situations, then we may differentiate both sides of (2.7) with respect to time, and use the Maxwell equations to replace $\partial \mathbf{B}/\partial t$ by the curl of the electric field vector $\mathbf{E}$. This gives the result

$$\text{curl}\left(\frac{\partial}{\partial t}(\Lambda \mathbf{j}_s) - \mathbf{E}\right) = 0 \tag{6.11}$$

Furthermore, the divergence of this same expression also vanishes in the superconducting volume: div ($\mathbf{E}$) is proportional to the electric charge density in the metal, which always vanishes (Abraham 49); div ($\mathbf{j}$) vanishes for the same reason; we also assume here that there is no conversion of "normal" to "superelectrons," so that div ($\mathbf{j}_n$) and div ($\mathbf{j}_s$) vanish separately.[6]

Since both the curl and the divergence of the bracketed expression in (6.11) vanish, it is tempting to set this expression equal to zero, identically; this would of course yield the second London equation, (6.1). However, this procedure is not strictly valid; if the superconducting volume is multiply connected (for example, a ring), there exist non-zero vector fields $\mathbf{v}_0(r)$ satisfying the conditions

$$\text{curl } \mathbf{v}_0 = 0 \tag{6.12a}$$

$$\text{div } \mathbf{v}_0 = 0 \tag{6.12b}$$

$$(\mathbf{v}_0)_\perp = 0 \quad \text{on the surface of } V \tag{6.12c}$$

Physically, these represent "streamline flow" of an incompressible fluid around the ring. If the volume V is doubly connected, there exists only *one* such field; we may normalize the vector function $\mathbf{v}_0(r)$ conveniently by the requirement

$$\int_V (\mathbf{v}_0)^2 \, dV = 1 \tag{6.12d}$$

[6] This assumption excludes consideration of phenomena analogous to "second sound" in liquid helium. These are unlikely to be observable, for a number of reasons (Bardeen 61, Thouless 61, Velibekov 62).

The proper mathematical deduction from the vanishing of both the curl and the divergence of the bracket in (6.11) is therefore

$$\frac{\partial}{\partial t}(\Lambda \mathbf{j}_s) - \mathbf{E} = k(t)\mathbf{v}_0(r) \tag{6.13}$$

where $k(t)$ is an arbitrary function of time. The function $\mathbf{v}_0(r)$ is defined by (6.12); it vanishes in a simply connected volume, is uniquely defined by (6.12) in a doubly connected volume; in a triply connected volume (two rings joined together) $\mathbf{v}_0$ is a sum of two linearly independent vector functions, etc.

Thus, the second London equation (6.1) "almost" follows from the first London equation (2.7); the additional step required is to set the arbitrary function $k(t)$ in (6.13) equal to zero.

Although this appears a very minor additional step, when put this way, it is actually a major assumption in its own right. The very existence of persistent ring currents depends entirely on this one term, proportional to $\mathbf{v}_0(r)$: In Section 2, we represented the supercurrent density $\mathbf{j}_s$ as the curl of a magnetization vector, Eq. (2.8). This representation of $\mathbf{j}_s$ fails if there is actual charge transport, as was shown at the end of Section 2 [see Eq. (2.26) and the discussion there]. However, Schafroth (56, 60) has shown that the general case can be represented as follows:

$$\mathbf{j}_s = c \text{ curl } \mathbf{M} + j_0(t)\mathbf{v}_0(r) \tag{6.14}$$

where $\mathbf{M}$ satisfies the boundary condition (2.23), and $\mathbf{v}_0$ has just been defined by (6.12). The decomposition (6.14) of the vector field $\mathbf{j}_s$ is always possible and unique, provided only that div $(\mathbf{j}_s) = 0$ and $\mathbf{j}_s$ is parallel to the surface at the boundary. It is clear that the coefficient $j_0(t)$ in (6.14) is not determined by (6.13) unless some assumption is made about the arbitrary function $k(t)$ in (6.13). However, the current I in the superconducting ring is determined entirely by $j_0(t)$, since the term c curl $\mathbf{M}$ gives rise to $I = 0$, according to (2.26)!

Thus, the "almost" derivation of the second London equation from the first one just misses deriving the really significant term. In spite of the suggestiveness of the argument, the second London equation is not only logically independent of the first one, but the new step required represents a far-reaching new physical assumption. What, then, is the physical meaning of this new assumption?

In a normal material, it is impossible to argue from the diamagnetic ring currents within atoms, to charge transporting electric currents through the specimen; the diamagnetic ring currents are *small* ring currents, and cannot lead to net charge transport through a complete cross section of a wire: the ring current always cuts such a cross section at two points, and for every bit of charge transported one way, an equal amount is transported the other way. This is the physical meaning of the vanishing of the net current in (2.26). However, in a superconductor obeying the London equations, the "diamagnetic" ring currents are not of this type at all; rather, they are huge rings, of size comparable to the entire superconductor. Let us start with a solid superconductor of cylindrical shape, to which we apply a magnetic field parallel to the surface. The Meissner effect means that shielding currents $\mathbf{j}_s$ are built up in the London penetration layer of thickness λ, and these currents shield the magnetic field to such an extent that there is substantially no field $\mathbf{B}$, and no current $\mathbf{j}_s$, in the interior of the cylinder. Physically, $\mathbf{j}_s$ represents a current of "superelectrons" running in rings around the cylinder, immediately adjacent to the cylindrical surface. The radius of the ring current is substantially the same as the radius of the cylinder.

Let us now bore a cylindrical hole into our superconducting cylinder, thereby making it into a cylindrical shell, of inner radius a and outer radius R, such that the thickness $R - a$ is much larger than the London penetration depth λ and other lengths characteristic of superconductivity (in particular, the Pippard coherence length ξ). According to the physical picture which we have just developed, the existence of this hole should make very little difference: at the position of the hole, the field $\mathbf{B}$ and the current density $\mathbf{j}_s$ would be very close to zero, anyway, even for the solid specimen. Once this is accepted, however, the supercurrent $\mathbf{j}_s$ in the cylindrical shell *does* lead to a net charge-transporting current I around the ring, quite obviously.

From this point of view, it is apparent that the supercurrent density $\mathbf{j}_s$, not the magnetization vector $\mathbf{M}$, must be considered as the directly given physical quantity, in the London theory. The above argument would have lost all plausibility if we had looked at $\mathbf{M}$ rather than at $\mathbf{j}_s$: unlike $\mathbf{j}_s$, $\mathbf{M}$ is by no means zero in the deep interior of a solid cylinder; rather, $\mathbf{M}$ is large and negative, in such a way that $\mathbf{B} = \mathbf{H} + 4\pi\mathbf{M}$ is close to zero ($\mathbf{H}$ having the same value inside the cylinder as outside; see Section 2). In the London theory, the decomposition (6.14), though mathematically always possible and unique, is artificial from the physical point of view.

The Pippard modifications to the London linear theory, discussed in Section 3, make no essential difference in this discussion. For very long wavelengths, i.e., for very small wave numbers q, the Pippard kernel $K(q)$, Eq. (3.11), reduces to the London kernel (3.9), except for a constant factor which does not alter the present argument. Since the argument given is concerned with the very long wavelength components in a Fourier analysis, the difference between (3.9) and (3.11) does not alter the discussion in essence.

The essential physical assumption implicit in the second London equation is, therefore, the same as in the first London equation: the "superelectrons" retain coherent properties over arbitrarily large distances (distances as large as the superconducting specimen itself). A theory of superconductivity is forced to explain, in the words of Casimir (55), the existence of "stable wave functions extending over a mile or so of dirty lead wire."

7. Flux Quantization

The need for coherent wave functions for the "superelectrons" extending all the way around the superconducting ring is shown particularly directly by the experiments on quantization of the magnetic flux through the hole in the ring (Deaver 61, Doll 61). Following upon a suggestion of London (50), these experimenters measured the magnetic flux through the hole of superconducting rings carrying a ring current, and found that this flux (more accurately, the fluxoid) appeared to be "quantized," the quantum unit of flux being

$$\Phi_1 = hc/2e \tag{7.1}$$

where $h = 2\pi\hbar$ is Planck's constant and e the charge on an electron.

A crude argument for flux quantization with a quantum of this general order of magnitude goes as follows (London 50): suppose the "superelectrons" have a wave function $\phi(r)$ in the ring, in the absence of any magnetic field. Consider a magnetic field $\mathbf{B} = \text{curl}\,\mathbf{A}$ with field lines through the hole in the ring, but no field lines through the material of the ring itself; as a result of the Meissner field expulsion, this may be considered a first approximation to the actual field distribution when the ring carries a current. Since $\mathbf{B} = \text{curl}\,\mathbf{A}$ is, by assumption, zero within the superconductor, and $\text{div}\,\mathbf{A} = 0$ is the, generally convenient, London gauge, we may represent the vector field $\mathbf{A}$ *within the superconducting volume* V as the gradient of a scalar function $f(r)$:

$$\mathbf{A} = \mathbf{grad}\, f \quad \text{for } \mathbf{r} \text{ inside } V \tag{7.2}$$

This, in turn looks very much like a gauge transformation on the vector potential, and we know how wave functions behave under a gauge transformation; the transformed wave function is (e^* is the charge of the "superelectron")

$$\phi'(r) = \exp\left[\frac{ie^*}{\hbar c} f(r)\right] \phi(r) \tag{7.3}$$

The scalar function $f(r)$, however, is not single-valued. Rather, $f(r)$ increases by a definite amount F in going once around the ring, and we now show that this quantity F just equals the flux Φ through the hole in the ring; let C be a closed curve within the wire, encircling the hole once; then

$$F = \oint_C (\mathbf{grad}\, f) \cdot \mathbf{ds} = \oint_C \mathbf{A} \cdot \mathbf{ds} = \int_S \mathbf{B} \cdot \mathbf{n}\, dS = \Phi \tag{7.4}$$

where S is the surface spanned by C. Thus, the new wave function ϕ' defined by (7.3) is multiple-valued unless the increase in the phase factor, $eF/\hbar c$, is an integral multiple of 2π. Solving for the flux $\Phi = F$, this gives the flux quantization condition

$$\Phi = n \frac{2\pi\hbar c}{e^*} \tag{7.5}$$

where n is an integer.

The argument just given is due to London (50), who put the charge e^* equal to an electronic charge. Comparison of the experimental flux quantum (7.1) with (7.5), however, shows that we must put

$$e^* = 2e \tag{7.6}$$

That is, *the "superelectrons" have a double electronic charge* (Onsager 61). We shall return to this point in Chapter III.

For the moment, we shall confine ourselves to two remarks: (1) The analogy between (7.2) and a gauge transformation is misleading; although curl $\mathbf{A}$ vanishes inside the volume V, it certainly does not vanish everywhere in space; in particular, $\mathbf{B}$ = curl $\mathbf{A}$ does not vanish within the hole in the ring. The replacement of $\mathbf{A} = 0$ by (7.2) is *not* a gauge transformation without physical significance, but rather is a real physical process, involving flux lines threading through the hole in the ring. Under these conditions, there is no *a priori* reason to insist that the new wave function ϕ' must be related to the old wave function ϕ by a mere gauge transformation (7.3). If, for example, the very existence of the flux through the ring changes the energy levels of the system, then a similarity transformation such as (7.3) is clearly insufficient to describe this change, since similarity transformations leave energy levels invariant.

Bohm and Aharanov (Aharanov 59) called attention to this consequence of quantum mechanics in another connection. Thus London's argument for the necessity of flux quantization is invalid, and an alternative explanation must be developed by the microscopic theory. The second remark is: (2) the London argument obviously requires a coherent wave function ϕ all the way around the circumference of the ring. However, we do not need London's argument to infer such a correlation from the flux quantization experiment. The flux through the hole in the ring is a property of the ring as a whole, not a "local" property of part of the ring; flux quantization thus provides immediate experimental proof that superconductors possess essentially non-localizable properties, consistent with a correlation distance in excess of the physical dimensions of the specimen.

Whereas the London phenomenology of the Meissner effect, based on the first London equation (2.7), contains the essence of the phenomenon, the existence of flux quantization shows that the second London equation (6.1) misses a phenomenon of prime importance. Flux quantization does not follow from Eq. (6.1), and it is hard to see how any "local" equation similar to (6.1) could ever lead to such a global consequence. The occurrence of Planck's quantum constant in the value of the flux unit (7.1) also indicates that we are here beyond the range of a purely classical phenomenological approach. London himself was fully aware of the limitations of any phenomenology, his own included; in fact, he himself suggested this flux quantization.

It is interesting that the Landau-Ginzburg phenomenological equations, which involve Planck's constant through (5.2), do yield flux quantization if one requires ψ_s to be single-valued.

Since flux quantization was discovered experimentally after the microscopic theory of superconductivity had been developed, this phenomenon played little part in the development of the theory. We shall return to it, as a consequence of the microscopic theory, at the appropriate point.

8. The Energy Gap

In many physical systems, the actual excitation energies E can be written approximately as sums of contributions of "elementary excitations":

$$E \simeq \sum_s n_s \epsilon_s \tag{8.1}$$

where $n_s = 0, 1, 2, \ldots$ is the number of elementary excitations of type s, and ϵ_s is the corresponding excitation energy. If the elementary excitation is "of Fermi type," n_s cannot exceed 1. Examples of elementary excitations in ordinary metals are electrons above the Fermi surface, holes below the Fermi surface (both these are of Fermi type), and phonons.

The system is said to have an *energy gap* Δ if there exists a volume-independent energy Δ such that

$$\epsilon_s \geqslant \Delta, \qquad \text{all} \quad s \tag{8.2}$$

This condition implies, but is not implied by, the fact that the lowest excitation energy E_1 of the system as a whole equals Δ, and therefore does not go to zero with increasing volume.

Normal metals do not have an energy gap in this sense, since electrons can be excited to states just above the Fermi surface, and holes can be created in states just below the Fermi surface. To the extent that phonon excitations are taken into account, superconductors also do not have an energy gap.

However, in the 1950's, there appeared increasingly strong evidence that Fermi-type excitations in superconductors do have an energy gap. Such a gap had been suggested first by Daunt and Mendelssohn (Daunt 46), and a discussion of the experimental evidence can be found in Biondi (58, 59), Richards (58), and Tinkham (58a, b).

The main evidence for the energy gap comes from (1) the low-temperature behavior of the specific heat, $c_v \sim \exp(-\Delta/kT)$, and (2) the frequency threshold ω_0 for a sudden onset of absorption of infrared photons by thin superconducting films. These experiments are consistent with a minimum energy Δ required to excite the superconducting specimen from its ground state. Completely similar effects, but with a very different value of Δ, are observed in insulators, where Δ is the "band gap" between the highest filled band and the empty band above it.

In insulators, Δ is of the order of a fraction of an electron volt; in superconductors, Δ is of the order of 10^{-3} ev, so that Δ is a few times kT_c, where T_c is the transition temperature. Quite apart from the difference in the order of magnitude of Δ, the very fact of electrical conduction shows that the energy gap in superconductors can not be an absolute absence of low-lying excited states, but is a more subtle phenomenon.

Conversely, the fact that insulators have an energy gap shows immediately that an energy gap, by itself, does not necessarily lead to superconductivity. In particular, it is impossible to deduce the Meissner effect from the existence of an energy gap [as attempted by Bardeen (55)] since then insulators would show a Meissner effect, contrary to experience.

9. Coulomb Interactions between Electrons

In Section 1, we saw that the onset of superconductivity is related to some form of increased order in the motion of the electrons in the metal. It is necessary, therefore, to take into account ways in which electrons interact with each other. In the ordinary band theory of solids, the electrons are treated as independent particles, in a self-consistent (Hartree-Fock) manner. The "bands" are sets of energy levels of such an independent-particle picture; each electron is described by its own Schrödinger wave equation, in a potential with the periodicity of the lattice; this potential includes the effect of the lattice ions, assumed stationary at their equilibrium positions, and the "smeared out" average potential due to all the other electrons.

The strongest, and immediately obvious, electron-electron interaction is the Coulomb repulsion between the electrons. If there were no other particles around, this would give rise to the electrostatic potential

$$V_{ij} = \frac{e^2}{|\mathbf{r}_i - \mathbf{r}_j|} \tag{9.1}$$

where e is the electronic charge. The presence of the other electrons, and of the lattice ions, results in profound modifications in this potential. First of all, the ions are deformable, i.e., can be polarized, and this effect can be taken into account, in a first approximation, by introducing a dielectric constant ϵ. Furthermore, however, there are strong dynamical correlations between different electrons, as a result of the electrostatic repulsion. Suppose we know that there is one electron at position $\mathbf{r}_i$. This knowledge tells us something about the probability P_{ij} of finding a second electron at a nearby point $\mathbf{r}_j$. If the electrons were purely classical particles, and did not interact with each other at all, the electrons would be statistically independent, and the probability P_{ij} would be equal to the single-electron probability P_j of finding one electron at point $\mathbf{r}_j$.

The Pauli exclusion principle introduces a *statistical correlation* between different electrons, even in the absence of dynamical interactions. Electrons with parallel spins tend to stay away from each other, by distances of the order of the average de Broglie wavelength of their relative motions.[7]

---

[7] The wave function is antisymmetric under exchange of $\mathbf{r}_1$ and $\mathbf{r}_2$, and therefore vanishes when $\mathbf{r}_1 = \mathbf{r}_2$. The de Broglie wavelength gives the minimum distance over which the wave function can change significantly.

The statistical correlations due to the Pauli exclusion principle do not affect the distribution function for pairs of electrons with opposite spins; their wave function is a product of a spin- and a space-part,

$$\phi(\mathbf{r}_1\zeta_1 , \mathbf{r}_2\zeta_2) = u(\mathbf{r}_1 , \mathbf{r}_2)\, \chi(\zeta_1 , \zeta_2) \tag{9.2}$$

and the spin wave function $\chi(\zeta_1 , \zeta_2)$ may itself be either symmetric[8]

$$\chi_{\text{sym}}(\zeta_1 , \zeta_2) = 2^{-1/2}[\alpha(\zeta_1)\beta(\zeta_2) + \beta(\zeta_1)\alpha(\zeta_2)] \tag{9.3}$$

or antisymmetric

$$\chi_{\text{anti}}(\zeta_1 , \zeta_2) = 2^{-1/2}[\alpha(\zeta_1)\beta(\zeta_2) - \beta(\zeta_1)\alpha(\zeta_2)] \tag{9.4}$$

The Pauli principle requires that the full wave function ϕ be antisymmetric under exchange of $(\mathbf{r}_1\zeta_1)$ and $(\mathbf{r}_2\zeta_2)$. This can be achieved by combining the symmetric spin function (9.3) with an antisymmetric space function $u(\mathbf{r}_1 , \mathbf{r}_2)$, or else by combining the antisymmetric spin function (9.4) with a symmetric space function $u(\mathbf{r}_1 , \mathbf{r}_2)$. Both symmetric and antisymmetric space functions $u(\mathbf{r}_1 , \mathbf{r}_2)$ are permitted, and there is no net correlation in space between electrons of opposite spin.

In addition to the effects of the exclusion principle, however, there also exist *dynamical correlations* induced by the forces between the electrons. The electrons repel each other electrostatically, no matter what their spin directions may be. Thus the conditional probability P_{ij} of finding an electron at $\mathbf{r}_j$, given that there is an electron at $\mathbf{r}_i$, is smaller than the single-electron probability P_j, for positions $\mathbf{r}_j$ close to $\mathbf{r}_i$. Around each electron, there exists a "correlation hole" in the distribution of all the other electrons. The "correlation hole" is more pronounced for electrons with parallel spin directions, since for such electrons the effects of the repulsive force and of the Pauli exclusion principle reinforce each other; but the "correlation hole" also exists between electrons of opposite spins.

The distribution of electrons around a given electron is a distribution of electrically charged particles, and therefore gives rise to an electric potential. The "correlation hole" in this electron distribution is equivalent to a distribution of charge, of opposite sign to the charge on the electron. The correlation hole thus produces an effective "screening" of the potential (9.1); this effect was introduced by Debye and Hueckel (Debye 23) in their theory of strong electrolytic solutions; Wigner and Seitz (Wigner 33, 34) discussed it in connection with their theory of

[8] We use $\alpha(\zeta)$ to denote the spin function for an electron with spin up, $\beta(\zeta)$ to denote the spin function for an electron with spin down.

the binding energy of solids; a more elaborate treatment was then given by Wigner (34a, 38); more recently, the problem has been taken up by a number of authors (Bohm 51, 53; Pines 52, 53; Bardeen 55a; Gell-Mann 57; Sawada 57, 57a; Wentzel 57[8a]). The recent work has led to an excellent asymptotic treatment of "high-density jellium," an imaginary substance defined as follows: a gas of negatively charged particles (electrons) is placed into a container of volume V, which also contains a fixed uniform distribution of positive charge; the positive charge distribution has no dynamical properties whatever, but is inserted merely to cancel out the main term in the electrostatic energy of the negative charges; all the dynamics resides in the negative charges; "high-density" means that the average distance between negative particles is small compared to the Bohr radius of one of these particles when bound to an infinitely heavy unit positive charge.

The basic physical ideas were put most clearly by Bohm; the screening of the long-range Coulomb interactions is associated with the possibility of "plasma oscillations" in the electron gas; these plasma oscillations are collective motions of the whole system, in which the density of the gas varies periodically in space and time. A periodic (in space) density fluctuation yields a corresponding charge fluctuation around the mean charge density, which latter is zero as a result of the background positive charge of the "jellium" system; this charge fluctuation in turn produces an electrostatic potential periodic in space; the associated electric field is such as to tend to reduce the density fluctuation; this leads to wave motion, since the electrons, accelerated by this field, get into a motion which overshoots the equilibrium, uniform, distribution. The plasma oscillation is a purely classical concept, and the frequency of these oscillations correspondingly does not contain Planck's constant. The classical result, due to Tonks and Langmuir (Tonks 29), is (N/V is the number density)

$$\omega_{\text{plasma}} = \left(\frac{4\pi N e}{V m}\right)^{1/2} \tag{9.5}$$

Bohm suggested that these plasma modes of freedom of the system should be eliminated first, by a transformation, before one can treat the electrostatic potential (9.1) as a small perturbation. The transformed potential, in first approximation, is just Debye's screened interaction

$$V_{ij} = \frac{e^2}{|\mathbf{r}_i - \mathbf{r}_j|} \exp(-r_{ij}/D) \tag{9.6}$$

[8a] Also Brout 57; Hubbard 57; DuBois 59; Sawada 60; Suhl 61; Roos 61, 62; Osaka 62; Werthamer 62; Dang 63.

where $r_{ij} = | \mathbf{r}_i - \mathbf{r}_j |$ and D is the Debye screening distance; in the classical case, D is

$$D = \frac{v_{\text{av}}}{\omega_{\text{plasma}}} = \frac{(\mathfrak{k}T/m)^{1/2}}{\omega_{\text{plasma}}} = \left(\frac{V\mathfrak{k}T}{4\pi Ne^2}\right)^{1/2} \quad \text{(classical)} \tag{9.7}$$

Here v_{av} is the average thermal velocity of the electrons (treated classically!), obtained by setting $\frac{1}{2}mv^2$ equal to the equipartition energy $\frac{1}{2}\mathfrak{k}T$. When the electrons are treated quantum-mechanically, allowing for the effects of the exclusion principle, the average velocity in the degenerate Fermi-Dirac gas becomes independent of the temperature, and of order of magnitude

$$v_{\text{av}} \sim (2\epsilon_F/m)^{1/2} = p_F/m$$

where $\epsilon_F = p_F^2/2m$ is the Fermi energy, p_F the electron momentum at the Fermi surface:

$$p_F = \hbar(3\pi^2 N/V)^{1/3} \tag{9.8}$$

Both the original treatment of Bohm (53) and the, mathematically superior, later work of Sawada (57, 57a) yield screening distances of the order of magnitude:

$$D \sim \frac{v_{\text{av}}}{\omega_{\text{plasma}}} \sim \frac{p_F/m}{\omega_{\text{plasma}}} = \left(\frac{9\pi}{64}\right)^{1/6} \left(\frac{\hbar}{m^{1/2}e}\right) \left(\frac{V}{N}\right)^{1/6} \quad \text{(quantum)} \tag{9.9}$$

Except for the numerical factor, which is of no significance, this screening distance is the geometric mean of the Bohr radius $\hbar^2/me^2$ and the mean distance between particles, $(V/N)^{1/3}$. For electrons in solids, both these are of the same order of magnitude, about 10^{-8} cm. Both the classical and the quantum treatments indicate that the "high-density" approximation is not valid in this situation; the condition of validity is

$$\left(\frac{V}{N}\right)^{1/3} << \frac{\hbar^2}{me^2} \quad \text{(high density)} \tag{9.10}$$

and this is strongly violated for electrons in solids, since $(V/N)^{1/3}$ is actually some 2 to 10 times as large as the Bohr radius.[9]

If one uses the theory uncritically for conduction electrons in solids (Pines 53), the screening distance D turns out to be *smaller* than the average distance between conduction electrons; since the screening

[9] N must be interpreted as the number of conduction electrons, of course, so that it is roughly equal to the number of nuclei, not to the, much larger, total number of electrons.

effect arises from a "correlation hole" in the distribution of the other electrons around the given electron, such a small screening distance indicates an inconsistency.

Nevertheless, although the quantitative results of the theory of "high-density jellium" are not applicable to electrons in solids, the qualitative picture is almost surely all right for the normal state of metals. What, then, about the superconducting state? Do the Coulomb forces between electrons, modified, perhaps, by the band structure of the lattice (which is ignored in the "jellium" model), give rise to a transition into a superconducting state at low enough temperatures?

For many years, this was the general belief, although no one had succeeded in demonstrating such a transition convincingly. Attempts in this direction were made by Slater (37, 37a), Heisenberg and Koppe (Koppe 50, 54, and references quoted there), Born and Cheng (Born 48, 48a, 48b; Cheng 49; Schachenmeier 51, 51a), Tisza (Tisza 50, 51, 52; Luttinger 50), and Bardeen (41, 51, 51a). These attempts are discussed by Schafroth (60); they were not successful.

10. The Fröhlich Interaction and the Isotope Effect

In 1950, Fröhlich initiated the modern approach to the theory of superconductivity, by focusing attention on an interaction between electrons which had been completely ignored up till then. Although Fröhlich's detailed theory has not stood the test of time, the ideas which he introduced have been so fruitful that, within less than a dozen years from the appearance of his paper, a broadly satisfactory microscopic theory of superconductivity has come into existence. For an authoritative history of this discovery, see Fröhlich (61); this review article includes an amusing (and thought-provoking) section on the "psychology of superconductivity".

To introduce Fröhlich's idea, let us recall the theory of electrical resistance in normal metals. In a perfect lattice (and in "jellium"), there is no electrical resistance at all, since there exist traveling wave solutions of the Schrödinger equation for the electron system, which are exact solutions as a result of translation invariance with respect to lattice translations. In the one-electron (independent-particle) approximation, these are the Bloch waves; but the argument from translation invariance is not restricted to the independent-particle picture. Thus any theory of electrical resistance must involve deviations from translation invariance.

In solids, there are two effects of main importance: (1) the presence of chemical impurities and other lattice imperfections, such as missing atoms, interstitial atoms, etc., and (2) the thermal motion of the lattice ions even in a perfect lattice. At room temperature, the second effect dominates and is responsible for the observed electrical resistance of metals. We shall restrict ourselves here to a quick, qualitative discussion; a detailed treatment can be found in the excellent text of Peierls (55).

The motion of the lattice ions is ignored altogether in zeroth approximation. This is physically plausible since the lattice ions are thousands of times more massive than the conduction electrons, and hence move much more slowly on the average. In the next approximation, the motion of the lattice ions is described by a superposition of quantized elastic waves propagating through the lattice; these quanta of elastic waves are called "phonons," since some of the elastic waves are involved in the propagation of sound through the metal.

The formal treatment of the lattice waves is very similar to the treatment of blackbody electromagnetic radiation. The average number $\bar{N}_q$ of "phonons" of wave number q, circular frequency ω_q, and quantum energy $\hbar\omega_q$, is given by the Planck law

$$\bar{N}_q = \frac{1}{\exp(\hbar\omega_q/\mathfrak{k}T) - 1} \tag{10.1}$$

To a crude first approximation (Debye 12) the frequency ω_q can be taken to be

$$\omega_q \cong u_s \,|\, \mathbf{q} \,|, \qquad |\, \mathbf{q} \,| < q_{\max} \tag{10.2}$$

where u_s is the velocity of propagation of the elastic wave in question (different for longitudinal and transverse waves) as computed from the known elastic constants, and $q_{\max}$ is the Debye cutoff on the wave number. This cutoff is determined in such a way that the number of different phonons exactly equals the number of degrees of freedom of the ionic motions, i.e., three degrees of freedom for each ion. This gives

$$q_{\max} = \left(\frac{6\pi^2 N'}{V}\right)^{1/3} \tag{10.3}$$

where N' is the number of lattice ions; except for a numerical factor of order unity, $q_{\max}$ is the inverse of the lattice spacing. The Debye temperature Θ is defined by

$$\mathfrak{k}\Theta = \hbar\omega_{\max} = \hbar\bar{u}q_{\max} \tag{10.4}$$

where $\bar{u}$ is a suitably chosen average of the three elastic velocities. Some typical Debye temperatures are: 90° for Pb, 210° for Ag, 400° for Al.

At room temperature, $\hbar\omega_q$ in (10.1) is smaller than or comparable to $\mathfrak{t}T$ for all $q < q_{\max}$, so that the average number of phonons is appreciable. The lattice motions are therefore excited significantly, the lattice deviates from mathematical lattice symmetry, and this results in scattering of the Bloch electron waves by this assembly of phonons.[9a] The "electron-phonon interaction" produces an electrical resistivity, since freely traveling electron waves are no longer exact solutions of the Schrödinger equation. The value of the resistivity depends on the strength of this interaction: a strong electron-phonon interaction leads to high electrical resistance.

Formally, the electron-phonon interaction appears in the theory as a matrix element for processes in which an electron either absorbs or emits a single phonon, the electron being scattered (changing its momentum) in the process.

Fröhlich pointed out that there is another effect of the electron-phonon interaction, of prime importance for the theory of superconductivity; if one electron gets scattered, emitting a phonon in the process, and another electron absorbs this very phonon, thereby itself being scattered, what we have in effect is an *electron-electron interaction transmitted by phonons*. This is the Fröhlich interaction; it is completely analogous to electromagnetic interactions between moving charges, which may be considered to be "mediated" by light quanta (photons) being emitted by one charge and absorbed by the other; or to the meson-theoretic interaction between nucleons, in which one nucleon emits a virtual meson, the other absorbs it. In all these cases, the average excitation of the fields is not a major factor; the field quanta traveling from one of the particles to the other, are actually emitted by one particle, absorbed by the other, i.e., they arise as a result of the particle-field interaction even at the absolute zero of temperature.

The Fröhlich interaction is quite unusual in detail, since the velocity of typical phonons (the speed of sound in the metal) is much smaller than the velocity of the typical electron (the Fermi velocity), the ratio being about 1 : 100. Thus the field quanta trail behind the particles, rather than outrunning them. We may visualize each electron as leaving behind it a narrow "wake" of phonons; other electrons interact with the first electron if they accidentally travel, at high speed, through this "wake." It is clear from this description that "retardation effects" are by no means minor corrections here.

[9a] A modern discussion of this process, with references to earlier work, is given by Nakajima (63).

In his paper, Fröhlich (50, 50a) suggested that the normal Fermi distribution of the electrons is *unstable* in the presence of the Fröhlich interaction. He obtained a lower energy from a quite different distribution of electrons over k-space, provided only that the strength of the Fröhlich interaction (proportional to the square of the "coupling constant" for the electron-phonon interaction) is sufficiently large. In this way Fröhlich obtained a criterion for distinguishing superconducting from normal metals. The criterion could be checked, since the electron-phonon scattering can be estimated from data on the normal resistivity of the metal at room temperature. The quantitative comparison was surprisingly good, considering the approximate nature of the treatment (e.g., the whole Debye theory is quite a rough approximation).

More important than the quantitative comparison, however, were the *qualitative* consequences of the Fröhlich theory: (1) superconductivity arises only in bad normal conductors (for only in those is the electron-phonon interaction strong), and (2) superconductivity involves the *motion*, i.e., the dynamics, of the lattice ions, and we must therefore expect significant effects on superconductivity of dynamical properties of the lattice ions, in particular their mass.

Both these consequences are in spectacular agreement with experiment. The best normal conductors, silver and copper, fail to become superconductors down to the lowest temperatures attained so far; bad conductors like tin, lead, and mercury all become superconducting. Furthermore, in the case of metals with a number of different isotopes (e.g., lead, mercury, tin), the superconducting transition temperature T_c shows a definite dependence on the isotopic mass of the nucleus. Not only the existence of this "isotope effect," but even the actual dependence (T_c proportional to $M^{-1/2}$), agrees with the Fröhlich theory. Whereas the distribution of superconductors in the periodic table was known, to some extent, before the work of Fröhlich, the isotope effect was not known to Fröhlich and thus constituted a significant theoretical prediction. The experimental discovery of the isotope effect (Reynolds 50, Maxwell 50) was made independently of Fröhlich's theoretical work, and almost at the same time. Fröhlich was indeed fortunate to have striking proof of his basic theoretical idea delivered to him within a few weeks of the inception of that idea.[10]

In spite of these successes, the Fröhlich theory taken as a whole was

---

[10] He was even more fortunate, in retrospect, in the actual metals used in the experiments. There is now strong evidence (Matthias 63, 63a) that the isotope effect occurs only in some superconducting metals, not in all; see Chapters V and X for further discussion.

not immune from criticism. Some of this criticism was purely technical (Bardeen 50, 50a), concentrating on Fröhlich's use of perturbation theory, and could be countered by showing that Fröhlich's procedure was a not unreasonable approximation (Buckingham 54). A criticism by Kohn (51) and Wentzel (51), to the effect that the Fröhlich criterion does not indicate the onset of superconductivity but rather the complete breakdown of the assumed lattice structure of the solid, could be diverted, but not really met, by more elaborate counterarguments (Fröhlich 52, Huang 51, Kitano 53), using analogies with the renormalization procedure in quantum field theory. But there remained some unanswerable objections: (1) the Meissner effect does not follow from the theory (Schafroth 51); (2) the energy difference between the superconducting and normal states, and the transition temperature, are both too large by roughly a factor 10 (this factor cancels out in the successful comparison between normal and superconducting metals); and (3) the Coulomb interaction between electrons, which Fröhlich ignored completely, is actually a larger contribution to the cohesive energy of the metal than the Fröhlich interaction. This is true even after one allows for the plasma screening effects discussed in the preceding section.[11]

These difficulties, however, do not detract in the least from the value of Fröhlich's contribution to the theory of superconductivity. On the contrary, the questions raised by his theory are themselves a vital part of his contribution. Once the right questions are being asked, the answers are not far behind. Only after Fröhlich was it possible to ask all the right questions. In concluding this chapter, let us now list the main questions:

(1) How does the, comparatively weak, Fröhlich interaction produce a thermodynamic transition to a superconducting state, whereas the strong Coulomb repulsion between electrons has no such *qualitative* effect?

(2) The characteristic temperature for phonons is the Debye temperature Θ. Why are superconducting transition temperatures lower than Θ by one or two orders of magnitude?

(3) What is the nature of the "ordering" of electron motions in the superconducting state? In particular, what is the interpretation of Pippard's "coherence length," and of the Landau-Ginzburg "wave function of the superelectrons"?

---

[11] Various alternative versions of the Fröhlich theory, with certain modifications, have been suggested by Ziman (55), Haug (56), Stumpf (56), and Nesbet (62, 62a). These theories are discussed critically by Englman (63), who concludes that all are substantially equivalent to Fröhlich's original proposal, and subject to the same objections.

(4) How does this ordering of electron motions result in a Meissner effect and in persistent currents?

(5) Are the persistent currents infinitely long-lived, and if not, what limits their life?

The first two questions depend completely on Fröhlich's work; the details of the third question could not be asked before about 1950. It is an indication of the striking progress the theory of superconductivity has made since Fröhlich's initial impetus, that only a decade later all but the fifth question had entirely satisfactory answers.

Of necessity, the treatment in this chapter has been restricted to essentials. For more detail, the reader may consult various review articles and books. Some of these are (in roughly historical order): Burton 34; Itterbeck 45; Justi 46; Ginzburg 46, 50b, 52a, 53; Mendelssohn 48, 55, 60; Meissner 48; Heisenberg 48; Gorter 49, 55, 61; Vick 49; London 50; Koppe 50; Bardeen 51b, 56, 61, 63; Laue 52; Shoenberg 52, 53, 60; Olsen 53; Eisenstein 54; Fröhlich 54a, 61; Pippard 54; Squire 55; Faber 55b; Serin 55, 56; Matthias 57; Bogoliubov 58; Kuper 59; Schafroth 60; Thouless 61a.

II

The Bose-Einstein Gas Model

It is often useful to find some simple theoretical model, which allows exact calculation and shows some of the essential features of the actual phenomenon. This is of course only a first step; the second step consists in showing the relation between the simple model and nature.

In the theory of superconductivity, the first step was taken, independently, in three places: in the U.S.S.R. by Ginzburg (52a, 53), in the U.S.A. by Feynman (53), and in Australia by Schafroth (54a, 55a). All three authors showed by explicit calculation that the ideal Bose-Einstein gas, below its Bose-Einstein condensation point, has the magnetic response characteristic of the Meissner effect. On more qualitative grounds, the Bose-Einstein condensation was suggested as an explanation for "superfluid" behavior (both in superconductors and in liquid helium) many years earlier, by London (38). In liquid helium, the idea of the Bose-Einstein model proved immediately attractive, since the helium atoms of the dominant isotope (He^4) obey Bose-Einstein statistics. The electrons in metals, however, are Fermi-Dirac particles, and for that reason the Bose-Einstein gas model of a superconductor was, mistakenly, considered a mere curiosity until the middle 1950's. In view of developments since then, however, it appears well worthwhile to devote a whole chapter to a detailed discussion of the Bose-Einstein model.

1. The Bose-Einstein Distribution, Normal State

Consider N non-interacting particles, contained in a box of volume V at a temperature T. Let us denote the single-particle energy levels in this box by ϵ_k, where k is a general index. In the special case of a cubic box of side length L and volume L^3, the general index k becomes a wave vector $\mathbf{k}$, and the energy levels are given by

$$\epsilon_k = \frac{\hbar^2 k^2}{2M} \tag{1.1}$$

where M is the particle mass; the components of $\mathbf{k}$ are related to the side length of the box by

$$(k_1, k_2, k_3) = \frac{2\pi}{L}(n_1, n_2, n_3) \tag{1.2}$$

The n_i are integers, negative or zero or positive (we assume periodic boundary conditions for simplicity). In much of what follows, we shall *not* use the explicit form (1.1); it is well known that there are interacting systems, e.g., electrons in solids, in which the independent-particle model is quite a good first approximation, provided the energies ϵ_k in that model are determined self-consistently and allowed to differ from the simple law (1.1).

The ϵ_k are single-particle energies. The quantum states of the gas as a whole are defined by stating the numbers of particles n_k actually in quantum state k (the occupation numbers). The set of these occupation numbers is called a "configuration"; we use the letter α to denote a configuration

$$\alpha = \{n_0, n_1, n_2, n_3, \ldots\} \tag{1.3}$$

These occupation numbers satisfy the obvious condition

$$\sum_k n_k = N \tag{1.4}$$

The energy of the configuration α of the N-particle system is[1]

$$E_\alpha = \sum_k n_k \epsilon_k \tag{1.5}$$

In statistical mechanics, the gas is not in any one quantum state α, but rather there is a distribution over all possible quantum states; the probability P_α of finding the system in state α is given by

$$P_\alpha = \exp\left(\frac{F - E_\alpha}{\mathfrak{k}T}\right) \tag{1.6}$$

[1] We use capital letters E for system energies, small Greek letters ϵ for single-particle energies. The distinction between these will be of considerable importance in what follows.

where $\mathfrak{k}$ is Boltzmann's constant, T is the absolute temperature, and F is the free energy, chosen so as to make the sum of all the P_α equal to unity:

$$\exp(-F/\mathfrak{k}T) = \sum_\alpha \exp(-E_\alpha/\mathfrak{k}T) \tag{1.7}$$

Although the probabilities (1.6) seem to favor energy levels in the immediate neighborhood of the ground state E_0, actually the density of states E_α increases so rapidly with increasing E_α that the probability per unit energy interval dE has a sharp maximum at some finite energy $\bar{E}(V, T, N)$, which depends on the volume V, the temperature T, and the particle number N; this is the thermodynamic "internal energy," which is proportional to V (if the particle number density N/V is kept fixed) for all reasonable systems.

Equations (1.6) and (1.7) apply to all systems. For independent-particle systems, Eqs. (1.4) and (1.5) show that it is convenient to introduce the concept of an "average occupation number" $\bar{n}_k$ defined by

$$\bar{n}_k = \sum_\alpha n_k^{(\alpha)} P_\alpha \tag{1.8}$$

where $n_k^{(\alpha)}$ is the occupation number n_k in the configuration α, Eq. (1.3). In terms of the average occupation numbers, we have

$$\sum_k \bar{n}_k = N \tag{1.9}$$

and

$$\sum_k \bar{n}_k \epsilon_k = \bar{E} \tag{1.10}$$

where $\bar{E}$ is, as before, the thermodynamic "internal energy" of the system. Equation (1.10) is a direct consequence of (1.5) and (1.8) and does not involve new assumptions.

So far, we have not introduced the "statistics" of the particles of our gas. The way the "statistics" enter here is in the determination of the possible values of the occupation numbers n_k, and hence of the permissible configurations α, Eq. (1.3). If the particles obey Fermi-Dirac statistics, the Pauli exclusion principle prevents us from putting more than one particle into any one single-particle state k,[2] so that

$$n_k = 0 \quad \text{or} \quad 1 \quad \text{only, for Fermi-Dirac statistics} \tag{1.11}$$

[2] Particles with Fermi-Dirac statistics always have half-integral spin; we shall include the spin index as part of the general index for the single-particle states; that is, we count $(\mathbf{k}, +)$ and $(\mathbf{k}, -)$ as two distinct states "k", each of which can be occupied by at most one electron.

This condition limits the permissible values of α in the sum (1.7), which sum in turn fixes everything else. The explicit evaluation in Appendix A gives the "Fermi-Dirac distribution" law:

$$\bar{n}_k = \frac{1}{\exp[\beta(\epsilon_k - \mu)] + 1} \qquad \text{(Fermi-Dirac)} \qquad (1.12)$$

where $\beta = 1/\mathfrak{k}T$ and μ is a new parameter, called the "chemical potential" or the Fermi energy. The actual value of μ for a given condition of the system (given V, T, N) is obtained by substituting (1.12) into (1.9), evaluating the sum, and solving the resulting equation for μ. We emphasize that (1.12) is, strictly speaking, *not* a distribution law, but is already an average over the full distribution, see (1.8).

When the particles of the gas obey Bose-Einstein statistics, the occupation numbers n_k range over all non-negative integers. The only restriction on the permissible configurations is condition (1.4). In Appendix A, the relevant summations are carried out explicitly, and we obtain the "Bose-Einstein distribution"

$$\bar{n}_k = \frac{1}{\exp[\beta(\epsilon_k - \mu)] - 1} \qquad \text{(Bose-Einstein)} \qquad (1.13)$$

Just as before, the chemical potential μ must be determined by substituting (1.13) into (1.9), and solving the resulting equation for μ.

Although the distribution laws (1.12) and (1.13) look similar, formally, they are quite different in a number of ways. The Fermi energy μ may become positive, and frequently does; it is always positive for a degenerate Fermi gas, e.g., electrons in metals. In the Bose-Einstein case, μ must remain below ϵ_0; a value of μ in excess of the single-particle ground state energy ϵ_0 would lead to a negative value for $\bar{n}_0$, which is absurd in view of Eq. (1.8): the average of a non-negative quantity must be non-negative. Qualitatively, the distributions (1.12) and (1.13) differ from the "classical" distribution law

$$\bar{n}_k = \exp[\beta(\mu - \epsilon_k)] \qquad \text{(classical)} \qquad (1.14)$$

in opposite directions: The Fermi-Dirac distribution limits the average occupation numbers to $\bar{n}_k \leqslant 1$; the Bose-Einstein distribution enhances occupation numbers which are already large in the classical approximation: "To him that hath shall be given." The Pauli principle leads effectively to a repulsion, in the Fermi-Dirac gas; the opposite requirement in the Bose-Einstein case (symmetrical wave functions) leads to an effective attraction, i.e., the "statistical correlations" (see Chapter I, Section 9) in the Bose-Einstein gas are attractive correlations.

It is instructive to evaluate the sums (1.9) and (1.10) explicitly, for the ordinary ideal gas, i.e., ϵ_k given by (1.1). We make the usual replacement of a sum over single-particle states k by an integral (Schiff 49):

$$\sum_k g_k \rightarrow \frac{V}{(2\pi)^3} \int d^3k \, g(k) \tag{1.15}$$

where d^3k indicates an integration over the three-dimensional $\mathbf{k}$-space. Since we are interested in the Bose-Einstein case, we may assume zero spin for the particles and shall do so henceforth.

Since ϵ_k depends only on the magnitude of the vector $\mathbf{k}$, the angle integrations are trivial, and (1.9) can be written in the form

$$\frac{N}{V} = \frac{1}{(2\pi)^3} \int_0^\infty \frac{4\pi k^2 \, dk}{\exp[\beta(\epsilon_k - \mu)] - 1} \equiv f(\beta, \mu) \tag{1.16}$$

where $f(\beta, \mu)$ is defined by the integral. Without explicit evaluation we see that in the permissible range of values of μ, i.e., $\mu \leqslant \epsilon_0$, $f(\beta, \mu)$ is a monotonically increasing function of μ.

Since $\mu < \epsilon_0$ and thus $\exp[\beta(\epsilon_0 - \mu)] > 1$, we can expand the fraction in (1.16) to get, after interchanging the order of summation and integration,

$$f(\beta, \mu) = \frac{1}{2\pi^2} \sum_{n=1}^{\infty} \exp[n\beta(\mu - \epsilon_0)] \int_0^\infty \exp[-n\beta(\epsilon_k - \epsilon_0)] \, k^2 \, dk \tag{1.17}$$

For the special energy levels (1.1), the integrals in (1.17) can be evaluated explicitly by introducing $x = \beta(\epsilon_k - \epsilon_0) = \beta\hbar^2 k^2/2M$ as a variable of integration. To write the result compactly, we introduce the "thermal wavelength"

$$\lambda\!\!\!^{-} = \left(\frac{2\pi\beta\hbar^2}{M}\right)^{1/2} = \left(\frac{2\pi\hbar^2}{M\mathfrak{k}T}\right)^{1/2} \tag{1.18}$$

Except for a numerical factor $(2\pi)^{1/2}$, $\lambda\!\!\!^{-}$ is the de Broglie reduced wavelength for a particle of mass M moving with kinetic energy $\frac{1}{2}\mathfrak{k}T$. $\lambda\!\!\!^{-}$ therefore represents the quantum-mechanical uncertainty in the position coordinate of the "typical" particle. The sum (1.17) then becomes

$$f(\beta, \mu) = \frac{1}{\lambda\!\!\!^{-3}} \sum_{n=1}^{\infty} \frac{\exp[n\beta(\mu - \epsilon_0)]}{n^{3/2}} \tag{1.19}$$

The "classical limit" is valid at high temperatures, when $\lambda\!\!\!^{-}$, Eq. (1.18), becomes small compared to the average distance between particles,

$(V/N)^{1/3}$. This limit is obtained from (1.19) by assuming that $\mu - \epsilon_0$ is large and negative, so that the first term of the sum suffices; combination of (1.16) and (1.19) then gives

$$\mu - \epsilon_0 \cong -\mathfrak{k}T \ln (V/N\lambda^3) \qquad \text{(classical)} \tag{1.20}$$

which is indeed large and negative for $V/N \gg \lambda^3$. Equation (1.20) is of course independent of the statistics, since both (1.12) and (1.13) reduce to the classical law (1.14) for large and negative $\mu - \epsilon_0$.

To obtain the internal energy $\bar{E}$, we apply (1.15) to (1.10) to get

$$\frac{\bar{E}}{V} = \frac{1}{(2\pi)^3} \int_0^\infty \frac{4\pi k^2 \epsilon_k \, dk}{\exp[\beta(\epsilon_k - \mu)] - 1} \equiv g(\beta, \mu) \tag{1.21}$$

The same expansion procedure leads to the evaluation

$$g(\beta, \mu) = \frac{3\mathfrak{k}T}{2\lambda^3} \sum_{n=1}^{\infty} \frac{\exp[n\beta(\mu - \epsilon_0)]}{n^{5/2}} \tag{1.22}$$

The internal energy per particle is obtained from (1.16) and (1.21):

$$\frac{\bar{E}}{N} = \frac{g(\beta, \mu)}{f(\beta, \mu)} \tag{1.23}$$

where $\mu = \mu(V, T, N)$ must be determined from (1.16); after this has been done, (1.23) gives the internal energy as a function of V, T, and N.

In the classical limit, the first term of the infinite sum in (1.22) suffices. The λ^3 drops out of the ratio (1.23) and we obtain the classical equipartition law:

$$\frac{\bar{E}}{N} = \frac{3}{2}\mathfrak{k}T \qquad \text{(classical)} \tag{1.24}$$

Quantum aspects become important when the de Broglie wavelength of the thermal motion, (1.18), becomes comparable to, or larger than, the average distance between particles:

$$\lambda \gtrsim (V/N)^{1/3} \qquad \text{(degeneracy condition)} \tag{1.25}$$

We then call the gas "degenerate." The degenerate Fermi-Dirac gas is well known in the theory of the solid state, since electrons in metals satisfy the condition (1.25) at all temperatures below the melting point of the metal. The degenerate Fermi-Dirac gas is characterized by a positive value of μ, the Fermi energy; $\bar{n}_k$ is nearly 1 for $\epsilon_k < \mu$, nearly 0 for $\epsilon_k > \mu$ [see Eq. (1.12)]. The degenerate Bose-Einstein gas behaves quite differently, however, and we now turn to an investigation of that case.

In the classical limit, the quantity $\lambda\!\!\!^{-}{}^3$ may be taken, to a first crude approximation, as the size of a "cell" in the volume V. There are $M = V/\bar{\lambda}^3$ such cells, and N particles to be distributed between cells. If we assume low density, so that the probability of multiple occupation of a cell is low, the number of ways in which N indistinguishable particles can be placed in M cells is

$$C = \frac{M!}{N!\,(M-N)!}$$

This is the number of "complexions," and the entropy S is given by

$$S = \mathfrak{k} \ln C$$

Using Stirling's approximation for the logarithm of a factorial, and expanding the result in the small quantity $y = N/M = N\bar{\lambda}^3/V$, we obtain

$$S = -N\mathfrak{k}[\ln y + O(1)] \tag{1.26}$$

where $O(1)$ means a term which remains constant when $\ln y$ becomes large. This result is in full agreement with the exact result

$$S = N\mathfrak{k}\left(\frac{5}{2} - \ln y\right) \tag{1.27}$$

which follows, with a little extra work, from (1.20) and (1.24), together with the thermodynamic relation that μ is the partial derivative of the free energy $F = \bar{E} - TS$ with respect to N, keeping V and T constant.

2. The Bose-Einstein Condensation

The degenerate Bose-Einstein gas, which satisfies condition (1.25), is obtained from the formulas of Section 1 by assuming that the chemical potential μ is close to its maximum value, $\mu = \epsilon_0$. Although the expansions (1.19) and (1.22) converge for all $\mu \leqslant \epsilon_0$, they become useless in this region, since the convergence is very slow.

We have remarked already that $f(\beta, \mu)$, defined by (1.16), is a monotonically increasing function of μ in the region $\mu \leqslant \epsilon_0$; $f(\beta, \mu)$ therefore assumes its maximum value at $\mu = \epsilon_0$, and this maximum value is

$$f_{\max}(\beta) = F(\beta, \epsilon_0) = \frac{1}{(2\pi)^3} \int_0^\infty \frac{4\pi k^2\, dk}{\exp\left[\beta(\epsilon_k - \epsilon_0)\right] - 1} \tag{2.1}$$

The crucial question is whether this integral converges, or is infinite. For large k, the exponential in the denominator is quite sufficient to ensure convergence; the behavior at small k, on the other hand, depends on ϵ_k in that region; the integrand, in that region, is proportional to $k^2/(\epsilon_k - \epsilon_0)$.[3] For the ordinary case, given by (1.1), $k^2/(\epsilon_k - \epsilon_0) = 2M/\hbar^2$ is a constant, and the integral (2.1) converges both at $k = 0$ and at $k \to \infty$. The value of $f_{\max}$ can be obtained from (1.19) by setting $\mu = \epsilon_0$; it is

$$f_{\max}(\beta) = \frac{1}{\lambda^3} \sum_{n=1}^{\infty} \frac{1}{n^{3/2}} = \frac{2.61}{\lambda^3} \tag{2.2}$$

Since $f(\beta, \mu)$ equals the number density N/V [see Eq. (1.16)], we therefore have the result

$$\frac{N}{V} \leqslant f_{\max} = \frac{2.61}{\lambda^3} \tag{2.3}$$

This result is absurd. Physically, imagine that we are keeping the temperature T fixed, so that the thermal wavelength λ remains fixed also, and we increase the number density by inserting more and more bosons into the volume V. What is there to prevent us from inserting more particles, even after N/V has reached the "limit" (2.3)? The particles are, by assumption, non-interacting, and, being Bose-Einstein particles, there is no exclusion principle effect (the actual statistical effect goes in the opposite direction!). Thus the upper limit (2.3) on the number density cannot be correct, and must arise from an error somewhere in the argument.

As Einstein (24, 25) showed long ago, the mathematical error lies in the replacement (1.15) of the sum over single-particle states k, by an integral over the three-dimensional **k**-space. Let us return to the original sum (1.9), and consider just the first term of that sum, corresponding to the single-particle ground state, $k = 0$. The average occupation number of that state is

$$\bar{n}_0 = \frac{1}{\exp\left[\beta(\epsilon_0 - \mu)\right] - 1} \tag{2.4}$$

The largest possible value of the chemical potential is $\mu = \epsilon_0$.

[3] We assume no energy gap here; that case will be discussed later.

For μ only slightly less than ϵ_0, we may expand (2.4) to obtain

$$\bar{n}_0 \cong \frac{1}{\beta(\epsilon_0 - \mu)} = \frac{\mathfrak{k}T}{\epsilon_0 - \mu} \qquad \text{for} \quad \epsilon_0 - \mu \ll \mathfrak{k}T \tag{2.5}$$

It follows from (2.5) that this single occupation number $\bar{n}_0$ can be made arbitrarily large, merely by choosing $\epsilon_0 - \mu$ to be sufficiently small. Since the very first term of the sum (1.9) can take arbitrarily large values, there cannot possibly be an upper limit to the value of the sum as a whole.

The term $\bar{n}_0$ of (1.9) gets lost in the approximation (1.15), since $\bar{n}_k$ is multiplied by the density of single-particle states k, which vanishes at $k = 0$. It turns out, however (and we shall show this explicitly in what follows), that an adequate approximation to (1.9) can be obtained by using (1.15) for all states *except* $k = 0$, i.e., we make the replacement[4]

$$\sum_k g_k \to g_0 + \frac{V}{(2\pi)^3} \int d^3k\, g(k) \tag{2.6}$$

If the general quantity g_k is taken to be the average occupation number $\bar{n}_k$, Eq. (2.6) gives an improved approximation for the number density N/V, to replace the inadequate (1.16), namely,

$$\frac{N}{V} = \frac{\bar{n}_0}{V} + f(\beta, \mu) \tag{2.7}$$

where $\bar{n}_0$ is given by Eq. (2.4) and $f(\beta, \mu)$ is defined by (1.16) and evaluated in (1.17) and (1.19).

Let us follow what happens when we keep the temperature, and hence the thermal wavelength λ, fixed, and increase N/V by inserting more particles into the box. Initially, when the gas is highly dilute, we are in the "classical" region, $(V/N)^{1/3} \gg \lambda$; μ is large and negative, so that $\bar{n}_0$, Eq. (2.4), is less than unity; the corresponding term of (2.7) is of order $1/V$ and thus of no consequence in statistical mechanics: all expressions in statistical mechanics must be understood to apply to the limit of very large volume V and large numbers N, the limit being taken in such a way that the number density N/V remains constant.

[4] We assume here that single-particle state ϵ_0 is non-degenerate. If it is degenerate (for example, if the bosons have spin 1, so that ϵ_0 is threefold degenerate), the analysis becomes more complicated, and the model of non-interacting particles becomes unrealistic. We mention, however, that such a gas would be ferromagnetic in the condensed state, if the spin 1 is associated with a magnetic moment. In the condensed state, the moments line up with each other, leading to a macroscopic permanent magnetization. The direction in space of this magnetization vector is determined by whatever slight external field there may be present, i.e., the system is ferromagnetic.

Now we put more particles into the box, so that N/V increases; the chemical potential μ increases as we do so, until μ comes close to its maximum value, $\mu = \epsilon_0$; this happens for $N/V = f_{\max}$, Eq. (2.3). At this stage, the first term on the right-hand side of (2.7) suddenly becomes appreciable, and *all additional particles go into the single-particle ground state*, $k = 0$. Since μ is very close to ϵ_0 at this stage, the replacement of (2.4) by (2.5) is permissible. Solving for μ, we obtain

$$\mu = \epsilon_0 - \frac{\mathfrak{k}T}{\bar{n}_0} \qquad \text{(condensation region)} \tag{2.8}$$

Since $\bar{n}_0$ must be a finite fraction of all particles N now, and N is proportional to the volume V, Eq. (2.8) shows that *the difference* $\epsilon_0 - \mu$ *is of order* $1/V$ *in the condensation region*. As far as the second term on the right-hand side of (2.7) is concerned, we can set $\mu = \epsilon_0$ in $f(\beta, \mu)$, so this term is just $f_{\max}$. The practical way to determine the Bose-Einstein distribution in this region is therefore as follows. We start from the given quantities V, T, N. We determine the thermal wavelength λ, and hence $f_{\max}$, (2.2). If N/V exceeds $f_{\max}$, the gas is condensed; we find the occupation number of the single-particle ground state from (2.7):

$$\bar{n}_0 = N - Vf_{\max} \qquad \text{(condensation region)} \tag{2.9}$$

Inserting this value of $\bar{n}_0$ into (2.8) then yields the chemical potential μ under the given conditions.

It remains to check that this procedure is consistent, i.e., that (2.6) is a sufficiently accurate formula for our purposes. To do so, let us split off the first *two* terms from the discrete sum over k, before replacing the remainder by an integral. This gives

$$N = \bar{n}_0 + \bar{n}_1 + Vf(\beta, \mu) \tag{2.10}$$

where $\bar{n}_1$ is the average occupation number of the single-particle state ϵ_1. The excitation energy $\epsilon_1 - \epsilon_0$ is usually infinitesimally small in the limit of large volume V (the exceptional case of an energy gap will be discussed later). We therefore can make, for $\bar{n}_1$, the same expansion which led from (2.4) to (2.5). This yields

$$\bar{n}_1 = \frac{1}{\exp[\beta(\epsilon_1 - \mu)] - 1} \cong \frac{\mathfrak{k}T}{\epsilon_1 - \mu} = \frac{\mathfrak{k}T}{(\epsilon_1 - \epsilon_0) + (\epsilon_0 - \mu)} \tag{2.11}$$

Our procedure is consistent provided $\bar{n}_1$ is of a lower order of magnitude than $\bar{n}_0$. This is indeed the case: the excitation energy $\epsilon_1 - \epsilon_0$ is proportional to $1/L^2$ where L is the side-length of the box [see Eqs. (1.1) and

(1.2)], whereas $\epsilon_0 - \mu$ is of order $1/V = 1/L^3$. Thus, in the last form on the right-hand side of (2.11), $\epsilon_0 - \mu$ can be ignored compared to $\epsilon_1 - \epsilon_0$, and we obtain the volume-dependences:

$$\bar{n}_1 \sim L^2 \qquad \text{compared to} \qquad \bar{n}_0 \sim L^3 \tag{2.12}$$

Looking at (2.7), we see that the significant quantity is the (intensive) particle number density N/V. According to (2.12), $\bar{n}_0/V$ is finite and must be retained, but $\bar{n}_1/V$ is of order $1/L$ and hence vanishes in the limit of large L; thus it is permissible to absorb $\bar{n}_1/V$ into the integral for $f(\beta, \mu)$, i.e., the approximation (2.6) is adequate. Q.E.D.[5]

So far, we have evaluated all formulas with the special assumptions (1.1) for the single-particle energies ϵ_k, and for a three-dimensional system [this latter assumption is contained in (2.6), of course]. Let us now see more generally under what conditions this Bose-Einstein condensation occurs.

The crucial condition is clearly that $f(\beta, \mu)$ has a finite maximum value, i.e., the integral (2.1), or its counterpart for a two-dimensional or one-dimensional system, should converge to a finite result. If and only if this is the case, do we get Bose-Einstein condensation as we increase the particle density at constant temperature.[6]

The crucial region in (2.1) is k near 0; different assumed energy level distributions ϵ_k, and/or different numbers of dimensions can give quite different results.

As an example, consider the Bose-Einstein gas *in two dimensions*. For a two-dimensional system, the factor $V/(2\pi)^3$ in (1.15) or (2.6) is replaced by $(L/2\pi)^2$, and the integration is over a two-dimensional rather than a three-dimensional k-space. Equation (2.7) is replaced by

$$\frac{N}{L^2} = \frac{\bar{n}_0}{L^2} + f_2(\beta, \mu) \qquad \text{(two dimensions)} \tag{2.13}$$

where

$$f_2(\beta, \mu) = \frac{1}{(2\pi)^2} \int_0^\infty \frac{2\pi k \, dk}{\exp[\beta(\epsilon_k - \mu)] - 1} \qquad \text{(two dimensions)} \tag{2.14}$$

---

[5] Strictly speaking, one should check that retaining the first cL terms (c = constant) of (1.9) explicitly, rather than merely the first two terms, does not produce a significant change from the approximation (2.6). This more detailed argument is given by London (38).

[6] In all practical cases, lowering the temperature at constant particle density also produces condensation, since any *finite* $f_{\max}(\beta)$ approaches 0 in the limit of zero temperature, $\beta \to \infty$; this follows from dimensional arguments applied to (2.1). The condensation process at constant temperature and variable particle density is easier to discuss, however.

The maximum value of f_2 is

$$f_{2,\max}(\beta) = \frac{1}{(2\pi)^2} \int_0^\infty \frac{2\pi k\, dk}{\exp[\beta(\epsilon_k - \epsilon_0)] - 1} \qquad \text{(two dimensions)} \qquad (2.15)$$

Equation (2.14) replaces (1.16); Eq. (2.15) replaces (2.1).

Near $k = 0$, the integrand in (2.15) is proportional to $k/(\epsilon_k - \epsilon_0)$. If we use the ordinary single-particle energies (1.1), this means the integrand is proportional to $1/k$, and the integral (2.15) diverges logarithmically. There is no finite upper limit to $f_2(\beta, \mu)$, and hence *the ordinary Bose-Einstein gas does not condense in two dimensions. Bose-Einstein statistics by itself if not a sufficient condition for Bose-Einstein condensation.*

The fact that the two-dimensional Bose gas fails to condense can be seen also from a comparison between $\bar{n}_0$ and $\bar{n}_1$ along the lines of (2.5), (2.11), and (2.12). We shall *assume* condensation, and derive a contradiction, as follows: according to (2.5), $\epsilon_0 - \mu$ must be of order $1/V$ in order that $\bar{n}_0$ should be of order V. In the two-dimensional case, V must be replaced by the area, L^2, of the box, thus we have $\epsilon_0 - \mu \sim 1/L^2$. Equation (2.11) remains correct, but now both $(\epsilon_1 - \epsilon_0)$ and $(\epsilon_0 - \mu)$ are of the same order of magnitude, i.e., of order of magnitude $1/L^2$. Thus, the assumption $\bar{n}_0 \sim L^2$ leads to the conclusion $\bar{n}_1 \sim L^2$, also. This is a contradiction to the initial assumption of Bose-Einstein condensation, since this would have meant $\bar{n}_1$ of lower order than $\bar{n}_0$. It is interesting, and will turn out useful, that the essential decision between condensation or no condensation can be obtained by such a limited consideration, in which we look at only the two lowest single-particle levels, ϵ_0 and ϵ_1, and their occupation numbers.

The simplest general way of stating the condition for Bose-Einstein condensation is to introduce the density of single-particle levels: let $\rho(\epsilon)\, d\epsilon$ be the number of single-particle levels ϵ_k in the range $\epsilon < \epsilon_k < \epsilon + d\epsilon$. In the "normal" region, above condensation, we can then replace the sum in (1.9) by an integral over ϵ, to get

$$N = \int_{\epsilon_0}^\infty \frac{\rho(\epsilon)\, d\epsilon}{\exp[\beta(\epsilon - \mu)] - 1} \qquad \text{(normal region)} \qquad (2.16)$$

The maximum permissible value of the chemical potential is $\mu = \epsilon_0$; this yields

$$N_{\max} = \int_{\epsilon_0}^\infty \frac{\rho(\epsilon)\, d\epsilon}{\exp[\beta(\epsilon - \epsilon_0)] - 1} \qquad (2.17)$$

If we restrict ourselves to simple functional forms for $\rho(\epsilon)$, then the condition for convergence of this integral, and hence for the occurrence of Bose-Einstein condensation, is

$$\lim_{\epsilon \to \epsilon_0} \rho(\epsilon) = 0 \qquad \text{(condition for condensation)} \tag{2.18}$$

It is easily seen that the examples given so far are consistent with (2.18).

We speak of an *energy gap* $\Delta\epsilon$ if the single-particle spectrum ϵ_k has a finite, volume-independent first excitation energy $\epsilon_1 - \epsilon_0 = \Delta\epsilon$. In that case, $\rho(\epsilon)$ is zero below $\epsilon = \epsilon_1$, and such a Bose gas condenses in any number of dimensions. We emphasize that this is a sufficient, but by no means a necessary, condition for Bose-Einstein condensation.

It is important at this stage to clarify the nature of the condensed state in the Bose-Einstein condensation, and in particular, to contrast this condensed state with the quantum-mechanical ground state E_0 of the N-particle gas. In terms of the "configurations" (1.3), the quantum-mechanical ground state E_0 of the N-particle system corresponds to the configuration

$$\alpha_0 = \{n_0\,, n_1\,, n_2\,, n_3\,, \ldots\} = \{N, 0, 0, 0, \ldots\} \tag{2.19}$$

This state is completely different from the "condensed state" in statistical mechanics, and has essentially vanishing probability of occurrence at any finite temperature, no matter how small.

The probability P_α of finding the state α at a given temperature is given by the general law (1.6). By looking at (1.7), we see that the free energy F is always less than the ground state energy E_0, but F approaches E_0 from below in the limit $T \to 0$. Furthermore, both F and E_0 are proportional to the size (the volume V) of the system, for all reasonable systems; i.e., F and E_0 are *extensive* quantities in the thermodynamic sense of that word. Hence the probability of finding the system in its true quantum-mechanical ground state E_0 is

$$P_0 = \exp\left[-\frac{E_0 - F(T)}{\mathfrak{k}T}\right] \tag{2.20}$$

where the exponent is negative and proportional to the volume of the system, at all finite temperatures T. Thus, in the limit of very large volume, the probability P_0 becomes exponentially small like $\exp(-aV)$, where a is independent of the volume.

Another way of saying the same thing is as follows. According to the equipartition principle, the average kinetic energy per degree of freedom is $\frac{1}{2}\mathfrak{k}T$. The equipartition law fails in quantum statistical mechanics, but we may still use it for order-of-magnitude estimates. The average energy $\bar{E}$ of an N-particle system is *not* of order $\mathfrak{k}T$, but of order $N\mathfrak{k}T$. Since N is of the order of Avogadro's number, $N \sim 10^{23}$, the factor N cannot be ignored. If we really want the system to have appreciable probability of being in its ground state E_0, the average excitation energy $\bar{E}$ must be less than, or of the order of, the first excitation energy $E_1 - E_0$. This implies the condition

$$N\mathfrak{k}T \lesssim E_1 - E_0 \tag{2.21}$$

However, $E_1 - E_0$ is independent of the size of the system (independent of N), so the relevant temperature T is proportional to $1/N$, and is vanishingly small for any system of reasonable size. This is true even if there is an energy gap ΔE; the relevant characteristic temperature is not $\Delta E/\mathfrak{k}$ but $\Delta E/N\mathfrak{k}$.

What, then, is the physical meaning of the "condensed state" in Bose-Einstein condensation? The configurations α which make the dominant contributions are *not* (2.19) or neighboring configurations, as we have just seen. Rather, the contributing configurations $\alpha = \{n_0, n_1, n_2, ...\}$ of the "condensed" state have the common property that *n_0 is a finite fraction of the total number N, whereas all other n_k are infinitesimally small compared to N.* If and only if this type of configuration dominates the probability distribution *at a finite, volume-independent temperature* do we speak of a Bose-Einstein condensation.

In the limit of strictly zero temperature, the ground state probability P_0 does become unity, of course, and the N-particle ground state E_0 is the actual state of the system. But what matters here is the way in which this limit is approached. In a two-dimensional Bose gas, the approach is "smooth": as the temperature decreases, the occupation numbers $\bar{n}_k$ for states with high ϵ_k decrease rapidly, the $\bar{n}_k$ for low ϵ_k increase, in such a way that N, Eq. (1.9), stays constant. At very low temperatures, all the occupation is concentrated in levels $\epsilon_k \lesssim \mathfrak{k}T$. But, there is no finite, volume-independent temperature at which the one single-particle level ϵ_0 is occupied unusually highly; the average occupation numbers $\bar{n}_0$ and $\bar{n}_1$ are of the same order of magnitude, at all temperatures.

By contrast, the three-dimensional Bose gas approaches the limit of strictly zero temperature quite differently. There is a finite, volume-independent temperature T_c at which $f_{\max}$, Eq. (2.2), equals the (constant) number density N/V. At all temperatures $T < T_c$, the average occupation number of the single-particle ground state, $\bar{n}_0$, is qualitatively dissimilar to all other occupation numbers. $\bar{n}_0$ is proportional to the volume $V = L^3$ (at constant N/V), whereas all the other $\bar{n}_k$ increase more slowly with volume. The Bose condensation means that *the Bose-Einstein statistics and the density of single-particle levels combine so as to select out the single-particle ground state* ϵ_0 *for especially high occupation, over a volume-independent, finite range of temperature* $0 < T < T_c$. The quantum-mechanical ground state E_0 is not involved here at all.

We stress the difference between states of interest in statistical mechanics, and the quantum mechanical ground state, because the two are so often confused. It is frequently true that quantum mechanical averages over the ground state wave function provide useful first approximations for statistical averages over the canonical ensemble, at sufficiently low temperatures. But this is by no means always true. For example, we shall see in Section 5 that the ground state of the Bose gas, in the presence of a sufficiently strong applied magnetic field, is qualitatively different from the relevant statistical ensemble even in the limit of zero temperature.

Superfluids are peculiar systems, and one must proceed with considerable caution. If one wishes to calculate with the ground state rather than with a statistical ensemble, this procedure must be motivated physically, at the very least. If it is merely an assumption, it is a very restrictive and severe assumption, which often amounts to assuming a property equivalent to the physical property it is desired to establish. This particular confusion has plagued many so-called theories of superconductivity in the past.

3. The Two-Fluid Model

We now show that the Bose-Einstein gas, below the condensation temperature T_c, behaves in a certain sense like a mixture of two fluids, a "normal fluid" and a "superfluid."

Let J be any quantum-mechanical operator representing a physical variable of the N-body system. The quantum-mechanical expectation value of J in quantum state α of the N-particle system is given by the $3N$-dimensional integral

$$J_{\alpha\alpha} = \int \Psi_\alpha^* J \Psi_\alpha \, d^{3N}r \tag{3.1}$$

In thermal equilibrium, the probability of finding the system in quantum state α is P_α, Eq. (1.6). Thus, the statistical expectation value of the quantity J is given by

$$\bar{J} = \sum_\alpha P_\alpha J_{\alpha\alpha} \tag{3.2}$$

For a completely general operator J, that is all that can be said. But most of the operators of practical interest are quite special in form, and for those operators the general expression (3.2) can be simplified considerably.

We shall be concerned very largely with *sums of one-particle operators*, i.e., N-particle operators J of the special form

$$J = \sum_{i=1}^{N} j_i \tag{3.3}$$

where the operator j_i acts only on the coordinates of particle number i in the wave function $\Psi_\alpha(r_1, r_2, ..., r_i, ..., r_N)$. Examples of sums of one-particle operators are

(i) The number operator: $j_i = 1$
(ii) The Hamiltonian of an independent-particle system: $j_i = (\mathbf{p}_i)^2/2M$
(iii) The electric current operator: $\mathbf{j}_i = e\mathbf{p}_i/M$

If the system is an independent-particle system, i.e., if H is a sum of one-particle operators, then there exist single-particle energy levels ϵ_k and corresponding single-particle wave functions $\phi_k(r)$. We define the matrix elements of the single-particle operator j by the usual equation:

$$j_{kk'} = \int \phi_k^* \, j \, \phi_{k'} \, d^3r \tag{3.4}$$

Unlike (3.1), the integration in (3.4) is over a three-dimensional space only. The subscripts k, k' refer to single-particle states, *not* to the number i of the particle, which is irrelevant here.

It can be shown (see Appendix A) that the general formula (3.2) reduces to

$$\bar{J} = \sum_k \bar{n}_k j_{kk} \tag{3.5}$$

under the stated conditions (independent-particle system, J a sum of one-particle operators). The $\bar{n}_k$ is given by (1.12), (1.13), or (1.14), depending upon the statistics of the particles. Equation (3.5) is remarkable in that only the *diagonal* elements j_{kk} enter, and in that it is formally the same for all statistics.

Let us apply the general expression (3.5) to the condensed Bose-Einstein gas. Then we must separate out the term $k = 0$ from the sum (3.5), before replacing the sum by an integral. The result is

$$\bar{J} = \bar{n}_0 j_{00} + \bar{J}_n \tag{3.6}$$

where $\bar{J}_n$ is the "normal" contribution, given by

$$\bar{J}_n = \frac{V}{(2\pi)^3} \int d^3k \; \bar{n}_k j_{kk} \tag{3.7}$$

and the other term may be considered the "superfluid" contribution and denoted by $\bar{J}_s$:

$$\bar{J}_s = \bar{n}_0 j_{00} \tag{3.8}$$

At temperatures above the condensation temperature, (3.8) is unimportant since $\bar{n}_0$ is of order 1 rather than of order N; for $T < T_c$, however, the contribution $\bar{J}_s$ is comparable to $\bar{J}_n$, i.e., both are proportional to the size of the system.

If J is the number operator, we have $j_i = 1$, $j_{kk'} = \delta_{kk'}$, $j_{00} = 1$, and thus

$$\bar{N}_s = \bar{n}_0 \tag{3.9}$$

The number of "superfluid" particles is the average occupation number of the single-particle ground state ϕ_0.

If J is the Hamiltonian of the independent-particle system, then $j_{kk'} = \epsilon_k \delta_{kk'}$ and $j_{00} = \epsilon_0$, the single-particle ground state energy. Without loss of generality, we may set $\epsilon_0 = 0$; this merely determines the zero of our energy scale. Under those conditions, (3.8) shows that *the "superfluid" makes no contribution to the internal energy $\bar{E}$ of the system.* Since the specific heat c_V equals $(\partial\bar{E}/\partial T)_{V,N}$, *the superfluid makes no contribution to the specific heat.* By inserting this result in the general expression for the entropy, Eq. (I,1.2), we conclude that *the superfluid makes no contribution to the entropy.*

These are exactly the properties postulated for the superfluid component in liquid helium II (Tisza 38, 38a, 40, 40a, 47, 49; Landau 41, 44, 47, 47a, 49), and this makes it very tempting to consider the condensed ideal Bose-Einstein gas a theoretical model for actual liquid helium II (London 38). The model has serious shortcomings, which we need not discuss here; but it contains a considerable element of truth, as is shown by the fact that the isotope He^3, which obeys Fermi-Dirac statistics, does not undergo a lambda-transition similar to the transition in He^4.

The relationship between the Bose-Einstein gas model and superconductivity is obviously more complex, since electrons in metals obey Fermi-Dirac statistics; we shall return to this problem in Chapter III.

For the moment, let us confine ourselves to the following useful observation: at temperatures $T \ll T_c$, the de Broglie thermal wavelength λ, Eq. (1.18), is much larger than the mean interparticle distance $(V/N)^{1/3}$. Under these conditions, the number of normal particles, $\bar{N}_n = Vf_{\max}$ [see Eq. (2.9)], approaches zero, and the number of superfluid particles, $\bar{N}_s = \bar{n}_0$, approaches the total particle number N. Inspection of (3.6) shows that the statistical expectation value $\bar{J}$ is given approximately by

$$\bar{J} \simeq Nj_{00} \qquad \text{for} \qquad T \ll T_c \tag{3.10}$$

This, however, is equal to the quantum-mechanical expectation value of the N-particle operator J over the N-particle ground state wave function Ψ_0:

$$J_{00} = \int \Psi_0^* \, J \, \Psi_0 \, d^{3N}r = Nj_{00} \tag{3.11}$$

Thus, *if* the system undergoes a Bose-Einstein condensation, and *if* the operator J of interest is a sum of single-particle operators, *then* the statistical expectation value $\bar{J}$ may be approximated, at temperatures $T \ll T_c$, by the quantum-mechanical expectation value J_{00}. This provides a very useful approximation.

Mathematically, the need for care here is dictated by the fact that the order of the two limiting processes: $T \to 0$ and $V \to \infty$, cannot be interchanged. In statistical mechanics, the limit of infinite volume must be performed *first*, i.e.,

$$\frac{\bar{J}}{V} \equiv \lim_{\substack{V\to\infty \\ N\to\infty \\ N/V=\text{constant}}} \left[\frac{\bar{J}(V,N,T)}{V}\right]$$

and the limit of zero temperature, if desired, must be performed *afterward*. *If* the system undergoes Bose-Einstein condensation, *then* the two limits can be interchanged with impunity. If the system fails to undergo Bose-Einstein condensation, then the two limits may, and often do, give very different results depending upon the order in which they are performed.

4. The Meissner Effect at Zero Temperature

Let us now investigate the magnetic response of the Bose-Einstein gas in the limit of zero temperature. As we have just seen, we may investigate this limiting case by employing the quantum-mechanical ground state, provided that there is Bose condensation. We shall assume so for the moment, but we shall have to check this condition for consistency later on, and we shall find that there exist cases (magnetic field above the critical value) in which this assumption of condensation is inconsistent.

The quantum-mechanical ground state Ψ_0 has the form

$$\Psi_0 = \phi_0(r_1)\,\phi_0(r_2)\,...\,\phi_0(r_N) \tag{4.1}$$

where $\phi_0(r)$ is the one-particle ground state wave function. This wave function is a solution of the one-particle Schrödinger equation

$$\frac{1}{2M}\left[-\,i\hbar\,\nabla - \frac{e}{c}\,\mathbf{A}(\mathbf{r})\right]^2 \phi_0 + e\Phi(\mathbf{r})\phi_0 = \epsilon_0\,\phi_0(\mathbf{r}) \tag{4.2}$$

where $\mathbf{A}(\mathbf{r})$ is the vector potential, $\Phi(\mathbf{r})$ the scalar potential at point $\mathbf{r}$, and ϕ_0 satisfies the boundary condition

$$\phi_0(\mathbf{r}) = 0 \qquad \text{on the surface of the volume} \quad V \tag{4.3}$$

The quantum-mechanical energy E_0 of the N-particle system is

$$E_0 = N\epsilon_0 \tag{4.4}$$

This energy does *not* include the energy contained in the electromagnetic field described by $\mathbf{A}$ and Φ, since the electromagnetic field is considered a given quantity here.

To avoid unessential complications, we consider the problem of the infinite slab; the volume V of the gas is contained between $x = a$ and $x = -\,a$; in the y- and z-directions there are periodic boundary conditions at $y = L$ and $z = L$; the effective volume is $V = 2aL^2$.

The simplest case is the gas subject to no electromagnetic fields, i.e., $\mathbf{A} = \Phi = 0$ in (4.2). The ground state wave function ϕ_0 is given by

$$\phi_0(\mathbf{r}) = (2/V)^{1/2} \cos(\pi x/2a) = (aL^2)^{-1/2} \cos(\pi x/2a) \tag{4.5}$$

and the single-particle ground state energy is

$$\epsilon_0 = \pi^2\hbar^2/8Ma^2 \tag{4.6}$$

Next, let us expose the system to the action of a prescribed magnetic field **B**, derived from a given vector potential **A**. Each gas particle is assumed to carry a test charge e, through which it responds to the magnetic field. The assumption of a "test" charge implies that the particles do not themselves create a significant electromagnetic field (we shall have to relax this assumption later on).

If we "switch on" the vector potential slowly, the N-particle ground state wave function Ψ_0 retains the *form* (4.1); *the N-particle system remains in its ground state.* What changes is not (4.1), but the details of the single-particle ground state wave function ϕ_0, which now satisfies (4.2) with $\mathbf{A} \neq 0$, but Φ still zero.

Before solving (4.2) explicitly for small fields, let us first establish gauge invariance. The gauge transformation

$$\mathbf{A}' = \mathbf{A} + \operatorname{grad} f \tag{4.7}$$

on the vector potential induces the associated transformation

$$\phi_0'(\mathbf{r}) = \exp\left[\frac{ie}{\hbar c} f(\mathbf{r})\right] \phi_0(\mathbf{r}) \tag{4.8}$$

on the single-particle ground state wave function. It is trivial to verify that (4.7) and (4.8), with arbitrary $f(\mathbf{r})$, lead to a Schrödinger equation for ϕ' of exactly the form (4.2), with $\mathbf{A}'$ replacing $\mathbf{A}$. Furthermore, the current density per particle is given by (Schiff 49)

$$\mathbf{j}(\mathbf{r}) = \frac{i\hbar e}{2M}\left[(\nabla\phi_0^*)\phi_0 - \phi_0^*(\nabla\phi_0)\right] - \frac{e^2}{Mc}\phi_0^*\phi\,\mathbf{A} = \mathbf{j}_1 + \mathbf{j}_2 \tag{4.9}$$

and this expression is also invariant under the gauge transformation (4.7), (4.8).

Having checked the gauge invariance of the theory, we are free to choose whatever gauge is most convenient for our purpose. This turns out to be the London gauge, $\operatorname{div} \mathbf{A}_1 = 0$. If the magnetic field **B** is in the z-direction, $B = B_z(x)$, then the London $\mathbf{A}_1$ is in the y-direction, and satisfies

$$dA/dx = B_z(x) \tag{4.10}$$

As a result of the London gauge condition $\operatorname{div} \mathbf{A}_1 = 0$, we also have the useful identity

$$\operatorname{div}(\mathbf{A}_1\phi_0) = \mathbf{A}_1 \quad \operatorname{grad} \phi_0 \tag{4.11}$$

Since the Meissner effect involves the response of the system to a *weak* magnetic field (i.e., we retain only terms which are *linear* in the applied field), we solve the Schrödinger equation (4.2) by the perturbation method, in the standard way (Schiff 49). The perturbing Hamiltonian is the operator[7]

$$H'\phi_0 = \frac{i\hbar e}{Mc} \mathbf{A}_1(\mathbf{r}) \cdot \text{grad}\, \phi_0 + \frac{e^2}{2Mc^2} [\mathbf{A}_1(\mathbf{r})]^2 \phi_0 \tag{4.12}$$

Since $\mathbf{A}_1$ is a vector in the y-direction and grad ϕ_0 a vector in the x-direction, the first term on the right vanishes. The second term is quadratic in $\mathbf{A}_1$. Thus, *to first order in the applied field, the ground state wave function of the bosons is unaltered, in London gauge.*

Thus, again to first order in the applied field, we may use the zero-order wave function (4.5) to compute the current (4.9). The term $\mathbf{j}_1$ vanishes identically, since ϕ_0 is a *real* function of $\mathbf{r}$. The second term, $\mathbf{j}_2$, makes a contribution. Multiplying by N to get the total current [we recall that (4.9) is the current per particle], we get the equation

$$\mathbf{j}_s(\mathbf{r}) = N\mathbf{j}_2(\mathbf{r}) = -\frac{Ne^2}{Mc} \mid \phi_0(\mathbf{r}) \mid^2 \mathbf{A}_1(\mathbf{r}) \tag{4.13}$$

This expression for the supercurrent density is highly reminiscent of the London equation, in the form (I,5.10). *If* we could replace the probability density $\mid \phi_0(r) \mid^2$ by its average value, $1/V$, we would obtain the London equation in the Ginzburg-Landau form (I,5.13), with m^* and e^* the mass and charge of the boson.

However, this procedure would be quite inconsistent. If we wish to make use of the peculiar features of the Bose-Einstein condensation to get a Meissner effect, we cannot ignore these features as soon as they become inconvenient. The strong x-dependence of the ground state wave function (4.5) is characteristic of the ground state in quantum mechanics, and it makes (4.13) very different from the London equation (I,5.13).

A formal way to obtain the conventional London equation from the ideal Bose gas model consists in altering the boundary condition (4.3) to read

$$\partial\phi_0/\partial n = 0 \qquad \text{on the surface of the volume } V \tag{4.3a}$$

With this altered boundary condition, the ground state wave function is no longer given by (4.5), but rather is a constant:

$$\phi_0(\mathbf{r}) = V^{-1/2} = (2aL^2)^{-1/2} \tag{4.5a}$$

The rest of the analysis goes through unaltered. Substitution of (4.5a) into the final formula (4.13) yields the London equation in the form (I, 5.13). We note for future reference that the ideal Bose gas, with

[7] We have used (4.11) to simplify (4.2).

either boundary condition (4.3) or (4.3a), has no gap in its energy spectrum.

Both boundary conditions are quite permissible, and one of them has led us to the desired result. However, the decisive influence of the choice of boundary condition in this model is a bit disconcerting, and makes this strictly formal treatment unconvincing physically. Within the confines of a model of strictly independent particles, there is no clear reason for preferring (4.3a) to the more usual (4.3).

However, the problem we have been considering so far is physically unrealistic. If the bosons are charged, as we had to assume to get any response to a magnetic field at all, then they repel each other by Coulomb forces, and these Coulomb forces cannot be ignored. The importance of the Coulomb effect was stressed by Schafroth (55a), who made a self-consistent calculation for "condensed Bose-Einstein jellium" along the following lines: The ground state wave function (4.1), together with the "background" uniform charge of the jellium model, generate a charge density

$$\rho(\mathbf{r}) = Ne\{\,|\,\phi_0(\mathbf{r})\,|^{\,2} - 1/V\} \tag{4.14}$$

This charge density acts as source for an electrostatic potential Φ which satisfies the Poisson equation:

$$\nabla^2 \Phi = -\,4\pi\rho \tag{4.15}$$

To achieve self-consistency, we must substitute *this* potential $\Phi(r)$ into the one-particle Schrödinger equation (4.2). The resulting set of equations is non-linear, of course, but it can be solved numerically. The numerical solution is simplified greatly by introducing dimensionless quantities. We define the length δ and the energy Δ by the two equations

$$\Delta = Ne^2\delta^2/V = \hbar^2/M\delta^2 \tag{4.16}$$

so that

$$\delta = \left(\frac{\hbar^2}{me^2}\cdot\frac{V}{N}\right)^{1/4} = (a_0 d^3)^{1/4} \tag{4.17a}$$

and

$$\Delta = \frac{Me^4}{\hbar^2}\left(\frac{a_0}{d}\right)^{3/2} \tag{4.17b}$$

where $a_0 = \hbar^2/Me^2$ is the Bohr radius of the boson, $d = (V/N)^{1/3}$ is the mean distance between bosons, and the dimensional factor in (4.17b) is the Rydberg energy for a boson. We now measure distances in units of δ, energies in units of Δ; we introduce the non-dimensional quantities

$$\zeta = x/\delta, \qquad \hat{\Phi} = e\Phi/\Delta, \qquad u = V^{1/2}\phi \tag{4.18}$$

In the absence of a magnetic field, these substitutions reduce (4.2) and (4.15) to the non-dimensional forms

$$-\frac{1}{2}\frac{d^2u}{d\zeta^2} + \hat{\Phi}u = (\epsilon_0/\varDelta)u \tag{4.19a}$$

and

$$\frac{d^2\hat{\Phi}}{d\zeta^2} = -4\pi(u^2 - 1) \tag{4.19b}$$

The boundary conditions are $u = \hat{\Phi} = 0$ for $\zeta = \pm a/\delta$, corresponding to $x = \pm a$, the surface of the slab. Since all coefficients in (4.19) are simple numbers, the numerical solution leads to an eigenvalue $\epsilon_0/\varDelta$ of the order of unity; furthermore, the "surface region" in which $u(\zeta)$ rises from 0 to its final value, 1, in the interior, is also of order unity in the ζ-variable; that means, *the "surface region" has a thickness of the order of* δ, Eq. (4.17a), and *the Coulomb energy per particle is of the order of* $\varDelta$, Eq. (4.17b).

For a first orientation, we may consider the boson mass M to be of the order of an electron mass, the boson charge e to be of the order of an electronic charge, and the number density of bosons, N/V, to be of the order of the number density of conduction electrons, say, 10^{22} cm^{-3}. With these assumptions, δ is of order 10^{-8} cm, very much smaller than the linear dimensions of interest in superconductivity (the London depth and the Pippard coherence length). Except for this extremely narrow transition region, the *self-consistent* ground state wave function $\phi_0(r)$ is constant inside the slab, and equal to $V^{-1/2}$. *The self-consistent treatment of the Coulomb repulsion between the bosons results in* (4.13) *becoming equivalent to the London equation almost everywhere within the volume*; the transition region, in which the variation of $\phi_0(x)$ cannot be ignored, is so narrow that it makes no practical difference.

The narrowness of the transition region is itself a consequence of the Bose condensation. It arises from the factor N in (4.14): the potential $\Phi(r)$ acting on any one boson has as its source the coherent charge density produced by all N bosons. Thus a small deviation of $\phi_0(r)$ from its average value, $V^{-1/2}$, produces an enormous net charge density in the jellium model, and such deviations can therefore occur only over very small regions.

It remains to check that a first-order treatment of the applied magnetic field still leads to (4.13); this check is straightforward, however, since the argument leading to (4.13) did not make any use of the specific dependence of $\phi_0(x)$ on x.

This concludes the proof that the ideal Bose gas in the condensed state has a Meissner effect, governed by the London equation.[7a] Before going on to study the critical field of the Bose gas model, let us make some observations: The crucial equation (4.13) arises from the general equation (4.9) through the fact that the ground state wave function ϕ_0 of the bosons is unaltered, in London gauge, upon switching on the magnetic field, to first order in the field strength. That is, the wave function is "stiff," in London's terminology. Since ϕ_0 for zero field must lead to zero current, a "stiff" ϕ_0 makes j_1 in (4.9) vanish, and only the "London term" j_2 survives. We must, however, be careful just what this "stiffness" really means. It does *not* mean that ϕ_0 is unaltered by switching on the vector potential **A**, no matter what the gauge of **A**. On the contrary, for all gauges other than the London gauge, ϕ_0 is far from "stiff," as is shown by (4.8). Thus, in all gauges other than the London gauge, both j_1 and j_2 in (4.9) contribute significantly; only their sum is gauge-invariant.

One could say that the one-particle wave function is "stiff" in one particular gauge (the London gauge), but such a statement has no physical meaning, since the gauge of the vector potential is not a physically meaningful concept at all; the vector potential, and with it its gauge, is a mathematical artifice; the physically measurable, and hence meaningful, quantity, is the magnetic field.

What, then, is meant by the term "stiff wave functions" applied to the condensed Bose gas? The one-particle wave function $\phi_0(r)$ must be "flexible," so what remains "stiff"? The answer is quite simple: The "stiffness" is a *many-particle* effect induced by the Bose-Einstein condensation; no matter how "flexible" $\phi_0(r)$ may be, the N-particle ground state wave function Ψ_0, Eq. (4.1), is certainly very "stiff": all factors $\phi_0(\mathbf{r}_i)$, $i = 1, 2, \ldots, N$, involve the *same* function ϕ_0. This is an extremely strong restriction on the permissible many-particle motions. What remains "stiff" is the *structure* of the N-particle wave function, i.e., the fact of Bose-Einstein condensation. This, and this alone, produces the "spectacular" magnetic response of the condensed Bose gas, given by (4.13).[8]

The second observation concerns the question of an energy gap.

[7a] For an alternative derivation of the Meissner effect in an interacting Bose gas, see Schwabl (62).

[8] The example worked out at the beginning of the section, ignoring the Coulomb repulsions between the bosons, shows immediately that even "stiff" wave functions do not necessarily lead to the detailed London equation. But the magnetic response (4.13) with ϕ_0 given by (4.5), though not identical with London's phenomenology, would certainly be considered spectacular; and alteration of the boundary condition from (4.3) to (4.3a), and ϕ_0 from (4.5) to (4.5a), does give the conventional London equation.

The simple energy spectrum (1.1) and (1.5) has no energy gap. Since we have just obtained the Meissner effect, we conclude that *an energy gap is not required to obtain a Meissner-Ochsenfeld effect.* This observation supplements the earlier remark (Chapter I, Section 8) that an energy gap is not a sufficient condition for the Meissner effect. Far from being causally connected, *the existence of an energy gap and of a Meissner effect are substantially unrelated to each other.* There exist energy-gap systems without Meissner effect (all insulators), and superconducting systems without energy gap (the ideal Bose gas).[9]

A look back over the derivation in this section shows clearly that (in the limit of zero temperature, for a condensed Bose gas) the Meissner effect is a property of the ground state wave function Ψ_0, and does not involve excited states at all. Thus the energies of excited states are irrelevant. The only way in which excited states enter is through the criterion for Bose condensation, Eq. (2.18). Once this is satisfied, we can forget about excited states.

The third observation concerns the relationship to the Ginzburg-Landau theory (Chapter I, Section 5). There is an obvious similarity between the results derived in this section and the linearized version of the Ginzburg-Landau theory, for example, compare (I,5.13) and the present (4.13). The similarity extends even to boundary conditions: Ginzburg and Landau use the condition $\partial\psi_s/\partial n = 0$ on the boundary, for their "wave function of the superelectrons." Since the self-consistent treatment of the Coulomb effect for the Bose gas has led us to an extremely narrow transition region at the surface, of order (4.17a) in thickness, the Bose particle wave function $\phi_0(x)$ may be replaced, for all practical purposes, by a constant, i.e., by a solution of the Schrödinger equation satisfying the same boundary condition as the Ginzburg-Landau assump-

[9] We apologize for stressing the obvious; but there is an enormous amount of confusion in the literature concerning this question.

Calculations based on an "elementary excitation" picture are often done, and are exceedingly useful. We shall consider them in Chapter VI. But the elementary excitations in superfluid systems provide a microscopic description of "normal fluid" properties in the two-fluid model; they do not describe, or explain, the true "superfluid" properties. Sometimes one may argue from superfluid properties to specific normal fluid properties: for example, from the Meissner effect at zero frequency, to a violation of the optical f-sum rule extended over non-zero frequencies. But an argument in the opposite direction, from normal fluid properties (e.g., the energy gap) to true superfluid properties, is of very doubtful validity and provides no physical understanding of the origin and cause of superfluidity. An energy gap in the normal excitation spectrum indicates that the normal fluid is hard to excite, and not likely to carry a supercurrent. The mechanism by which the supercurrent *is* carried, or the conditions under which we have a superconductor rather than an insulator, cannot be understood by looking at elementary excitations.

tion. More fundamentally, the essential assumption of the Ginzburg-Landau theory, that all "superelectrons" can be described by one and the same wave function ψ_s, corresponds precisely to the Bose condensation in the Bose-Einstein gas model.

5. The Critical Field of the Ideal Bose Gas, at Zero Temperature

In order to study the question of a critical field for the Bose gas model, we now turn our attention to the *normal* state of the Bose gas in the presence of a magnetic field. *If* a sufficiently high magnetic field $\mathfrak{H}_c$ destroys superconductivity in the Bose gas, *then* for fields $\mathfrak{H} > \mathfrak{H}_c$ the Bose gas must be in the normal, uncondensed state, even in the limit of very low temperature. We shall make this assumption, and check its consistency afterwards.

We begin by studying the energy levels of a single particle in the presence of a constant "acting" magnetic field $\mathfrak{H}'$. The "acting field" is the applied field, modified by the average field due to all the other particles (Becker 33). We shall return to the relationship between $\mathfrak{H}'$ and the applied field $\mathfrak{H}$ later. Unlike the condensed Bose gas, with its Meissner effect, the normal Bose gas allows the presence of a constant (in space) magnetic field, so there is nothing inconsistent in that assumption.

We follow closely the treatment of Peierls (55) for the analogous case of the Fermi gas. We consider the infinite slab, bounded by $x = a$ and $x = -a$, with periodic boundary conditions in the y- and z-directions, periodicity distance L. The constant acting field $\mathfrak{H}'$ in the interior of the slab, assumed to be in the z-direction, can be represented as the curl of the vector potential:

$$A_x = 0, \qquad A_y = \mathfrak{H}'x \qquad A_z = 0 \tag{5.1}$$

The Schrödinger equation becomes

$$-\frac{\hbar^2}{2M}\left[\frac{\partial^2}{\partial x^2} + \left(\frac{\partial}{\partial y} - \frac{ie}{\hbar c}\,\mathfrak{H}'x\right)^2 + \frac{\partial^2}{\partial z^2}\right]\phi(x, y, z) = \epsilon\,\phi(x, y, z) \tag{5.2}$$

Unlike the case of the condensed Bose gas, it is not necessary here to include the self-consistent effect of the Coulomb repulsion to ensure a substantially uniform charge density; we shall get a uniform charge density from the normal Bose distribution, anyway.

The differential equation (5.2) is invariant under displacements in the y- and z-directions. We therefore may assume eigenfunctions ϕ of the form

$$\phi(x, y, z) = u(x) \exp(ik_2 y + ik_3 z) \tag{5.3}$$

The periodic boundary conditions imply

$$k_2 = n_2(2\pi/L), \qquad k_3 = n_3(2\pi/L) \tag{5.4}$$

where n_2 and n_3 are positive or negative integers. Substitution of (5.3) into (5.2) gives

$$-\frac{\hbar^2}{2M}\frac{d^2u}{dx^2} + \frac{(e\mathfrak{H}')^2}{2Mc^2}(x - x_0)^2 u = \left[\epsilon - \frac{(\hbar k_3)^2}{2M}\right] u \tag{5.5}$$

where

$$x_0 = \hbar c k_2 / e\mathfrak{H}' \tag{5.6}$$

Equation (5.5) is the wave equation of a simple harmonic oscillator, centered at $x = x_0$, with spring constant

$$K = (e\mathfrak{H}')^2/Mc^2 \tag{5.7}$$

The classical frequency of the oscillator is

$$\omega = (K/M)^{1/2} = e\mathfrak{H}'/Mc \tag{5.8}$$

which is twice the Larmor frequency. The boundary conditions on $u(x)$ make (5.5) different from an ordinary oscillator; the boundary conditions are

$$u(a) = u(-a) = 0 \tag{5.9}$$

However, as long as x_0, Eq. (5.6), lies well inside the slab, the natural falloff of the oscillator wave functions is sufficient to make (5.9) very nearly satisfied even if condition (5.9) is not imposed explicitly. Under these circumstances, then, the energy eigenvalues are

$$\epsilon_{mk} \cong \frac{(\hbar k_3)^2}{2M} + (m + \tfrac{1}{2})\hbar\omega = \frac{(\hbar k_3)^2}{2M} + (2m + 1)\mu_0 \mathfrak{H}' \tag{5.10}$$

where m takes the values 0, 1, 2, 3, ..., and μ_0 is the Bohr magneton

$$\mu_0 = e\hbar/2Mc \tag{5.11}$$

As indicated by the notation ϵ_{mk}, the quantum number k_2, corresponding to the motion in the y-direction, does not appear in ϵ at all. The energy eigenvalues are therefore highly degenerate. The degeneracy is equal to the number of distinct values of k_2 consistent with $-a < x_0 < a$, For x_0 outside that range, the oscillator solutions are quite inconsistent with (5.9), and in fact no eigensolutions exist. Using (5.4) and (5.6), we get the result

$$\text{Degeneracy of } \epsilon_{mk} \cong 2aL \frac{e\mathfrak{H}'}{2\pi\hbar c} \tag{5.12}$$

We are now in a position to do statistical mechanics; at this point, our problem becomes different from the one discussed in Peierls' book, since we have Bose-Einstein rather than Fermi-Dirac statistics. The total particle number is given by (1.9); since the oscillator quantum number m is a discrete variable, we write (1.9) in the form

$$N = \sum_{m=0}^{\infty} N_m \tag{5.13}$$

where (using $V = 2aL^2$)

$$N_m = \frac{V}{(2\pi)^2} \frac{e\mathfrak{H}'}{\hbar c} \int_{-\infty}^{+\infty} dk \frac{1}{\exp\{\beta[(2m+1)\mu_0\mathfrak{H}' + \hbar^2k^2/2M - \mu]\} - 1} \tag{5.14}$$

As before, μ is the chemical potential; but now the maximum value of μ is not 0, but the single-particle ground state energy [$m = k_3 = 0$ in Eq. (5.10)]

$$\mu_{\max} = \mu_0\mathfrak{H}' \tag{5.15}$$

The complicated sums and integrals in (5.13) and (5.14) have been carried out for the general case by Schafroth (55a). For the special case of very low temperatures, however, we can obtain the essential results more simply, as follows. In the limit of zero temperature, the condition

$$\mathfrak{k}T \ll \mu_0\mathfrak{H}' \tag{5.16}$$

is certainly satisfied. Inspection of (5.14) shows that, under that condition, we have $N_0 \gg N_1 \gg N_2 \gg \cdots$, so that we may restrict ourselves to the very first term in the sum (5.13). Introducing the variable of integration $v = \beta\hbar^2k^2/2M$ and the quantity

$$v_0 = \beta(\mu - \mu_0\mathfrak{H}') = \frac{\mu - \mu_0\mathfrak{H}'}{\mathfrak{k}T} \tag{5.17}$$

we obtain from (5.14), for $m = 0$,

$$N_0 = \frac{Ve\mathfrak{H}'}{2\pi^{3/2}\hbar c\lambda} \int_0^{\infty} dv \frac{1}{v^{1/2}[\exp(v - v_0) - 1]} \tag{5.18}$$

Let us first check that the assumption of a normal, uncondensed system is consistent. When μ approaches its maximum value, Eq. (5.15), the quantity v_0, Eq. (5.17), becomes zero. In that case, the integral (5.18) diverges at $v = 0$ like $\int dv/v^{3/2}$; thus, by allowing v_0 to approach zero, the number N_0, Eq. (5.18), can be made arbitrarily large. *In the presence of a constant magnetic field $\mathfrak{H}'$ within the volume V, the ideal Bose-Einstein gas does not condense at all.*

This result justifies our initial assumption of a "normal" Bose-Einstein distribution.[10] Contrary to first impression, this result does *not* invalidate the calculation of Section 4, in which Bose-Einstein condensation was assumed. If there is Bose-Einstein condensation, the resulting Meissner effect prevents a *constant* magnetic field $\mathfrak{H}'$. Conversely, if there *is* a constant magnetic field, the occurrence of discrete quantum numbers in the energy levels (5.10) means that a perturbation expansion in powers of the applied field, such as was used in Section 4, is invalid. Thus the treatments of Section 4 and the present section supplement each other, being valid in different regions of the $\mathfrak{H}$-T plane; they do not contradict each other.

Let us now determine the region of consistency of the present treatment, in the limit of very low temperatures. For very low T, the thermal wavelength $\lambda\!\!\!^{-}$, Eq. (1.18), becomes large. To maintain a constant number of particles, (5.18) shows that the integral must become large as well. This can happen only if v_0 approaches zero; thus (5.17) leads to the limiting result:

$$\lim_{T\to 0} \mu = \mu_0 \mathfrak{H}' \tag{5.19}$$

The chemical potential μ is the free energy per particle; thus the free energy approaches the limit $N\mu_0\mathfrak{H}'$. The total magnetic moment is the negative derivative of the free energy with respect to the acting magnetic field; the magnetization $\mathbf{M}$ (magnetic moment per unit volume) is therefore given by

$$\lim_{T\to 0} \mathbf{M} = -\frac{N}{V}\mu_0 \mathbf{e}' \tag{5.20}$$

where $\mathbf{e}'$ is a unit vector in the direction of $\mathfrak{H}'$.

At this stage, we must determine the relation between the acting field $\mathfrak{H}'$ and the applied field $\mathfrak{H}$. By definition, $\mathfrak{H}'$ is the field acting on a given particle, due to the combined effect of an external agency and all the other particles. Under ordinary conditions $\mathfrak{H}'$ is *not* equal to

[10] This assumption appears in the calculation leading to (5.14), implicitly, through the use of asymptotic expressions for energy levels, degeneracies, etc. If there were Bose-Einstein condensation, the ground state wave function in the magnetic field $\mathfrak{H}'$ would have had to be determined in much more detail.

the average field as seen by a probe; the space average of the microscopic field $\mathfrak{h}$ is the magnetic flux density vector B (for a discussion, see Becker 33):

$$\langle \mathfrak{h} \rangle_{\text{av}} = \mathbf{B} = \mathfrak{H} + 4\pi \mathbf{M} \tag{5.21}$$

Under normal conditions, the field due to the particle under consideration makes a significant contribution to $\langle \mathfrak{h} \rangle_{\text{av}}$, and this contribution must be excluded when calculating the "acting field" $\mathfrak{H}'$. The conventional calculation (Becker 33) gives the Lorenz-Lorentz relation

$$\mathfrak{H}' \cong \mathfrak{H} + \frac{4\pi}{3} \mathbf{M} \tag{5.22}$$

However, the conditions are not "normal" even in the uncondensed Bose gas, and (5.22) does not apply here. Let us determine the size of the typical particle orbit in the magnetic field $\mathfrak{H}'$. In the limit of low temperature, this size is the mean spread of the oscillator wave function $u_0(x)$ corresponding to $m = 0$ in (5.10). Setting $\frac{1}{2} K(\Delta x)^2$ equal to the quantum energy $\frac{1}{2}\hbar\omega$, and using (5.7) and (5.8), we obtain the estimate

$$\Delta x \sim \left(\frac{\hbar c}{e \mathfrak{H}'}\right)^{1/2}$$

We assert that this orbit size is much larger than the average distance between particles, for reasonable number densities N/V and magnetic fields. The condition $\Delta x \gg (V/N)^{1/3}$ is equivalent to

$$\mathfrak{H}' \ll \frac{\hbar c}{e} \left(\frac{N}{V}\right)^{2/3} \tag{5.23}$$

With e of the order of an electronic charge and N/V of order 10^{22} cm^{-3}, Eq. (5.23) means $\mathfrak{H}' \ll 10^7$ gauss, a condition which is easily satisfied.

Since the orbit of the particle under consideration is much larger than the distance between particles, the field produced by this particle is a quite negligible fraction of the field, at the momentary position of the particle, produced by all the *other* particles. Thus, contrary to the usual situation, the simple space average (5.21) is an excellent approximation for the acting field $\mathfrak{H}'$, giving

$$\mathfrak{H}' \cong \mathfrak{H} + 4\pi \mathbf{M} = \mathbf{B} \tag{5.24}$$

Combination of (5.20) and (5.24) yields the magnetization law for the uncondensed ideal Bose gas in the limit of zero temperature:

$$\mathbf{B} = \mathfrak{H} - 4\pi \frac{N}{V} \mu_0 \mathbf{e}' \tag{5.25}$$

where $\mathbf{e}'$ is a unit vector in the common direction of $\mathfrak{H}$ and $\mathbf{B}$.

Equation (5.25) is consistent provided $\mathfrak{H}$ exceeds the value

$$\mathfrak{H}_c = 4\pi \frac{N}{V} \mu_0 \tag{5.26}$$

For $\mathfrak{H} > \mathfrak{H}_c$, the direction of **B** in (5.25) agrees with the common direction of $\boldsymbol{\mathfrak{H}}$ and $\mathbf{e}'$, and the magnitude of B is $\mathfrak{H} - \mathfrak{H}_c$. The uncondensed ideal Bose gas, in the limit of zero temperature, has a permanent magnetic moment **M**; only the direction of **M**, not its magnitude, depends on the applied field. In this respect, the Bose gas behaves similarly to a ferromagnetic material; however, the direction of the permanent magnetic moment is opposite to the applied field direction, whereas in the ferromagnet it is parallel to the applied field.

On the other hand, for $\mathfrak{H} < \mathfrak{H}_c$, Eq. (5.25) becomes inconsistent, since either direction for $\mathbf{e}'$ results in the wrong direction for **B**.

We conclude that $\mathfrak{H}_c$, Eq. (5.26), is the critical field value for the charged ideal Bose gas. For $\mathfrak{H} > \mathfrak{H}_c$, the calculation leading to (5.25) applies, the Bose gas is uncondensed (though far from "normal" in its response to an applied magnetic field), and allows a constant field to penetrate into its interior. For $\mathfrak{H} < \mathfrak{H}_c$, the Bose gas is condensed, and the Meissner effect (as calculated in Section 4) prevents the establishment of a constant field. The calculation of Section 4 assumes a *small* field applied to the outside of the slab, and to that extent there still remains an intermediate region ($\mathfrak{H}$ not small, but still $\mathfrak{H} < \mathfrak{H}_c$) which has not been covered by the calculation; however, it is reasonable to assume that the Meissner field expulsion, established for $\mathfrak{H} \ll \mathfrak{H}_c$, does not change qualitatively for larger values of $\mathfrak{H} \lesssim \mathfrak{H}_c$. In any case, the Bose gas model is sufficiently different from real superconductors to make a detailed calculation of the intermediate region of $\mathfrak{H}$ uninteresting.

Let us now compare the Bose gas model, in the limit of zero temperature, with actual superconductors. Qualitatively, there is agreement for the condensed (superconducting) state, i.e., we do get a Meissner effect, but disagreement for the normal state; actual superconductors exposed to fields $\mathfrak{H} > \mathfrak{H}_c$ behave to all extents and purposes like normal metals, showing an extremely weak magnetic susceptibility, nothing like the permanent magnetic moment of the Bose gas model. Thus, *if superconductivity in metals is to be related, somehow, to the existence of quasi-bosons, these quasi-bosons cannot be permanent constituents of the substance, but must disappear* (*or at least, most of them must disappear*), *in the normal state of the metal.* The number N of bosons must itself be a variable, not a fixed number.

In spite of the disagreement for the normal state, there is surprisingly good agreement between the Bose gas model and actual superconductors, in the limit of zero temperature. The experimental quantities with which we may compare, at this stage, are two: The critical field $\mathfrak{H}_c(0)$ may be compared with (5.26), and the London penetration depth λ may be compared with the formula deduced from (I,2.16), (I,5.10), and (II,4.13), the latter with the self-consistent value $|\phi_0(r)|^2 = 1/V$; this Bose gas result for the penetration depth is:

$$\lambda = c\left(\frac{VM}{4\pi Ne^2}\right)^{1/2} \tag{5.27}$$

Using (5.26) and (5.27), we deduce the following two useful formulas for the Bose gas:

$$\lambda^2\mathfrak{H}_c = \frac{\hbar c}{2e} \tag{5.28}$$

and

$$(\lambda\mathfrak{H}_c)^2 = \frac{\pi\hbar^2}{M}\frac{N}{V} \tag{5.29}$$

Equation (5.28) allows a direct determination of the boson charge e. If we use the zero-temperature values for tin, $\mathfrak{H}_c = 306$ gauss (Serin 55) and $\lambda = 5.3 \times 10^{-6}$ cm (Pippard 53), we obtain about 4 electronic charges for e from (5.28); for the product $(\lambda\mathfrak{H}_c)^2$ the experimental value is 2.8×10^{-6}; using the electronic mass for M and a number density $N/V \sim 10^{22}$ cm^{-3}, the Bose gas value of this same quantity is rather larger, about 3.5×10^{-5}. Detailed quantitative agreement is out of the question, both in view of the qualitative difference for the "normal" state (which enters into the determination of $\mathfrak{H}_c$) and in view of the fact that whatever the "boson" may be, it certainly is not a simple conduction electron. The quantitative comparison just made shows, however, that "bosons" with reasonable charges, masses, and number densities, are quite consistent with observations on superconductors at zero temperatures.

6. Magnetic Behavior of the Bose Gas Model at Non-Zero Temperature

The calculations for non-zero temperature are very lengthy and complicated (Schafroth 55a). We shall therefore content ourselves with quoting the results, and commenting on those results.

The transition temperature of the Bose gas in the absence of magnetic fields is given by equality in Eq. (2.3); solving for T_c gives[11]

$$\mathfrak{k}T_c = \frac{2\pi\hbar^2}{M}\left(\frac{N}{2.61V}\right)^{2/3} \tag{6.1}$$

This result appears terrible at first sight: for reasonable values of M and N/V, of the order of the ones just found from (5.28) and (5.29), T_c is of the order of the Fermi degeneracy temperature of the metal, i.e., some 10^4 °K.

How can we get around this difficulty? A clue is provided by the qualitative discussion of the degeneracy condition [see (1.25)]. The thermal wavelength $\lambdabar$ depends only on the mass of the "boson" and the temperature. The "boson" mass can hardly change much with temperature, thus $\lambdabar$ is essentially a known quantity. However, the number density N/V *can* change with temperature, provided that the "bosons" are not elementary particles but composite entities of some sort. Even at zero temperature, the "bosons" must be capable of disappearance in order to give *qualitative* agreement with metallic properties for $\mathfrak{H} > \mathfrak{H}_c$. It is not surprising, therefore, that "bosons" may disappear also when one raises the temperature. Since T_c in metals is about 10^{-3} of the Fermi degeneracy temperature, and since it involves $(N/V)^{2/3}$, the "boson" density would have to decrease by about a factor 10^5 between $T = 0$ and $T = T_c$. With such a tremendous temperature variation, the model is clearly unreliable for any comparison with experiment: the mechanism, whatever it may be, by which "bosons" are produced and destroyed, must have a major influence on the thermodynamics of the system.

It is most surprising, therefore, that the *magnetic* response of the Bose gas model is still in qualitative agreement with experimental facts on superconductors, provided one does the obvious scaling of tempera-

[11] This is the transition temperature in the absence of self-consistent Coulomb effects. The "jellium" system with Bose-Einstein statistics has not yet been treated adequately at non-zero temperature. If the lowest excitation levels correspond invariably to density fluctuations (phonons) in the gas, then there is an energy gap: phonons go over into plasma waves in "jellium"; the resulting transition temperature would be rather higher than (6.1), even more in disagreement with experiment. For treatments of the Bose system in the presence of interactions, see Bogoliubov 47; Penrose 51, 56; Lee 57, 57a, 58, 59, 59a, 59b, 60, 60a, 60b; Huang 57, 57a, 60; Zubarev 58; Foldy 61; Wu 61; Kromminga 62; Luban 62; Parry 62; Yang 62; and Lieb 63.

Although the Bose gas represents merely a first, crude model for actual metallic superconductors, it is possible that alpha-particles in the interior of a white dwarf star may undergo a Bose-Einstein condensation which is modified only slightly by interactions (Ninham 63); if so, the evolution of such stars may be influenced profoundly.

tures, i.e., expresses temperature-dependent quantities as functions of the reduced temperature

$$t = T/T_c \tag{6.2}$$

The ideal Bose gas in a weak magnetic field is condensed for $T < T_c$ ($t < 1$), with average occupation number for the single-particle ground state given by

$$\bar{n}_0 = N(1 - t^{3/2}) = \bar{N}_s \tag{6.3}$$

The remaining particles form the "normal fluid" component

$$\bar{N}_n = Nt^{3/2} \tag{6.4}$$

The magnetic response, as defined by the general equation (I.3.8), is governed by the kernel

$$K(q) = K_s(q) + K_n(q) \tag{6.5a}$$

where $K_s(q)$ comes from the "superfluid" and is given by

$$K_s(q) = -\frac{\bar{N}_s}{V}\frac{e^2}{Mc^2}\frac{1}{q^2} \tag{6.5b}$$

and $K_n(q)$ comes from the uncondensed particles

$$K_n(q) = -\frac{\Pi}{8}\frac{e^2}{Mc^2}\frac{1}{\bar{\lambda}^2 q} + \text{terms independent of } q \tag{6.5c}$$

For large depth of penetration into the specimen (low values of q), the leading terms given here suffice. The temperature dependence of (6.5b) arises from (6.3), that of (6.5c) from $\bar{\lambda}$, Eq. (1.18). The fact that the "normal" contribution to $K(q)$ is still singular, although only like $1/q$, should not be too surprising in view of the highly "abnormal" behavior of the uncondensed Bose gas at zero temperature (see Section 5). In practice, K_n is quite negligible compared to K_s for all values of T and q of interest. The form K_s is just the one appropriate to the London equation [see (I,3.9)] with a temperature-dependent Λ and hence a temperature-dependent penetration depth:

$$\lambda(T) = \frac{\lambda(0)}{(1 - t^{3/2})^{1/2}} \tag{6.6}$$

The experimental results show rather less dependence of λ on T, except in the immediate neighborhood of the transition temperature; there, however, the type of divergence, proportional to $(1 - t)^{-1/2}$, is in agreement with experiment.

The critical field of the Bose gas has the temperature dependence:

$$\mathfrak{H}_c(T) = \mathfrak{H}_c(0)\,(1 - t^{3/2}) \tag{6.7}$$

This again differs in detail from the experimental values for superconductors, but the disagreement is quantitative rather than qualitative. In fact, the Bose gas result obtained from (6.6) and (6.7)

$$\lambda^2 \mathfrak{H}_c = \text{independent of temperature} \tag{6.8}$$

is in surprisingly good agreement with experiment, much better agreement than either (6.6) or (6.7) taken separately. Since this quantity depends only on the charge of the boson, and not on the number of bosons [see Eq. (5.28)], this agreement is a favorable point for the Bose gas model.

7. Other Properties of the Bose Gas Model

Besides the response to a magnetic field, the condensed ideal Bose-Einstein gas displays "superfluid" behavior in other ways. In this section, we give a short discussion of these other properties.

In his theory of superfluid helium, Landau (41) suggested the "rotating bucket" experiment as a test of true superfluid behavior: The liquid is placed in a cylindrical bucket with vertical axis; the bucket is set into rotation with angular velocity ω, and one measures the angular momentum L of the liquid. A classical liquid gives

$$L = I\omega \tag{7.1}$$

with moment of inertia I equal to the classical value I_0:

$$I_0 = \tfrac{1}{2} NMR^2 \tag{7.2}$$

where N is the number of particles, M the mass per particle, and R the radius of the bucket. If the liquid is a mixture of "superfluid" and "normal fluid," the superfluid component does not take part in the rotation (at least for sufficiently small angular velocities), and only the normal fluid component rotates along with the bucket. We then expect a non-classical moment of inertia

$$I = (\bar{N}_n/N) I_0 \tag{7.3}$$

where $\bar{N}_n$ is the number of particles in the "normal fluid." After the war, such experiments were carried out by Andronikashvili (46), and gave the predicted result. The fraction of normal fluid, $\bar{N}_n/N$, showed a strong temperature dependence in agreement with Landau's predictions; the fraction becomes zero at the absolute zero of temperature, and becomes unity at the transition temperature.

An investigation of the condensed ideal Bose-Einstein gas in a rotating cylindrical bucket (Blatt 54, 55a) shows that the system obeys (7.3), with $\bar{N}_n$ given by (6.4), at very small angular velocities $\omega < \omega_1$, where

$$\omega_1 = 4.45 \frac{\hbar}{MR^2} \tag{7.4}$$

For normal laboratory dimensions $R \sim 1$ cm and M the mass of a helium atom, (7.4) is of the order of 10^{-3} rad/sec, much too small to fit in with observed critical flow velocities in liquid helium. For angular velocities ω in excess of ω_1, the behavior of the Bose gas model depends on whether full thermodynamic equilibrium is maintained during the experiment. The equilibrium situation is as follows: as soon as ω exceeds ω_1, Eq. (7.4), the single-particle quantum state most favorable for condensation is no longer the lowest state with angular momentum quantum number $m = 0$, but rather the lowest state with $m = 1$. There is thus, at $\omega = \omega_1$, a thermodynamic transition of first order, in which the angular momentum L of the whole system increases discontinuously by an amount $\Delta L = \bar{N}_s\hbar$. Another thermodynamic first-order transition occurs at $\omega = \omega_2$, a somewhat higher angular velocity, where the condensed particles shift from the lowest $m = 1$ state to the lowest $m = 2$ state; and so on. The ratio L/ω rapidly increases, and soon *exceeds* the classical value (7.2), even though the slope $dL/d\omega$ remains equal to the value (7.3) within the continuous sections of the discontinuous curve of L versus ω. Since one would measure, not the slope, but rather L/ω, the condensed Bose-Einstein gas in this region would be an "infrafluid" rather than a "superfluid."

However, if full thermal equilibrium is *not* maintained, then the condensed particles may remain in the lowest $m = 0$ state, even though a lower free energy can be had by transferring them to some other state; this depends on the time-scale of the experiment in relation to the relevant relaxation times, which are very hard to estimate reliably. Assuming that the relaxation times are sufficiently long, the Bose gas would maintain the low moment of inertia (7.3) well beyond $\omega = \omega_1$, as a metastable superfluid state.

The relationship of this calculation to actual liquid helium is not clear at this time; it is usually maintained that rotating liquid helium, with the normal fluid rotating and the superfluid at rest, is in full thermodynamic equilibrium at angular velocities well in excess of ω_1, Eq. (7.4); if so, the analogy to the ideal Bose-Einstein gas is insufficient to explain the observations.

Since our interest here is not in liquid helium, but in superconductivity, we need merely observe that the ideal Bose gas model shows superfluid properties under rotation, as well as under the influence of an applied magnetic field.

Of more interest for superconductivity is the behavior of the Bose gas in a ring-shaped volume, with magnetic flux through the hole in the ring. This will be discussed in detail in Chapter IX; suffice it here to state that the calculation (Blatt 61, 61a; Bloch 62) yields persistent currents with infinite lifetime, and flux quantization with quantum value $2\pi\hbar c/e$, e being the charge of the boson.

In conclusion, it appears that the ideal Bose gas in its condensed state has many properties in common with superconductors and superfluid liquid helium. The correspondence is not perfect, but is certainly highly suggestive. The uncondensed Bose gas, however, is qualitatively different from a normal metal in a number of ways. Also, the fact that the superconducting transition temperature T_c is some 1000 times smaller than the "natural" degeneracy temperature of a gas of Bose particles of reasonable mass, charge and number density indicates that the simple Bose gas model is defective in some essential respect. The most likely explanation of this discrepancy is that the number of "bosons" in a metal is not a fixed quantity, independent of temperature, magnetic field, and other conditions, but rather the "bosons" are composite entities of some sort which can be formed, and can disappear again, depending upon circumstances. If this explanation be accepted, then the data on superconducting metals indicate that the "boson" number is comparable to the number of conduction electrons at zero temperature and in the absence of a magnetic field. The "boson" number drops rapidly as the temperature is increased, by a large factor (perhaps of order 10^5), between $T = 0$ and $T = T_c$. The "boson" number also becomes negligible if the applied magnetic field exceeds the critical field.

III

The Quasi-Chemical Equilibrium Theory

1. The Concept of Electron Pairs as Quasi-Molecules

The simple theoretical model of a condensed Bose-Einstein gas has many features in common with actual superconductors, but there are enough differences and discrepancies to make any immediate application to actual superconductors out of the question. The most obvious and striking discrepancy is of course the Fermi-Dirac statistics of electrons in solids.

Thus, although the idea of Bose-Einstein condensation as an explanation of superconductivity had been mooted as early as 1938 by London, it was not until the middle 1950's that serious attempts were made in this direction, and even then it took several years before there was anything like general acceptance of the idea. Surprisingly enough, although the Meissner effect of the ideal Bose gas was calculated independently by three people, two of them considered it a mathematical curiosity of no relevance to actual superconductors.

Schafroth, however, decided to take the Bose-Einstein model seriously. If the "free electrons" in solids are unsuitable for the purpose, as they obviously are, then we must *construct* our bosons somehow. This point of view is strengthened by the comparison of the Bose gas model with actual superconductors, which shows that the "bosons," whatever they are, must be capable of appearing and disappearing; their number cannot be constant.

If one looks at the theory of solids from this point of view, it is immediately apparent that "constructed" bosons already exist. The two best-known examples are phonons and excitons. Phonons (quanta of sound waves) satisfy Bose-Einstein statistics; they are unsuitable for superconductivity because they carry no charge, and because we know, from purely thermodynamic considerations (see Chapter I, Section 1), that the electrons must be involved in the "ordering" of the superconducting state. Excitons are bound states of a conduction electron and a positively charged hole in a filled band, rather similar to positronium. Let $\phi(\mathbf{x}, \mathbf{y})$ be the wave function of an exciton, $\mathbf{x}$ and $\mathbf{y}$ being the coordinates of the electron and hole, respectively. Since this is a bound state, the distance $|\mathbf{x} - \mathbf{y}|$ remains small, of order d, say. Let us consider a situation in which there are *two* excitons present, with wave functions $\phi_a(\mathbf{x}, \mathbf{y})$ and $\phi_b(\mathbf{x}, \mathbf{y})$, respectively. The product wave function $\phi_a(\mathbf{x}_1, \mathbf{y}_1)\,\phi_b(\mathbf{x}_2, \mathbf{y}_2)$ is not permissible, because it is not antisymmetric under exchange of identical particles. We antisymmetrize in the usual fashion, by carrying out a sum over permutations of identical particles. The result is

$$\begin{aligned}\psi(\mathbf{x}_1, \mathbf{y}_1; \mathbf{x}_2, \mathbf{y}_2) &= \phi_a(\mathbf{x}_1, \mathbf{y}_1)\,\phi_b(\mathbf{x}_2, \mathbf{y}_2) + \phi_a(\mathbf{x}_2, \mathbf{y}_2)\,\phi_b(\mathbf{x}_1, \mathbf{y}_1) \\ &\quad - \phi_a(\mathbf{x}_1, \mathbf{y}_2)\,\phi_b(\mathbf{x}_2, \mathbf{y}_1) - \phi_a(\mathbf{x}_2, \mathbf{y}_1)\,\phi_b(\mathbf{x}_1, \mathbf{y}_2) \end{aligned} \tag{1.1}$$

Let us compute the normalization integral for that function! If we assume that the wave functions ϕ_a and ϕ_b are already normalized, the "direct" terms give 1. There are also two kinds of exchange integrals, namely,

$$E = \int \phi_a^*(\mathbf{x}, \mathbf{y})\,\phi_b(\mathbf{x}, \mathbf{y})\, d^3x\, d^3y$$

and

$$S = \int \phi_a^*(\mathbf{x}_1, \mathbf{y}_1)\,\phi_b^*(\mathbf{x}_2, \mathbf{y}_2)\,[\phi_a(\mathbf{x}_1, \mathbf{y}_2)\,\phi_b(\mathbf{x}_2, \mathbf{y}_1) + \phi_a(\mathbf{x}_2, \mathbf{y}_1)\,\phi_b(\mathbf{x}_1, \mathbf{y}_2)]\, d^6x\, d^6y$$

The normalization integral is

$$\int |\psi(\mathbf{x}_1\ \mathbf{y}_1; \mathbf{x}_2, \mathbf{y}_2)|^2\, d^6x\, d^6y = 4(1 + |E|^2 - S)$$

The two exchange integrals which enter here represent two different kinds of "exchange." The term $|E|^2$ arises from a joint exchange, in which both the electron and the hole are interchanged, i.e., the two excitons are exchanged as units. The term S arises from exchanges of only one of the two particles making up each exciton.

There exists a physical situation in which $|E|^2$ and S have very different orders of magnitude: when we write $\phi(\mathbf{x}, \mathbf{y})$ as a product of an internal wave function, depending upon $|\mathbf{x} - \mathbf{y}|$, and of a center-of-gravity wave function, depending upon $\frac{1}{2}(\mathbf{x} + \mathbf{y})$, then suppose that the "spread" of the internal wave function, d, is much smaller than the "spread" of the center-of-gravity wave function D, and that this latter spread D is of the same order of magnitude as the distances between the centers of gravities of the excitons in states ϕ_a and ϕ_b. In that case, $|E|^2$ is comparable to unity, but S is small, of order $\exp(-D/d)$. We may then neglect S altogether. Looking back at (1.1), we see that this is equivalent to working with just the first line of (1.1), i.e., with the simpler wave function

$$\psi_m(\mathbf{x}_1, \mathbf{y}_1; \mathbf{x}_2, \mathbf{y}_2) = \phi_a(\mathbf{x}_1, \mathbf{y}_1)\,\phi_b(\mathbf{x}_2, \mathbf{y}_2) + \phi_a(\mathbf{x}_2, \mathbf{y}_2)\,\phi_b(\mathbf{x}_1, \mathbf{y}_1) \tag{1.2}$$

This, however, is a symmetrical combination of the wave functions of the two separate "molecules." To this extent, then, a molecule composed of two Fermi-Dirac atoms obeys Bose-Einstein statistics.

This is, of course, only a crude approximation, since, in general, *all* overlap integrals must be taken into account. Nonetheless, in some sort of first approximation, we may consider excitons to obey Bose-Einstein statistics; and the number of excitons is clearly a variable quantity, depending upon temperature and other conditions.

Thus, there already exists, in the standard theory of solids, a composite particle, built up from electrons, and obeying Bose-Einstein statistics. It is not known at this time whether excitons can undergo Bose-Einstein condensation; they cannot do so in full thermodynamic equilibrium, since the number of excitons is quite small under equilibrium conditions; but excitons might perhaps condense if a high concentration of them is maintained artificially (Blatt 62a). However, interesting as such a condensation would undoubtedly be, it would not be condensation into a superconducting state, for the simple reason that excitons are electrically neutral.

It is not unreasonable, however, to consider another kind of composite "particle" at this stage, composed not of an electron and a hole, but of two electrons. That is, we consider the "chemical reaction"

$$e + e \rightleftarrows e_2 \tag{1.3}$$

in analogy to ordinary chemical reactions such as the one forming hydrogen molecules from hydrogen atoms:

$$\mathrm{H} + \mathrm{H} \rightleftarrows \mathrm{H}_2 \tag{1.4}$$

In ordinary chemical reactions, low temperatures favor the presence of large numbers of molecules, i.e., the reaction equilibrium shifts away from single atoms towards a larger fraction of molecules. By analogy, we might expect to find mostly single electrons in solids at room temperature, but increasing numbers of electron pairs e_2 as the temperature decreases. *If* enough electron pairs are formed, *if* these pairs can undergo a condensation similar to Bose-Einstein condensation, and *if* the condensate retains enough similarity to the ideal Bose-Einstein condensed system to yield a Meissner effect and persistent currents, *then* we have the beginning of a theory of superconductivity.

This is a formidable array of if's, and more difficulties can easily be found. However, sometimes one can make progress by ignoring all the difficulties initially, continuing as if they did not exist, and then returning to clear up the difficult points at a later stage. This is what we shall do here; it is what was actually done in the development of the theory of superconductivity.

The history of the idea of Bose-condensed electron pairs is not untypical of the history of new ideas in physics, generally. Since I was involved personally in the development, let me drop the impersonal "we" for a few reminiscences.

The first suggestion of electron pairs and their Bose condensation was made by Ogg (46), on the basis of experiments on, of all things, very dilute solutions of alkali metals in liquid ammonia! Under certain special conditions, Ogg claimed persistent ring currents at temperatures as high as −180°C, and even higher. Ogg had some indirect evidence for the existence of trapped electron pairs in his solutions, and he estimated their Bose-Einstein degeneracy temperature, on the basis of their approximately known concentration and a guess at their effective mass, to be in the right range. However, Ogg's experiments proved to be difficult to repeat, his theoretical ideas were discounted by the theorists, and the whole thing was forgotten. In retrospect, I can recall listening to a seminar lecture by Ogg, right after the war, and to that extent the Sydney work may well have been influenced by Ogg. But this influence was entirely subconscious—neither Schafroth, Butler, nor I remembered anything of Ogg's work at the time the Sydney work started, in the latter half of 1954. Only much later, in 1961, was I reminded of Ogg's work in the course of a correspondence with P.T. Landsberg; it would be a pity, though, to omit Gamow's limerick on Ogg's bi-electron theory, quoted by Landsberg:

"In Ogg's theory it was his intent
That the current keep flowing, once sent;
So to save himself trouble,
He put them in double,
And instead of stopping, it went."

There is no doubt about Ogg's priority of the suggestion, and we all owe him apologies for forgetting so completely about it in our publications.

Ogg's ideas were discounted so strongly, I am convinced, largely because they were phrased in the language of an experimental chemist, rather than that of a theoretical physicist. It is easy to see that a truly stable, bound "dielectron" is quite out of the question inside a solid, and thus ordinary chemical concepts need considerable modification.[1]

The Sydney group consisted of theoretical physicists, and thus our language, at least, was more in line with that of others. Schafroth, in his initial Letter to the Editor (Schafroth 54), already pointed out that the electron pairs must be considered metastable, "resonance" states, not truly bound states, the resonance energy lying just below the Fermi energy of the ordinary free electron gas in the metal. Only the presence of all the other electrons, which inhibits most of the otherwise possible transitions to lower states, makes the metastable resonance state effectively stable. We realized very quickly that there was no *statistical mechanical* treatment of chemical equilibrium known at that time, even though the thermodynamics of chemical reactions was an old subject, and even though an approximate statistical mechanics had been worked out in analogy to the thermodynamic treatment. This, however, was clearly inadequate for dealing with resonance states in a Fermi gas background. Thus, we devoted considerable effort to finding a theory of ordinary chemical equilibrium, starting from fundamental statistical mechanics. Although published much later, most of this was finished in late 1955, by Schafroth in Princeton[1a] (he visited the Institute for Advanced Study for one term) and by Butler and myself in Sydney. But, in spite of our using the theorists' language, and in spite of Schafroth's missionary activities in Princeton, it took a long while before electron pairs became respectable.

The lack of acceptance of the Sydney work, at that time, was by no means just prejudice and conservatism. There were any number of serious difficulties, stressed particularly in Casimir's perceptive discussion (Casimir 55), and the quasi-chemical equilibrium theory was not developed far enough to deal with these objections.

An apparently different line of development then commenced elsewhere, starting with a self-consistent calculation of the effective binding energy of an electron pair by Cooper (56). This was followed quickly by the BCS theory [Bardeen (57); see Bardeen (61) for a complete review], the work of Valatin (58), and of Bogoliubov and co-workers (Bogoliubov 57, 58, 58a, 58b), all

[1] The same objection applies to some suggestions made much later (Pitzer 56, Weyerer 58, Rumer 60).

of them nearly simultaneously. By 1958, electron pairing of some kind was generally accepted as the explanation of superconductivity. We shall discuss these theories and their interrelationships in Chapters V through VII.

2. Thermodynamics of Chemical Equilibrium

If a binary chemical reaction, such as (1.4), takes place in the gaseous state, and the gas is sufficiently dilute, we may consider the gas as a mixture of two ideal gases, one of atomic hydrogen, the other of molecular hydrogen. However, the number of particles in each of these subsystems is not fixed. Rather, N_1, the number of atoms, and N_2, the number of diatomic molecules, are related by

$$N_1 + 2N_2 = N \tag{2.1}$$

where N is the given number of protons in the system. We would like to know the values of N_1 and N_2 which correspond to complete (chemical as well as thermal) equilibrium.

If the system is kept at constant volume and temperature, the equilibrium condition is that the free energy F be a minimum. In the approximation of dilute gases, the free energy is a sum of two terms: the free energy $F_1(N_1)$ of a gas of N_1 hydrogen atoms, and the free energy $F_2(N_2)$ of a gas of N_2 hydrogen molecules. In this problem, it is important to keep the zero of energy consistent. We shall define zero energy to mean the energy of an assembly of stationary atoms, far removed from each other. In that case, the free energy of the molecular gas, $F_2(N_2)$, does not approach zero in the limit of zero temperature, but rather approaches $-N_2B$, where B is the molecular binding energy. The total free energy is

$$F = F_1(N_1) + F_2(N_2) = F_1(N - 2N_2) + F_2(N_2) \tag{2.2}$$

where we have used (2.1). We differentiate this with respect to N_2 to get the equilibrium condition[2]

$$\frac{dF}{dN_2} = -2F_1'(N - 2N_2) + F_2'(N_2) = 0$$

The derivative of the free energy with respect to the particle number,

[2] The symbol F_1' means the derivative of F_1 with respect to its argument.

which occurs here, is called the "chemical potential" and denoted by the Greek letter μ; it is the same μ which occurs in the Fermi-Dirac and Bose-Einstein distributions of Chapter II, and which is called the "Fermi energy" in solid state physics. Stating explicitly the quantities kept constant during the differentiation, the definition of μ is

$$\mu = \left(\frac{\partial F}{\partial N}\right)_{V,T} \tag{2.3}$$

In terms of the chemical potentials, the equilibrium condition reads

$$\mu_2 = 2\mu_1 \tag{2.4}$$

The physical meaning of this condition is clear: if two hydrogen atoms combine to form one hydrogen molecule, we lose $2\mu_1$ free energy from the atomic gas, and gain μ_2 free energy from the molecular gas. In chemical equilibrium, the gain and loss must balance.

The factor 2 in (2.4) is the same 2 which occurs in (2.1). In the chemical equilibrium between ozone (O_3), ordinary molecular oxygen (O_2), and atomic oxygen, the condition on the numbers is

$$N_1 + 2N_2 + 3N_3 = N \tag{2.5}$$

where N is the number of oxygen nuclei. The chemical equilibrium conditions are

$$\mu_2 = 2\mu_1 \qquad \text{and} \qquad \mu_3 = 3\mu_1 \tag{2.6}$$

Conversely, if there is no chemical combination, but only simple particle transfer (for example, consider a junction between two metals, with electrons able to move from one metal to the other), the equilibrium condition is simple equality of the chemical potentials μ. In the case of a metallic junction, the chemical potential μ is just the Fermi energy, so that the condition of equilibrium means that the Fermi levels in both metals must be the same.

To get some feeling for what is involved, consider the chemical equilibrium between hydrogen atoms and hydrogen molecules in the gaseous state, at normal temperatures. We may then use the classical approximation (II, 1.20) for the chemical potential of the atomic gas

$$\mu_1 = -\, \mathfrak{k}T \ln (V/N_1\lambda_1^3) = \mathfrak{k}T \ln (y_1) \tag{2.7}$$

where we have introduced the non-dimensional quantity

$$y_1 = N_1\lambda_1^3/V \tag{2.8}$$

The physical meaning of y is a "reduced concentration": if we imagine the volume V divided up into cells of size λ_1^3, where λ_1 is the atomic thermal de Broglie wavelength, (II, 1.18), then the number of "cells" is $C = V/\lambda_1^3$, and y_1 is the average number of particles per cell. For atomic hydrogen at room temperature, λ_1 is of the order of 10^{-7} cm; if the atomic hydrogen were at normal temperature and (partial) pressure, the distance between hydrogen atoms would be of the same order of magnitude; in fact, however, the hydrogen is almost entirely in the form of molecules, so that y_1 is very small indeed.

The chemical potential of the hydrogen molecular gas is, allowing for the different zero of energy and ignoring excited states of the molecule,[3]

$$\mu_2 = -\mathfrak{k}T \ln (V/N_2\lambda_2{}^3) - B = \mathfrak{k}T \ln (y_2) - B \tag{2.9}$$

The equilibrium condition (2.4) then assumes the form

$$(y_1)^2 = y_2 \exp (-B/\mathfrak{k}T) \tag{2.10}$$

This must be solved subject to the condition (2.1) on the numbers. We make use of the relation [see (II, 1.18) and note $M_2 = 2M_1$]

$$\lambda_2 = \left(\frac{2\pi\hbar^2}{2M\mathfrak{k}T}\right)^{1/2} = \frac{\lambda_1}{\sqrt{2}} \tag{2.11}$$

and the definitions

$$y_0 = N\lambda_1{}^3/V, \qquad \alpha = 2^{-7/2} \exp (-B/\mathfrak{k}T) \tag{2.12}$$

to put (2.10) in the form

$$(y_1)^2 = 2\alpha(y_0 - y_1) \tag{2.13}$$

with the solution

$$y_1 = \alpha\left[\left(1 + \frac{2y_0}{\alpha}\right)^{1/2} - 1\right] \tag{2.14}$$

This is the explicit solution for the reduced concentration of atomic hydrogen, as a function of temperature and number density of protons.

At normal temperature and pressure, y_0 differs from unity by only a few orders of magnitude, but α is exceedingly small, since molecular binding energies are of the order of electron volts, whereas $\mathfrak{k}T$ is a small fraction of an electron volt. Thus (2.14) can be approximated by

$$y_1 \cong (2\alpha y_0)^{1/2} = (y_0)^{1/2}\, 2^{-5/4} \exp (-B/2\mathfrak{k}T) \ll y_0 \tag{2.15}$$

Correspondingly, the gas is practically entirely in molecular form, $N_2 \cong \frac{1}{2}N$.

[3] See later in this section concerning this correction. λ_2 is the *molecular* thermal de Broglie wavelength [see Eq. (2.11)].

The opposite situation is reached only at very high temperatures or exceedingly low concentrations. For $2y_0 \ll \alpha$ we obtain $y_1 \simeq y_0$, i.e., $N_1 \simeq N$, so that the gas is substantially in atomic form. The crossover occurs for $2y_0 = \alpha$, which we can rewrite in the form

$$d = 2^{3/2}\,\lambda\!\!\!^{-}_1 \exp(B/3\mathfrak{k}T) \tag{2.16}$$

where $d = (V/N)^{1/3}$ is the average distance between protons under atomic conditions. For reasonable number densitites (and hence values of d), the crossover occurs at temperatures of the order of 10^4 °K. Conversely, at ordinary temperatures, the distance d between atoms is enormous before the chemical equilibrium favors atoms over molecules.

Note, however, that complete dissociation of the molecule is more likely than mere excitation to an excited molecular quantum state with excitation energy E^* just below the binding energy B. Letting N_2^* be the number of excited molecules, and $y_2^* = N_2^* \lambda\!\!\!^{-}{}_2^3/V$ be their reduced concentration, the Maxwell distribution gives

$$y_2^* = y_2 \exp(-E^*/\mathfrak{k}T) \tag{2.17}$$

which must be compared with the reduced concentration (2.15) for the atoms. For E^* comparable to B, Eq. (2.15) is enormously larger than Eq. (2.17), primarily because of the exponential factors. This state of affairs justifies the usual approximation of ignoring highly excited states altogether in calculations of chemical equilibrium.

There is one correction, however, which is necessary to get accurate results. Most molecules have a large number of low-lying, rotational levels, with excitations energies less than, or of the order of, $\mathfrak{k}T$. These levels are populated significantly, of coursc, and cannot be ignored. The correction for this effect is simple, although a detailed derivation of it would take too much space here. Let us define the "internal free energy per molecule," f, by the equation

$$\exp(-\beta f) = \sum_s \exp(-\beta\eta_s) \tag{2.18}$$

where η_s is the internal energy (not including center-of-gravity motion) of molecular level number s. Letting $s = 0$ be the molecular ground state, we have $\eta_0 = -B$, and, from (2.18),

$$\lim_{T\to 0} f = \lim_{\beta\to\infty} f = -B$$

The correction in question consists simply in replacing $-B$, wherever it occurs in the formulas so far, by this new quantity f. It is obvious from (2.18) that molecular levels η_s with excitation energy $\eta_s - \eta_0 \gg \mathfrak{k}T$ make only very small contributions to f.[4]

To obtain an order of magnitude for the correction involved here, let us restrict ourselves to the low-lying rotational levels of the molecule. The quantum number "s" is then the pair of quantum numbers j, m, where j is the rotational angular momentum, and there is a $(2j + 1)$-fold degeneracy of level η_j, due to $m = j_z$; we write, ignoring spins,

$$\eta_s = \eta_j = -B + \frac{\hbar^2 j(j+1)}{2I} \tag{2.19}$$

Only alternate values of j are permitted, depending upon the statistics of the atoms. If the lowest rotational excitation energy is much less than $\mathfrak{k}T$ (this is the usual case), the sum in (2.18) can be replaced by an integral over a continuous variable j and evaluated easily. We supply a factor $\frac{1}{2}$ since half the levels j are forbidden by the symmetry requirements, to obtain the result

$$f \cong -B + \tfrac{1}{2}\mathfrak{k}T \ln (\hbar^2/2I\mathfrak{k}T) \tag{2.20}$$

This quantity must be substituted into (2.10) and (2.12) in place of $-B$; with this substitution, (2.14) is correct as it stands. The additional term in (2.20) is only of minor importance; the logarithm is negative (since, by assumption, $\hbar^2/2I \ll \mathfrak{k}T$), but, being a logarithm, is not large numerically. Thus, we are replacing, in effect, B by a quantity slightly larger than B, the difference being of the order of $\mathfrak{k}T$. For hydrogen molecules at room temperature, we have $\hbar^2/2I\mathfrak{k}T \cong \frac{1}{4}$, and B is replaced by $B + (0.7)\mathfrak{k}T$, a very small correction indeed. The direction of the effect is to favor molecules even more, compared to atoms.

Excited atomic levels invariably have excitation energies much larger than $\mathfrak{k}T$, and can be ignored.

The relative favoring of dissociation of a molecule, compared to mere excitation [see Eqs. (2.15) and (2.17)] is understandable in terms of the extra entropy (disorder) gained by having twice as many particles to distribute over the "cells" of size $\lambda\!\!\!^{-3}$ mentioned at the end of Chapter II, Section 1. This effect is particularly strong when the gas is highly diluted; it becomes less important as the concentration increases, for then a larger fraction of the available "cells" is already filled even if there are only molecules, no atoms.

---

[4] This is fortunate for this rudimentary theory, for otherwise we would be in serious trouble in principle: among the excited molecular states are not only bound states, $\eta_s < 0$, but also scattering states with $\eta_s > 0$. Scattering states, however, are presumably to be counted, not as molecules, but as part of the *atomic* gas, the scattering leading to a correction from ideal gas behavior. We shall return to this point in Section 6 of this chapter.

Thus an increasing density of particles favors molecules over atoms, in the chemical equilibrium, whereas extreme dilution always leads to a purely atomic gas. But (2.16) shows that this dilution must be extreme indeed to outweigh the effect of the molecular binding energy, at room temperature.

3. Electron Pairs and the Three-Fluid Model

We now resume the qualitative discussion of Section 1, in which we suggested a Bose-Einstein-like condensation of electron pairs as an explanation for superconductivity. Although the simple theory of chemical equilibrium, developed in Section 2, is clearly inadequate to deal with electrons in metals, it may suffice to give us at least some qualitative information. In particular, we would like to know whether a thermodynamic transition is possible in such a system, and whether the number of electron pairs can vary strongly with temperature, becoming essentially negligible above the transition temperature.

We therefore apply the simple theory of chemical equilibrium to the reaction between single electrons and electron pairs, Eq. (1.3). Since the electron gas in metals is a degenerate Fermi-Dirac gas, the quantum effects on the thermodynamics must be taken into account. Thus, although Eq. (2.4) is still the correct chemical equilibrium condition, we must not use the classical approximation (2.7) for the relation between chemical potential and particle density.

The gas of unpaired electrons has energy levels ϵ_k, and the total number of unpaired electrons is given by the Fermi distribution

$$N_1 = \sum_k \bar{n}_k = \sum_k \frac{1}{\exp\,[\beta(\epsilon_k - \mu)] + 1} \tag{3.1}$$

where $\mu = \mu_1$ is the chemical potential of the single electrons.

In addition to levels for single electrons, we must now also consider levels for electron pairs. We shall denote these by an index α, their occupation numbers by ν_α, and their energies by η_α. If we assume simple Bose statistics for electron pairs (this is of course only a crude, preliminary approximation), the total number of pairs is given by a Bose distribution

$$N_2 = \sum_\alpha \bar{\nu}_\alpha = \sum_\alpha \frac{1}{\exp\,[\beta(\eta_\alpha - \mu_2)] - 1} \tag{3.2}$$

When we combine the chemical equilibrium condition $\mu_2 = 2\mu$ with the condition that the total number of electrons must equal N, we obtain the determining equation for μ:

$$N = N_1 + 2N_2$$
$$= \sum_k \frac{1}{\exp\,[\beta(\epsilon_k - \mu)] + 1} + 2\sum_\alpha \frac{1}{\exp\,[\beta(\eta_\alpha - 2\mu)] - 1} \tag{3.3}$$

For any given $\beta = 1/\mathfrak{k}T$ and number density N/V, Eq. (3.3) is an implicit equation for μ; once μ has been found, N_1 and N_2 follow.

We follow the work of Schafroth (54), making the simplest assumptions at every stage, so as to obtain a first orientation. Thus, the single-particle spectrum is assumed to be

$$\epsilon_k = \hbar^2 k^2/2M \tag{3.4}$$

and the electron pair spectrum is taken as

$$\eta_\alpha = \eta_0 + \frac{\hbar^2 K_\alpha{}^2}{2(2M)} \tag{3.5}$$

where η_0 is the ground state energy of a pair, and K_α is the center-of-gravity momentum of the pair in state α. Excited internal states of a pair are ignored. Assumptions (3.4) and (3.5) allow us to work with the well-known formulas for ideal Fermi-Dirac and Bose-Einstein gases.

Since the average pair occupation numbers $\bar{\nu}_\alpha$ cannot be negative, Eq. (3.3) immediately gives the condition

$$2\mu < \eta_0 \tag{3.6}$$

Let us first dispose of the case of really bound electron pairs, $\eta_0 < 0$. In that case, Eq. (3.6) implies that the chemical potential μ is negative, also. For a Fermi-Dirac gas, a negative μ means a nondegenerate gas (we recall that the degenerate Fermi-Dirac gas has μ positive and close to the zero-temperature Fermi energy ϵ_F). Thus, *if bound pairs can be formed, the gas of single electrons is always non-degenerate*. Since we know that the electron gas in normal metals is highly degenerate, bound pairs are excluded immediately.

This conclusion is, of course, in full agreement with the fact that the main interaction between electrons, the screened Coulomb interaction, is repulsive, and the Fröhlich interaction, though not purely repulsive, is quite weak. Thus, there is no physical mechanism by which fully bound electron pairs could be formed.

We therefore make the opposite assumption, $\eta_0 > 0$, corresponding to a positive energy "resonant state," rather than a truly bound state of the electron pairs. Such resonant states can occur only quite close to the Fermi surface, since they are basically scattering resonances, and electrons deep inside the Fermi sea are prevented from undergoing scattering collisions, by the Pauli exclusion principle. We thus assume that the energy per particle in the pair, $\frac{1}{2}\eta_0$, is very close to the zero-temperature Fermi energy:

$$\eta_0 = 2\epsilon_F(1 - \delta) \tag{3.7}$$

with $|\delta| \ll 1$.

At very low temperatures, only electrons in the immediate neighborhood of the Fermi surface are free to scatter each other; electrons deep inside the Fermi sea cannot undergo scattering collisions, because one or both of the final states would be already occupied. If the interaction between electrons at the Fermi surface is purely repulsive, we expect to find scattering resonances only above the Fermi energy, i.e., $\delta < 0$ in (3.7). We now show that such scattering resonances, though undoubtedly present, are of no importance for a theory of superconductivity, i.e., they do not give rise to a thermodynamic transition or to a Bose-Einstein condensation.

The proof goes as follows: If we ignore pairs altogether, i.e., set $N_1 = N$ in (3.1), and then go to the limit of zero temperature, the chemical potential μ approaches the zero-temperature Fermi energy

$$\epsilon_F = \frac{\hbar^2}{2M}\left(\frac{6\pi^2 N}{V}\right)^{2/3} \tag{3.8}$$

Since the number of single electrons, N_1, is always less than or equal to N, we conclude that $\mu \leqslant \epsilon_F$ also when the pairs are included, i.e., $\mu \leqslant \epsilon_F$ in (3.3) at all temperatures. For a scattering resonance above the Fermi surface, $\delta < 0$, we have $\eta_0 > 2\epsilon_F$, and thus the exponents $\beta(\eta_\alpha - 2\mu)$ never vanish. The gas of pairs never undergoes a Bose condensation.

If we accept the result that a purely repulsive interaction produces scattering resonances, if any, only above the Fermi surface, then we have an immediate answer to the first question at the end of Chapter I: *The Coulomb interaction between electrons, by itself, cannot lead to superconductivity, because this interaction is purely repulsive.* Weak though the Fröhlich interaction undoubtedly is, the Fröhlich interaction is not a pure repulsion, and may, under favorable conditions, lead to an electron pair resonance below the Fermi surface. Such a resonance, with $\delta > 0$ in (3.7), does lead to a condensation, as we shall now show.

In the limit of low temperatures, the sum (3.1) for N_1 may be approximated by standard methods (Landau 58) to give

$$\frac{N_1}{N} = \left(\frac{\mu}{\epsilon_F}\right)^{3/2} \left[1 + \frac{1}{8}\left(\frac{\pi \mathfrak{k} T}{\mu}\right)^2 + \ldots\right] \tag{3.9}$$

If we set μ equal to its maximum value, according to (3.6), and use (3.7), we get the *maximum* number of single electrons from (3.9):

$$N_{1,\max} = N(1 - \delta)^{3/2}[1 + \cdots] \cong N(1 - \tfrac{3}{2}\delta) \tag{3.10}$$

Here we have ignored the correction term in the square bracket of (3.9); the ratio $\mathfrak{k}T/\mu \simeq \mathfrak{k}T/\epsilon_F$ is of order 10^{-3}, so this correction is of order 10^{-6}, at temperatures of interest for superconductors. We have also assumed $\delta \ll 1$.

Since (3.10) is the maximum number of single electrons at low temperatures, the remaining electrons must be contained in the pairs. Whereas the number of singles, N_1, has only a very weak explicit temperature dependence (and a correspondingly weak variation of μ with temperature if N_1 is to remain constant), the Bose distribution for the pairs gives a strong explicit temperature dependence to N_2. The critical situation is obtained by setting μ equal to its limit, Eq. (3.6). When this happens, the lowest exponent in the pair distribution, $\beta(\eta_0 - 2\mu)$, vanishes; i.e., the pair gas undergoes a Bose-like condensation. At this transition point, the number density of the pairs, N_2/V, is given by $f_{\max}$, (II,2.3), i.e.,

$$\frac{N_2}{V} = \frac{2.61}{\lambda_2^3} \tag{3.11}$$

where λ_2 is the thermal de Broglie wavelength for the pairs, i.e., we use a mass $2M$ in (II,1.18),

$$\lambda_2 = \left(\frac{2\pi\hbar^2}{2M\mathfrak{k}T}\right)^{1/2} \tag{3.12}$$

Since $N_2 = \frac{1}{2}(N - N_1)$ and N_1 is fixed by (3.10) for this value of μ, Eq. (3.11) determines the condensation temperature T_c. Using (3.10), this condition is

$$\frac{3}{4}\delta\frac{N}{V} \cong \frac{2.61}{\lambda_2^3} \tag{3.13}$$

Since δ is very small, and since λ_2 increases with decreasing temperature, the condensation condition (3.13) leads to a much lower transition temperature than we would get for a Bose gas of mass $2M$ and boson number $\frac{1}{2}N$. The pair gas is highly dilute and correspondingly undergoes Bose condensation only at very low temperatures.

Conversely, we may estimate δ from the known electron density N/V and known transition temperatures for superconductors. This gives $\delta < 10^{-4}$, i.e., a resonance energy extremely close to the Fermi surface.

The present treatment is greatly oversimplified, and must be taken only in a qualitative sense. The inhibition of electron-electron collisions, and hence electron-electron resonances, by the Pauli exclusion principle, makes it reasonable to expect such resonances only very close to the Fermi surface. However, this also means that the calculation should be carried out in some self-consistent fashion, in which this "quenching" effect of the Pauli principle is included explicitly. Furthermore, we have assumed a very simple energy spectrum for the pairs, Eq. (3.5). If there is an "energy gap," both the spectrum of singles and the spectrum of pairs are altered, and the details of the calculation lose all meaning.

However, once it is understood that the discussion in this section is only qualitative, the qualitative picture which emerges is illuminating indeed. What we have is a *three-fluid model*, the three fluids being the unpaired electrons, the uncondensed (normal) part of the pair distribution, and the condensed (superfluid) pairs. The total number N of electrons is divided up among these, as follows:

$$N = N_1 + 2N_{2n} + 2N_{2s} \tag{3.14}$$

At temperatures above T_c, $N_{2s} \cong 0$ and N_{2n} is small compared to N_1, so that the band theory of solids (in which the pairs are ignored completely) is an excellent approximation. This remains true right down to $T = T_c$, since the density of "normal" pairs remains quite small throughout. Below $T = T_c$, we get increasing numbers of condensed pairs, N_{2s}. These condensed pairs are responsible for superconductivity in the metal.

The possible criticisms of this picture are many and cogent. Let us concentrate on the most important points (Casimir 55). The discussion centers on the *internal size* of these "pairs." They are either small (pair size comparable to the average distance between electrons) or large, and we encounter difficulties either way. If the pair size is small, the energy associated with forming a pair can be estimated from the uncertainty principle. This estimate leads to an energy of the order of electron volts, completely out of line with the known small energy differences between the superconducting and normal states. On the other hand, if the pair size is *large*, we are in even more trouble: between the two electrons forming the pair, there are large numbers of unpaired electrons cruising about, and if the pairs are really large, then even the pair wave functions of different pairs overlap, so that the Bose statistics becomes a very questionable concept.

It is important to realize that there are two distinct aspects to this last criticism:

(1) If the pairs are large, what is the justification for the restriction to pairs? Should one not also include triplets, quadruplets, etc.?

(2) If the pairs are large and there are enough of them that the pair size exceeds the average distance between centers of pairs, the pair wave functions must overlap appreciably. What, then, is the justification for calling the pairs Bose-Einstein particles? Certainly the simple argument of Section 1 fails under those conditions.

It is clear that these arguments cannot be answered on the basis of the oversimplified theory of this section. We now turn to the development of a more satisfactory theoretical framework.

4. Correlation Functions and the Virial Expansion: Classical Treatment

The formation of stable molecules out of atoms is an extreme manifestation of forces between atoms. In order to prepare the way for the statistical mechanics of chemical equilibrium, let us, in this section, review the well-known theory of the way in which interatomic forces modify the perfect gas law

$$pV = N\mathfrak{k}T \tag{4.1}$$

As a result of interatomic forces, Eq. (4.1) is only an approximation for actual gases. It is conventional to write the equation of state of the gas in the form of the "virial expansion"

$$pV = N\mathfrak{k}T\left[1 + \frac{B(T)}{V} + \frac{C(T)}{V^2} + \ldots\right] \tag{4.2}$$

where $B(T)$ is called the "second virial coefficient," $C(T)$ the "third virial coefficient," and so on.

It is easiest to start with a purely classical treatment, ignoring quantum effects. In classical statistical mechanics, the probability of finding the system in a state in which particle i has momentum between p_i and $p_i + dp_i$, position between r_i and $r_i + dr_i$, is proportional to

$$\exp\left[-\frac{H_N(r_1, p_1, r_2, p_2, \ldots, r_N, p_N)}{\mathfrak{k}T}\right] d^3r_1\, d^3p_1\, d^3r_2\, d^3p_2 \ldots d^3r_N\, d^3p_N$$

where H_N is the classical Hamiltonian of the N-particle system. If there are forces between pairs of particles only, H_N is given by

$$H_N = \sum_{i=1}^{N} \frac{(\mathbf{p}_i)^2}{2M} + \sum_{i<j=1}^{N} v(r_{ij}) \tag{4.3}$$

where $r_{ij} = |\mathbf{r}_i - \mathbf{r}_j|$ is the distance between particles i and j, and v is the interparticle potential.

The free energy F_N, in the classical approximation to (II,1.6) and (II,1.7), is given by the integral

$$\exp\left(-\frac{F_N}{\mathfrak{k}T}\right) = \frac{1}{N!}\int \frac{d^{3N}r\, d^{3N}p}{(2\pi\hbar)^{3N}} \exp\left(-\frac{H_N}{\mathfrak{k}T}\right) \tag{4.4}$$

The factor $N!$ comes from the identity of the particles: integration over the positions and momenta of all particles, independently, counts the same physical situation $N!$ times, instead of once. The factor $(2\pi\hbar)^{3N}$ makes (4.4) dimensionless, as it must be; the actual factor involves Planck's constant, and can be derived by performing a limiting process on the quantum-mechanical equation (II,1.7), see ter Haar (54).

We follow ter Haar's discussion of the virial expansion, in the main. The free energy F_N defined by (4.4) depends on the number N of particles. It is convenient to introduce a generating function, depending on a parameter z, by performing the sum

$$Q(z) = \sum_N z^N \exp(-F_N/\mathfrak{k}T) \tag{4.5}$$

Clearly, if $Q(z)$ is known, we can recover the N-particle free energy by a contour integration in the complex z-plane[5]

$$\exp(-\beta F_N) = \frac{1}{2\pi i}\oint \frac{Q(z)}{z^{N+1}}\, dz \tag{4.6}$$

where the contour encircles the origin once, counterclockwise.

---

[5] In practice, it is even simpler than that, as we shall see.

The parameter z is called the "activity"; it is conventional to introduce the "chemical potential" μ by the relation

$$z = \exp(\beta\mu) \tag{4.7}$$

and the "grand canonical potential" $\Omega = \Omega(V, T, \mu)$ by

$$Q(V, T, z) = Q(V, T, e^{\beta\mu}) = \exp(-\beta\Omega) \tag{4.8}$$

The use of these quantities is discussed in more detail in Appendix A. Ter Haar (54) uses the notation q for our $-\beta\Omega$, and ν for our $\beta\mu$.

Combining the equations so far, we obtain the following definition of the grand canonical potential:

$$\exp[-\beta\Omega(V, T, \mu)] = \sum_{N=1}^{\infty} \frac{1}{N!} \int \frac{d^{3N}r\, d^{3N}p}{(2\pi\hbar)^{3N}} \exp[\beta(N\mu - H_N)] \tag{4.9}$$

Let us consider the first few terms of the sum, and the corresponding integrands. For $N = 1$ the integrand is, apart from constant factors,

$$W_1(r, p) = \exp(-\beta H_1) = \exp\left[-\frac{\beta p^2}{2M}\right] \tag{4.10}$$

The integrand for $N = 2$ is, again apart from constant factors,

$$W_2(r_1, p_1, r_2, p_2) = \exp(-\beta H_2) = \exp\left[-\beta\left(\frac{p_1^2}{2M} + \frac{p_2^2}{2M} + v(r_{12})\right)\right] \tag{4.11}$$

In practice, the interatomic potential $v(r_{12})$ approaches zero rapidly for r_{12} in excess of atomic dimensions (the van der Waals force between atoms gives a potential decreasing like $1/r^6$); thus, unless the two particles are rather close together (form a "cluster"), W_2 is the product of separate factors W_1, one for each particle. We therefore define a correlation function (Ursell 27) U_2 by

$$\begin{aligned} U_2(r_1, p_1, r_2, p_2) &= W_2(r_1, p_1, r_2, p_2) - W_1(r_1, p_1)\, W_1(r_2, p_2) \\ &= \exp\left[-\beta\left(\frac{p_1^2}{2M} + \frac{p_2^2}{2M}\right)\right]\left\{\exp[-\beta v(r_{12})] - 1\right\} \end{aligned} \tag{4.12}$$

and write the integrand W_2 in the abbreviated form

$$W_2(1, 2) = W_1(1)\, W_1(2) + U_2(1, 2) \tag{4.13}$$

In a similar way, the three-particle Ursell function $U_3(1, 2, 3)$ is defined by the equation

$$\begin{aligned} W_3(1, 2, 3) = {} & W_1(1)\, W_1(2)\, W_1(3) + W_1(1)\, U_2(2, 3) + W_1(2)\, U_2(3, 1) \\ & + W_1(3)\, U_2(1, 2) + U_3(1, 2, 3) \end{aligned} \tag{4.14}$$

It is apparent from this construction [and a proof can be found in ter Haar (54)] that the general Ursell function $U_n(1, 2, 3, \ldots, n)$ vanishes unless the n particles form a "cluster," in the sense that any particle can be reached from any other particle by traversing a chain of intermediate particles, with no link of the chain longer than the range of the potential function v.

Let us now consider the *integral* for the term $N = 3$ in (4.9); except for constants, this is just the integral of W_3, Eq. (4.14). The second term on the right-hand side of (4.14) gives the contribution

$$\int \frac{d^9r\, d^9p}{(2\pi\hbar)^9} W_1(1)\, U_2(2, 3)$$

$$= \int \frac{d^3r_1\, d^3p_1}{(2\pi\hbar)^3} W_1(1) \int \frac{d^3r_2\, d^3p_2\, d^3p_3}{(2\pi\hbar)^6} U_2(2, 3) = I_1 I_2 \tag{4.15}$$

where

$$I_n = \int \frac{d^{3n}r\, d^{3n}p}{(2\pi\hbar)^{3n}} U_n(1, 2, \ldots, n) \tag{4.16}$$

(we define $U_1 = W_1$ for convenience of notation). The integral I_1 can be evaluated immediately from (4.10) and (4.16) and is

$$I_1 = V/\lambda\!\!\!^{-3} \tag{4.17}$$

where $\lambda\!\!\!^{-}$ is the familiar thermal de Broglie wavelength (II,1.18).

To evaluate I_2, we substitute (4.12) into (4.16); the p-space integrations are trivial. In the integrations over $\mathbf{r}_1$ and $\mathbf{r}_2$, it is convenient to introduce the vector difference $\mathbf{r}_2 - \mathbf{r}_1$ as dummy variable of integration. With the exception of values of $\mathbf{r}_1$ extremely close to the boundary of the volume V, the integral over $\mathbf{r}_2 - \mathbf{r}_1$ is independent of $\mathbf{r}_1$ and emerges as a constant factor in the subsequent integration over $\mathbf{r}_1$. Thus there is only *one* factor volume, V, in the integral I_2. We define "b-coefficients" b_n by

$$I_n = \frac{V}{\lambda\!\!\!^{-3}} n!\, b_n \tag{4.18}$$

to obtain, for $n = 2$,

$$b_2 = \frac{1}{2!} \int \frac{d^3r}{\lambda\!\!\!^{-3}} \{\exp[-\beta v(r)] - 1\} \tag{4.19}$$

In general, except for surface terms which are of no importance, there is only one factor volume in each I_n, and the coefficients b_n in (4.18) are volume-independent, intensive quantities. This rule holds for all low-order b_n, but fails when n becomes comparable to the total number of particles, N. Such large values of n are of importance for the liquid state (liquid drops), but do not contribute in the gaseous state.

The expansion (4.9) then assumes the form (we write the integrals for $N = 1, 2, 3$ explicitly)

$$\exp[-\beta\Omega(V, T, \mu)] = \exp(\beta\mu) I_1 + \frac{\exp(2\beta\mu)}{2!}(I_1^2 + I_2) + \frac{\exp(3\beta\mu)}{3!}(I_1^3 + 3I_1I_2 + I_3) + \cdots$$

$$= \exp\left[\sum_{n=1}^{\infty} (1/n!) \exp(n\beta\mu) I_n\right] \tag{4.20}$$

The proof that the terms can be rearranged to give the last line of (4.20) is given in ter Haar (54). Combining (4.18) and (4.20), we obtain the following formula for the grand canonical potential:

$$\Omega(V, T, \mu) = -\mathfrak{k}T \frac{V}{\lambda^3} \sum_{n=1}^{\infty} b_n e^{n\beta\mu} \tag{4.21}$$

At this stage, we make use of some of the thermodynamic relations proved in Appendix A. It is shown there that the number of particles in the gas, for given V, T, and μ, is obtained by differentiation of Ω with respect to μ, as follows:

$$N = -\left(\frac{\partial\Omega}{\partial\mu}\right)_{V,T} = \frac{V}{\lambda^3} \sum_{n=1}^{\infty} nb_n e^{n\beta\mu} \tag{4.22}$$

and that the pressure p is given by

$$pV = -\Omega(V, T, \mu) \tag{4.23}$$

If one eliminates the chemical potential μ between Eqs. (4.21) and (4.22), Eq. (4.23) gives the equation of state.

We shall do this elimination here in a low approximation, for a dilute gas. Retaining only the term $n = 1$ yields the ideal gas law (4.1). Keeping the first two terms, $n = 1$ and $n = 2$, gives for N:

$$N = \frac{V}{\lambda^3}(e^{\beta\mu} + 2b_2 e^{2\beta\mu} + \cdots) \tag{4.24}$$

In this, classical, region, μ is negative and $\exp(\beta\mu)$ is small, so we solve (4.24) approximately to obtain

$$\exp(\beta\mu) = \frac{\lambda^3 N}{V} - 2b_2\left(\frac{\lambda^3 N}{V}\right)^2 + \cdots \tag{4.25}$$

Substitution into (4.21) and (4.23) gives

$$pV = N\mathfrak{k}T\left[1 - b_2 \frac{\lambda^3 N}{V} + \cdots\right] \tag{4.26}$$

where the omitted terms involve higher powers of the number density N/V. Comparison of (4.2) and (4.26) then gives the standard formula for the second virial coefficient:

$$\begin{aligned} B(T) &= - N\lambda^3 b_2 \\ &= - \tfrac{1}{2} N \int d^3r \, \{\exp[- \beta v(r)] - 1\} \end{aligned} \tag{4.27}$$

This is the classical result, which is used in practice to determine the interparticle potential $v(r)$ from experimental measurements on the gas.

5. Correlation Functions and the Virial Expansion: Quantum Treatment

Our starting point for the quantum treatment is the expression (II,1.7) for the free energy, which replaces the classical expression (4.4). Since (II,1.7) is a sum over all eigenstates E_α (of the proper symmetry) of the N-particle Hamiltonian H_N, this sum is also the trace of the operator $\exp(-\beta H_N)$

$$\exp(- \beta F_N) = \sum_\alpha{}' \exp(- \beta E_\alpha) = \text{Trace}' \exp(- \beta H_N) \tag{5.1}$$

The prime on the sum over α, and on the Trace, indicates restriction to states of the appropriate symmetry, i.e., purely symmetric states for Bose-Einstein statistics, purely antisymmetric states for Fermi-Dirac statistics.

The exponential of an operator may be defined by means of the power series for the exponential function, but this is rather inconvenient to apply in practice. A more useful procedure is to differentiate $\exp(-\beta H_N)$ with respect to β, thereby obtaining the *Bloch differential equation*:

$$\frac{d}{d\beta} \exp(- \beta H_N) = - H_N \exp(- \beta H_N) \tag{5.2}$$

This must be solved subject to the initial condition

$$\lim_{\beta \to 0} \exp(- \beta H_N) = I \tag{5.3}$$

where I stands for the unit operator.

Since the trace of an operator is the sum of all diagonal elements, in all representations, it is not necessary to find the eigenvalues E_α of H_N explicitly, in order to evaluate (5.1). We may, for example, work in configuration space, in spite of the fact that H_N is not diagonal in that space. As a useful first example, consider the operator $\exp(-\beta H_1)$, where $H_1 = p_1^2/2M$ is the kinetic energy operator for one particle. The Bloch equation (5.2) then assumes the form

$$\frac{d}{d\beta} \langle \mathbf{r} \mid \exp(-\beta H_1) \mid \mathbf{r}' \rangle = -\frac{\hbar^2}{2M} \nabla_r^2 \langle \mathbf{r} \mid \exp(-\beta H_1) \mid \mathbf{r}' \rangle \tag{5.4}$$

subject to the initial condition

$$\lim_{\beta \to 0} \langle \mathbf{r} \mid \exp(-\beta H_1) \mid \mathbf{r}' \rangle = \delta(\mathbf{r} - \mathbf{r}')$$

It is easily verified that the following expression satisfies both the differential equation and the initial condition:

$$\langle \mathbf{r} \mid \exp(-\beta H_1) \mid \mathbf{r}' \rangle = \frac{1}{\lambda^3} \exp\left[-\frac{\pi(\mathbf{r} - \mathbf{r}')^2}{\lambda^2}\right] \tag{5.5}$$

As another example, consider H_2^0, the Hamiltonian for two particles without interactions. Since there is no crossterm between particles 1 and 2, the exponential of the sum is the product of the exponentials:

$$\exp(-\beta H_2^0) = \exp[-\beta H_1(1) - \beta H_1(2)] = \exp[-\beta H_1(1)] \exp[-\beta H_1(2)]$$

or, in terms of matrix elements,

$$\langle \mathbf{r}_1 \mathbf{r}_2 \mid \exp(-\beta H_2^0) \mid \mathbf{r}_1' \mathbf{r}_2' \rangle = \frac{1}{\lambda^6} \exp\left\{-\frac{\pi}{\lambda^2} [(\mathbf{r}_1 - \mathbf{r}_1')^2 + (\mathbf{r}_2 - \mathbf{r}_2')^2]\right\} \tag{5.6}$$

The restriction on the trace (5.1), to states of appropriate symmetry only, is inconvenient in practice. To avoid this restriction, one can use operators which give 1 when applied to a function of the right symmetry, and give 0 for all other functions. Letting $\epsilon = +1$ for Bose-Einstein statistics, and $\epsilon = -1$ for Fermi-Dirac statistics, the appropriate projection operators are sums over all permutations of the N particles:

$$T = \frac{1}{N!} \sum_P (\epsilon)^P P$$

where $(-1)^P$ is -1 for odd permutations P, and equals $+1$ for even permutations P. We then get the identity, for any operator Q,

$$\text{Trace}'(Q) = \text{Trace}(QT)$$

where the trace on the right-hand side is unrestricted.

We now apply this to (5.1). Let $r_1 r_2 \cdots r_N$ be an arbitrary configuration; we denote the "permuted" configuration by $r_{1P} r_{2P} \cdots r_{NP}$; for example, if P is the interchange of particles 1 and 2, then $r_{1P} = r_2$ and $r_{2P} = r_1$. With this notation, Eq. (5.1) becomes

$$\exp(-\beta F_N) = \frac{1}{N!} \int d^3r_1 \cdots d^3r_N \sum_P (\epsilon)^P \langle \mathbf{r}_1 \mathbf{r}_2 \cdots \mathbf{r}_N | \exp(-\beta H_N) | \mathbf{r}_{1P} \mathbf{r}_{2P} \cdots \mathbf{r}_{NP} \rangle \tag{5.7}$$

For $N = 2$ this becomes

$$\exp(-\beta F_2) = \tfrac{1}{2} \int d^3r_1\, d^3r_2 [\langle \mathbf{r}_1 \mathbf{r}_2 | \exp(-\beta H_2) | \mathbf{r}_1 \mathbf{r}_2 \rangle + \epsilon \langle \mathbf{r}_1 \mathbf{r}_2 | \exp(-\beta H_2) | \mathbf{r}_2 \mathbf{r}_1 \rangle] \tag{5.7a}$$

For the special case of no interactions between the particles, substitution of (5.6) in (5.7a) yields

$$\exp(-\beta F_2^0) = \frac{1}{2!} \int d^3r_1\, d^3r_2 \left\{ \frac{1}{\lambda^6} + \epsilon \frac{1}{\lambda^6} \exp\left[-\frac{2\pi(\mathbf{r}_1 - \mathbf{r}_2)^2}{\lambda^2} \right] \right\} \tag{5.8}$$

The second term of the integrand arises from the quantum statistics of the particles, and represents a *statistical correlation* between the particles. This statistical correlation produces deviations from the classical ideal gas law (4.1), even in the complete absence of interparticle forces.

We now define W_1, W_2, and the correlation function U_2, in analogy to the classical treatment of the preceding section. Thus

$$U_1(\mathbf{r}_1) = W_1(\mathbf{r}_1) = \langle \mathbf{r}_1 | \exp(-\beta H_1) | \mathbf{r}_1 \rangle = 1/\lambda^3 \tag{5.9}$$

$$W_2(\mathbf{r}_1, \mathbf{r}_2) = \langle \mathbf{r}_1 \mathbf{r}_2 | \exp(-\beta H_2) | \mathbf{r}_1 \mathbf{r}_2 \rangle + \epsilon \langle \mathbf{r}_1 \mathbf{r}_2 | \exp(-\beta H_2) | \mathbf{r}_2 \mathbf{r}_1 \rangle \tag{5.10}$$

$$U_2(\mathbf{r}_1, \mathbf{r}_2) = W_2(\mathbf{r}_1, \mathbf{r}_2) - W_1(\mathbf{r}_1)\, W_1(\mathbf{r}_2) \tag{5.11}$$

In the absence of interparticle forces, the Ursell correlation function U_2 is the purely statistical correlation function

$$U_2^0(\mathbf{r}_1, \mathbf{r}_2) = \frac{\epsilon}{\lambda^6} \exp\left[-\frac{2\pi(\mathbf{r}_1 - \mathbf{r}_2)^2}{\lambda^2} \right] \tag{5.12}$$

We see that the statistical correlations have a "range" equal to the thermal de Broglie wavelength λ. This is the quantum-mechanical uncertainty in the position of a particle with kinetic energy of the order of $\mathfrak{k}T$.

From here on, the quantum treatment is very similar to the classical treatment; we refer to ter Haar (54) for details. The quantum analog of the classical integral I_n, Eq. (4.16), is

$$I_n = \int d^{3n}r \; U_n(\mathbf{r}_1, \mathbf{r}_2, \ldots, \mathbf{r}_n) \tag{5.13}$$

and the coefficients b_n are defined as before, by (4.18). All the work leading to the expression for the second virial coefficient $B(T)$ goes through as before, the only difference being that b_2 is no longer given by the classical form (4.19).

In particular, in the absence of interactions, we have

$$I_2^0 = \int d^3r_1 \, d^3r_2 \; U_2^0(r_1, r_2) = \epsilon \frac{V}{\lambda^3} \frac{1}{2^{3/2}} = \frac{V}{\lambda^3} 2! \, b_2^0 \tag{5.14}$$

so that the second virial coefficient becomes

$$B^0 = -N\lambda^3 b_2^0 = -\epsilon \frac{N\lambda^3}{2^{5/2}} \tag{5.15}$$

This, substituted into the virial series (4.2), gives the leading term in an expansion of pV for the ideal Bose-Einstein (or Fermi-Dirac) gas, valid at high temperatures.

When there are forces acting between the particles of the gas, there appear *dynamical correlations* in addition to the statistical correlations. The Ursell pair correlation function $U_2(\mathbf{r}_1, \mathbf{r}_2)$ can be written in the form

$$U_2(\mathbf{r}_1, \mathbf{r}_2) = U_2^0(\mathbf{r}_1, \mathbf{r}_2) + U_2'(\mathbf{r}_1, \mathbf{r}_2) \tag{5.16}$$

where the additional term U_2' represents the dynamical correlation function. In (5.16), U_2 is given by (5.11) and U_2^0 by (5.12).

A corresponding separation can be made in the integral $I_2 = I_2^0 + I_2'$, Eq. (5.13), and thus in the second virial coefficient $B(T)$:

$$B(T) = B^0 + B' = -N\lambda^3(b_2^0 + b_2') \tag{5.17}$$

The evaluation of the dynamical part of the virial coefficient, B', is beyond the scope of this book. We refer the reader to ter Haar (54) and to the original papers (Uhlenbeck 36, 37; Gropper 36, 37, 39; Green 52; Blatt 56, 56a). Here we shall state the result, comment on it, and attempt to make it plausible. The result is (ignoring spins)

$$B' = -2^{3/2} N\lambda^3 \sum_l{}' (2l+1) \left\{ \sum_i \exp(-\beta\eta_{li}) + \frac{1}{\pi} \int_0^\infty dk \frac{d\delta_l}{dk} \exp\left[-\frac{(k\lambda)^2}{2\pi} \right] \right\} \tag{5.18}$$

where the prime on the sum over l means only even l for Bose-Einstein statistics, only odd l for Fermi-Dirac statistics; the η_{li} are the bound state energies of the two-particle Hamiltonian H_2 with orbital angular momentum l (note $\eta_{li} < 0$), and $\delta_l(k)$ is the scattering phase shift of the two-particle scattering with angular momentum l.

The statistics of the particles enter into (5.18) only through the choice of even or odd values of the angular momentum quantum number l; the change of sign (factor ϵ) in the statistical part of the second virial coefficient, Eq. (5.15), has no direct analog in the dynamical part, Eq. (5.18).

If the two-particle Hamiltonian *has* bound states η_{li}, then their contribution dominates the second virial coefficient at sufficiently low temperatures: the bound states give increasing exponentials

$$\exp(-\beta\eta_{li}) = \exp\left[+\frac{B_{li}}{fT}\right]$$

where $B_{li} = -\eta_{li}$ is the binding energy of bound state number i with angular momentum l. All other terms in (5.15) and (5.18) have much weaker temperature dependence.

Although a detailed derivation of (5.18) would take us too far afield, the following semiderivation of the bound-state terms in (5.18) may be helpful. By combining earlier formulas in this section, the integral I_2, Eq. (5.13), can be written as

$$\begin{aligned} I_2 &= \text{Trace}\,[\exp(-\beta H_2)\,T] - \text{Trace}\,\exp(-\beta H_2^0) \\ &= 2!\,\text{Trace}'\,\exp(-\beta H_2) - [\text{Trace}\,\exp(-\beta H_1)]^2 \end{aligned} \tag{5.19}$$

Since the second term on the right does not involve any bound states, the bound state terms must arise purely from the first term. We let I_{2b} be the bound state contribution to I_2; a double prime denotes restriction to bound states $\eta_\alpha < 0$ of the appropriate symmetry. The bound state with internal energy η_α gives rise to the energy levels

$$\eta_\alpha + \frac{\hbar^2 K^2}{2(2M)}$$

where K is the center-of-gravity momentum; we thus have

$$I_{2b} = 2! \sum_K \sum_\alpha{}'' \exp\left[-\beta\eta_\alpha - \frac{\beta\hbar^2 K^2}{4M}\right] \tag{5.20}$$

The sum over K separates out as a factor, and is given by

$$\sum_K \exp\left(-\frac{\beta\hbar^2K^2}{4M}\right) = \frac{V}{(2\pi)^3}\int d^3K \exp\left[-\frac{\beta\hbar^2K^2}{4M}\right] = \frac{V}{(\lambda_2)^3} \tag{5.21}$$

where λ_2 is the thermal de Broglie wavelength of the molecular motion, Eq. (3.12). The corresponding contribution to $b_2 = \frac{1}{2}(\lambda^3/V)\, I_2$ is

$$(b_2)_b = \frac{\lambda^3}{2V} I_{2b} = 2^{3/2} \sum_\alpha{}'' \exp(-\beta\eta_\alpha) \tag{5.22}$$

where we have used the relation (2.11). The second virial coefficient is $B = -N\lambda^3 b_2$, and the bound state contribution to B is therefore

$$(B)_b = -N\lambda^3(b_2)_b = -2^{3/2} N\lambda^3 \sum_\alpha{}'' \exp(-\beta\eta_\alpha) \tag{5.23}$$

This agrees precisely with the bound state terms in (5.18); we need merely remember that a bound state with angular momentum l is $(2l+1)$-fold degenerate.

6. Correlation Functions and Chemical Equilibrium

In the conventional approach to the discussion of chemical equilibrium, one considers the system to consist of two interpenetrating gases: a gas of atoms and a gas of molecules. The total pressure is the sum of the partial pressures of the two gases:

$$p = p_1 + p_2 \tag{6.1}$$

Since the grand canonical potential $\Omega(V, T, \mu)$ equals $-pV$, generally, Eq. (6.1) can also be written as

$$\Omega = \Omega_1(V, T, \mu_1) + \Omega_2(V, T, \mu_2) \tag{6.2}$$

where μ_1 is the chemical potential of the atoms, μ_2 the chemical potential of the molecules. They are related by the chemical equilibrium condition (2.4), $\mu_2 = 2\mu_1$. The number of particles is, quite generally, the negative derivative of Ω with respect to μ, at constant V and T (see Appendix A); thus the number of atoms is

$$N_1 = -\left(\frac{\partial\Omega_1}{\partial\mu_1}\right)_{V,T} \tag{6.3}$$

and the number of molecules is

$$N_2 = -\left(\frac{\partial \Omega_2}{\partial \mu_2}\right)_{V,T} \tag{6.4}$$

The total number of particles, including those bound within the molecules, is

$$N = -\left(\frac{\partial \Omega}{\partial \mu}\right)_{V,T} = N_1 + 2N_2 \tag{6.5}$$

where we have used the chemical equilibrium condition (2.4).

Although the procedure outlined here is quite neat, formally, it is most unsatisfactory from a fundamental point of view. We do *not* have two non-interacting, interpenetrating gases in our box. We have *one* gas, whose atoms may, and occasionally do, combine to form molecules. The same interatomic forces which lead to the formation of molecules, also lead to deviations from the ideal gas law in the atomic gas. In what way, then, is the formation of molecules distinguished from ordinary effects which are contained in the virial expansion?

One conceivable answer to this question is: molecules are *bound* states of two atoms, whereas the "ordinary" deviations from the ideal gas law for the atomic gas are connected with *scattering* states of two, or more, atoms. The trouble with this answer is that the bound states *do* contribute to the second virial coefficient, as we have just seen.

In fact, the entire distinction between bound states and scattering states is difficult in an interacting system such as a gas. If two particles are by themselves in a large box, the distinction is clear and unambiguous. But in a gas, there are always other particles about, other molecules and other atoms, which collide with the supposedly stable molecule, and may do so with sufficient energy to lead to molecular break-up. If we, in our imagination, follow the path of a "stable" molecule through our gas, after some time this molecule will undergo a break-up collision, and will not be there any more! Of course, for any one molecule which breaks up, there is on the average another molecule which gets formed in an encounter between atoms, and in this way the total number of molecules stays constant in chemical equilibrium. But no one molecule is "stable" in the sense of remaining around indefinitely.

Does this imply that there are no such things as molecules? Of course it does not; but it does imply that the concept "molecule" must be considered a bit more carefully. Suppose a Laplace demon wishes to determine the number of molecules in the gas; how should he go about it? He may start by taking a "snapshot" of the gas, and he will notice, in this snapshot, that certain pairs of atoms are very close together, compared to the mean distance $(V/N)^{1/3}$. However, not all these pairs are really molecules. Even in a gas of hard spheres, occasionally two hard spheres will be close together; but no molecules are formed. Our Laplace demon, therefore, must take additional snapshots, at slightly later times. He can now classify the close pairs of atoms in his first snapshot into two classes: pairs which stay together for some time, and pairs which separate rapidly. The demon then counts as "molecules" those pairs which stay together for times much longer than the characteristic time $\tau = a/v$ (a = diameter of atom, v = average thermal velocity); the other pairs are accidental, and should not be counted as molecules.

Thus, in a gas, a molecule is not a bound state; rather, *a diatomic molecule is an attractive pair correlation.* More precisely, it is an attractive *dynamical* pair correlation; purely statistical pair correlations, corresponding to the $U_2^0(r_1, r_2)$ in (5.16), must not be counted as molecules.

Since the theory of the second virial coefficient involves these pair correlations, we may expect a close relationship between the virial expansion and the theory of chemical equilibrium. This is indeed the case. We shall first establish the formal connection and discuss the physical connection afterward.

The virial expansion (4.2) amounts to a rearrangement of the terms of the series (4.21) for the grand canonical potential. If we define the chemical potential of n-particle correlations by

$$\mu_n = n\mu_1 \tag{6.6}$$

then the series (4.21) can be rewritten as

$$\Omega = \sum_{n=1}^{\infty} \Omega_n(V, T, \mu_n) \tag{6.7}$$

where[6]

$$\Omega_n(V, T, \mu_n) = -kT \frac{V}{\lambda^3} b_n \exp(\beta\mu_n) \tag{6.8}$$

If we restrict ourselves to $n = 1$ and 2, Eq. (6.7) is precisely the law of partial pressures, Eq. (6.2); and Eq. (6.6) with $n = 2$ is just the chemical equilibrium condition (2.4).

Furthermore, Eq. (4.24) for the number of particles is equal to (6.5), provided we make the identifications[6]

$$N_1 = \frac{V}{\lambda^3} \exp(\beta\mu_1) \tag{6.9}$$

[6] These identifications are only preliminary, and require correction; this will be done later.

and

$$N_2 = \frac{V}{\lambda^3} b_2 \exp(\beta\mu_2) = \frac{V}{\lambda^3} b_2 \exp(2\beta\mu_1) \tag{6.10}$$

Equation (6.9) is equivalent to the classical approximation (2.7) for the chemical potential of the atomic gas. Before considering refinements, let us first check that the dominant (bound state) terms in b_2 lead to the expected expression (2.9) for the molecular gas.

The bound state contribution to b_2, called $(b_2)_b$, is given by (5.22). When we substitute (5.22) into (6.10), we obtain the bound state contribution to the number of molecules, $(N_2)_b$. We make use of the relation $\lambda_2 = 2^{-1/2}\lambda$ for the thermal de Broglie wavelength of the molecules, to get

$$(N_2)_b = \frac{V}{\lambda_2^3} \sum_{\alpha}'' \exp[\beta(\mu_2 - \eta_\alpha)] \tag{6.11}$$

We recall that the double prime on the sum indicates a restriction to bound states, i.e., $\eta_\alpha < 0$, all of the appropriate symmetry to fit the Bose-Einstein or Fermi-Dirac statistics of the atoms.

If we ignore all excited states of the molecule, and put $\eta_0 = -B$, we get exactly (2.9). Equation (6.11) is more accurate than that, however, since it includes the correction for excited (but still bound) molecular states: In fact, we have

$$(N_2)_b = \frac{V}{\lambda_2^3} \exp(\beta\mu_2) \sum_{\alpha}'' \exp(-\beta\eta_\alpha) = \frac{V}{\lambda_2^3} \exp[\beta(\mu_2 - f)] \tag{6.12}$$

where f is the internal free energy of the molecule, defined by (2.18). Thus, the "$-B$" in Eq. (2.9) has been replaced by f, just as was stated at the end of Section 2.

Hence, subject to corrections to be discussed below, *we obtain a statistical mechanical theory of chemical equilibrium from the theory of the virial expansion, simply by taking Eq.* (4.21) *seriously, as it stands, even in cases where some of the "correction" terms become large.* In particular, we should not restrict ourselves to situations in which N_2, Eq. (6.10), is small compared to N_1, Eq. (6.9), but should permit N_2 to become comparable to N_1 or even exceed N_1 significantly.

At first sight, one tends to be sceptical of the validity of stopping with the terms N_1 and N_2 of an expansion, in which N_2 is much larger than N_1. However, this is just the physical meaning of a chemical equilibrium between binary molecules and single atoms! Our Laplace demon, with his successive snapshots of the gas, detects an enormous number of highly persistent, attractive pair correlations. But if the chemical bond saturates with binary molecules (as it does for hydrogen, for example: there is no stable H_3 molecule), then the Laplace demon does not see any persistent, attractive triplet correlations at all.

If there is actually chemical equilibrium between atoms and binary molecules, then N_1 and N_2 are the leading, significant terms; the first "correction" term is N_3; this term includes the effects, not only of triple collisons between atoms, but also of binary collisions between one atom and one molecule. If there are no ternary molecules, N_3 contains *only* corrections for deviations of the atomic and molecular gases from perfect gas behavior. The term $n = 4$ contains the effects of encounters between four atoms, of encounters between two atoms and one molecule, and of encounters between two molecules; and so on for $n = 5, 6$, etc.

We emphasize that the low-order terms, considered here, have nothing whatever to do with condensation of the gas into the liquid state. Formally, the series (4.21) contains the full equation of state of the material, including the phase transition to the liquid and solid states. However, the transition to the liquid state involves the sudden coming into prominence of very high order terms, with n comparable to the total number of particles N. These terms represent "liquid drops," and cannot be treated by the methods applicable to gases. As has been stressed particularly by Fierz (51), the transition to the liquid state has nothing to do with the importance, or lack of importance, of low terms in the virial series. If the vapor is sufficiently dilute, it behaves substantially like a perfect gas right down to the condensation temperature.

Having discussed the general approach, let us now retrace our steps more carefully, so as to obtain more accurate identifications of "atomic" and "molecular" terms in the virial expansion.

First, we must distinguish between statistical and dynamical correlations, and their respective contribution to the virial series. Recalling (5.16) and (5.17), we may rewrite Ω_n, Eq. (6.8), as

$$\Omega_n = \Omega_n^0 + \Omega_n' \tag{6.13}$$

where Ω_n^0 is the nth term in the series (4.21) for the *ideal*, non-interacting gas. We shall show in Section 7 that this is given by[7]

$$\Omega_n^0(V, T, \mu) = \frac{V}{\lambda^3} b_n^0 \exp(n\beta\mu) = (\epsilon)^{n+1} \frac{V}{\lambda^3} \frac{\exp(n\beta\mu)}{n^{5/2}} \tag{6.14}$$

[7] As before, $\epsilon = +1$ for Bose-Einstein statistics, $\epsilon = -1$ for Fermi-Dirac statistics, of the atoms.

This part of Ω_n must be included as pact of the "atomic" contribution. *Only dynamical correlations can give rise to molecules.*

Next, consider the remaining, dynamical contribution to the "number of molecules" N_2, Eq. (6.10), i.e., we consider

$$N'_2 = \frac{V}{\lambda^3} b'_2 \exp(2\beta\mu) = \tfrac{1}{2} I'_2 \exp(2\beta\mu)$$
$$= \exp(2\beta\mu) \operatorname{Trace}' [\exp(-\beta H_2) - \exp(-\beta H_2^0)] \qquad (6.15)$$

where the prime on the trace indicates restriction to states of the appropriate symmetry, and H_2^0 is the two-particle Hamiltonian without interactions.

It is helpful to separate out the center-of-gravity motion, i.e., we write

$$H_2 = -\frac{\hbar^2}{2(2M)} \nabla_R^2 + \left[-\frac{\hbar^2}{2(\frac{1}{2}M)} \nabla_r^2 + v(r)\right] = H_R + H_r \qquad (6.16)$$

where $\mathbf{R} = \frac{1}{2}(\mathbf{r}_1 + \mathbf{r}_2)$ and $\mathbf{r} = \mathbf{r}_2 - \mathbf{r}_1$. Furthermore, the force-free Hamiltonian is

$$H_2^0 = H_R + H_r^0 = H_R - \frac{\hbar^2}{2(\frac{1}{2}M)} \nabla_r^2 \qquad (6.17)$$

Both H_2 and H_2^0 contain the same center-of-gravity part, and thus this part may be treated separately. Eigenfunctions of H_R have the form

$$\psi_{K,\alpha} = \exp(i\mathbf{K}\cdot\mathbf{R})\, \phi_\alpha(\mathbf{r}) \qquad (6.18)$$

where $\phi_\alpha(\mathbf{r})$ is a function of $\mathbf{r} = \mathbf{r}_2 - \mathbf{r}_1$. The corresponding eigenvalue of H_R is equal to $\hbar^2K^2/4M$.

Equation (6.15) shows that we are not actually interested in the eigenvalues and eigenfunctions of either H_2 or H_2^0, but rather in the eigenvalues of the operator

$$U'_2 = \exp(-\beta H_2) - \exp(-\beta H_2^0)$$
$$= \exp(-\beta H_R) [\exp(-\beta H_r) - \exp(-\beta H_r^0)] \qquad (6.19)$$

The factorization in (6.19) is possible because H_R commutes with both H_r and H_r^0, and because the center-of-gravity factor $\exp(i\mathbf{K}\cdot\mathbf{R})$ is symmetric under exchange of 1 and 2. Application of this operator to a function of the form (6.18) gives

$$[\exp(-\beta H_2) - \exp(-\beta H_2^0)]\, \psi_{K,\alpha}$$
$$= \exp\left(-\frac{\beta\hbar^2K^2}{4M}\right) \left[\exp(-\beta H_r) - \exp(-\beta H_r^0)\right] \psi_{K,\alpha} \qquad (6.20)$$

The operator in the square brackets on the right-hand side of (6.20) acts only on the factor $\phi_\alpha(r)$ of the wave function (6.18). We are therefore interested in the eigenvalues $\mathfrak{v}_\alpha$ defined by

$$[\exp(-\beta H_r) - \exp(-\beta H_r^0)]\, \phi_\alpha(r) = \mathfrak{v}_\alpha\, \phi_\alpha(r) \tag{6.21}$$

In terms of these eigenvalues, Eq. (6.15) becomes [see Eq. (5.21) for the evaluation of the sum over K]

$$N_2' = \exp(2\beta\mu) \frac{V}{\lambda_2^3} \sum_\alpha{}' \mathfrak{v}_\alpha \tag{6.22}$$

where the prime on the sum over α has the usual significance, of restriction to the appropriate symmetry of the eigenfunctions $\phi_\alpha(r)$.

If we can make the identification

$$\mathfrak{v}_\alpha = \exp(-\beta\eta_\alpha) \tag{6.23}$$

then (6.22) reduces to the expression for the number of binary molecules in the theory of chemical equilibrium [see Eq. (6.12)]. This identification is not exact, however, for two reasons:

(1) Even for low-lying eigenvalues η_α of H_r, the corresponding eigenfunction is *not* equal to the $\phi_\alpha(r)$ in (6.21), and the eigenvalue $\mathfrak{v}_\alpha$ is *not* given by (6.23). The difference $\exp(-\beta H_r) - \exp(-\beta H_r^0)$ is not equivalent to $\exp(-\beta H_r)$, taken by itself.

However, for these low-lying levels, the operator $\exp(-\beta H_r^0)$ can be treated as a small perturbation, and thus (6.23) becomes a reasonable first approximation. We can, if we wish, define "effective energy levels" $\hat{\eta}_\alpha$ of the molecules by the equation

$$\mathfrak{v}_\alpha = \exp[-\beta\hat{\eta}_\alpha(\beta)] \tag{6.24}$$

but, as the notation indicates, these effective energy levels are themselves somewhat temperature-dependent.

(2) For high-lying levels, the eigenvalue spectrum of the operator in (6.21) differs qualitatively from the eigenvalue spectrum of $\exp(-\beta H_r)$. The eigenvalues of $\exp(-\beta H_r)$ are all positive numbers, and there is both a discrete spectrum (corresponding to bound states of the Hamiltonian H_r) and a continuous spectrum; this is indicated schematically in Fig. 3.1. By contrast, the eigenvalue spectrum $\mathfrak{v}_\alpha$ in (6.21) is entirely a discrete spectrum (Schafroth 57), and in general there are negative eigenvalues as well as positive eigenvalues. It is apparent that a negative $\mathfrak{v}_\alpha$ can never be written in the form (6.24).

The positive, discrete values of $\mathfrak{v}_\alpha$ can be written, at least formally, in the form (6.24), and we may think of positive values of $\hat{\eta}_\alpha(\beta)$ as temperature-dependent "resonant states" of the molecule. All positive $\mathfrak{v}_\alpha$ make positive contributions to the number of molecules, Eq. (6.22).

However, negative eigenvalues $\mathfrak{v}_\alpha$ are obviously not to be identified with molecules, since the corresponding contribution to N_2', Eq. (6.22), would give a negative number of such "molecules."

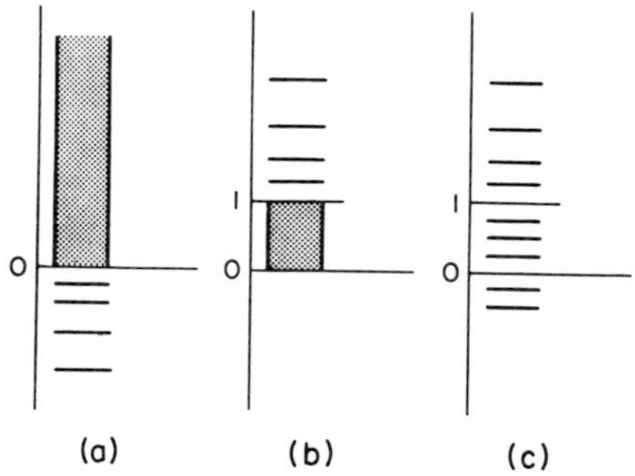

FIG. 3.1. Schematic comparison of the eigenvalue spectra of the operators:

(a) H_r, the Hamiltonian of the two-particle system for center-of-gravity momentum $\mathbf{K} = 0$. Note the discrete states for negative energies, the continuous spectrum for positive energies.

(b) $\exp(-\beta H_r)$ where $\beta = 1/\mathfrak{k}T$. Note that all eigenvalues are positive. The continuous spectrum corresponds to eigenvalues less than 1, the discrete spectrum appears as eigenvalues larger than 1.

(c) The correlation matrix, $\exp(-\beta H_r) - \exp(-\beta H_r^0)$, where H_r^0 is the kinetic energy of relative motion. Note that the *entire* spectrum of this operator is discrete, and that negative eigenvalues are possible.

This distinction makes good sense, physically: The operator in (6.21) is an average pair correlation, in thermal equilibrium, for pairs with center-of-gravity momentum $K = 0$. Positive $\mathfrak{v}_\alpha$ correspond to attractive pair correlations, and can be identified with molecules. Negative $\mathfrak{v}_\alpha$ are *repulsive* pair correlations and have nothing to do with molecule formation—these terms are corrections to the equation of state of the *atomic* gas.

Thus, finally, the terms in the virial series (4.21) which are to be identified with binary molecules are given by

$$\Omega_2'' = -\mathfrak{k}T \frac{V}{\lambda^3} 2^{3/2} \sum_\alpha{}''' \mathfrak{v}_\alpha \exp(2\beta\mu) \tag{6.25}$$

where the $\mathfrak{v}_\alpha$ are eigenvalues of the operator in (6.21), and the triple prime on the sum α indicates restriction to *positive* eigenvalues of the correct symmetry. The factor $2^{3/2}$ comes from the ratio of λ^3 to $(\lambda_2)^3$.

7. Statistical Correlations in the Degenerate Ideal Fermi Gas

The Ursell correlation function U_n, introduced in Sections 4 and 5, can be evaluated particularly easily if the gas is ideal, i.e., if the correlations in question are purely statistical correlations.

The function $U_1(r_1) = W_1(r_1)$, Eq. (5.9), is the diagonal element of the operator $\exp(-\beta H_1)$. Let us now introduce the general matrix element

$$\langle \mathbf{r} \mid U_1 \mid \mathbf{r}' \rangle = \langle \mathbf{r} \mid \exp(-\beta H_1) \mid \mathbf{r}' \rangle \tag{7.1}$$

The explicit formula for this quantity is given in Eq. (5.5).

We now notice that the purely statistical pair correlation function U_2^0, Eq. (5.12), can be expressed immediately in terms of the matrix U_1. Comparison of (5.5) and (5.12) leads to the identity

$$U_2^0(\mathbf{r}_1, \mathbf{r}_2) = \epsilon \langle \mathbf{r}_1 \mid U_1 \mid \mathbf{r}_2 \rangle \cdot \langle \mathbf{r}_2 \mid U_1 \mid \mathbf{r}_1 \rangle \tag{7.2}$$

where $\epsilon = -1$ for Fermi-Dirac statistics.

To find the generalization of (7.2) for arbitrary order n, we use the fact that H_n^0 is a sum of mutually commuting, single-particle Hamiltonians H_1, so that the exponential is a product of exponentials

$$\begin{aligned} \langle r_1 r_2 \cdots r_n \mid \exp(-\beta H_n^0) \mid r_1' r_2' \cdots r_n' \rangle \\ = \langle r_1 \mid U_1 \mid r_1' \rangle \langle r_2 \mid U_1 \mid r_2' \rangle \cdots \langle r_n \mid U_1 \mid r_n' \rangle \end{aligned} \tag{7.3}$$

Looking at (5.7), (5.9), and (5.10), the general formula for the integrand W_n is

$$W_n(r_1 r_2 \cdots r_n) = \sum_P (\epsilon)^P \langle r_1 r_2 \cdots r_n \mid \exp(-\beta H_n) \mid r_{1P} r_{2P} \cdots r_{nP} \rangle \tag{7.4}$$

where the P denote permutations of the n indices.

If we substitute (7.3) into (7.4), there is considerable simplification. It is well-known that permutations can be written as products of mutually commuting, cyclical permutations. If the permutation P breaks up into a product of two (or more) cycles, the corresponding term in W_n^0 becomes a product of two (or more) "unlinked" factors. It is easy to see, in this way, that the only contributions to the Ursell function U_n^0 arise from those permutations P which cannot be broken up into smaller cycles, i.e., the cyclical permutations of n things [for example, with $N = 3$, we have the 2 distinct cycles (1, 2, 3) and (1, 3, 2)]. All these permutations are even if n is odd, and odd if n is even, so that

$$(\epsilon)^P = (\epsilon)^{n-1} \qquad \text{for cyclic permutations of order } n.$$

We thus have the general formula (Kahn 38)

$$U_n^0(r_1, r_2, \ldots, r_n) = (\epsilon)^{n-1} \sum_{P \text{ cyclic}} \langle r_1 \mid U_1 \mid r_{1P}\rangle \langle r_2 \mid U_1 \mid r_{2P}\rangle \cdots \langle r_n \mid U_1 \mid r_{nP}\rangle \tag{7.5}$$

in which the sum contains $(n - 1)!$ terms.

By substituting (7.1) into (7.5), we obtain the explicit form

$$U_n^0(r_1, r_2, \ldots, r_n) = \frac{(\epsilon)^{n-1}}{\lambda^{3n}} \exp\left\{-\frac{\pi}{\lambda^2}[(\mathbf{r}_1 - \mathbf{r}_2)^2 + (\mathbf{r}_2 - \mathbf{r}_3)^2 + \cdots + (\mathbf{r}_{n-1} - \mathbf{r}_n)^2 + (\mathbf{r}_n - \mathbf{r}_1)^2]\right\} + \text{terms from the other cyclic permutations of order } n \tag{7.6}$$

This shows the nature of these statistical linkings explicitly. The linking is in the form of chains which must close in on themselves, with each link of the chain being of order λ in length. If n is large, the average linear dimension of such a chain is of order $(n)^{1/2}\lambda$. For electrons in metals, λ is itself quite large compared to a lattice distance, and furthermore, large values of n make important contributions. The statistical chains are therefore of enormous dimensions.

When we evaluate the integral (5.13) for I_n with (7.5) or (7.6), it is easily seen that all the $(n - 1)!$ cyclic permutations give rise to the same value (they amount to a permutation of the dummy variables of integration). We thus have

$$I_n^0 = (\epsilon)^{n-1}(n-1)! \int \frac{d^{3n}r}{\lambda^{3n}} \exp\left\{-\frac{\pi}{\lambda^2}[(\mathbf{r}_1 - \mathbf{r}_2)^2 + (\mathbf{r}_2 - \mathbf{r}_3)^2 + \cdots + (\mathbf{r}_{n-1} - \mathbf{r}_n)^2 + (\mathbf{r}_n - \mathbf{r}_1)^2]\right\} \tag{7.7}$$

It will turn out instructive to go through the evaluation of the integral explicitly. We introduce the variables

$$\mathbf{y}_1 = \mathbf{r}_1 - \mathbf{r}_2, \quad \mathbf{y}_2 = \mathbf{r}_2 - \mathbf{r}_3, \ldots, \mathbf{y}_{n-1} = \mathbf{r}_{n-1} - \mathbf{r}_n, \quad \mathbf{y}_n = \mathbf{r}_n - \mathbf{r}_1 \tag{7.8}$$

These are not all independent, but rather they satisfy the relation

$$\mathbf{y}_1 + \mathbf{y}_2 + \cdots + \mathbf{y}_n = 0 \tag{7.9}$$

In addition to the $\mathbf{y}_i$, we need one more variable, which we can take to be $\mathbf{r}_1$, say. We thus get

$$I_n^0 = (\epsilon)^{n-1}(n-1)! \int d^3r_1 \int \frac{d^{3n}y}{\lambda^{3n}} \,\delta(\mathbf{y}_1 + \cdots + \mathbf{y}_n) \exp\left[-\frac{\pi}{\lambda^2}(\mathbf{y}_1{}^2 + \cdots + \mathbf{y}_n{}^2)\right] \tag{7.10}$$

The delta-function in (7.10) takes care of the restriction (7.9).

The integral over r_1 gives rise to a factor V.[8] We now make use of the identity

$$\delta(\mathbf{Y}) = \frac{1}{V} \sum_{\mathbf{k}} \exp(i\mathbf{k} \cdot \mathbf{Y}) \tag{7.11}$$

We substitute (7.11) into (7.10), and interchange the order of summation over $\mathbf{k}$ and integration over the $\mathbf{y}_i$ to obtain the result

$$I_n^0 = (\epsilon)^{n-1}(n-1)! \sum_{\mathbf{k}} (\hat{\mathrm{u}}_{\mathbf{k}})^n \tag{7.12}$$

where[9]

$$\hat{\mathrm{u}}_{\mathbf{k}} = \int d^3y \exp(i\mathbf{k} \cdot \mathbf{y}) \langle \mathbf{r} + \mathbf{y} \mid U_1 \mid \mathbf{r} \rangle = \int \frac{d^3y}{\lambda^3} \exp\left[i\mathbf{k} \cdot \mathbf{y} - \frac{\pi y^2}{\lambda^2}\right]$$
$$= \exp\left(-\frac{k^2\lambda^2}{4\pi}\right) = \exp\left(-\frac{\beta\hbar^2k^2}{2M}\right) \tag{7.13}$$

Let us comment on this result. The plane waves $\exp(i\mathbf{k} \cdot \mathbf{r})$ are eigenfunctions of the one-particle Hamiltonian H_1, and hence also of the operator $U_1 = \exp(-\beta H_1)$. The identity (7.11) for the delta-function amounts to introduction of these eigenfunctions, and the $\hat{\mathrm{u}}_k$ are the eigenvalues of the operator U_1, so that we have

$$\hat{\mathrm{u}}_k = \exp(-\beta\epsilon_k) \tag{7.14}$$

[this follows from (7.13)] and also

$$\langle k \mid U_1 \mid k' \rangle = \delta_{k,k'}\, \hat{\mathrm{u}}_k \tag{7.15}$$

i.e., U_1 is diagonal in the k-representation.

Let us now return to the calculation of I_n from Eq. (7.12). We have

$$\sum_k \exp(-n\beta\epsilon_k) = \frac{V}{(2\pi)^3} \int d^3k \exp(-n\beta\epsilon_k) = \frac{V}{n^{3/2}\lambda^3} \tag{7.16}$$

and thus

$$I_n^0 = (\epsilon)^{n-1}(n-1)!\ \frac{V}{n^{3/2}\lambda^3} \tag{7.17}$$

[8] For very large cycles, n of order N, this statement becomes untrue because the actual value of r_1 affects the limits of integration for the y-variables. However, such cycles are of concern to us only beyond the Bose condensation point, and as we shall see right away, the series expansion (4.21) needs modification there, anyway.

[9] The matrix element $\langle \mathbf{r} + \mathbf{y} \mid U_1 \mid \mathbf{r} \rangle$ is independent of $\mathbf{r}$, because of translation invariance.

Relation (4.18) defines the b-coefficients and gives

$$b_n^0 = \frac{(\epsilon)^{n-1}}{n^{5/2}} \tag{7.18}$$

When we substitute this into the virial series (4.21) and (4.22), we get the expansions for the ideal gas:

$$\Omega^0(V, T, \mu) = -\mathfrak{k}T \frac{V}{\lambda^3} \sum_{n=1}^{\infty} \frac{(\epsilon)^{n-1} \exp(n\beta\mu)}{n^{5/2}} \tag{7.19}$$

and

$$N = \frac{V}{\lambda^3} \sum_{n=1}^{\infty} \frac{(\epsilon)^{n-1} \exp(n\beta\mu)}{n^{3/2}} \tag{7.20}$$

We note that this last result agrees exactly with the Bose gas expansion (II,1.19): we put $\epsilon = +1$ and recall that the f in (II,1.19) equals N/V, see (II,1.16).

In the ideal Bose gas, μ is non-positive, and thus (7.19) and (7.20) converge, although the convergence is too slow to be practically useful for μ near 0. In the degenerate ideal Fermi gas, however, μ is large and positive, close to the zero temperature Fermi energy ϵ_F, and thus *both* (7.19) *and* (7.20) *are divergent series.*

In order to obtain useful expressions, we must reorder these series. Looking at (4.20), we can write $-\beta\Omega$ in the form

$$-\beta\Omega^0(V, T, \mu) = \sum_{n=1}^{\infty} \frac{I_n^0 \exp(n\beta\mu)}{n!}$$

$$= \epsilon \sum_k \sum_{n=1}^{\infty} \frac{[\epsilon \hat{u}_k \exp(\beta\mu)]^n}{n} \tag{7.21}$$

where we have made use of (7.12) and have interchanged orders of summation.

For each k, the sum over n is recognizable as the expansion of a logarithm, and thus

$$\Omega^0(V, T, \mu) = \epsilon \mathfrak{k}T \sum_k \ln[1 - \epsilon \hat{u}_k \exp(\beta\mu)] \tag{7.22}$$

This is a standard expression in statistical mechanics; from it we can deduce the, more familiar, average occupation numbers $\bar{n}_k$ by carrying out the differentiation in (4.22):

$$N = -\left(\frac{\partial\Omega}{\partial\mu}\right)_{V,T} = \sum_k \frac{\hat{u}_k \exp(\beta\mu)}{1 - \epsilon\hat{u}_k \exp(\beta\mu)} = \sum_k \bar{n}_k \tag{7.23}$$

When we substitute (7.14) into (7.23), we get (II,1.12) for $\epsilon = -1$ and (II,1.13) for $\epsilon = +1$, as we should.

The conclusion from this derivation is that, in the degeneracy region, the series (4.21) cannot be used as it stands, but must be summed selectively. If the series diverges even for the ideal gas, as it does, it must not be used uncritically for an interacting gas.

There is an important and useful generalization of this approach, however, which we must discuss before going on. A look at the derivation just completed, shows that we did not need to use the explicit expression (5.5) for U_1, or (7.13) for the eigenvalues $\hat{u}_k$ of U_1. In particular the reduction of the statistical n-particle correlation matrix U_n^0 to an expression involving only the matrix U_1, i.e., our equation (7.5), is valid no matter what we assume about U_1. To some extent, this is clear, since the single-particle Hamiltonian and its eigenvalues ϵ_k are completely at our disposal, formally speaking.

The degree of generality is higher than that, however. The form (7.13) for $\hat{u}_k$ is quite restrictive, even if the ϵ_k are an arbitrary single-particle spectrum. According to (7.13), all the $\hat{u}_k$ are independent of the chemical potential μ, and have an exponential dependence on $\beta = 1/\mathfrak{k}T$. The derivation of (7.22), however, goes through even if $\hat{u}_k$ is a completely arbitrary function of β and μ.

It turns out that, in an interacting system, the independent-particle approximation can be improved enormously by allowing U_1 to have this degree of flexibility, i.e., by considering an "effective" single-particle Hamiltonian which itself depends on temperature and particle density. With this assumption, the eigenvalues ϵ_k are themselves functions of β and μ, and $\hat{u}_k = \exp[-\beta\epsilon_k(\beta, \mu)]$ becomes quite arbitrary.

The idea of a self-consistent effective single-particle spectrum is an old one. It underlies the Hartree-Fock approximation (Schiff 49), and has undergone extensive development recently as a result of the work of Brueckner and others (Brueckner 58, Bell 61, and references quoted there). In statistical mechanics, the analog of the Hartree-Fock approximation has been worked out by Husimi (40), and Brueckner-type methods have been worked out by Bloch and Dominicis (Bloch 59, 59a).

The common core of all these methods is an improvement of the single-particle approximation, by including a large fraction of the correlation effects between particles within an "effective" single-particle spectrum, and effective single-particle wave functions. These latter represent, not "bare" particles, but "dressed" particles, that is, they represent elementary excitations of the system which can, in some approximation, be superposed as if they were independent entities. The very great success of the band theory of solids, and the smaller, but still considerable, success of the nuclear shell model, indicate to how large an extent such a program is possible. In the theory of solids, in particular, the independent-particle model (the band theory of solids) is so spectacularly successful that it amounts almost to a miracle. Although qualitative arguments have been advanced for many years, only very recently (Kohn 57) has there been any formal justification for the independent-particle model of electrons in solids; and it is doubtful whether, even now, we fully understand just why the model is as good as it appears to be, experimentally. The screening of the Coulomb force between electrons by plasma effects (see Chapter I, Section 9) is by no means sufficient explanation. Even if we ignore all small-angle (say, less than 90 degrees) Coulomb scattering of electrons by electrons, the cross section is still of the order of 10^{-16} cm^2, and a straightforward calculation of the mean free path of an electron gives something of the order of one lattice distance! The Pauli exclusion principle is essential here: in the degeneracy region, most of the electron-electron collisions are forbidden by the Pauli exclusion principle, since one or both of the electrons would land in already occupied final states. Only electrons in the immediate neighborhood of the Fermi surface can collide with each other effectively. All other collisions are "quenched" out. In a first, crude approximation, one may describe this "quenching" by replacing the transition probability $\langle k_1k_2 \mid T \mid k_1'k_2'\rangle$ from initial states k_1', k_2' to final states k_1, k_2, by the "quenched" transition probability

$$\langle k_1k_2 \mid \tilde{T} \mid k_1'k_2'\rangle = (1 - \bar{n}_{k_1})(1 - \bar{n}_{k_2})\langle k_1k_2 \mid T \mid k_1'k_2'\rangle \tag{7.24}$$

The "quenching factors" $1 - \bar{n}_k$ are the probabilities that the final state is not already occupied by another particle of the Fermi distribution; $\bar{n}_k$ is the average occupation number, Eq. (7.23).[10] When these quenching factors are substituted in the calculation of the mean free path of an electron against electron-electron collisions, the result becomes much more reasonable.

In terms of our present discussion of correlation effects, we may rephrase this as follows: the very pronounced statistical correlations in the Fermi gas have the effect of largely "quenching" the dynamical correlations. Even quite strong interparticle *forces* result in only weak dynamical *correlations* between the particles. Thus, while it would be hopeless to expand in powers of the interparticle *potential*, it is by no

[10] Since, in the general case, $\hat{u}_k$ is itself a function of μ, it is no longer true that $N = -(\partial\Omega/\partial\mu)_{V,T}$. However, it can be shown (see Appendix B) that the remainder of Eq. (7.23) is still valid.

means hopeless to expand in the order of the interparticle dynamical correlations. Although the introduction of "effective" particles can account for much of the actual correlation effects, this method cannot encompass all the consequences of correlations. Among the missing phenomena, which cannot be explained on any independent-particle model, no matter how elaborate, is superconductivity.

8. Quasi-Chemical Equilibrium in a Degenerate Fermi Gas

The theory of chemical equilibrium in gases, as given in Section 6, provides an adequate treatment under ordinary conditions. But if the "atomic" gas is a degenerate Fermi gas, such as the electron gas in a metal, the theory needs modification. As we have just seen, by considering the *ideal* Fermi gas, it would be utterly unrealistic to restrict ourselves to just the first two terms of the series (4.21); the high-order terms of that series are of major importance, so much so that the series does not even converge as it stands, but requires re-ordering of terms.

However, the formation of molecules is described by the *dynamical* part U_2' of the pair correlation $U_2 = U_2^0 + U_2'$; the divergence of the series (4.21) for the ideal gas is associated purely with the *statistical* correlation functions U_n^0. Furthermore, we have seen, at the end of Section 7, that the very importance of these statistical correlations serves to "quench out" the dynamical correlations. Thus, it becomes a reasonable approximation to take into account statistical correlation terms U_n^0 of *all* orders n, but break off the series in the dynamical correlations U_n' with some low order $n = m$, in practice usually $m = 2$.[11] If we ignore dynamical correlations altogether ($m = 1$), we have just the independent-particle model. This model can be improved by working with "effective" particles, and we shall assume that this has been done. But even then, true dynamical correlation effects exist, incorporated in correlation matrices U_2', U_3', etc. *The quasi-chemical equilibrium approximation consists in retaining* U_2' *and all powers of* U_2', *but discarding* U_3', U_4', *etc.*

If binary "molecules," i.e., attractive binary correlations between electrons, exist in appreciable numbers, then this approximation includes the effect of the basic Fermi-Dirac statistics of the electrons on the chemical equilibrium between single electrons and electron pairs.

[11] This suggestion is due to S. T. Butler.

Thus, we should obtain the quenching factors in (7.24) automatically, as a result of the theory. Furthermore, this approximation allows us to investigate to what extent these electron pairs are similar to Bose-Einstein particles, and in particular, whether the electron pair gas can undergo a Bose-Einstein-like condensation. All this can be done within the confines of the basic approximation.

Another question, going beyond this approximation, concerns the neglect of triplet, quadruplet, and higher dynamical correlations. We shall discuss this a bit farther on.

For the moment, however, let us illustrate the quasi-chemical equilibrium approximation on the specific example of the term $N = 4$ in (4.5); we shall use the abbreviated notation of eqs. (4.13) and (4.14), and we shall indicate by dots a set of terms obtained from the given term by performing permutations of the particle coordinates. In this notation, the integrand for the evaluation of exp $(-F_4/\mathfrak{k}T)$ is, apart from constant factors,

$$\begin{aligned} W_4(1, 2, 3, 4) = &\ U_1(1)\, U_1(2)\, U_1(3)\, U_1(4) \\ &+ U_2(1, 2)\, U_1(3)\, U_1(4) + \cdots \\ &+ U_2(1, 2)\, U_2(3, 4) + \cdots \\ &+ U_3(1, 2, 3)\, U_1(4) + \cdots \\ &+ U_4(1, 2, 3, 4) \end{aligned} \tag{8.1}$$

In the *complete virial series* (4.21), all terms of (8.1) contribute; the nth term in that series arises from an integral over the function U_n. In the theory of *ordinary chemical equilibrium*, Section 6, we retain U_1 and U_2, but discard all higher U_n. In (8.1), this means the last two lines are discarded. In the *quasi-chemical equilibrium* approximation, the higher U_n are not discarded completely, but rather are replaced by their purely statistical components U_n^0. Thus (8.1) becomes:

$$\begin{aligned} W_4(1, 2, 3, 4) \simeq &\ U_1(1)\, U_1(2)\, U_1(3)\, U_1(4) \\ &+ U_2(1, 2)\, U_1(3)\, U_1(4) + \cdots \\ &+ U_2(1, 2)\, U_2(3, 4) + \cdots \\ &+ U_3^0(1, 2, 3)\, U_1(4) + \cdots \\ &+ U_4^0(1, 2, 3, 4) \end{aligned} \tag{8.2}$$

Furthermore, and this is an essential aspect of the quasi-chemical equilibrium theory, we must recognise that the series, of which (8.2) yields the fourth term, does not converge as it stands, but requires a re-ordering of terms. In this process of re-ordering, each factor $U_2(i, j)$

is split into its statistical and dynamical parts, $U_2 = U_2^0 + U_2'$ and the U_2^0 terms are combined with the other U_n^0 in the selective summation of terms.

The mathematical treatment of this approximation is complicated, and is relegated to Appendix B. Here, we confine ourselves to stating the main results. The grand canonical potential of the system, in the quasi-chemical equilibrium theory, is given by

$$\exp(-\beta\Omega) = \sum_{n=1}^{\infty} \frac{\exp(N\beta\mu)}{N!} \operatorname{Trace}(W_N) \tag{8.3}$$

where W_N is taken as consistent with the approximation, for example, W_4 is the expression (8.2), and the trace is unrestricted.

It is highly convenient to work with the eigenvalues $\hat{\mathfrak{u}}_k$ and eigenfunctions ϕ_k of the U_1 matrix, so that U_1 assumes the form

$$\langle k \mid U_1 \mid k' \rangle = \hat{\mathfrak{u}}_k \delta_{kk'} \tag{8.4}$$

Furthermore, since factors U_1 always occur in conjunction with factors $\exp(\beta\mu)$, we introduce the notation

$$\mathfrak{u}_k = \hat{\mathfrak{u}}_k \exp(\beta\mu) = \exp[\beta(\mu - \hat{\epsilon}_k)] \tag{8.5}$$

where the $\hat{\epsilon}_k$ are "effective" single-particle energies, i.e., $\hat{\epsilon}_k$ may itself be a function of β and μ.

Summation of the series (8.3) leads to a result which can be written in the form

$$\Omega(V, T, \mu) = \Omega^0 + \Omega' \tag{8.6}$$

where Ω^0 is the grand canonical potential of the unpaired electrons, given by (7.22) with $\epsilon = -1$ for Fermi statistics; in our present notation (8.5), we have

$$\Omega^0(V, T, \mu) = \mathfrak{k}T \sum_k \ln(1 + \mathfrak{u}_k) \tag{8.7}$$

To write down the other contribution, Ω', which of course represents the electron pairs, it is desirable to introduce, in place of the dynamical pair correlation matrix U_2', a "quenched" version of this matrix, defined by

$$\langle kl \mid \tilde{U}_2 \mid k'l' \rangle = \frac{\langle kl \mid U_2' \mid k'l' \rangle}{[(1 + \mathfrak{u}_k)(1 + \mathfrak{u}_l)(1 + \mathfrak{u}_{k'})(1 + \mathfrak{u}_{l'})]^{1/2}} \tag{8.8}$$

Since the average occupation number of single-particle state k is given by [see Eq. (7.23)]

$$\bar{n}_k = \frac{\mathrm{u}_k}{1 + \mathrm{u}_k} \tag{8.9}$$

we have the identity

$$\frac{1}{1 + \mathrm{u}_k} = 1 - \bar{n}_k \tag{8.10}$$

The factors on the right-hand side of (8.8) are therefore exactly the "quenching" factors introduced ad hoc into (7.24), symmetrized as between initial and final states. The symmetrization is understandable, since this is an equilibrium calculation, and in thermal equilibrium inverse processes occur equally often. It is not unexpected, but pleasing, to find that these quenching factors enter the final result of the quasi-chemical equilibrium theory in a natural and direct manner, without any further assumptions beyond the ones already made.

With this notation, the "molecular" grand canonical potential Ω' in (8.6) is given by

$$\exp(-\beta\Omega') = \sum_{N=0,2,4,6,\cdots} \frac{\exp(N\beta\mu)}{N!!} \sum_P (-1)^P \sum_{k_1,k_2,\cdots,k_N}$$

$$\langle k_1k_2 \mid \tilde{U}_2 \mid k_{1P}k_{2P}\rangle \langle k_3k_4 \mid \tilde{U}_2 \mid k_{3P}k_{4P}\rangle \cdots \langle k_{N-1}k_N \mid \tilde{U}_2 \mid k_{N-1,P}k_{N}\rangle \tag{8.11}$$

where N represents the number of particles which are paired up, i.e., twice the number of pairs; $N!! = 2 \cdot 4 \cdot 6 \cdot \cdots \cdot N$ for even N; P is a permutation of N things, $(-1)^P$ is -1 for odd permutations, $+1$ for even permutations; and k_{1P} is an index depending upon the permutation P: for example, if P replaces 1 by 5, then k_{1P} means k_5.

We have written down the formula (8.11) in order to indicate the complexity of this type of calculation, not because (8.11) is by itself a useable result. The sum over permutations "links together" all the indices $k_1, \ldots, k_N$, and thus allows for the influence of the Fermi-Dirac statistics of the electrons which form the pairs, on the thermodynamic functions of the electron pair gas.[12] The detailed evaluation of (8.11) for an arbitrary matrix $\tilde{U}_2$ is extremely difficult, and has not been carried out. It is possible, however, to discuss certain useful special cases.

For this purpose, it is desirable to introduce the eigenvalues and

[12] The influence of the unpaired electrons is contained in the quenching factors in (8.8), and appears in (8.11) only implicitly.

eigenfunctions of the quenched pair correlation matrix $\tilde{U}_2$.[13] Let us call the eigenvalues $\hat{\mathfrak{v}}_\alpha$ and the eigenfunctions $\phi_\alpha(k_1, k_2)$:

$$\sum_{k_1'k_2'} \langle k_1k_2 \mid \tilde{U}_2 \mid k_1'k_2' \rangle \, \phi_\alpha(k_1', k_2') = \hat{\mathfrak{v}}_\alpha \phi_\alpha(k_1, k_2) \tag{8.12}$$

so that

$$\langle k_1k_2 \mid \tilde{U}_2 \mid k_1'k_2' \rangle = \sum_\alpha \hat{\mathfrak{v}}_\alpha \phi_\alpha(k_1k_2) \, \phi_\alpha^*(k_1'k_2') \tag{8.13}$$

In a crude way of speaking, $\phi_\alpha(k_1k_2)$ is the wave function of an electron pair in pair state α, and $\hat{\mathfrak{v}}_\alpha$ is related to a pair energy $\hat{\eta}_\alpha$ by the correspondence (6.24).[14] As we have pointed out there, however, this correspondence is far from perfect, and in particular, some of the $\hat{\mathfrak{v}}_\alpha$ may be negative, indicating repulsive pair correlations associated with the pair function ϕ_α. Such pair correlations cannot be considered analogous to binary molecules.

In analogy to (8.5), we define quantities $\mathfrak{v}_\alpha$ by

$$\mathfrak{v}_\alpha = \hat{\mathfrak{v}}_\alpha \exp(2\beta\mu) \tag{8.14}$$

The exponential factor contains the chemical potential of the pairs, $\mu_2 = 2\mu$, as is to be expected.

The special cases which can be treated are best described in terms of these quantities $\mathfrak{v}_\alpha$. Let us, to start with, be quite naive and make the identification (6.24) for $\hat{\mathfrak{v}}_\alpha$ in terms of a pair energy; the $\mathfrak{v}_\alpha$ is then to be identified with[14]

$$\mathfrak{v}_\alpha \sim \exp[\beta(2\mu - \hat{\eta}_\alpha)] = \exp[\beta(\mu_2 - \hat{\eta}_\alpha)] \tag{8.15}$$

where we have avoided writing an equality sign since the correspondence is only a rough, qualitative one, and there exist negative eigenvalues $\mathfrak{v}_\alpha$, for which (8.15) fails completely.

If we think of our pairs as Bose-Einstein particles with an energy spectrum $\hat{\eta}_\alpha$, the condition for the "no overlap" approximation to hold is that μ_2 be large and negative, i.e.,

$$\mathfrak{v}_\alpha \ll 1 \qquad (\text{all } \alpha) \qquad \text{"no overlap" approximation} \tag{8.16}$$

[13] Eigenfunctions and eigenvalues of the original, dynamical pair correlation matrix U_2' play no role whatsoever in the theory. The "quenching" must be done first, before obtaining any physically meaningful results (May 59c).

[14] Note, however, that our index α here, and the energy $\hat{\eta}_\alpha$, *include* the center-of-gravity motion of the pair, whereas this was excluded in the discussion at the end of Section 6. Furthermore, $\mathfrak{v}_\alpha$ of Section 6 corresponds to $\hat{\mathfrak{v}}_\alpha$ here.

The extreme opposite case, of Bose-Einstein condensation, is achieved when μ_2 moves up to equal the lowest pair energy η_0, i.e., when

$$\mathfrak{v}_0 = \text{Max}\,(\mathfrak{v}_\alpha) = 1 \qquad \text{condensation} \tag{8.17}$$

Clearly, condition (8.17) can never be fulfilled if *all* the $\mathfrak{v}_\alpha$ are negative, i.e., if all the pair correlations are purely repulsive in nature. *If we identify the Bose-like condensation of electron pairs with the onset of superconductivity, this shows that purely repulsive correlations cannot lead to superconductivity.*

The simplest case in which (8.11) can be evaluated to a good approximation is the "no overlap" case, Eq. (8.16). The result is (Schafroth 57)

$$\Omega' = -\mathfrak{k}T \sum_\alpha \ln\,(1 - \mathfrak{v}_\alpha) \qquad \text{"no overlap" approximation} \tag{8.18}$$

Apart from notation, this is exactly the ideal Bose gas result (7.22), with $\epsilon = +1$. The number of pairs, N_2, is given by

$$N_2 = \sum_\alpha \frac{\mathfrak{v}_\alpha}{1 - \mathfrak{v}_\alpha} \qquad \text{"no overlap" approximation} \tag{8.19}$$

in agreement with (7.23) for $\epsilon = +1$. Thus, the average number of electron pairs in pair state α is given by

$$\bar{\nu}_\alpha = \frac{\mathfrak{v}_\alpha}{1 - \mathfrak{v}_\alpha} \qquad \text{"no overlap" approximation} \tag{8.20}$$

If we ignore the limitations of this approximation for a moment, and substitute $\mathfrak{v}_\alpha = 1$ into (8.20), we get an infinite number of this type of pair, i.e., we have reached the condensation region [see Eq. (8.17)]. Actually, however, (8.18)—(8.20) are completely invalid in that region, and a much more detailed investigation is required. We shall discuss this problem in Section 9. Suffice it to say, now, that overlap integrals can certainly *not* be ignored, or even assumed to be small, once there is condensation. For the condensed pairs have a center-of-gravity motion with a de Broglie wavelength comparable to the size of the specimen, and thus the condensed pairs most certainly have overlapping wave functions. Since (8.18)—(8.20) are obtained from the general (8.11) by neglecting overlap integrals, these formulas are not applicable in the condensation region.

We close this section with some remarks concerning the neglect of triplet, quadruplet, and higher dynamical correlations. Such higher correlations can obviously be neglected if the pair correlations have short range, i.e., if the "molecules" are small compared to the distance between molecules, and also small compared to the distance between "atoms" (unpaired electrons).

This condition, though certainly *sufficient*, fortunately is *not necessary*. Triplet and higher dynamical conditions can be neglected under much less stringent conditions, *as a result of the quenching of dynamical correlations by the Pauli exclusion principle*. We have discussed this effect at the end of Section 7, in relation to pair correlations. Let us now extend the discussion to correlations of higher order.

In order to get an effective triplet correlation, for example, all *three* final states must lie close to the Fermi surface. The fraction of states within a distance $\mathfrak{k}T$ of the Fermi surface is of order $\mathfrak{k}T/\epsilon_F$, where ϵ_F is the Fermi energy. To get an effective n-particle correlation, the fraction of n-particle states which are "available" is only

$$f_n = (\mathfrak{k}T/\epsilon_F)^n \tag{8.21}$$

and this gives an order-of-magnitude estimate for the "quenching" of n-particle dynamical correlations by the Pauli exclusion principle.

Since $\mathfrak{k}T/\epsilon_F$ is of order 10^{-3} to 10^{-4} at temperatures of interest in superconductors, we see that successive orders n of correlations are ever more strongly quenched out, by very appreciable factors. Since we know, from the success of the band theory of normal metals, that even the pair correlations ($n = 2$) are almost completely quenched, we can conclude that triplet and higher correlations are likely to be utterly negligible.[15]

Furthermore, even if triplet correlation did make some small contribution, this contribution would be of no interest for a theory of superconductivity; for triplets made up of Fermions are themselves Fermions [see Blatt (58), Section 5, for a precise statement of this], and thus triplets cannot Bose-condense.

The first possible competition for pair correlations is provided by quadruplet correlations. These are down, compared to pair correlations, by a factor $(\mathfrak{k}T/\epsilon_F)^2$, of order 10^{-6} to 10^{-8}. Thus, if there is a Bose-like condensation of electron complexes at all, it is very much more likely to occur with pairs of electrons rather than with quadruplets, even if the pairs are "large." The extremely strong statistical correlations associated with the Pauli exclusion principle mean that the straightforward classical estimates of regions of validity in terms of the "size" of the system are not valid.

[15] We recall that we are working with "effective" single particles here, i.e., the vast majority of the true correlations between electrons is already contained in the effective single-particle Hamiltonian. The correlations under consideration here are the residual correlations, which cannot be included in any single-particle Hamiltonian.

At the time the quasi-chemical equilibrium theory was being developed, this was all that could be said. By now, however, there is the strongest possible experimental evidence that the "superelectrons" are electron pairs, rather than more complex structures. The experimental evidence arises from the value of the flux quantum (I,7.1), which corresponds to a charge $e^* = 2e$ for the "superelectrons". Since the only other constants entering the flux quantum are the universal constants h and c, there can be little doubt indeed that the "superelectrons" are electron pairs (Onsager 61).

9. Schafroth Condensation of Electron Pairs

The "no overlap" approximation (8.16) and (8.18)–(8.20) fails in the condensation region, defined by (8.17). In this latter region, overlap integrals between pair wave functions cannot be neglected. The electron pairs are by no means the same as elementary Bose-Einstein particles, and the pair partition function (8.11) is not even approximately similar to the Bose gas result (8.18).[16] We shall call the condensation phenomenon for such pairs a "Schafroth condensation," to distinguish it from ordinary Bose-Einstein condensation.

Fortunately, we do not *need* a full correspondence to a simple Bose-Einstein gas. The only thing we really need, for a theory of superconductivity, is the existence of a *condensation* similar to the Bose-Einstein condensation. In this section, we shall discuss this Schafroth condensation phenomenon qualitatively, showing in particular that the Pauli exclusion principle is not a hindrance to condensation of electron pairs, but is actually a positive, helping influence. The detailed mathematical discussion is given in the original papers (Schafroth 57, Blatt 58, Matsubara 60) and is summarized in Appendix B.

The first remark is that the Fermi statistics of electrons does *not* prevent us from putting many electron pairs into one pair state $\phi(k_1, k_2)$. For example, consider two electron pairs, i.e., four electrons altogether. The product wave function $\phi(k_1, k_2)\ \phi(k_3, k_4)$ is not antisymmetric, and is therefore forbidden by the Pauli principle. Consider, however, the wave function:

$$\psi(1, 2, 3, 4) = \phi(1, 2)\,\phi(3, 4) - \phi(1, 3)\,\phi(2, 4) + \phi(1, 4)\,\phi(2, 3) \tag{9.1}$$

[16] A more mathematical treatment, contained in Appendix B, shows the same thing; the creation and destruction operators for these pairs do *not* obey Bose-Einstein commutation rules, not even approximately.

Direct calculation shows that this function is antisymmetric with respect to any exchange of particles, and is therefore a permissible wave function, describing two electron pairs, both in the quantum state ϕ.[17]

The function (9.1) is built up from the one-pair function $\phi(k_1, k_2)$, and in that sense we may say that both electron pairs are "in state ϕ." However, it is *not* true that expectation values of operators, over the wave function ψ, are directly related to expectation values of the same operators over the one-pair function ϕ. For example, let v_{12} be the potential energy of interaction between particles 1 and 2. There is then no *simple* relation between the integrals

$$\int \phi^*(1, 2)\, v_{12}\, \phi(1, 2)\, d^3r_1\, d^3r_2 = (\phi, v\phi) \tag{9.2}$$

and

$$\int \psi^*(1, 2, 3, 4)\, v_{12}\, \psi(1, 2, 3, 4)\, d^3r_1\, d^3r_2\, d^3r_3\, d^3r_4 = (\psi, v\psi) \tag{9.3}$$

A relationship does exist, of course, but it is complicated due to the overlap integrals which occur when (9.1) is substituted into (9.3).

It is instructive to consider the special case in which the operator v_{12} is given by

$$v_{12} = \delta(\mathbf{r}_1 - \boldsymbol{\rho}_1)\, \delta(\mathbf{r}_2 - \boldsymbol{\rho}_2) \tag{9.4}$$

Then (9.3) is the probability, over the wave function ψ, of finding particle 1 at position $\boldsymbol{\rho}_1$, particle 2 at $\boldsymbol{\rho}_2$, and the remaining two particles anywhere at all. (Since ψ is antisymmetric, this probability is symmetric under the interchange of 1 and 2). This probability is *not* equal to the absolute square of the basic "pair" wave function $\phi(\boldsymbol{\rho}_1, \boldsymbol{\rho}_2)$; that is, ϕ is *not* the "two-particle projection" of the four-particle wave function ψ, even though ψ is "built up from" factors ϕ.

We emphasize that by the term "pair state" we mean a full wave function $\phi(k_1, k_2)$ for a pair of particles. We do *not* mean a "pair of" single-particle states, say k_1' and k_2'. Putting one particle into single-particle state k_1 and the other particle into single-particle state k_2 corresponds to the unphysical pair wave function

$$\phi(k_1, k_2) = 2^{-1/2}(\delta_{k_1k_1'}\, \delta_{k_2k_2'} - \delta_{k_1k_2'}\, \delta_{k_2k_1'})$$

[17] This expression differs from the function (1.1) for a pair of excitons, because here all four particles are identical, whereas in the exciton, the electron and hole are inequivalent particles. In showing the antisymmetry of (9.1), it is of course necessary to remember that $\phi(1, 2) = -\phi(2, 1)$, for example.

It is easily verified that this function, when substituted into (9.1), causes ψ to vanish identically. This is of course just a complicated way of saying a simple thing: once the single-particle states k_1' and k_2' are occupied, with occupation probability equal to 1, then no further particles can be accommodated in these states.

But if we put a pair of particles into a physically reasonable pair wave function $\phi(k_1, k_2)$, then the probability of finding one member of the pair in single-particle state k_1', the other member of the pair in single-particle state k_2', is given by $|\phi(k_1', k_2')|^2$, and this is infinitesimally small for any reasonable pair wave function ϕ.[18] Thus, we exclude only a very tiny fraction of the available phase space by putting one pair of particles into state ϕ, and there is ample phase space left to put more pairs into that same pair wave function.

Having shown that we can put many pairs of particles into any one pair state ϕ, it becomes interesting to investigate the limiting case where *all pairs* are in one and the same pair state ϕ, i.e., the case of complete Schafroth condensation of pairs. Since the pair wave functions ϕ_α are the eigenfunctions of the quenched pair correlation matrix $\tilde{U}_2$ [see Eqs. (8.12) and (8.13)] this means that the quenched pair correlation matrix assumes the factored (separable) form:

$$\langle k_1 k_2 | \tilde{U}_2 | k_1' k_2' \rangle = \hat{\mathfrak{v}}_0 \, \phi(k_1, k_2) \, \phi^*(k_1', k_2') \tag{9.5}$$

A look at (8.8) shows that the "unquenched" dynamical pair correlation matric U_2' is then also separable

$$\langle k_1 k_2 | U_2' | k_1' k_2' \rangle = \hat{\mathfrak{v}}_0 \, \chi(k_1, k_2) \, \chi^*(k_1', k_2') \tag{9.6}$$

where

$$\chi(k_1, k_2) = (1 + \mathfrak{u}_{k_1})^{1/2} (1 + \mathfrak{u}_{k_2})^{1/2} \, \phi(k_1, k_2) \tag{9.7}$$

That is, *a complete Schafroth condensation of pairs corresponds, in this formalism, to a "separable" dynamical pair correlation matrix.*

Following (8.14), we introduce the quantity $\mathfrak{v}_0$ by

$$\mathfrak{v}_0 = \hat{\mathfrak{v}}_0 \exp(2\beta\mu) \tag{9.8}$$

and we wish to evaluate the grand canonical potential for the pairs, $\Omega'(V, T, \mu)$, from the general formula (8.11) for the special assumption (9.5) and for various values of the parameter $\mathfrak{v}_0$. We only quote the result here.

---

[18] This probability is of order $(1/V)^2$ in the general case, and of order $1/V$ if there are selection rules, in particular conservation of center-of-gravity momentum. We are not concerned with the precise order of magnitude here, but require merely that it be infinitesimal in the limit of large volume V.

We recall [see Eqs. (8.16) and (8.17)] that the "no overlap" region corresponds to $\mathfrak{v}_0 < 1$, and that we expect a phenomenon similar to Bose condensation when $\mathfrak{v}_0 = 1$. It is therefore not surprising that the detailed calculation yields values of Ω' which are completely different for $\mathfrak{v}_0 < 1$ and for $\mathfrak{v}_0 > 1$.

For $\mathfrak{v}_0 < 1$, we get substantially the ideal Bose gas result

$$\Omega' = +\mathfrak{k}T \ln (1 - \mathfrak{v}_0) \qquad \text{for} \quad \mathfrak{v}_0 < 1 \tag{9.9}$$

The average number of pairs in this pair state is equal to

$$\bar{\nu}_0 = \frac{\mathfrak{v}_0}{1 - \mathfrak{v}_0} \qquad \text{for} \quad \mathfrak{v}_0 < 1 \tag{9.10}$$

These formulas are correct except for terms of relative order $1/V$; they are also just what we would obtain from the "no overlap" approximations (8.18) and (8.19), by omitting all pair states $\alpha \neq 0$ from the sums over α.

Since (9.10) is of order unity, not of order volume, this one pair state makes no significant contribution to the thermodynamic and other properties of the substance for $\mathfrak{v}_0 < 1$. In this region, one would have to allow for all pair states α, i.e., perform the sums in (8.18) and (8.19), in order to obtain significant (of order V) contributions.

However, the situation changes completely when $\mathfrak{v}_0$ exceeds 1. Formulas (9.9) and (9.10) are then invalid, and the values of Ω' and $\bar{\nu}_0$ depend on the details of the pair wave function ϕ. The formulas get very complicated if ϕ is a completely general function. Here, we shall quote the results if there is conservation of center-of-gravity momentum, i.e.,

$$\phi(k_1, k_2) = \phi(\mathbf{k}_1 s_1, \mathbf{k}_2 s_2) = 0 \qquad \text{unless} \qquad \mathbf{k}_1 + \mathbf{k}_2 = \mathbf{K} \tag{9.11}$$

The ground state of the pairs corresponds to $\mathbf{K} = 0$, i.e., $\mathbf{k}_1 = -\mathbf{k}_2$. As a further simplification, we may ignore spin-orbit coupling, so that the pair wave function is a product of a function of the vectors $\mathbf{k}_1$, $\mathbf{k}_2$ and a function of the spin variables s_1, s_2. For two electrons, the spin function may be either a singlet function, (I,9.4) or a triplet function, (I,9.3). The singlet spin function is antisymmetric, and must be combined with a symmetric space function; if the pair correlation is basically attractive (as we are assuming here), then a symmetric space function is more favourable than an antisymmetric space function.[19] In the preferred,

[19] This is opposite to the situation in atoms, where the correlations between electrons are basically repulsive, and the antisymmetric space function gives rise to a lower energy. This is the origin of Hund's rule for atoms: "High multiplicity, low term." For attractive correlations, we have an inverted Hund's rule, the lowest term having also the lowest multiplicity ($S = 0$). The inverted form of Hund's rule is also applicable to light nuclei.

singlet, spin state, $s_1 = -s_2$, and we therefore have the selection rule

$$\phi(\mathbf{k}_1, s_1; \mathbf{k}_2, s_2) = 0 \qquad \text{unless} \qquad \mathbf{k}_1 = -\mathbf{k}_2 \quad \text{and} \quad s_1 = -s_2 \tag{9.12}$$

Furthermore, $\phi(\mathbf{k}, +1; -\mathbf{k}, -1) = -\phi(\mathbf{k}, -1; -\mathbf{k}, +1)$ for a singlet spin state. To save writing, we shall use the notation

$$\begin{aligned}\phi(\mathbf{k}, +1; -\mathbf{k}, -1) &= -\phi(-\mathbf{k}, -1; +\mathbf{k}, +1) = \phi(-\mathbf{k}, +1; +\mathbf{k}, -1)\\ &= -\phi(\mathbf{k}, -1; -\mathbf{k}, +1) = \phi_{\mathbf{k}}\end{aligned} \tag{9.13}$$

By rotation symmetry, the ground state of the pairs is an S-state (orbital angular momentum zero), i.e., $\phi_{\mathbf{k}}$ is independent of the direction of the vector $\mathbf{k}$. Finally, by time reversal invariance (Wigner 32), we may take $\phi_{\mathbf{k}}$ to be real, without loss of generality. We also note that the sum over the formal index $\mathbf{k}$ includes a sum over the two values of the spin index, as well as a sum over vectors $\mathbf{k}$. In all practical cases, we have the equivalence

$$\sum_k \to 2 \sum_{\mathbf{k}} \tag{9.14}$$

We now state results: let x be a c-number parameter, and define $g_{\mathbf{k}}(x)$ by

$$g_{\mathbf{k}}(x) = \frac{2x \mid \phi_{\mathbf{k}} \mid^2}{1 + 2x \mid \phi_{\mathbf{k}} \mid^2} \tag{9.15}$$

Let t be the solution of the equation

$$t = \sum_{\mathbf{k}} g_{\mathbf{k}}(\mathfrak{v}_0 t) \tag{9.16}$$

As we shall see on hand of a specific example, later, t is proportional to the volume V if $\mathfrak{v}_0$ exceeds 1. We then have

$$\Omega' = \mathfrak{k}T \left[t - \sum_{\mathbf{k}} \ln\,(1 + 2\mathfrak{v}_0 t \mid \phi_{\mathbf{k}} \mid^2)\right] \tag{9.17}$$

and

$$\bar{\nu}_0 = t = \sum_{\mathbf{k}} g_{\mathbf{k}} \tag{9.18}$$

These formulas still require evaluation of the sums over $\mathbf{k}$, and solution of the implicit Eq. (9.16) for $\bar{\nu}_0 = t$. To illustrate these expressions, and to establish orders of magnitude, we now make a very simple assumption concerning the pair wave function, which allows completely explicit solutions, and which is not unreasonable physically.[20] The function chosen here is the ground state wave function for a truly bound pair (i.e., a real molecule, not a quasi-molecule), held together by a force of vanishing range:

$$\phi(\mathbf{r}_1 s_1, \mathbf{r}_2 s_2) = C \frac{\exp(-\gamma r_{12})}{r_{12}} \frac{\alpha(s_1)\,\beta(s_2) - \beta(s_1)\,\alpha(s_2)}{2^{1/2}} \tag{9.19}$$

where C is a normalization constant, $r_{12} = |\, \mathbf{r}_1 - \mathbf{r}_2 \,|$, the constant γ is related to the binding energy of the pair in the usual way, and $\alpha(s)$ is the spin-function for an up-spin, $\beta(s)$ the spin-function for a down-spin. The normalization constant is

$$C = \left(\frac{\gamma}{2\pi V}\right)^{1/2} \tag{9.20}$$

and the k-space transform of (9.19) yields, in the notation (9.13),

$$\phi_{\mathbf{k}} = \frac{4\pi C}{2^{1/2}(k^2 + \gamma^2)} \tag{9.21}$$

The sum on the right-hand side of (9.16) becomes an elementary integral over k-space, upon using (II, 1.15); not only can the integral be done, but the resulting equation (9.16) allows an explicit solution for t, namely,

$$t = \bar{\nu}_0 = \frac{V\gamma^3 \mathfrak{v}_0(\mathfrak{v}_0^2 - 1)}{2\pi} \qquad \text{for} \quad \mathfrak{v}_0 > 1 \tag{9.22}$$

Substitution of this value of t into (9.17), and evaluation of the sum over $\mathbf{k}$, then gives

$$\Omega' = -\mathfrak{k}T \frac{V\gamma^3}{6\pi} (\mathfrak{v}_0 + 2)(\mathfrak{v}_0 - 1)^2 \qquad \text{for} \qquad \mathfrak{v}_0 > 1 \tag{9.23}$$

We now discuss these results. The first, crucial remark is that both (9.22) and (9.23) are extensive quantities, proportional to the volume V of the system. Thus, for $\mathfrak{v}_0 > 1$, this one pair state ϕ accommodates a number of pairs, $\bar{\nu}_0$, proportional to the size of the system, and makes an extensive, significant contribution to the free energy. It is furthermore obvious from (9.22) that we can make the number density of pairs, $\bar{\nu}_0/V$, arbitrarily large by letting $\mathfrak{v}_0$ become arbitrarily large. Thus any

[20] The correct pair wave function ϕ has to be determined self-consistently, of course, and does not turn out to be of the form (9.19); this, however, does not affect the present discussion.

number of electrons can be crowded into this one pair state. *The Fermi-Dirac statistics of the single electrons does not prevent infinite occupation of an electron pair state.*

The statistics does have an effect, however. In the ideal gas of elementary bosons (see Chapter II), the condensed particles make no contribution at all to the pressure of the system. We recall the thermodynamic identity $\Omega = -pV$, and the fact that Ω' is the contribution of the condensed pairs to Ω; we then see from (9.23) that it does take additional pressure to force more and more particles into this condensed pair state; $\mathfrak{v}_0$ cannot stay equal to 1, but must increase beyond 1, in order to get a finite number density of these pairs. To this extent, then, the Pauli principle does affect the condensation: it requires actual pressure to make the pairs condense, whereas it took no pressure at all to allow any number of elementary bosons to occupy the condensed state.

The value of the pressure required depends on the internal wave function ϕ of the pairs. In effect, we are using the "tail" (in k-space) of the pair wave function ϕ in order to accommodate more and more particles without violating the exclusion principle. The wave function ϕ_k, Eq. (9.21), drops off for large k. By putting in more and more of these pairs, we "fill up" the low values of k; but there always remains room for more particles in the region of very high k. We pay for exploiting this high-k region by getting a higher average energy—and this shows itself thermodynamically as a pressure, which increases with the number density of the pairs.

So far we have shown that the Fermi-Dirac statistics does not prevent infinite occupation of any one pair state. It remains to show, however, that there is actual condensation into the lowest pair state (lowest in energy, highest in $\mathfrak{v}_\alpha$). We recall that, in the ideal Bose gas of elementary bosons, we were able to obtain the essential results (condensation or no condensation) by considering only the lowest *two* states of one boson in the box [see Eqs. (II,2.11) and (II,2.12) for the three-dimensional Bose gas, and the later argument for the two-dimensional Bose gas (which does not condense)].

In the pair gas, the calculation of Ω' from (8.11) is so complicated that it has been carried through, completely, only for this case of just two quantum states of the pairs, a lowest state ϕ_0 and a translational state ϕ_1. That is, we assume, instead of (9.5), the form

$$\langle k_1 k_2 \mid \tilde{U}_2 \mid k_1' k_2' \rangle = \hat{\mathfrak{v}}_0 \, \phi_0(k_1, k_2) \, \phi_0^*(k_1', k_2') + \hat{\mathfrak{v}}_1 \, \phi_1(k_1, k_2) \, \phi_1^*(k_1', k_2') \tag{9.24}$$

and we work with the quantities

$$\mathfrak{v}_0 = \hat{\mathfrak{v}}_0 \exp(2\beta\mu), \qquad \mathfrak{v}_1 = \hat{\mathfrak{v}}_1 \exp(2\beta\mu) \tag{9.25}$$

ϕ_0 is taken to be the function ϕ introduced previously, i.e., given by (9.12) and (9.13). ϕ_1 is taken to correspond to a given, non-zero center-of-gravity momentum $\mathbf{K}$, i.e.,

$$\phi_1(\mathbf{k}_1 , s_1 , \mathbf{k}_2 , s_2) = 0 \quad \text{unless} \quad \mathbf{k}_1 + \mathbf{k}_2 = \mathbf{K} \quad \text{and} \quad s_1 = - s_2 \tag{9.26}$$

With these assumptions, Ω' can be evaluated completely for infinitesimal (of order $1/L$) values of K; these are the interesting values, since they are most likely to spoil the condensation. The calculation (Matsubara 60) is exceedingly complex, and cannot be reproduced here; we shall quote the essential result, and we shall then do our best to make this result plausible.

The occupation numbers $\bar{\nu}_0$ and $\bar{\nu}_1$ of these two pair states depend upon $\mathfrak{v}_0$ and $\mathfrak{v}_1$, and it is necessary to have some information about these two quantities, to obtain meaningful results. In principle, one should carry out a self-consistent calculation, which would then yield self-consistent pair wave functions ϕ_0 and ϕ_1, as well as self-consistent values of $\mathfrak{v}_0$ and $\mathfrak{v}_1$. Fortunately, the main result can be obtained more simply, by making only one, very reasonable, assumption about the ratio

$$q \equiv \frac{\mathfrak{v}_0 - \mathfrak{v}_1}{\mathfrak{v}_0} = \frac{\hat{\mathfrak{v}}_0 - \hat{\mathfrak{v}}_1}{\hat{\mathfrak{v}}_0} \tag{9.27}$$

Let us be naive and relate the $\mathfrak{v}_\alpha$ to pair energies η_α by (8.15). The ratio q is then related to the pair energies by

$$q \sim 1 - \exp\,[\beta(\eta_0 - \eta_1)] \cong \frac{\eta_1 - \eta_0}{\mathfrak{k}T} \tag{9.28}$$

Let us make the most unfavorable assumption for condensation, i.e., that the energy difference $\eta_1 - \eta_0$ is as small as possible; even then, one concludes from the uncertainty principle that this energy difference is at least of order $\hbar^2/ML^2$, where L is the side length of the box. We therefore *assume*

$$q \gtrsim O(1/L^2) \tag{9.29}$$

and we evaluate the occupation numbers $\bar{\nu}_0$ and $\bar{\nu}_1$ under this assumption.

The assumption (9.29) is very conservative; the self-consistent calculations give an energy gap Δ for excitation of a pair, with Δ being independent of the size of the box, and thus enormously much larger than the estimate (9.29). However, these calculations (see Chapter IV) start out from the assumption of a Schafroth condensation, and the resulting argument is not conclusive. By contrast, if we get a Schafroth condensation of the pairs even with the very conservative assumption

(9.29), then the possibility of condensation of electron pairs may be considered established.

This is precisely what happens. There are three regions of interest:

Region 1: $\mathfrak{v}_1 < \mathfrak{v}_0 < 1$ "No overlap" region, both $\bar{\nu}_0$ and $\bar{\nu}_1$ of order 1.
Region 2: $\mathfrak{v}_1 < 1 < \mathfrak{v}_0$ $\bar{\nu}_0$ of order V, $\bar{\nu}_1$ of order 1, as expected.
Region 3: $1 < \mathfrak{v}_1 < \mathfrak{v}_0$ $\bar{\nu}_0$ of order V, $\bar{\nu}_1$ of order 1!

This last result establishes the Schafroth condensation of the pairs. If pair state ϕ_1 were the *only* pair state available, then the preceding calculation, with $\phi = \phi_1$, would apply, and would give $\bar{\nu}_1$ of order V. But the fact that a lower (in energy; higher in $\mathfrak{v}_\alpha$) pair state ϕ_0 exists, leads to preferential occupation of this lower pair state, at the expense of the occupation number $\bar{\nu}_1$.

Furthermore, the detailed calculation shows that the Pauli principle not only does not interfere with Schafroth condensation of electron pairs, but actually aids this condensation.

Let us now try to understand this state of affairs qualitatively (Blatt 62). Suppose we have $2N$ electrons to distribute over the possible states k, and suppose we distribute them according to one or both of the pair wave functions $\phi_0(k_1, k_2)$ and $\phi_1(k_1, k_2)$. Let us start by putting the first pair of electrons into states $k_0 = (\mathbf{k}_0, +1)$ and $-k_0 = (-\mathbf{k}_0, -1)$, i.e., we have occupied one of the combinations of single-particle states permitted by the ground state pair wave function ϕ_0, Eq. (9.13).

Having done this, a certain amount of phase space is excluded for the next electron pair. We now point out that this exclusion is *worse* for a ϕ_1 type pair than for another ϕ_0 type pair! For another ϕ_0 type pair, the only set of indices excluded is the set $(k_0; -k_0) = (\mathbf{k}_0, +1; -\mathbf{k}_0, -1)$. For a type ϕ_1 pair, however, it follows from (9.26) that *two* sets of indices are excluded, namely,

$$(\mathbf{k}_0, +1; -\mathbf{k}_0 + \mathbf{K}, -1) \quad \text{and} \quad (\mathbf{k}_0 + \mathbf{K}, +1; -\mathbf{k}_0, -1)$$

The first is excluded because the state $(\mathbf{k}_0, +1)$ is already occupied; the second, because the state $(-\mathbf{k}_0, -1)$ is already occupied.

Thus, we minimize the effects of the exclusion principle by putting the second pair into equal-and-opposite momentum states, say $(k_1; -k_1) = (\mathbf{k}_1, +1; -\mathbf{k}_1, -1)$; i.e., by making the second pair a ϕ_0 type pair, rather than a ϕ_1 type pair.

Having two pairs, we now "fill in" the third pair. The same argument clearly goes through, and we are on our way to a situation in which *all* N electron pairs are ϕ_0 type pairs, i.e., the Schafroth condensed state of the pairs. *Exclusion principle effects occur in any case, whether the particles are all put into one pair state, or distributed more evenly over the available pair states. But the exclusion effects are minimized, and energy is thus saved, by putting all the pairs into the same pair state. The exclusion principle not only does not forbid, but positively favors, Schafroth condensation of Fermion pairs.*

This explains why we are able to obtain a condensation of Bose-Einstein type even for highly "overlapping" pairs. The deviations from strict Bose-Einstein behavior occasioned by the non-zero overlap integrals actually help the condensation. The bigger the pairs, the more easily do they condense! The competition is not between condensed and non-condensed pairs, but between a condensed pair state of the system and the ordinary, single-particle Fermi sea. If the condensed pair state has lower energy than the Fermi sea state of singles, then nothing is gained (at zero temperature) by introducing non-condensed pairs.

In passing, we might mention that the qualitative argument just given, unlike the detailed calculation, is not restricted to very large numbers of particles. Thus, from the qualitative argument one would conclude that a Schafroth-condensed pair state may be a useful approximation even for small numbers of particles. This appears to be true in fact, as is shown by the success of such pairing wave functions in nuclear physics, where the number of nucleons outside closed shells is by no means very large (Bohr 58; Bogoliubov 58c; Soloviev 58, 60, 61, 61a, 61b, 62, 62a, 62b; Belyaev 59; Migdal 59; Arvieu 60; Baranger 60, 61; Bayman 60; Griffin 60; Kisslinger 60, 62; Marumori 60; Mottelson 60; Thouless 60b; Kerman 61; Nilsson 61; Tamura 61; Wada 61; Yoshida 61; Gallagher 62; Katz 59; Khanna 62; Nagasaki 62; Pavlikovski 62; Terasawa 62; Volkov 62; Voros 62; Mikhailov 63; Sano 63; Bremond 63; Lang 63; Pal 63).

IV

Self-consistent Treatment of the Ground State

Although the quasi-chemical equilibrium theory is simple conceptually, it is quite complicated practically. The need for self-consistent calculations was recognized quite early, but considerable time elapsed before the quasi-chemical equilibrium theory could be brought to the necessary state of technical development.

Thus the first self-consistent calculations in the theory of superconductivity were done by other methods; Bardeen, Cooper, and Schrieffer (Bardeen 57) used a highly intuitive approach which yielded excellent agreement with experiment in a number of ways; soon thereafter, basing themselves on a method developed by Bogoliubov for systems of interacting bosons (Bogoliubov 47), Bogoliubov and his collaborators developed a beautiful and ingenious technique for Fermion systems (Bogoliubov 58, 58a), which allows explicit inclusion of the electron-phonon coupling; simultaneously Valatin (58) arrived at the same transformation.

The BCS and Bogoliubov approaches are complementary to the quasi-chemical equilibrium approach, in the sense that they allow easy calculation but are much harder to motivate. Furthermore, in both the BCS and Bogoliubov techniques, the use of creation and destruction operators is absolutely essential, since in their wave function the number of particles is not a constant of motion.

We therefore deviate from the historical order at this stage, and devote the present chapter to a discussion of the self-consistent ground state wave function, starting from the physical concept of Schafroth condensation of electron pairs, without explicit use of advanced methods.[1] In

[1] Since the number of particles is conserved at all stages, the use of creation and

Chapter V, we then state and discuss the BCS and Bogoliubov theories, still at zero temperature, and establish their relation to each other and to the quasi-chemical equilibrium theory. Non-zero temperatures require extension of the formalism, and are treated in Chapters VI and VII. For non-zero temperatures, the historical order is also preferable didactically, and thus Chapter VI contains the BCS and Bogoliubov treatment of the thermodynamics, Chapter VII the quasi-chemical equilibrium treatment.

1. Variational Derivation of the BCS-Bogoliubov Integral Equation

The simplest case, which nevertheless contains most of the essential elements of the full theory, is obtained by assuming a complete Schafroth condensation of electron pairs, that is, no pairs in excited pair states, and no unpaired electrons at all. Furthermore, we shall assume that the interaction between electrons can be represented, in some approximation, by an operator v_{ij} acting directly on the electron wave function; that is, we assume that the phonons which transmit the Fröhlich interaction can be eliminated. We shall discuss these assumptions later. For the moment, let us see where they lead us.

We have shown in Chapter II that properties of the condensed system at small but finite temperatures can be deduced validly by calculating with the quantum mechanical ground state, rather than with a statistical ensemble. If all N electrons (N even) are paired up, and all $N/2$ pairs occupy the same pair state φ, then the N-particle ground state wave function is an antisymmetrized product of pair wave functions $\varphi(i, j) = -\varphi(j, i)$:

$$\Psi_N = C_N \sum_P (-1)^P P[\varphi(1, 2) \varphi(3, 4) \varphi(5, 6) \cdots \varphi(N-1, N)] \tag{1.1}$$

where C_N is a normalization constant, and the sum extends over all $N!$ permutations of the N particles.

---

destruction operators is not *necessary*, although it is a great convenience. The technical difficulties of the theory are concerned with the explicit evaluation of expectation values of the relevant operators over the condensed-pair wave function (1.1). This technical work is done in Appendix B.

We now consider (1.1) as a *trial function* in the Rayleigh-Ritz variation principle

$$\frac{\int \Psi_N^* H_N \Psi_N \, d\tau}{\int \Psi_N^* \Psi_N \, d\tau} = \text{Minimum} \tag{1.2}$$

where the N-particle Hamiltonian is given by

$$H_N = \sum_{i=1}^{N} \frac{\mathbf{p}_i^2}{2M} + \sum_{i<j=1}^{N} v_{ij} \tag{1.3}$$

The variational parameters at our disposal are the complex numbers $\varphi(k_1, k_2)$ which define the basic pair wave function. The normalization constant C_N drops out of (1.2), of course.

The discussion is simplified enormously by making use of invariance properties, since thereby we restrict the Ansatz for $\varphi(k_1, k_2)$. We have gone through this discussion already, in Chapter III [see (III, 9.11) to (III, 9.13)]; and we merely repeat the final result: the independent components of the pair wave function are real numbers φ_k equal to

$$\varphi_k = \varphi(\mathbf{k}, +1; -\mathbf{k}, -1) = -\varphi(-\mathbf{k}, -1; +\mathbf{k}, +1) \tag{1.4}$$

and φ_k is independent of the direction of the vector $\mathbf{k}$. All other components $\varphi(\mathbf{k}_1, s_1; \mathbf{k}_2, s_2)$ vanish identically, as a result of the various symmetries.

In (III, 9.15), the wave function components φ_k enter not by themselves, but multiplied by an adjustable constant $(2x)^{1/2}$. It turns out that this is also true here, and we therefore define an un-normalized wave function χ_k by

$$\chi_k = (2x)^{1/2} \varphi_k \tag{1.5}$$

φ_k (but not the χ_k) are subject to the normalization condition

$$\sum_k | \varphi_k |^2 = 1 \tag{1.6}$$

The expectation values of one-particle operators (such as the kinetic energy) and of two-particle operators [such as the potential energy in (1.3)] over the wave function (1.1) can be worked out completely generally (Blatt 60a).

The general results simplify considerably for pair wave functions $\varphi(k_1, k_2)$ with the "*simple pairing*" *property*: there exists a one-to-one correspondence $k \leftrightarrow \hat{k}$ between pairs of indices k, such that $\varphi(k_1, k_2)$ vanishes unless $k_2 = \hat{k}_1$. The wave function (1.4) has this property, the one-to-one correspondence being $(\mathbf{k}, s) \leftrightarrow (-\mathbf{k}, -s)$. We quote the simple results here (see Appendix B for the derivation).

Expectation values of one-particle operators involve the quantity $g_k(x)$, Eq. (9.15); in our present notation this is given by (we use the fact that χ_k is real)

$$g_k(x) = \frac{2x \mid \varphi_k \mid^2}{1 + 2x \mid \varphi_k \mid^2} = \frac{\chi_k^2}{1 + \chi_k^2} \tag{1.7}$$

Let J be a sum of single-particle operators, see Chapter II, Section 3, and let $j_{kk'}$ be the matrix element of j between single-particle states k and k'. Then the expectation value of J over the condensed pair wave function (1.1) is given by[2]

$$J = \frac{\int \Psi_N^* \, J \Psi_N \, d\tau}{\int \Psi_N^* \Psi_N \, d\tau} = \sum_k j_{kk} g_k(x) + \text{terms of relative order } 1/N \tag{1.8}$$

The parameter x is determined by the condition that the number N of particles must be the value of the number operator $J = N = \Sigma_{i=1}^N 1$, i.e., $j_{kk'} = \delta_{kk'}$. Thus we must have

$$N = \sum_k g_k(x) = \sum_k \frac{\chi_k^2}{1 + \chi_k^2} \tag{1.9}$$

which is an implicit equation for x.[2] The expectation value of the kinetic energy operator is obtained from (1.8) by setting $j_{kk'} = \epsilon_k \, \delta_{kk'} = (\hbar^2 k^2/2M) \, \delta_{kk'}$.

We also need the expectation value of two particle operators such as v_{ij}. Let $\langle k, l \mid v \mid k', l' \rangle$ be the matrix element of v_{12} between the states: particle 1 in state k, particle 2 in state l; and: particle 1 in state k', particle 2 in state l'. Furthermore, define ψ_k by

$$\psi_k = \frac{(2x)^{1/2} \, \varphi_k}{1 + 2x \mid \varphi_k \mid^2} = \frac{\chi_k}{1 + \chi_k^2} \tag{1.10}$$

Then the expectation value of the potential energy operator over the wave function (1.1) is given by [we use the notation $-k$ for $(-\mathbf{k}, -s)$]

$$\begin{aligned} \frac{\int \Psi_N^* \, \Sigma_{i<j} \, v_{ij} \, \Psi_N \, d\tau}{\int \Psi_N^* \Psi_N \, d\tau} &= \tfrac{1}{2} \sum_{k,k'} \langle k, -k \mid v \mid k', -k' \rangle \, \psi_k \psi_{k'} \\ &+ \tfrac{1}{2} \sum_{k,l} [\langle k, l \mid v \mid k, l \rangle - \langle k, l \mid v \mid l, k \rangle] \, g_k g_l \\ &+ \text{terms of relative order } 1/N \end{aligned} \tag{1.11}$$

[2] The sums over the formal index k, here, include summation over the spin index with its two values. Thus (1.9) for the number of *particles* agrees with (III,9.18) for the number of *pairs*, the factor 2 coming from the different summation indices.

We consider χ_k as the independent parameters for our variation; the only subsidiary condition is (1.9); we multiply (1.9) by the Lagrange multiplier μ (it will turn out to be equal to the chemical potential), and are thus interested in minimizing the quantity

$$\langle H_N \rangle - \mu N = \sum_k (\epsilon_k - \mu) g_k + \tfrac{1}{2} \sum_{k,k'} \langle k, -k \mid v \mid k', -k' \rangle \psi_k \psi_{k'}$$

$$+ \tfrac{1}{2} \sum_{k,l} [\langle k, l \mid v \mid k, l \rangle - \langle k, l \mid v \mid l, k \rangle] g_k g_l \tag{1.12}$$

We use the equations

$$\frac{dg_k}{d\chi_k} = \frac{2\chi_k}{(1 + \chi_k^2)^2} \tag{1.13}$$

and

$$\frac{d\psi_k}{d\chi_k} = \frac{1 - \chi_k^2}{(1 + \chi_k^2)^2} \tag{1.14}$$

and the definitions[3]

$$L_k = \tfrac{1}{4} \sum_l [\langle kl \mid v \mid kl \rangle - \langle kl \mid v \mid lk \rangle + \langle lk \mid v \mid lk \rangle - \langle lk \mid v \mid kl \rangle] g_l \tag{1.15}$$

$$\Delta_k = -\tfrac{1}{2} \sum_{k'} [\langle k, -k \mid v \mid k', -k' \rangle + \langle k', -k' \mid v \mid k, -k \rangle] \psi_{k'} \tag{1.16}$$

Straightforward differentiation of (1.12) with respect to χ_k, and removal of a common factor $\frac{1}{2}(1 + \chi_k^2)^2$ from all terms, gives

$$(\epsilon_k - \mu + 2L_k) \chi_k + \tfrac{1}{2} (\chi_k^2 - 1) \Delta_k = 0 \tag{1.17}$$

Following Bogoliubov (57, 58, 58a), we consider (1.17) as a quadratic equation for χ_k, which can be solved immediately. We introduce the notation

$$\xi_k = \epsilon_k + 2L_k - \mu \tag{1.18}$$

and obtain

$$\chi_k = -\frac{\xi_k}{\Delta_k} \pm \sqrt{1 + \left(\frac{\xi_k}{\Delta_k}\right)^2} \tag{1.19}$$

For large values of k, we should choose the plus sign in (1.19), for the following reason: For large k, the term ϵ_k dominates (1.18), and ξ_k is thus positive. If the interaction is attractive, Δ_k is also positive. The

---

[3] The minus sign in (1.16) has the effect that Δ_k is positive if the interaction is predominantly attractive (i.e., if v is negative).

plus sign in (1.19) then ensures that χ_k becomes small for large k, so that χ_k is square integrable, as it must be.[4]

When we substitute (1.19) into (1.7) and (1.10), we obtain functional forms for g_k and ψ_k, in terms of ξ_k and Δ_k. These are

$$g_k = \frac{1}{2}\left\{1 \mp \frac{(\xi_k/\Delta_k)}{[1 + (\xi_k/\Delta_k)^2]^{1/2}}\right\} \tag{1.20a}$$

and

$$\psi_k = \pm \frac{1}{2[1 + (\xi_k/\Delta_k)^2]^{1/2}} \tag{1.20b}$$

The signs in (1.19) and (1.20) are correlated, in the sense that one should choose either the upper sign in all equations, or the lower sign in all equations. In the simple case of a purely attractive interaction, the upper sign should be chosen throughout.

In order to gain a qualitative understanding of the expressions so far, let us observe that sums over k-space, such as (1.15) and (1.16), are slowly varying functions of k compared to ξ_k, Eq. (1.18). Furthermore, if μ lies close to the usual Fermi energy,[5] and if L_k is negative [as it must be for an attractive interaction, see Eq. (1.15)], then there exists a value of k such that $\xi_k = 0$. Let k_0 be this value, i.e.,

$$\xi_{k0} = \epsilon_0 + 2L_0 - \mu = 0 \tag{1.21a}$$

$$\epsilon_0 = \hbar^2 k_0^2/2M \tag{1.21b}$$

$$L_0 = (L_k)_{k=k_0} \tag{1.21c}$$

Near $k = k_0$, we may approximate ξ_k, (1.18), by[6]

$$\xi_k \cong \epsilon_k - \epsilon_0 \qquad \text{for} \quad k \cong k_0 \tag{1.22}$$

If the interaction is attractive everywhere, so that Δ_k does not change sign, we may neglect the variation of Δ_k with k for a first orientation.

[4] This argument depends on a basically attractive interaction. If the interaction changes sign (for example, a combination of a Fröhlich interaction and an electrostatic repulsion) the sign in (1.19) is not fixed so easily, and different signs may be appropriate in different regions of k-space. χ_k may then pass through zero for certain values of k.

[5] This must be so, from physics: μ is the energy required to remove one electron from the solid (more precisely, half the energy to remove one pair of electrons); this energy is known to be about the same in the superconducting and normal states.

[6] The essential assumption here is that L_k varies much more slowly than ϵ_k. We could include the variation of L_k with k, by multiplying the right-hand side of (1.22) by the correction factor $1 + 2(dL_k/d\epsilon_k)_0$. This correction produces no qualitative effects.

The formulas (1.19), (1.20a), and (1.20b) then lead to the behavior shown schematically in Fig. 4.1. We see that $\chi_k = 1$ and $g_k = \psi_k = \frac{1}{2}$ at $k = k_0$. For $k \ll k_0$, χ_k becomes large and positive, g_k approaches unity, and ψ_k approaches zero; for $k \gg k_0$, all three quantities approach zero. The width, in energy, of the transition region is determined by Δ_k.

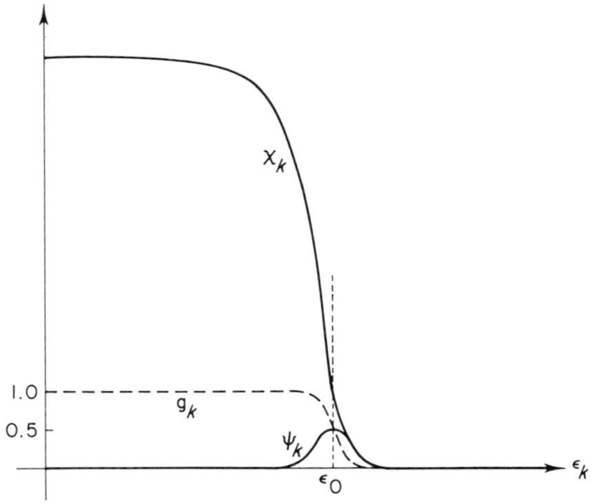

FIG. 4.1. Schematic drawing of quantities characteristic of the Schafroth-condensed pair state. The pair wave function χ_k is large near $k = 0$, passes through 1 at the Fermi energy ϵ_0, and drops to zero at higher energies like $\Delta/2(\epsilon - \epsilon_0)$. The analytic formula is Eq. (1.19). The quantities g_k, Eq. (1.20a), and ψ_k, Eq. (1.20b), are auxiliary quantities constructed from the pair wave function χ_k.

It is also useful to observe that the expectation values of operators over the usual Fermi sea state can be obtained by a simple, formal limiting procedure from the expectation values (1.8) and (1.11) over the condensed pair state (1.1). The limit in question is as follows: we assume that χ_k, Eq. (1.5), is constant for $k < k_F$ and zero for $k > k_F$, and we then let the constant approach infinity. We thus obtain, from (1.7) and (1.10),

$$g_k \to g_k^0 = 1 \quad \text{for} \quad k < k_F \tag{1.23a}$$

$$= 0 \quad \text{for} \quad k > k_F$$

$$\psi_k \to \psi_k^0 = 0 \quad \text{for all } k \tag{1.23b}$$

When we substitute these limiting values into (1.8), (1.9), and (1.11), we obtain

$$\langle J \rangle = \sum_k j_{kk} g_k^0 = \sum_{k<k_F} j_{kk} \tag{1.24a}$$

$$N = \sum_{k<k_F} 1 \tag{1.24b}$$

$$\begin{aligned}\left\langle \sum_{i<j} v_{ij} \right\rangle &= \tfrac{1}{2} \sum_{k,l} [\langle kl \mid v \mid kl \rangle - \langle kl \mid v \mid lk \rangle]\, g_k^0 g_l^0 \\ &= \tfrac{1}{2} \sum_{k,l<k_F} [\langle kl \mid v \mid kl \rangle - \langle kl \mid v \mid lk \rangle] \end{aligned} \tag{1.24c}$$

Equation (1.24b) shows that the quantity k_F is exactly the conventional Fermi momentum; Eqs (1.24a) and (1.24c) are then the usual Hartree-Fock expectation values over the Fermi sea state.

In this same limit, the quantity Δ_k, Eq. (1.16), vanishes identically. It is therefore reasonable to expect Δ_k to be small in the actual case. If Δ_k is indeed small, the transition region in Fig. 4.1 is narrow, and g_k differs from g_k^0 only over this narrow region. The essential difference between (1.11) and the Hartree-Fock limit (1.24c) is therefore not the difference between g_k and g_k^0, but rather the appearance of the first sum in (1.11), which has no counterpart in (1.24c).

We therefore concentrate our attention on the corresponding term in the variational equation, i.e., on the quantity Δ_k, Eq. (1.16). We substitute the functional form (1.20b) into (1.16), and we make use of the fact that for interactions invariant under time reversal

$$\langle k, -k \mid v \mid k', -k' \rangle = \langle k', -k' \mid v \mid k, -k \rangle \tag{1.25}$$

This leads to the *BCS-Bogoliubov integral equation*

$$\Delta_k = \mp \sum_{k'} \frac{\langle k, -k \mid v \mid k', -k' \rangle}{2[1 + (\xi_{k'}/\Delta_{k'})^2]^{1/2}} \tag{1.26}$$

If v is an attractive potential (negative), the upper sign should be chosen in (1.26). If we replace ξ_k by the approximate form (1.22), which contains only one unknown constant (ϵ_0), then (1.26) is a *non-linear integral equation for* Δ_k. Once this equation has been solved, the un-normalized pair wave function is given by (1.19), g_k and ψ_k are given by (1.20), and the constant ϵ_0 is determined by (1.9). Thus, the problem of self-consistency for the extreme case of a Schafroth condensate of electron pairs has been reduced to the solution of the integral Eq. (1.26).

We note that only a special class of matrix elements of the interaction enters the integral equation (1.26). These particular matrix elements are called the "reduced interaction," and in the BCS theory (see Chapter V) all other matrix elements are discarded from the start.

However, it is clear from the present derivation that the "model" assumption is not an assumption on the Hamiltonian at all [we used the full Hamiltonian (1.3)], but rather it is an assumption concerning the trial wave function in the variation principle. This trial wave function was chosen on the basis of a Schafroth condensate of electron pairs, i.e., the wave function (1.1). *The restriction to the "reduced" interaction is a consequence of condensation of electron pairs, not vice versa.* The "pairing interaction" consists of those matrix elements of the full interaction which matter if one chooses a "pairing wave function." The choice of the wave function is primary.[7]

At this stage, we are also in a position to relax the initial assumption of no unpaired electrons. Physically, one would expect that the correlated pairs are formed from components in k-space close to the Fermi momentum; k-values well inside the Fermi surface should not be involved in the pairing. Thus, a more reasonable wave function than (1.1) would consist of a filled Fermi sea, filled up to some momentum k_m slightly lower than the conventional Fermi momentum k_F, with pairs sitting on top of this. This wave function is easily written down in second quantization formalism (see Appendix B) but is awkward to write otherwise. However, all we need here are expectation values, and these can be obtained from our present expectation values by a limiting process analogous to the limit we used to get (1.23): We assume that χ_k is constant and large for all $k < k_m$, and we let this constant value approach infinity. This gives

$$g_k = 1, \qquad \psi_k = 0 \qquad \text{for all} \quad k < k_m \qquad \text{(Fermi sea below pairs)} \qquad (1.27)$$

With such a wave function, the quantity k_m becomes a variational parameter, and the quantities χ_k for $k < k_m$ cease being variational parameters.

2. Solution of the Integral Equation

We now proceed to solve the integral Eq. (1.26) for Δ_k, in which ξ_k is defined by (1.22). Since this integral equation is non-linear in the

[7] The restriction to the reduced interaction is not an absolute restriction in any case: the quantity L_k, Eq. (1.15), contains quite different matrix elements, and L_k is necessary for the complete set of self-consistent equations, namely (1.15), (1.16), (1.18), and (1.20). The reduction from this complicated set to the single equation (1.26) is only an approximation, albeit a very good one. Even with that approximation, however, L_k is necessary to find the relation between ϵ_0 of (1.22), and the chemical potential μ. This relation is given by (1.21). If one takes the reduced interaction literally, L_k vanishes identically.

unknown quantity Δ_k, there is little general theory available concerning the existence of solutions, their uniqueness, or other properties. We shall obtain unique solutions of known form by the artifice of using an extremely schematic assumption about the interaction matrix elements in (1.26), namely, the assumption of a "separable" reduced interaction:

$$\langle \mathbf{k}, s, -\mathbf{k}, -s \mid v \mid \mathbf{k}', s', -\mathbf{k}', -s' \rangle = -\delta_{ss'} \frac{\gamma_k \gamma_{k'}}{V} \tag{2.1}$$

The minus sign ensures that this interaction is basically attractive. The factor $\delta_{ss'}$ is appropriate for a spin-independent interaction; both the Fröhlich interaction and the Coulomb interaction are spin-independent to a good approximation. The quantities γ_k are real numbers which may be chosen so that (2.1) represents, in some crude sense, the actual matrix elements which one should use. The factor $1/V$ makes γ_k independent of the volume.

We emphasize that (2.1) is *not* physically reasonable, and this assumption is made only for the sake of mathematical convenience. The hope is that the solutions of (1.26) obtained with this assumption are not qualitatively different from the solutions of (1.26) with the correct matrix elements. This hope is supported by some numerical calculations with more realistic interactions (see Chapter V, Section 6).

Substitution of (2.1) into (1.26), and summation over the spin index, yields[8]

$$\Delta_{\mathbf{k},+} = S\gamma_k \tag{2.2}$$

where S is a constant, given by

$$S = \frac{1}{V}\sum_{\mathbf{k}} \frac{\gamma_k}{2[1 + (\xi_k/\Delta_k)^2]^{1/2}} = \frac{1}{V}\sum_{\mathbf{k}} \frac{\gamma_k}{2[1 + (\xi_k/S\gamma_k)^2]^{1/2}} \tag{2.3}$$

We write (2.3) as an implicit equation for the unknown constant S, namely,

$$\frac{1}{V}\sum_{\mathbf{k}} \frac{\gamma_k}{2[S^2 + (\xi_k/\gamma_k)^2]^{1/2}} = 1 \tag{2.4}$$

Since γ_k is assumed to be known, and ξ_k is given by (1.22), the sum on the left of (2.4) is, at least in principle, a known function of S, and (2.4) determines the self-consistent value (or values) of S. Once S has been determined from (2.4), $\Delta_{\mathbf{k},+} = -\Delta_{-\mathbf{k},-}$ is given by (2.2) and the other quantities of the theory follow from (1.19) and (1.20).

---

[8] Since (2.1) is purely attractive, we choose the upper sign in (1.26). Note that $\Delta_{-\mathbf{k},-} = -\Delta_{\mathbf{k},+}$, for all $\mathbf{k}$.

At this stage, we follow Bardeen, Cooper, and Schrieffer (Bardeen 57) in making the extreme simplification

$$\begin{aligned} \gamma_k &= \gamma \quad \text{(a constant)} \quad \text{for} \quad |\xi_k| = |\epsilon_k - \epsilon_0| < \hbar\omega_c \\ &= 0 \quad \text{otherwise} \end{aligned} \tag{2.5}$$

This is meant to be a schematic representation of the Fröhlich interaction; thus $\hbar\omega_c$ is taken to be of the order of magnitude of the Debye cutoff energy of lattice waves, as given by (I,10.3) and (I,10.4). Since Debye temperatures are typically of the order of 10^2 °K, whereas the Fermi degeneracy temperature of the electron gas in a metal is of the order of 10^4 °K, $\hbar\omega_c$ is about 100 times smaller than the Fermi energy ϵ_F. The energy ϵ_0 turns out to be extremely close to the Fermi energy, so that $\hbar\omega_c \ll \epsilon_0$.

With the assumption (2.5), the sum over k in (2.4) goes only over values of k in the immediate neighborhood of the Fermi surface, and we may therefore ignore the k-independence of the density of states:

$$\rho_E = \frac{VMk}{2\pi^2\hbar^2} \cong \frac{VMk_0}{2\pi^2\hbar^2} \tag{2.6}$$

Replacing the sum over $\mathbf{k}$ by an integral over ξ_k, and using (2.6), Eq. (2.4) becomes

$$\frac{\frac{1}{2}\rho_E\gamma}{V}\int_{-\hbar\omega_c}^{+\hbar\omega_c}\frac{d\xi}{[S^2 + (\xi/\gamma)^2]^{1/2}} = \frac{\gamma^2\rho_E}{V}\operatorname{arc\,sinh}\left(\frac{\hbar\omega_c}{\gamma S}\right) = 1 \tag{2.7}$$

We introduce the non-dimensional "coupling constant" ρ by[9]

$$\rho = \gamma^2\rho_E/V \tag{2.8}$$

and obtain the solution

$$\Delta_{\mathbf{k},+} = -\Delta_{-\mathbf{k},-} = \frac{\hbar\omega_c}{\sinh(1/\rho)} \cong 2\hbar\omega_c \exp(-1/\rho) \tag{2.9}$$

where the last approximation refers to the "weak coupling" case in which $\rho \ll 1$.

Since ρ is essentially a coupling constant, and since the function $\exp(-1/\rho)$ is non-analytic at $\rho = 0$, *the result* (2.9) *cannot be obtained*

[9] This quantity is called $N(0)V$ by Bardeen (57). Since ρ_E is propertional to V [see (2.6)], ρ is volume-independent.

by any perturbation calculation, to any finite order, which starts from the Fermi sea state as the zero-order wave function.[10]

This is as expected: the transition from the normal to the superconducting state is a phase transition in the thermodynamic sense; in general, the properties of one phase cannot be deduced by analytic continuation of the properties of an adjacent phase. There is a line of analytical discontinuity separating the two regions.[11]

In the case of superconductivity, the cause of the thermodynamic transition is the Schafroth condensation of pairs of electrons. The condensed pair wave function (1.1), from which everything else is derived, cannot be obtained from the Fermi sea state by any *finite* perturbation procedure. We may of course express (1.1) as a sum over unperturbed, independent-particle type, wave functions of the N-body system, i.e., as a sum over "configurations" (II,1.3). However, this requires a huge number of terms, a number which becomes infinite as the number of particles becomes infinite.[12] The perturbation approach fails.

---

[10] Historically, the first model in which the coupling constant enters in the form $\exp(-1/\rho)$ was proposed by Fröhlich (54), who considered a "one-dimensional superconductor." This model is summarized by Schafroth (60).

In spite of the undoubted historical importance of this work of Fröhlich, it is probable that the instability of the one-dimensional Fermi sea state in the presence of electron-phonon coupling, found by Fröhlich, does not indicate the onset of a superconducting state, but rather the transition to a non-metallic (insulating) state of the linear chain. As Peierls (55) points out, a one-dimensional lattice is never metallic. It is possible to make slight, periodic distortions of the original chain, so as to form a "superlattice" whose periodicity distance is an integral multiple of the original periodicity distance. There is always at least one such superlattice in which the Fermi energy lies inside one of the new forbidden bands introduced by the lowering of the symmetry. The Fermi sea state of the superlattice has lower energy than the original Fermi sea state, and since the Fermi energy now lies inside a forbidden band, the superlattice state is an insulator, not a metal. It is highly likely that this form of instability, rather than a transition to a superconducting state, lies at the basis of Fröhlich's one-dimensional model.

We also observe that "perturbation theory" is meant here in the primitive sense of a power series expansion in a coupling constant such as ρ, which can never lead to $\exp(-1/\rho)$. It is possible, however, to perform selective summation of specific terms of the perturbation series, so as to obtain non-analytic functions. This has been done by Katz (63). The prescription used for retaining certain diagrams and discarding other diagrams, then takes the place of the choice of a trial wave function in the approach used here.

[11] For a finite system, with N particles, the free energy F_N is determined by $\exp(-\beta F_N) = \text{Trace}\exp(-\beta H_N)$. The trace is an analytic function of β, and there is no analytic discontinuity in F_N. However, in statistical mechanics we must take the limit: $N \to \infty$, $V \to \infty$, $N/V =$ constant. The limit of an analytic function of β need not itself be an analytic function of β. At phase transition points β_c, the analytic form of the limit function is different to the right and to the left of β_c.

[12] Mathematically, the difficulty may be expressed as follows: The expectation value of an operator over a wave function which is a linear combination of configurations, is

With the simple form (2.5), the solution (2.9) is valid only for $|\xi_k| < \hbar\omega_c$. For all other values of k, Δ_k vanishes identically, and with it ψ_k. Equations (1.7) and (1.10) imply that the quantity g_k is related to ψ_k quite generally through

$$\psi_k = \pm [g_k(1 - g_k)]^{1/2} \tag{2.10}$$

Thus $\psi_k = 0$ implies that g_k is either 1 or 0. In accordance with the phsysical idea of a filled Fermi sea below the pairs, we choose

$$g_k = 1 \quad \text{for} \quad \xi_k < -\hbar\omega_c; \qquad g_k = 0 \quad \text{for} \quad \xi_k > \hbar\omega_c \tag{2.11}$$

This choice of g_k is not only physically reasonable, but is also the only possible choice here: If we chose $g_k = 1$ for $\xi_k > \hbar\omega_c$, the number of particles (1.9) would become infinite; if we chose $g_k = 0$ for $\xi_k < -\hbar\omega_c$, the number of particles (1.9) would be severely limited, to a value much smaller than needed in a metal.

The energy ϵ_0 in (1.22) is expected to be close to the Fermi energy ϵ_F. We now show that, to the accuracy of our present approximations, $\epsilon_0 = \epsilon_F$. To see this, we observe that g_k, Eq. (1.20), has considerable symmetry around $y \equiv \xi_k/\Delta_k = 0$. To be precise, $g_k - \frac{1}{2}$ is an antisymmetric function of y. When we replace the sum (1.9) by an integral in the usual fashion, we obtain

$$N = \int_0^\infty 2g_k\rho_E \, dE = \sum_{\mathbf{k}<k_m} 2 + \int_{-\hbar\omega_c}^{+\hbar\omega_c} 2g(\xi)\,\rho_E(\xi)\,d\xi \tag{2.12}$$

The factor 2 comes from the sum over spins; k_m is defined by $\hbar^2k_m^2/2M = \epsilon_0 - \hbar\omega_c$; and ρ_E is the density of states (2.6). If we neglect the variation of ρ_E over the narrow energy interval involved here, then $g(\xi)$ may be replaced by $\frac{1}{2}$ in the integrand, and we obtain

$$N = \sum_{\mathbf{k}<k_0} 2 \tag{2.13}$$

which is precisely the condition that k_0 be the conventional Fermi momentum!

Thus we have obtained a reasonable physical result: Most of the electrons make up a Fermi sea, of maximum momentum k_m slightly smaller than the usual Fermi momentum k_F. The remaining electrons,

---

an integral over a sum of terms. If the sum is finite, we may, and commonly do, interchange the orders of summation and integration. In the present case, this is not permissible: The sum of integrals is a power series in the coupling constant ρ, whereas the integral of the sum leads to the non-analytic function $\exp(-1/\rho)$.

in the immediate neighborhood of the Fermi surface, are organized into attractively coupled pairs, the pair wave function χ_k being large for $\epsilon_k < \epsilon_F$ and dropping down rapidly as we pass outside the conventional Fermi sphere. The electrons bound together in pairs form a small fraction, of the order of $\hbar\omega_c/\epsilon_F$, of the total number of electrons.

Unfortunately, this simple and reasonable physical picture becomes spoiled when we use a more realistic assumption for the electron-electron interaction. For example, instead of taking $\gamma_k = 0$ for $|\xi_k| > \hbar\omega_c$, we may let γ_k become very small, but non-zero, in that region. Equation (2.2) shows that Δ_k is then non-zero for all k, i.e., we have lost the filled Fermi sea, and all electrons are paired up. The pair wave function χ_k is given by (1.19) for all k; it is still normalizable, but its physical meaning for k deep inside the Fermi sea is a bit doubtful. It is possible, however, that this vast extension of the pair wave function over all of k-space is due to the excessive simplifications introduced into the theory, rather than being a true consequence of the theory. We shall return to this point in Chapter VII.

At this stage, we can also investigate the consequence of choosing a purely repulsive interaction. Within the limitation to "separable" interactions, we get a repulsive interaction from (2.1) by inverting the sign, i.e., we now choose:

$$\langle \mathbf{k}, s, -\mathbf{k}, -s \mid v \mid \mathbf{k}', s', -\mathbf{k}', -s' \rangle = + \delta_{ss'}\gamma_k \gamma_{k'}/V \tag{2.14}$$

With this assumption, (2.2) remains correct, but the opposite sign is necessary in (2.3). Equation (2.4) becomes

$$\frac{1}{V}\sum_{\mathbf{k}} \frac{\gamma_k}{2[S^2 + (\xi_k/\gamma_k)^2]^{1/2}} = -1 \tag{2.15}$$

If $\gamma_k \geqslant 0$ everywhere [for example, we may choose (2.5)], Eq. (2.15) cannot be satisfied. Thus, under these admittedly rather special conditions, a purely repulsive interaction does not give rise to a self-consistent solution of Schafroth condensed pair type.

3. The Critical Field at Zero Temperature

The simple theory presented so far is quite incomplete; we have discussed only the ground state, and we have not specified just what should be chosen for the interaction matrix elements. Nevertheless there are already two significant consequences which can be deduced. The first of these is a right order of magnitude for the energy difference between the superconducting and normal states.

According to the discussion of Chapter I, Section 4, this energy difference determines the critical field H_c at zero temperature; in this limit, all free energies reduce to ordinary energies (the entropies vanish). Let E_{0s} be the ground state energy of the superconducting material, of volume V, and let E_{0n} be the energy of the lowest "normal" state, in which the magnetic field can penetrate the material. Both E_{0s} and E_{0n} depend somewhat on the external magnetic field H applied to the surface of the cylinder. The critical field H_c is given by the relation

$$V \frac{H_c^2}{8\pi} = E_{0n}(H_c) - E_{0s}(H_c) \tag{3.1}$$

Although we have allowed for a dependence of E_{0s} and E_{0n} on the applied field, these dependences are actually very slight. In the superconducting state, the field does not penetrate (Meissner effect, see Chapter VIII), and thus E_{0s} depends on H only through the energy of the field within the London penetration depth. This contribution becomes negligible in the limit of large volume V. The dependence of E_{0n} on H arises from the normal magnetic susceptibility of the metal, which is an extremely small quantity.[13]

We shall neglect both these effects, in order to gain a quick orientation. We shall identify the "normal" state as the Fermi sea state, and the superconducting state as the condensed pair state (1.1), and shall calculate their energies with $H = 0$. This procedure is quite rough and ready, since there is no condensation in the normal state, and thus the system should be described by a statistical ensemble at all non-zero temperatures, rather than by a single quantum state such as the Fermi sea state. The proof, that our procedure gives the correct limit of the free energy as T approaches zero, would take us too far afield here.

For the normal, Fermi sea state, we use (1.24) to obtain

$$E_{0n} = \sum_k (\epsilon_k + L_k^0)\, g_k^0 = \sum_{k<k_F} (\epsilon_k + L_k^0) \tag{3.2}$$

[13] In the ideal Bose gas model, Chapter II, Section 5, the "normal" magnetic susceptibility is itself abnormally large, so that E_{0n} depends very strongly indeed on the applied field H. This peculiarity of the Bose gas model makes it quite different from ordinary metals, in the "normal" state.

where we have used the notation (1.23a), and where we have defined L_k^0 by analogy to (1.15) through

$$L_k^0 = \tfrac{1}{4}\sum_l [\langle kl \mid v \mid kl\rangle - \langle kl \mid v \mid lk\rangle + \langle lk \mid v \mid lk\rangle - \langle lk \mid v \mid kl\rangle]\, g_l^0$$

$$= \tfrac{1}{2}\sum_{l<k_F} [\langle kl \mid v \mid kl\rangle - \langle kl \mid v \mid lk\rangle] \tag{3.3}$$

In the second step, we have used (1.23a) as well as the natural symmetry of the matrix elements of the interaction. The expression (3.3) is one half of the "average potential" seen by a particle in state k, due to all the other particles in the Fermi sea. The term $l = k$, representing the self-interaction, automatically vanishes in (3.3). The factor $\frac{1}{2}$ is well known in the Hartree-Fock theory: without it, we would be counting each pair of interacting particles twice over, instead of once only.

For the superconducting state, we use (1.8), (1.11), (1.15), and (1.16) to get

$$E_{0s} = \sum_k [(\epsilon_k + L_k)\, g_k - \tfrac{1}{2}\psi_k \Delta_k] \tag{3.4}$$

The energy difference in (3.1) therefore assumes the form

$$E_{0n} - E_{0s} = \sum_k \epsilon_k(g_k^0 - g_k) + \sum_k (L_k^0 g_k^0 - L_k g_k) + \tfrac{1}{2}\sum_k \psi_k \Delta_k$$

$$= \Delta E_1 + \Delta E_2 + \Delta E_3 \tag{3.5}$$

We discuss these three contributions in turn. To simplify ΔE_1, we note that

$$\sum_k g_k = \sum_k g_k^0 = N \tag{3.6}$$

so that we may replace ϵ_k by $\epsilon_k - \epsilon_0 = \epsilon_k - \epsilon_F = \xi_k$ without altering the value of ΔE_1 (we ignore the very slight difference between ϵ_0 and ϵ_F). Making use of the fact that $g_k - \frac{1}{2}$ is an odd function of ξ_k [see Eq. (1.20a)], we obtain

$$\Delta E_1 = \sum_k \xi_k(g_k^0 - g_k) = -2\sum_{k>k_F} \xi_k g_k$$

$$= -4\rho_E \int_0^{\hbar\omega_c} \xi g(\xi)\, d\xi$$

$$= \rho_E \{\hbar\omega_c\, [(\hbar\omega_c)^2 + \Delta^2]^{1/2} - (\hbar\omega_c)^2 - \Delta^2 \operatorname{arc\,sinh}\,(\hbar\omega_c/\Delta)\} \tag{3.7}$$

Next, we show that ΔE_2 in (3.5) is a higher-order term which may be neglected. To see this we write

$$\Delta E_2 = \tfrac{1}{2}\sum_{k,l} [\langle kl \mid v \mid kl\rangle - \langle kl \mid v \mid lk\rangle] (g_k^0 g_l^0 - g_k g_l) \tag{3.8}$$

We use the symmetry of the matrix elements of v, together with the identity

$$g_k^0 g_l^0 - g_k g_l = \tfrac{1}{2}(g_k^0 - g_k)(g_l^0 + g_l) + \tfrac{1}{2}(g_k^0 + g_k)(g_l^0 - g_l)$$

to write

$$\Delta E_2 = \sum_k (L_k^0 + L_k)(g_k^0 - g_k)$$

We have already had occasion to introduce a symbol L_0 for the value of L_k at the Fermi surface [see Eq. (1.21c)]. Similarly, we define L_0^0 to be the value of L_k^0 at the Fermi surface, and we use the identity (3.6) in order to obtain the final form

$$\Delta E_2 = \sum_k [(L_k^0 - L_0^0) + (L_k - L_0)] (g_k^0 - g_k) \tag{3.9}$$

In this form, it is easily seen that ΔE_2 is a higher-order term: according to its definition, Eq. (1.15), L_k is a slowly varying function of k. Thus the difference $L_k - L_0$ is small near the Fermi surface, and the same is true of $L_k^0 - L_0^0$. The region near the Fermi surface is the only contributing region, however, since $g_k^0 - g_k$ vanishes for $|\xi_k| > \hbar\omega_c$.

It should be emphasized that the term $\Sigma_k L_k g_k$ in (3.4), and its counterpart in (3.2), are not themselves negligible. However, these contributions are so nearly the same in the normal and superconducting states, that their difference ΔE_2 can be ignored.

Last, we evaluate ΔE_3 in (3.5). In the simple theory, Δ_k is a constant, Δ, for $|\xi_k| < \hbar\omega_c$; and Δ_k vanishes elsewhere. We use (2.2) and (2.3) to obtain

$$\Delta E_3 = \tfrac{1}{2}\sum_k \psi_k \Delta_k = \sum_{\mathbf{k}} \psi_{\mathbf{k},+}\Delta_{\mathbf{k},+} = \Delta \sum_{\mathbf{k}} \tfrac{1}{2}[1 + (\xi_k/\Delta_k)^2]^{-1/2}$$
$$= \frac{\Delta S V}{\gamma} = \frac{\Delta^2 V}{\gamma^2} \tag{3.10}$$

We are now in a position to combine our results. We neglect (3.9), and use (3.7) and (3.10) to obtain $E_{0n} - E_{0s}$. When we employ (2.9), we find that two terms cancel. The net result is

$$E_{0n} - E_{0s} = \rho_E \hbar\omega_c \{[(\hbar\omega_c)^2 + \Delta^2]^{1/2} - \hbar\omega_c\} = \frac{2\rho_E(\hbar\omega_c)^2}{\exp(2/\rho) - 1} \cong \tfrac{1}{2}\rho_E \Delta^2 \tag{3.11}$$

The last, approximate form of (3.11) is valid in the weak-coupling case, where $\Delta \ll \hbar\omega_c$.

We emphasize that the exponential factor in (3.11) is the *square* of the exponential which enters in (2.9). As we shall now show, this exponential is a very significant factor, and this difference must be kept in mind.

Let us insert numbers so as to get order of magnitude estimates. We shall use tin as an example. The critical field at zero temperature is $H_c = 306$ gauss; the number density of ions is 3.7×10^{22} cm^{-3}; we assume (Pines 58) that the ions are quadrivalent, so that the number density of the conduction electrons is 1.5×10^{23} cm^{-3}. From this and Eq. (I,9.8) we deduce a Fermi wave number $k_F = 1.6 \times 10^8$ cm^{-1}; insertion of this value into (2.6) yields $\rho_E/V = 6.9 \times 10^{33}$. The Debye temperature of tin is 195°K; we assume, arbitrarily, that the cutoff energy $\hbar\omega_c$ is half the Debye energy, i.e., $\hbar\omega_c = \frac{1}{2}k\Theta_D = 1.3 \times 10^{-14}$ erg. The magnetic energy per unit volume is $H_c^2/8\pi = 3.7 \times 10^3$ erg/cm^3. Setting this equal to $(E_{0n} - E_{0s})/V$, Eq. (3.11) yields.

$$\frac{H_c^2}{8\pi} \cong \frac{\rho_E}{2V}\Delta^2$$

and thus

$$\Delta \cong 10^{-15} \text{ erg}, \qquad \Delta/k \cong 7.3\ ^\circ\text{K} \qquad (\text{tin}) \tag{3.12}$$

Substitution of this value into (2.9) gives an estimate for the coupling constant ρ, namely,

$$\rho \cong \frac{-1}{\ln(\Delta/2\hbar\omega_c)} \cong 0.3 \qquad (\text{tin}) \tag{3.13}$$

An estimate of the importance of the exponential terms in this theory is obtained by observing that

$$\exp(1/\rho) \cong 26, \qquad \exp(2/\rho) \cong 670 \qquad (\text{tin}) \tag{3.14}$$

It is obvious from these numbers that the precise value of ρ is quite critical; and it is therefore rather hopeless to *predict* critical fields, and other quantities involving such exponentials, directly from first principles. Indeed, the first attempt in this direction by Pines (58) led to ridiculous values (for tin: $\rho = 0.07$ instead of 0.3). A careful and elaborate calculation by Morel and Anderson (Morel 62) gives very much better agreement; for tin, they obtain 0.24 for the "theoretical" value, and 0.25 (instead of our 0.3) for the "experimental" value. The difference in the "experimental" values arises from a more careful evaluation of the density of states, and of the effective cutoff energy $\hbar\omega_c$ (for which we only made a crude guess). It should be observed that quite a bit of theory enters into the "experimental" value of ρ. In general, the "experimental" and "theoretical" values of Morel (62) are reasonably close, although the agreement is not always as good as it is for tin.

All "experimental" values of ρ are in the range $0.1 < \rho < 0.5$. This is not at all accidental, and in fact represents a major success of the theory. The lower limit of this observed range is experimental in origin: for $\rho < 0.1$, the critical fields become exceedingly small (other things being equal, we would get $H_c = 0.8$ gauss for tin, with $\rho = 0.1$). Thus the slightest admixture of magnetic impurities spoils the superconductivity and forces the material back into the normal state. Just this effect was responsible for the incorrect assignment of molybdenum and iridium as non-superconducting metals; extreme purification of these metals showed that both are really superconductors with quite small critical fields, e.g., $H_c(0) \cong 20$ gauss for iridium (Hein 62, Geballe 62).

The upper limit, $\rho < 0.5$, can be deduced theoretically: Kohn (51) and others have called attention to the fact that a value of ρ in excess of 0.5 indicates instability of the assumed lattice structure: Some of the phonon modes acquire unbounded amplitudes of zero-point vibration. We shall give the calculation in Chapter V, Section 5.

This limitation of ρ has a number of consequences, only one of which we can deduce at this stage: the exponential factor in (3.11) is *at least* $\exp(2/\rho) > \exp(4) = 54$. This is the smallest factor by which the energy difference (3.11) differs from the order-of-magnitude estimate in the Fröhlich (50) theory. Without such a factor, there would not be even order-of-magnitude agreement between theory and experiment. As we shall see later, the same limitation $\rho < 0.5$ also explains why there are no room-temperature superconductors.

4. The Pippard Coherence Distance

The non-analytic dependence on the coupling constant, and the right order of magnitude for the critical field are two important results of the microscopic theory at zero temperature. A third result emerges from a study of the *electron-electron correlation function* implied by the theory.

Let $\mathbf{r}$ be a given displacement vector, and let us determine the probability of finding two electrons displaced from each other by precisely this amount. Since electrons are indistinguishable, we must symmetrize between them, and we therefore consider the two-particle operator

$$W(\mathbf{r}) = \sum_{i<i=1}^{N} w_{ij}(\mathbf{r}) \tag{4.1}$$

with

$$w_{ij}(\mathbf{r}) = \tfrac{1}{2}[\delta(\mathbf{r}_i - \mathbf{r}_j - \mathbf{r}) + \delta(\mathbf{r}_j - \mathbf{r}_i - \mathbf{r})] \tag{4.2}$$

If the distance r is much larger than all correlation distances, then the expectation value of $W(r)$ can be computed by the simple device of replacing the square of the N-particle wave function by its average value, V^{-N}. Each term w_{ij} then gives $1/V$ upon integration, and thus

$$\lim_{r\to\infty} W(r) = \frac{N(N-1)}{2V} \tag{4.3}$$

Deviations from this value are to be interpreted as correlation effects.

It is highly instructive to be more specific about the spin directions of the two electrons. The displacement vector $\mathbf{r}$ has a beginning and an end point; we specify that the electron at the beginning of the vector $\mathbf{r}$ must have spin down, that at the end of $\mathbf{r}$ must have spin up. Recalling the standard definition of the Pauli spin matrix σ_z, we observe that the operator $\frac{1}{2}(1 + \sigma_z)$ selects electrons with spin up, the operator $\frac{1}{2}(1 - \sigma_z)$ selects electrons with spin down. We are therefore interested, not in $W(\mathbf{r})$, but in the more specific operator

$$T(\mathbf{r}) = \sum_{i<j=1}^{N} t_{ij}(\mathbf{r}) \tag{4.4}$$

where

$$t_{ij}(\mathbf{r}) = (1/8)\,[(1 + \sigma_{zi})(1 - \sigma_{zj})\,\delta(\mathbf{r}_i - \mathbf{r}_j - \mathbf{r}) + (1 + \sigma_{zj})(1 - \sigma_{zi})\,\delta(\mathbf{r}_j - \mathbf{r}_i - \mathbf{r})] \tag{4.5}$$

Of all the electrons contributing to the asymptotic result (4.3), we have picked out one fourth. Thus we expect, and shall find,

$$\lim_{r\to\infty} T(r) = \frac{N(N-1)}{8V} \cong \frac{N^2}{8V} \tag{4.6}$$

where we have neglected a term of relative order $1/N$ in the last step.

The operator (4.4) is a typical two-particle operator. Its expectation value over the Schafroth-condensed pair wave function (1.1) is given by (1.11), with v_{ij} replaced by t_{ij}. In order to use this equation, we need the matrix elements of t_{ij} between plane wave states. A straightforward calculation gives

$$\langle \mathbf{k}_1 s_1 \mathbf{k}_2 s_2 \mid t_{12} \mid \mathbf{k}_1' s_1' \mathbf{k}_2' s_2' \rangle = (2V)^{-1} \delta_{s_1 s_1'} \delta_{s_2 s_2'} \delta(\mathbf{k}_1 + \mathbf{k}_2 - \mathbf{k}_1' - \mathbf{k}_2')$$
$$[\delta_{s_1,+} \delta_{s_2,-} \exp(-i\boldsymbol{\varkappa} \cdot \mathbf{r}) + \delta_{s_1,-} \delta_{s_2,+} \exp(+i\boldsymbol{\varkappa} \cdot \mathbf{r})] \tag{4.7}$$

where

$$\boldsymbol{\varkappa} = \tfrac{1}{2}(\mathbf{k}_1 - \mathbf{k}_2 - \mathbf{k}_1' + \mathbf{k}_2') \tag{4.8}$$

We substitute (4.7) into (1.11) and perform the sum over spin indices explicitly. Writing the two terms in (1.11) separately, we have

$$T(r) = T_1(r) + T_2(r) \tag{4.9}$$

with

$$T_1(r) = \frac{1}{2V} \sum_{\mathbf{k},\mathbf{k}'} \cos[(\mathbf{k} - \mathbf{k}') \cdot \mathbf{r}] \, \psi_{\mathbf{k},+} \psi_{\mathbf{k}',+} \tag{4.10}$$

$$T_2(r) = \frac{1}{2V} \sum_{\mathbf{k},l} g_{\mathbf{k}} g_l = \frac{N^2}{8V} \tag{4.11}$$

The result (4.11) follows from the fact that the sum of all $g_{\mathbf{k}}$ equals $\frac{1}{2}N$ (factor $\frac{1}{2}$ because the sum over spin indices is missing). This term is therefore independent of the distance r, and equal to the limiting value (4.6). In the Fermi sea state, this is the *only* contribution, since $\psi_{\mathbf{k}}$ vanishes in the Fermi sea state. Electrons of antiparallel spin are uncorrelated in the Fermi sea state. This is as expected from the Pauli principle, since electrons with opposite spins are distinguishable particles.

In the Schafroth-condensed pair state (1.1), however, electrons of antiparallel spin are correlated dynamically, and this shows itself in the contribution $T_1(r)$, (4.10). We use the exponential form of the cosine function to write

$$T_1(\mathbf{r}) = \frac{1}{2V} \operatorname{Re} \sum_{\mathbf{k},\mathbf{k}'} \exp[i(\mathbf{k} - \mathbf{k}') \cdot \mathbf{r}] \, \psi_{\mathbf{k}} \psi_{\mathbf{k}'} = \tfrac{1}{2} |\psi(\mathbf{r})|^2 \tag{4.12}$$

where

$$\psi(\mathbf{r}) = V^{-1/2} \sum_{\mathbf{k}} \exp(i\mathbf{k} \cdot \mathbf{r}) \, \psi_{\mathbf{k},+} \tag{4.13}$$

Thus, the Fourier transform of the quantity $\psi_{\mathbf{k}}$ plays the role of a "wave function" for the correlations of electrons with opposite spins.

Equation (4.13) is difficult to evaluate in general, but there are two limiting cases, $r = 0$ and large r, which can be handled without trouble, and which suffice for our purposes.

For $r = 0$, we have $\exp(i\mathbf{k} \cdot \mathbf{r}) = 1$, and we can use (2.2) and (2.3) together with the identity $\epsilon_F \rho_E = 3N/4$ to obtain

$$\psi(0) = \frac{\Delta V^{1/2}}{\gamma^2} = \frac{N}{V^{1/2}} \frac{3\Delta}{4\epsilon_F} \frac{1}{\rho} \tag{4.14}$$

We substitute (4.14) into (4.12) to obtain

$$T_1(0) = \frac{N^2}{8V} \left(\frac{3\Delta}{2\epsilon_F \rho} \right)^2 \tag{4.15}$$

This is the additional pair correlation between electrons of opposite spin, produced by the formation of Schafroth-condensed pairs, in the limit of zero distance between the two particles of the pair. Since Δ/ϵ_F is of order 10^{-3}, and ρ is of order 0.3, the factor multiplying $N^2/8V$ is very small, of order 10^{-5} to 10^{-4}. Thus this additional correlation is quite weak, numerically, important though it is for the physical properties of superconductors.

Though weak, the correlation $T_1(r)$ has quite a long range. To see this, we now evaluate $\psi(r)$, Eq. (4.13), in the limit of large r. First of all, we replace the sum over k by an integral, in the usual way, and we perform the angle integrations explicitly. The integral of $\exp(i\mathbf{k} \cdot \mathbf{r})$ over the full solid angle is $4\pi \sin(kr)/(kr)$. Next, we note that there is a cutoff in the simple theory, i.e., $\psi_k = 0$ for $|\xi_k| = |\epsilon_k - \epsilon_F| > \hbar\omega_c$. The cutoff energy $\hbar\omega_c$ is much smaller than the Fermi energy, so that the integration is confined to the immediate neighborhood of the Fermi surface. We therefore approximate as follows:

$$\xi_k = \epsilon_k - \epsilon_F = \frac{\hbar^2}{2M}(k^2 - k_F^2) \cong \frac{\hbar^2 k_F}{M}(k - k_F)$$
$$k - k_F \cong \frac{M\xi}{\hbar^2 k_F} \tag{4.16}$$

We can replace k by k_F everywhere except inside the argument of the $\sin(kr)$. Using the explicit form (1.20) for ψ_k, all these approximations taken together yield

$$\psi(r) \cong \frac{\rho_E}{V^{1/2}} \frac{1}{k_F r} \int_{-\hbar\omega_c}^{+\hbar\omega_c} d\xi \, \frac{\sin[(k_F + M\xi/\hbar^2 k_F) r]}{2[1 + (\xi/\Delta)^2]^{1/2}}$$
$$= \frac{\rho_E}{V^{1/2}} \frac{\sin(k_F r)}{k_F r} \int_0^{+\hbar\omega_c} d\xi \, \frac{\cos(q\xi)}{[1 + (\xi/\Delta)^2]^{1/2}} \tag{4.17}$$

where $q = Mr/\hbar^2 k_F$.

Clearly, for large enough r, and hence large enough q, the cosine factor ensures that the main contribution to the integral comes from small values of ξ, allowing us to extend the limit of integration from $\hbar\omega_c$ to infinity. We define the distance r_1 by (we use the notation $\lambda_F = 1/k_F$)

$$r_1 = \frac{\hbar k_F}{M\omega_c} = \frac{2\epsilon_F}{\hbar\omega_c} \lambda_F \tag{4.18}$$

The upper limit of integration in (4.17) can be extended to infinity if the cosine varies rapidly when $\xi = \hbar\omega_c$. This condition is equivalent to $r \gg r_1$. Numerically, the ratio $\epsilon_F/\hbar\omega_c$ is of the order of 10^2 while $\lambda_F \sim 10^{-8}$ cm. Thus this approximation is valid for distances r in excess of 10^{-6} cm.

We use the result (Watson 48, p. 172)

$$K_0(qy) = \int_0^\infty dx \frac{\cos(qx)}{(x^2 + y^2)^{1/2}} \tag{4.19}$$

where $K_0(z)$ is a Hankel function of order zero and imaginary argument, with the asymptotic form

$$K_0(z) \sim (\pi/2z)^{1/2} \exp(-z) \qquad \text{for} \quad z \gg 1 \tag{4.20}$$

It is convenient to define another characteristic distance r_P by

$$r_P = \frac{2\epsilon_F}{\Delta} \lambda_F = \frac{\hbar\omega_c}{\Delta} r_1 \sim 10^{-4} \text{ cm} \tag{4.21}$$

We now combine (4.12), (4.17), (4.19), and (4.21), and separate out a factor $N^2/8V$ as before. The result is

$$T_1(r) = \frac{N^2}{8V} \left[\frac{3\Delta}{2\epsilon_F} \frac{\sin(k_F r)}{k_F r} K_0\left(\frac{r}{r_P}\right)\right]^2 \qquad \text{for} \quad r \gg r_1 \tag{4.22}$$

Using the asymptotic form (4.20) in (4.22), we see that the "coherence distance" for electrons with opposite spin, due to the Schafroth-condensate of pairs, is the distance r_P, Eq. (4.21). This distance is precisely the right order of magnitude for the coherence distance of Pippard. *Pippard's coherence length is the average distance between members of the Schafroth-condensed pairs.* No such distance appears in the ideal Bose-Einstein gas model of Chapter II, simply because the ideal gas is a gas of elementary bosons, without internal structure. The Pippard distance is a measure of the size of the "pseudo-boson" in solids.

It should be noted that there are two kinds of coherence or correlation distances for a system of pairs of particles: one is the internal size of each pair, the other is a correlation distance for center-of-gravity motion of different pairs. The Pippard coherence distance is the internal size of the pair, i.e., the distance between two electrons belonging to the same pair. The other "correlation length," for center-of-gravity motion of different pairs, is *infinite* (as large as the volume V) in a Schafroth-condensed pair state. The center-of-gravity wave function of the pairs spreads through the entire volume, and is by no means confined to distances of the order of 10^{-4} cm.

The Pippard coherence distance is the range of the *wave function* for relative motion of the pair of electrons. It is not equal to, and is in fact much larger than, the range of the *force* responsible for the pairing. The effective range of the Fröhlich interaction is not easy to estimate, but is almost surely less than 10^{-6} cm, more likely of the order of 10^{-7} cm. The range of the Coulomb repulsion between electrons is even shorter, because of the Debye screening. The fact that the wave function has a range much larger than the range of the force is familiar from ordinary quantum mechanics of weakly bound systems (e.g., the deuteron). The effect is particularly large in the electron system, because the pair binding is so very weak, compared to the Fermi energy.

This very large internal size of the pairs, compared to the average distance between electrons in the metal (of order 10^{-8} cm), means that overlap integrals between wave functions of different pairs are of crucial importance. The basic pair wave function from which the whole theory is built up is the $\varphi(i, j)$ in (1.1); except for normalization, its k-space form is χ_k, Eq. (1.5). Yet the pair correlation quantity $T_1(r)$ involves, not the Fourier transform of χ_k, but rather the Fourier transform $\psi(r)$ of ψ_k, Eq. (1.10). The difference between ψ_k and χ_k is profound, see Fig. 4.1 for example. This difference is caused by the requirements of anti-symmetrization in (1.1); it is the tribute which we are required to pay to the Fermi-Dirac statistics of the particles which make up our pseudo-bosons. In terms of the discussion at the beginning of Chapter III, Section 1, the two types of overlap integrals, S and E, are of completely comparable magnitude here, and a naive picture of Bose condensation of well-separated pairs of electrons is out of the question. However, as was shown in Chapter III, Section 9, the antisymmetry requirements actually *help* the Schafroth-condensation of pairs. Thus the large internal size of the pairs in the theory of superconductivity does not disturb the basic concept of a condensation of these pairs.

With pairs of such large size, however, there is a difficulty of quite another sort: why should one restrict oneself to *pair* correlations? Triplet, quadruplet, and higher correlations might well be expected to play an important role. Certainly, there is no separation in space between different pairs. Arguments for neglecting higher order correlations have been given at the end of Chapter III; they are based on the "quenching" of all *dynamical* correlations by the Pauli exclusion principle, the quenching being more important, the higher the order of the correlation. *Statistical* correlations are included to all orders (they are caused by the antisymmetrization in (1.1), and we have just seen that they are indeed very important).

Before leaving the subject of electron-electron correlations in superconductors, let us investigate the correlation between electrons of parallel spin. That is, we replace the operator t_{ij}, Eq. (4.5), by another operator which requires both electron spins to point up:

$$t_{ij}^{+}(r) = \tfrac{1}{8}[(1 + \sigma_{zi})(1 + \sigma_{zj})\,\delta(\mathbf{r}_i - \mathbf{r}_j - \mathbf{r}) + (1 + \sigma_{zj})(1 + \sigma_{zi})\,\delta(\mathbf{r}_j - \mathbf{r}_i - \mathbf{r})] \tag{4.23}$$

Equation (4.7) is modified in the obvious way, by requiring that both s_1 and s_2 must be $+\tfrac{1}{2}$. This apparently minor modification produces major changes in (4.9)—(4.11). The contribution T_1^+ vanishes identically, and we have instead the appearance of an "exchange" term to supplement the "direct" term in $T_2^+(r)$:

$$T^+(r) = T_2^+(r) = \frac{1}{2V} \sum_{\mathbf{k},\mathbf{l}} \{1 - \cos[(\mathbf{k} - \mathbf{l}) \cdot \mathbf{r}]\}\, g_{\mathbf{k}} g_{\mathbf{l}} \tag{4.24}$$

The "1" again gives rise to the asymptotic contribution $N^2/8V$. Using the same technique as in (4.12), the other term can be written as the square of the Fourier transform of g_k:

$$T^+(r) = \frac{N^2}{8V} - \tfrac{1}{2}\,|\,g(\mathbf{r})\,|^2 \tag{4.25}$$

where

$$g(\mathbf{r}) = V^{-1/2} \sum_{\mathbf{k}} \exp(i\mathbf{k} \cdot \mathbf{r})\, g_{\mathbf{k}} \tag{4.26}$$

In the Fermi gas state, g_k becomes g_k^0, Eq. (1.23a), and the integral is easily evaluated to give the standard result (Wigner 33): We define $G(x)$ by

$$G(x) = \frac{3}{x^3}[\sin(x) - x\cos(x)] \tag{4.27}$$

and obtain:

$$T^+(r) = \frac{N^2}{8V}\{1 - [G(k_F r)]^2\} \qquad \text{(Fermi gas)} \tag{4.28}$$

This gives the correlation for electrons of parallel spin in the Fermi sea state. This correlation is entirely "statistical" in the sense discussed in Chapter III. Note, however, that the function $[G(k_F r)]^2$ is qualitatively different from the statistical linking factors (III, 5.5) which may be thought as underlying (4.28). The range of the correlation term in (4.28) is of the order of $\lambda_F = 1/k_F \sim 10^{-8}$ cm. By contrast, the "statistical links" in (III, 5, 5) have a range equal to the thermal de Broglie wavelength λ, which is of the order of 10^{-6} cm at superconducting temperatures, and becomes infinite in the limit of zero temperature. Thus the over-all correlation, averaged over the entire Fermi sea, is quite different from the statistical correlation between just two electrons in an otherwise empty box.

In view of the general similarity of g_k and g_k^0 (see Fig. 4.1), we may be tempted to use (4.28) also for the superconducting ground state (1.1). This, however, would be incorrect. Although (4.26) appears to involve an integral over the entire interior of the Fermi sea, it is easily seen[14] that the effective contribution comes only from the immediate neighborhood of the Fermi surface. By methods similar to the ones used already, we can show that $G(k_F r)$ in (4.28) must be replaced, for $r \gg r_1$, by

$$\begin{aligned} H(r) &\cong G(k_F r) - \frac{3\Delta}{4\epsilon_F} \frac{\cos(k_F r)}{k_F r} K_1\left(\frac{r}{r_P}\right) \\ &\cong G(k_F r)\left[1 + \frac{r}{r_P} K_1\left(\frac{r}{r_P}\right)\right] \qquad \text{for} \quad r \gg r_1 \end{aligned} \tag{4.29}$$

The correction is by no means small; indeed, for $r_1 < r < r_P$, it amounts to a factor 2 in this function, and hence to a factor 4 in the correlation term of (4.28).

Of more importance than the actual analytic expression, is the qualitative statement that the correlation between pairs of electrons with parallel spin is altered significantly, and that the Fourier transform of g_k plays the same role for parallel spin, as the Fourier transform of ψ_k plays for antiparallel spin. Neither g_k nor ψ_k are basic quantities from which everything else can be derived. The basic quantity is the pair wave function $\varphi(i, j)$ which appears in (1.1), and its k-space form φ_k (or multiple thereof, χ_k). The g_k and ψ_k are intermediate quantities useful for calculational purposes, but are not "pair wave functions." This will appear even more clearly in Chapter VIII.

[14] Replace the sum by an integral, integrate over angles, then integrate by parts so as to exhibit a factor dg_k/dk in the integrand, explicitly.

So far, then, the basic assumption of a Schafroth condensation of electron pairs, Eq. (1.1), has led to the right order of magnitude for the critical field at zero temperature, to a physical understanding of Pippard's coherence length, and to the occurrence of a function of the coupling constant which is non-analytic at zero (and hence explains the failure of perturbation treatments).

V

The BCS and Bogoliubov Theories, at Zero Temperature

In this chapter, we state and discuss the theories of Bardeen, Cooper, and Schrieffer (Bardeen 57, 61) and of Bogoliubov (58), to the extent that they are applied to the ground state of the system. The thermodynamic properties will be treated in Chapters VI and VII.

Since these methods require a knowledge of the use of creation and destruction operators, we give a short introduction to this method in Section 1. More detail, and proofs of the commutation relations in their full generality, can be found in books on quantum mechanics (e.g., Schiff 49 or Landau 59).

1. Creation and Destruction Operators

The use of creation and destruction operators, also called the technique of "second quantization," is a convenient way of handling the quantum mechanics of many-particle systems. It is *not* a basic extension of the concepts and ideas of quantum mechanics, but merely a technique of simplifying the writing of equations for many-body systems.

For definiteness, we start with a system of identical particles obeying Fermi-Dirac statistics. Let q denote the full set of coordinates (position as well as spin) of a particle, and let $u_k(q)$, $k = 1, 2, 3, \cdots$, be a complete orthonormal set of one-particle wave functions. This set need not be related in any way to the Hamiltonian of the N-particle system, even for $N = 1$.

The most general wave function of the N-particle system can be expanded as a sum of Slater determinants; each Slater determinant is associated with a "configuration" $\{n_1, n_2, n_3, \cdots\}$ where $n_k = 1$ if the function $u_k(q)$ occurs in the determinant, and $n_k = 0$ otherwise. The values $n_k = 0$ and 1 are the only possible values, since the Slater determinant vanishes if any function $u_k(q)$ occurs more than once. This is just a complicated way of stating the well-known Pauli exclusion principle: no more than one particle in any one state.

As a particular example, consider the antisymmetric state built up from the particular functions $u_k(q)$ and $u_l(q)$. With proper normalization to unity, it is

$$\Phi_{kl}(q_1, q_2) = 2^{-1/2}[u_k(q_1)\, u_l(q_2) - u_k(q_2)\, u_l(q_1)] \tag{1.1}$$

We note that the order of the indices k and l matters to the extent of a sign:

$$\Phi_{kl}(q_1, q_2) = -\Phi_{lk}(q_1, q_2) \tag{1.2}$$

The relation (1.2) shows immediately that Φ_{kl} vanishes identically for $k = l$, as it must according to the Pauli principle.

The most general antisymmetric wave function $\varphi(q_1, q_2) = -\varphi(q_2, q_1)$ of a pair of particles can be written as a superposition of functions (1.1), as follows: We introduce Fourier coefficients φ_{kl} in the usual way[1]:

$$\varphi_{kl} = \iint dq_1 dq_2\, u_k^*(q_1)\, u_l^*(q_2)\, \varphi(q_1, q_2) \tag{1.3}$$

Since $\varphi(q_1, q_2)$ is antisymmetric, Eq. (1.3) implies antisymmetry of the Fourier coefficients:

$$\varphi_{kl} = -\varphi_{lk} \tag{1.4}$$

We can expand $\varphi(q_1, q_2)$ as follows:

$$\begin{aligned} \varphi(q_1, q_2) &= \sum_{k,l} \varphi_{kl}\, u_k(q_1)\, u_l(q_2) \\ &= \tfrac{1}{2} \sum_{k,l} \varphi_{kl}[u_k(q_1)\, u_l(q_2) - u_k(q_2)\, u_l(q_1)] \\ &= 2^{-1/2} \sum_{k,l} \varphi_{kl}\, \Phi_{kl}(q_1, q_2) \end{aligned} \tag{1.5}$$

We shall now introduce "creation" operators a_k^+ which allow us to build up a general configuration one particle at a time, starting from

[1] Integration over q_i denotes both an integration over the space coordinates and a sum over the spin coordinate of particle number i.

the vacuum state $|0\rangle$ in which there are no particles present at all. Thus, $a_k^+ |0\rangle$ denotes a one-particle state, with wave function $u_k(q_1)$. If we operate on this state with the creation operator a_l^+, we get a two-particle state with the "second particle" in state l, i.e.,

$$a_l^+ a_k^+ |0\rangle = \Phi_{kl}(q_1, q_2) \tag{1.6}$$

Equation (1.2) implies the anticommutation rule:

$$a_l^+ a_k^+ = -a_k^+ a_l^+ \tag{1.7}$$

The Pauli principle says that no more than one particle can be put into any one state k. Thus repeated application of the creation operator a_k^+ must give a vanishing result:

$$(a_k^+)^2 = 0 \tag{1.8}$$

This is in fact an immediate consequence of the anticommutation rule (1.7).

The more general pair wave function (1.5) can also be written in terms of creation operators acting on the vacuum state. We introduce the "pair creation operator" b^+ associated with the pair wave function $\varphi(q_1, q_2)$ by the definition

$$b^+ = 2^{-1/2} \sum_{k,l} \varphi_{kl} a_l^+ a_k^+ \tag{1.9}$$

Then a comparison of (1.6) and (1.5) shows immediately that

$$b^+ |0\rangle = \varphi(q_1, q_2) \tag{1.10}$$

So far, there seems to be little advantage in these creation operators. The real simplification occurs only when we consider larger numbers of particles. For example, the Schafroth-condensed pair wave function (IV, 1.1) can be obtained by applying the pair creation operator b^+ to the vacuum state, $\frac{1}{2}N$ times in succession. Unlike a_k^+, the square of b^+ does not vanish, see the discussion at the beginning of Section 9 of Chapter III.

The resulting N-particle wave function is given by

$$\Psi_N = C_N (b^+)^{N/2} |0\rangle \tag{1.11}$$

where C_N is a normalization constant. Comparison of the simple form (1.11) with the complicated equation (IV, 1.1) gives an indication of the very real power of the method of creation operators. All the tedious sums over permutations are now unnecessary; the antisymmetry of

(1.11) is assured by the anticommutation rule (1.7) for the creation operators.

Creation operators alone suffice to write down the most general, antisymmetric wave function for N particles. But if we want to write expressions for linear operators acting on such N-particle wave functions (e.g., the Hamiltonian), we also need the Hermitean conjugates of the creation operators, called destruction operators. The destruction operator a_k has the effect of diminishing the occupation number n_k of state k by one unit. That is, if the configuration $\{n_1, n_2, \cdots n_k, \cdots\}$ has $n_k = 1$, application of the operator a_k results in the corresponding configuration with $n_k = 0$. If the configuration in question already has $n_k = 0$ to start with, application of the destruction operator a_k leads to a vanishing result.

Since n_k can be at most unity, repeated application of the operator a_k always gives zero, i.e., we have the operator identity

$$(a_k)^2 = 0 \tag{1.12}$$

Formally, this identity can be deduced from (1.8) by taking Hermitean conjugates on both sides (the Hermitean conjugate of the zero operator is the zero operator).

Taking Hermitean conjugates on both sides of the more general equation (1.7), we obtain the anticommutation rule for destruction operators:

$$a_l a_k = -a_k a_l \tag{1.13}$$

Equation (1.12) is the special case $k = l$ of Eq. (1.13).

Next, let us consider the effect of the operator $a_k^+ a_k$. Let us operate first on an N-particle configuration in which state k is occupied, i.e., $n_k = 1$. Application of the operator a_k leads to an $(N - 1)$-particle configuration with $n_k = 0$; but subsequent application of a_k^+ restores the Nth particle, again in state k. Hence nothing has changed at all, and $a_k^+ a_k$ is equivalent to 1 for such a configuration. On the other hand, if the N-particle configuration has no particle in state k, i.e., $n_k = 0$, the very first step (application of the destruction operator a_k) gives a vanishing result, and there is nothing for the operator a_k^+ to act on.[2] Thus we have obtained the result

$$\begin{aligned} a_k^+ a_k &= 1 \quad \text{when acting on a configuration with} \quad n_k = 1 \\ &= 0 \quad \text{when acting on a configuration with} \quad n_k = 0 \end{aligned} \tag{1.14}$$

[2] It is necessary to distinguish clearly between the vacuum state $|\,0\rangle$, which is a possible quantum-mechanical state normalized to unity, and a vanishing wave function (which is not normalizable at all). Application of a_k to a configuration with $n_k = 0$ leads to a vanishing wave function, *not* to the vacuum state.

We conclude that the operator $a_k^+ a_k$ is the operator analog of the occupation number n_k, i.e.,

$$(n_k)_{\mathrm{Op}} = a_k^+ a_k \tag{1.15}$$

Next, consider the operator $a_k a_k^+$. By the same chain of reasoning as before, we see that application of this operator to a configuration with $n_k = 0$ restores the same configuration, whereas application to a configuration with $n_k = 1$ yields a vanishing result. Thus

$$a_k a_k^+ = 1 - (n_k)_{\mathrm{Op}} = 1 - a_k^+ a_k \tag{1.16}$$

Next, consider a scattering event in which a single particle of the configuration is transferred from state k' to state k. In order that such an event may happen, it is necessary that $n_k = 0$ and $n_{k'} = 1$ in the initial configuration. The final configuration coincides with the initial one, except that $n_k = 1$ and $n_{k'} = 0$ in the final state. If k and k' are different states, as we shall assume for the moment, this transfer can be produced by either one of the two operators $a_{k'} a_k^+$ and $a_k^+ a_{k'}$. Both these operators have the effect of reducing $n_{k'}$ from one to zero, and raising n_k from zero to one. Both operators give vanishing results if the initial configuration violates the conditions $n_k = 0$ and $n_{k'} = 1$. A more detailed argument, which we omit here, shows that the two operators are equivalent to each other except for a sign, i.e.,

$$a_k^+ a_{k'} = -a_{k'} a_k^+ \qquad \text{for} \qquad k \neq k' \tag{1.17}$$

The rules (1.16) and (1.17) can be combined conveniently into the single anticommutation rule

$$a_k^+ a_{k'} + a_{k'} a_k^+ = \delta_{kk'} \tag{1.18}$$

The anticommutation rules (1.7), (1.13), and (1.18) essentially determine the set of operators a_k and a_k^+ in the algebraic sense: Let α_k and α_k^+ be another set of operators satisfying (1.7), (1.13), and (1.18). Then it can be shown that the α_k differ from the a_k only by a similarity transformation, i.e., there exists a Hermitean operator S such that

$$\alpha_k = e^{iS}\, a_k\, e^{-iS} \qquad \text{and} \qquad \alpha_k^+ = e^{iS}\, a_k^+\, e^{-iS} \qquad (\text{all } k) \tag{1.19}$$

We need expressions for the linear operators which enter the theory, in particular the linear operators which enter the Hamiltonian. These operators are of two types: sums of single-particle operators, and sums of two-particle operators. Let J be a sum of single-particle operators

$$J = \sum_{i=1}^{N} j(i) \tag{1.20}$$

where $j(i)$ is an operator acting on the coordinates of particle i only. Let $\langle k \mid j \mid k' \rangle$ be the matrix element of the operator j between the single-particle states $u_k(q)$ and $u_{k'}(q)$, i.e.,[3]

$$\langle k \mid j \mid k' \rangle = \int dq\, u_k^*(q)\, j_{\text{Op}}\, u_{k'}(q) \tag{1.21}$$

Then it can be shown (Landau 59, paragraph 63) that application of the operator J, Eq. (1.20), to an arbitrary N-particle configuration is equivalent to application of the operator

$$J = \sum_{k,k'} \langle k \mid j \mid k' \rangle\, a_k^+ a_{k'} \tag{1.22}$$

Physically, this equivalence is understandable: the operator $a_k^+ a_{k'}$ leads to a "scattering" from state k' to state k, and this process is associated with the matrix element (1.21).

Although the operators a_k and a_k^+ by themselves do not conserve the number of particles (a_k destroys one particle, a_k^+ creates one particle), the product $a_k^+ a_{k'}$ leaves the number of particles invariant, as it must. We repeat that (1.22) contains no physical idea or generalization, compared to (1.20); (1.22) is merely a more convenient method of writing the effect of (1.20) when applied to the most general configuration of N particles.

One convenience of (1.22) is immediately obvious: there is no longer an explicit reference to the number N of particles. When we consider two systems with different numbers N and N' of particles, the *form* of the operator (1.20) alters (there are N terms in the sum for the first system, N' terms for the second system). By contrast, the *form* of (1.22) remains the same.

A particular example of (1.22) is the operator for the number of particles in the system. This operator can be written in the form (1.20) by the simple device of putting $j(i) = 1$. The matrix elements of this particular operator j are equal to $\delta_{kk'}$, from (1.21) and the orthonormality of the $u_k(q)$. Thus (1.22) yields

$$N_{\text{Op}} = \sum_k a_k^+ a_k \tag{1.23}$$

in full agreement with our identification (1.15): the total number of particles equals the sum of all occupation numbers.

[3] The number i of the particle is irrelevant for (1.21), since $q = q_i$ is just a dummy variable of integration.

Having disposed of sums of single-particle operators, let us now turn to sums of two-particle operators, of type

$$T = \sum_{i<j=1}^{N} t(i,j) \tag{1.24}$$

where $t(i, j)$ acts on the coordinates of particles i and j only. Examples of such operators are the Coulomb repulsion energy between electrons, and the operators (IV, 4.1), (IV, 4.4), and (IV, 4.23), which were used to study correlations between electrons.

We define the matrix element of t in the usual way:

$$\langle kl \mid t \mid k'l' \rangle = \iint dq_1\, dq_2\, u_k^*(q_1)\, u_l^*(q_2)\, t_{\mathrm{Op}}(1, 2)\, u_{k'}(q_1)\, u_{l'}(q_2) \tag{1.25}$$

Then it can be shown (Landau 59) that application of T, Eq. (1.24), to a general N-particle configuration is equivalent to application of the operator

$$T = \tfrac{1}{2} \sum_{k,l,k',l'} \langle kl \mid t \mid k'l' \rangle\, a_k^+ a_l^+ a_{l'} a_{k'} \tag{1.26}$$

Note the order of the indices on the destruction operators: this matters, because of (1.13).

Physically, the operator $a_k^+ a_k^+ a_{l'} a_{k'}$ describes an event in which two particles interact in such a way that they are moved from initial states k' and l' to final states k and l. This operator automatically gives zero unless the initial configuration satisfies the conditions: $n_k = n_l = 0$ and $n_{k'} = n_{l'} = 1$. It is reasonable that this operator should be multiplied by the matrix element (1.25). The factor $\frac{1}{2}$ in front, and the detailed order of indices in (1.26), are more specific properties which need a detailed derivation (Landau 59).

In view of the fact that the sums in (1.26) go over all values of all indices, and in view of the anticommutation rules (1.7) and (1.13), we note that we may replace the matrix element of t in (1.26), without loss of generality, by the "antisymmetrized" matrix element

$$\langle kl \mid t \mid k'l' \rangle \to \tfrac{1}{4} [\langle kl \mid t \mid k'l' \rangle - \langle lk \mid t \mid k'l' \rangle - \langle kl \mid t \mid l'k' \rangle + \langle lk \mid t \mid l'k' \rangle] \tag{1.27}$$

The replacement (1.27) makes no difference to the value of the sum (1.26).

This concludes this brief survey of the method of creation and destruction operators for systems of Fermions. Last, we state the corresponding results for systems of Bosons; we shall need these results for the study of the electron-phonon coupling, since the phonons obey Bose-Einstein statistics.

It turns out (Landau 59) that the expressions (1.15), (1.22), and (1.26) remain unchanged; in (1.27) all terms have plus signs (symmetrization instead of antisymmetrization); the main change is that the anticommutation rules (1.7), (1.13), (1.18) are replaced by the *commutation* rules

$$a_k^+ a_l^+ = a_l^+ a_k^+ \qquad \text{(Bose statistics)} \tag{1.28a}$$

$$a_k a_l = a_l a_k \qquad \text{(Bose statistics)} \tag{1.28b}$$

$$a_k a_{k'}^+ - a_{k'}^+ a_k = \delta_{kk'} \qquad \text{(Bose statistics)} \tag{1.28c}$$

As before, it can be shown that these operators a_k and a_k^+ are determined by (1.28) up to a similarity transformation (1.19). But unlike the Fermi case, the operator $a_k^+ a_k$ now has the eigenvalues 0, 1, 2, 3, $\cdots$, instead of merely 0 and 1.

2. The BCS Theory at Zero Temperature

A. The BCS Wave Function

Historically, the first self-consistent calculation based on the electron pair idea was done by Cooper (56) for a *single pair* of electrons imbedded in a Fermi sea of other electrons. Cooper found that such an electron pair can form a stable, bound state, no matter how weak the electron-electron attraction (there is no bound state if there is only repulsion). The binding energy of the pair involves the factor exp $(-1/\rho)$, where ρ is proportional to the coupling constant of the attractive force. The concept of a "bound" state requires some care here: the binding energy is taken with respect to the Fermi energy of the surrounding electron gas, and the two particles of the pair stay together only in the presence of such a surrounding Fermi sea. Without the other electrons, there would not be any binding. The internal size of the pair turns out to be of order 10^{-4} cm, in agreement with the Pippard coherence length. But the details of the pair wave function are rather different from that of χ_k, Eq. (IV, 1.19)]. In particular, the Cooper wave function, considered in x-space, decreases with the interparticle distance much more slowly than the exponential fall-off of correlations found in Chapter IV, Section 4.

These differences are due to the fact that the Cooper calculation is for only one pair, whereas in fact there are very many pairs, forming a Schafroth condensate. It was realized immediately that, once the Fermi sea state is shown to be unstable against formation of a single pair, there will be a tendency to form large numbers of pairs. The problem then becomes a technical one, of how a state with many pairs is to be treated. The obvious answer is to write down the wave function (1.11) [or its equivalent, (IV,1.1)] and use it as the trial function in a variation calculation. This is what we did in Chapter IV. However, such a program requires explicit evaluation of the expectation values of one-particle and two-particle operators over this wave function, i.e., the formulas (IV, 1.8), and (IV, 1.11). These formulas were not known at the time, and thus the historical development of the theory did not follow this course.

Instead, Bardeen, Cooper, and Schrieffer (Bardeen 57) introduced a rather different concept of "pairing," namely a *pairing of single-electron states*, rather than a quantum state of an electron pair. The difference is best explained in terms of the operators creating the two kinds of "pair." The pair creation operator of the quasi-chemical equilibrium theory is the operator b^+, Eq. (1.9), and the N-particle wave function composed of $\frac{1}{2}N$ pairs is given by (1.11). By contrast, BCS introduce a whole set of "pair" creation operators, one for each value of the momentum vector $\mathbf{k}$. The "pair" so created consists of an electron with momentum $\mathbf{k}$ and spin up, together with another electron of momentum $-\mathbf{k}$ and spin down. Letting $\beta_{\mathbf{k}}^+$ be this creation operator, its definition is

$$\beta_{\mathbf{k}}^+ = a_{\mathbf{k},+}^+ a_{-\mathbf{k},-}^+ \tag{2.1}$$

BCS obtain their many-particle wave function by considering different momentum vectors $\mathbf{k}$ to be independent, and assigning, for each $\mathbf{k}$, a probability amplitude $v_{\mathbf{k}}$ to find the "$\mathbf{k}$-pair" occupied, and a probability amplitude $u_{\mathbf{k}}$ to find this "$\mathbf{k}$-pair" unoccupied.[4] Configurations in which the state $(\mathbf{k}, +)$ is occupied but the state $(-\mathbf{k}, -)$ is empty, are not used in the BCS wave function. Thus each momentum vector $\mathbf{k}$ is associated, in the BCS wave function, with the operator factor

$$u_{\mathbf{k}} + v_{\mathbf{k}} \beta_{\mathbf{k}}^+$$

and the wave function as a whole is given by

$$\Psi_{\text{BCS}} = \left\{ \prod_{\mathbf{k}} (u_{\mathbf{k}} + v_{\mathbf{k}} \beta_{\mathbf{k}}^+) \right\} | 0 > \tag{2.2}$$

[4] We use probability amplitudes, rather than probabilities, to avoid the continual occurrence of square roots. BCS use the notation $h_{\mathbf{k}}$ for our $(v_{\mathbf{k}})^2$. The numbers $u_{\mathbf{k}}$ should not be confused with the one-particle wave functions $u_k(q)$ of Section 1.

where $|0>$ is the vacuum state. The different momentum vectors **k** in (2.2) are quite independent. The corresponding β^+-operators commute:

$$\beta_{\mathbf{k}}^+\beta_l^+ = \beta_l^+\beta_{\mathbf{k}}^+ \tag{2.3}$$

This formula follows directly from the basic anticommutation rule (1.7) and the definition (2.1).

The wave function (2.2) is normalized correctly to unity provided that $u_{\mathbf{k}}$ and $v_{\mathbf{k}}$ satisfy the condition[5]

$$(u_{\mathbf{k}})^2 + (v_{\mathbf{k}})^2 = 1 \qquad (\text{all } \mathbf{k}) \tag{2.4}$$

In addition to the simple commutation rule (2.3), there is a special identity

$$(\beta_{\mathbf{k}}^+)^2 = 0 \qquad (\text{all } \mathbf{k}) \tag{2.5}$$

This is an operator identity, which follows directly from the Pauli exclusion principle. If a "**k**-pair" is already occupied, creation of a second "**k**-pair" would lead to double occupation of the states (**k**, +) and (−**k**, −), which is forbidden. More formally, (2.5) follows from (1.7) and (1.8).

It is convenient to introduce the Hermitean conjugate of (2.1)

$$\beta_{\mathbf{k}} = a_{-\mathbf{k},-}a_{\mathbf{k},+} \tag{2.6}$$

This is the destruction operator for the "**k**-pair." Use of the relations (1.7), (1.13), and (1.17) yields

$$\beta_{\mathbf{k}}\beta_l^+ = \beta_l^+\beta_{\mathbf{k}} \qquad \text{for} \quad \mathbf{k} \neq l \tag{2.7}$$

This relation, which holds only for $\mathbf{k} \neq l$, establishes the complete independence of these "**k**-pairs."

[5] The calculation goes as follows: The normalization integral is

$$(\Psi_{\mathrm{BCS}}, \Psi_{\mathrm{BCS}}) = \langle 0 \mid \prod_{\mathbf{k}} (u_{\mathbf{k}} + v_{\mathbf{k}}\beta_{\mathbf{k}}) \prod_{\mathbf{k}'} (u_{\mathbf{k}'} + v_{\mathbf{k}'}\beta_{\mathbf{k}'}^+) \mid 0\rangle$$

Because of the straightforward commutation rules (2.3) and (2.7), we are allowed to work with one **k**-value at a time, i.e., we obtain

$$(\Psi_{\mathrm{BCS}}, \Psi_{\mathrm{BCS}}) = \prod_{\mathbf{k}} \langle 0 \mid (u_{\mathbf{k}} + v_{\mathbf{k}}\beta_{\mathbf{k}})(u_{\mathbf{k}} + v_{\mathbf{k}}\beta_{\mathbf{k}}^+) \mid 0\rangle$$

$$= \prod_{\mathbf{k}} \{u_{\mathbf{k}}^2 + u_{\mathbf{k}}v_{\mathbf{k}}(\langle 0 \mid \beta_{\mathbf{k}} \mid 0\rangle + \langle 0 \mid \beta_{\mathbf{k}}^+ \mid 0\rangle) + v_{\mathbf{k}}^2 \langle 0 \mid \beta_{\mathbf{k}}\beta_{\mathbf{k}}^+ \mid 0\rangle\}$$

The vacuum expectation values of $\beta_{\mathbf{k}}$ and $\beta_{\mathbf{k}}^+$ are zero, and the remaining vacuum expectation value is unity. The normalization integral therefore is an infinite product of factors (2.4), each of which is maintained equal to one.

The simplicity of the operator relations (2.3), (2.5), and (2.7) makes the wave function (2.2) very much easier to handle than the function (1.11). Yet (2.2) still incorporates, in a sense to be elucidated in more detail later, the basic idea of electron pairing. To quote Cooper (59): "The basic approximation of the BCS theory of superconductivity rests in their assumption that it is the two-body correlations that are responsible for the qualitative features of superconductivity and that of the two-body correlations there is a very strong preference for singlet zero momentum pairs—so strong that one can get an adequate description of superconductivity by treating these correlations alone."

Of course, one pays a price for the great simplicity of (2.2): the number N of electrons is no longer uniquely defined. The *average* number of electrons is given by

$$\langle N \rangle_{\text{av}} = (\Psi_{\text{BCS}}, N_{\text{Op}}\, \Psi_{\text{BCS}}) = 2 \sum_{\mathbf{k}} (v_{\mathbf{k}})^2 \tag{2.8}$$

but Ψ_{BCS} is not an eigenfunction of the number operator. However, the mean deviation of N from the average value (2.8) is small compared to N, of relative order $N^{-1/2}$, and thus the wave function (2.2) represents an approximately constant number of particles.

The argument used by BCS to justify (2.2) is based on a consideration of the interaction between the electrons. They write down an effective electron-electron interaction produced by Fröhlich's electron-phonon coupling, namely (we again use $\hbar\omega_q$ for the phonon energy)

$$\langle k_1 k_2 \mid V_{\text{ph}} \mid k_1' k_2' \rangle = \frac{\hbar\omega_q \mid M_q \mid^2}{(\epsilon_{k1}' - \epsilon_{k1})^2 - (\hbar\omega_q)^2} + \frac{\hbar\omega_q \mid M_q \mid^2}{(\epsilon_{k2}' - \epsilon_{k2})^2 - (\hbar\omega_q)^2} \tag{2.9}$$

This is the perturbation-theoretic result from the interchange of one phonon, between electrons in plane wave states. BCS observe, following Fröhlich (50), that this interaction is attractive (negative) if the energy difference between the electron states is less than $\hbar\omega_q$. BCS make maximum use of the attractive portions of the interaction (2.9), by constructing a wave function in which all these attractive matrix elements enter with equal phases, so as to reinforce each other. In fact, they start from a *reduced Hamiltonian* which contains only those matrix elements connecting the ($\mathbf{k}$, $+$; $-\mathbf{k}$, $-$) "pairs" to each other. All other matrix elements are discarded from the start. This leads to the form

$$H_{\text{red}} = \sum_{\mathbf{k},s} \epsilon_{\mathbf{k}} a_{\mathbf{k}s}^{+} a_{\mathbf{k}s} + \tfrac{1}{2} \sum_{\mathbf{k},s,\mathbf{k}',s'} \langle \mathbf{k}, s, -\mathbf{k}, -s \mid v \mid \mathbf{k}', s', -\mathbf{k}', -s' \rangle\, a_{\mathbf{k}s}^{+} a_{-\mathbf{k}-s}^{+} a_{-\mathbf{k}'-s'} a_{\mathbf{k}'s'}$$

If we make the (usual) assumption that the interaction is spin-independent, then the matrix elements of v contain a factor $\delta_{ss'}$ and are independent of s. We thus obtain

$$H_{\text{red}} = \sum_{\mathbf{k},s} \epsilon_{\mathbf{k}} a_{\mathbf{k}s}^{+} a_{\mathbf{k}s} + \sum_{\mathbf{k},\mathbf{k}'} \langle \mathbf{k}, +, -\mathbf{k}, - \mid v \mid \mathbf{k}', +, -\mathbf{k}', - \rangle \beta_{\mathbf{k}}^{+} \beta_{\mathbf{k}} \tag{2.9a}$$

It can be shown that, with suitably chosen $u_{\mathbf{k}}$ and $v_{\mathbf{k}}$, the state (2.2) is very close to an eigenstate of this reduced Hamiltonian.

From here on, the going is easy. If we make the identification $g_{\mathbf{k}} = (v_{\mathbf{k}})^2$, formula (IV, 1.8) is an immediate consequence of (2.2).[6] When it comes to the interaction operator, the BCS formula is even simpler than (IV, 1.11), since the second sum on the right of (IV, 1.11) (the "Hartree-Fock" term) vanishes identically for the BCS reduced Hamiltonian. The quantity $\psi_{\mathbf{k}}$ in (IV, 1.11) must be identified with the product $u_{\mathbf{k}} v_{\mathbf{k}}$. With these basic formulas under control, the minimization of the energy proceeds as in Chapter IV, Sections 1 and 2; the results of Chapter IV were first obtained by BCS, in that fashion.

B. The Energy Gap

An additional result, of considerable importance, can be derived particularly easily from the BCS theory: the energy gap. Let us suppose that one BCS "pair" has been broken up into two uncorrelated electrons, one with momentum vector $\mathbf{k}'$, the other with momentum vector $\mathbf{k}''$. The spin directions of the two uncorrelated electrons are not important; we shall assume that the electron with momentum $\mathbf{k}'$ has spin up, that with momentum $\mathbf{k}''$ has spin down, mostly for the sake of definiteness. The essential point is that these two electrons do not have "partners" in the BCS sense, i.e., there is no electron with momentum $-\mathbf{k}'$ and spin down, and there is no electron with momentum $-\mathbf{k}''$ and spin up. The wave function for this excited state is given by

$$\Psi_{\text{exc}} = \left\{ \prod_{\mathbf{k} \neq \mathbf{k}', -\mathbf{k}''} (u_{\mathbf{k}} + v_{\mathbf{k}} \beta_{\mathbf{k}}^{+}) \right\} a_{\mathbf{k}',+}^{+} a_{\mathbf{k}'',-}^{+} \mid 0 \rangle \tag{2.10}$$

This wave function is orthogonal to (2.2), and it is easy to compute the

[6] We remind the reader that the formal index k in (IV, 1.8) includes not only the momentum vector $\mathbf{k}$, but also the spin index s. This accounts for the extra factor 2 in (2.8), for example.

expectation value of the reduced Hamiltonian over (2.10). The result is

$$(\Psi_{\text{exc}}, H_{\text{red}}\Psi_{\text{exc}}) = \sum_{\mathbf{k}\neq\mathbf{k}',-\mathbf{k}''} 2\epsilon_k(v_{\mathbf{k}})^2 + \epsilon_{\mathbf{k}'} + \epsilon_{\mathbf{k}''} + \sum_{\mathbf{k}\neq\mathbf{k}',-\mathbf{k}''}\ \sum_{\boldsymbol{l}\neq\mathbf{k}',-\mathbf{k}''} \langle \mathbf{k}, -\mathbf{k} \mid v \mid \boldsymbol{l}, -\boldsymbol{l}\rangle \psi_{\mathbf{k}}\psi_{\boldsymbol{l}} \tag{2.11}$$

The ground state expectation value differs from (2.11) in three ways:

(i) The sums over $\mathbf{k}$ and $\boldsymbol{l}$ are unrestricted.

(ii) The extra terms $\epsilon_{\mathbf{k}'} + \epsilon_{\mathbf{k}''}$ are not present.

(iii) The quantities $u_{\mathbf{k}}$ and $v_{\mathbf{k}}$ (and hence also $\psi_{\mathbf{k}} = u_{\mathbf{k}}v_{\mathbf{k}}$) are slightly different for the two states, in order to give the same average number of particles, N.

If it were not for (iii), the energy difference would be simple to compute. In order to get around this difficulty, we notice that the $u_{\mathbf{k}}$ and $v_{\mathbf{k}}$ are determined, for the ground state, by making $(\Psi_0, H_{\text{red}}\Psi_0)$ a minimum, subject to the condition that the average value of N, Eq. (2.8), is kept fixed. With the usual Lagrange multiplier μ, this means we minimize the expectation value of

$$W_{\text{red}} = H_{\text{red}} - \mu N_{\text{Op}} \tag{2.12}$$

where N_{Op} is the number operator.

As a result of this minimization, the expectation value of W_{red}, but *not* the expectation value of H_{red}, is variationally correct with respect to small variations of the $u_{\mathbf{k}}$ and $v_{\mathbf{k}}$, if the chemical potential μ is kept fixed.

We now keep μ fixed and compute the difference between the expectation values of W_{red} for the ground state and the excited state (2.10). The terms $-\mu N_{\text{Op}}$ then give contributions which cancel each other, if the calculation is done correctly. By keeping these terms in, however, we are allowed to do the calculation more crudely, i.e., as a result of the variational property, we may now ignore the difference (iii) between the two states, and use the same $u_{\mathbf{k}}$, $v_{\mathbf{k}}$ for both states. Furthermore, we note that $\mu = \epsilon_F$ in the BCS theory, since the Hartree-Fock term (by which μ differs from ϵ_F in general) is zero for the reduced interaction.

Using the definition (IV, 1.16) of $\Delta_{\mathbf{k}}$, the excitation energy is therefore

$$W_{\mathbf{k}'\mathbf{k}''} - W_0 = (\epsilon_{\mathbf{k}'} - \epsilon_F)(1 - 2v_{\mathbf{k}'}^2) + (\epsilon_{\mathbf{k}''} - \epsilon_F)(1 - 2v_{\mathbf{k}''}^2) + 2(\psi_{\mathbf{k}'}\Delta_{\mathbf{k}'} + \psi_{\mathbf{k}''}\Delta_{\mathbf{k}''}) \tag{2.13}$$

We now use (IV, 1.18), (IV, 1.20b), and (IV, 1.20a) for $g_{\mathbf{k}} = v_{\mathbf{k}}^2$ to get

$$W_{\mathbf{k}'\mathbf{k}''} - W_0 = (\xi_{\mathbf{k}'}^2 + \Delta_{\mathbf{k}'}^2)^{1/2} + (\xi_{\mathbf{k}''}^2 + \Delta_{\mathbf{k}''}^2)^{1/2} \tag{2.14}$$

This shows the existence of the *energy gap*: in the BCS theory $\Delta_{\mathbf{k}} = \Delta$ is a constant, and the energy difference (2.14) is at least 2Δ in magnitude, no matter what $\mathbf{k}'$ and $\mathbf{k}''$ we choose.

The reason for the energy gap is the restriction on the range of summation of the last term (the double sum) in (2.11). The other terms in (2.11) would not give a gap. We may say that the different "pairs" $(k, -k)$ reinforce each other, and we decrease the binding energy of the state as a whole, by "killing off" the momenta $\mathbf{k}'$ and $\mathbf{k}''$ as far as "pair" formation is concerned.

It should be noted that we have proved only that there is an energy gap for this particular kind of excited state. It is possible (Anderson 58, Bogoliubov 58) to show the existence of other kinds of excited states, called "collective excitations," which do not have an energy gap if the range of the electron-electron interaction is finite (for example, a Fröhlich interaction plus a screened Coulomb interaction). These "collective" excited states correspond to density fluctuations in the electron system. If the Coulomb interaction is unscreened (i.e., taken to have an infinite range), then these states are plasma wave states with quite high excitation energy ($\hbar\omega_{\mathrm{pl}}$ is of the order of 15 to 20 ev), i.e., the gap is restored. This still leaves open the question whether there are other "collective" excitations without an energy gap. So far, none have been found, but it is not certain that all low-lying excitations have been investigated.

The BCS theory establishes the existence of an energy gap for a particular type, and clearly an important type, of excitations of the system. The energy gap is of the right order of magnitude: we can find the value of Δ from the critical field at zero temperature, see Eq. (IV, 3.12). The energy gap is twice Δ, i.e., it corresponds to about 15.2°K. A quantity directly comparable with the energy gap is the value of $\hbar\omega_0$ where ω_0 is the critical (circular) frequency for onset of absorption of electromagnetic radiation by thin films (see Chapter I, Section 8). The early data are summarized by Biondi (58); the experiments of Tinkham and Glover (Glover 56, Tinkham 56, Glover 57, Tinkham 58) are certainly consistent with a cutoff at that frequency (the corresponding wavelength is about 0.1 cm, i.e., the very short microwave region); but the data are insufficient to define the precise cutoff frequency closely enough to be significant for a quantitative comparison, even though they do indicate the existence of a cutoff frequency as such. Later experiments (Ginsberg 60) define the cutoff frequency more precisely, and are in good agreement with the value predicted from the critical field at zero temperature.

C. The Relation between the BCS and Quasi-Chemical Equilibrium Theories

The BCS theory has been successful in a large number of calculations of experimentally observable properties of superconductors. Not only the energy gap, but the temperature dependence of this quantity, the temperature dependence of the critical field, the specific heat curve, ultrasonic absorption, etc.—all these have been calculated on the BCS theory, and agree at least as well with experiment as the rough initial assumptions of this theory could lead one to expect; in fact, the agreement is remarkably good. Thus, the BCS theory was the first quantitative theory of the superconducting state, and a decisive step forward in the field.[6a]

The theory has difficulties, however, of a different nature: it is impossible to accept the justification given by BCS for their initial use of the wave function (2.2), and of the "reduced interaction." One way would be to accept the reduced interaction, in which case (2.2) represents a very accurate approximation to the exact solution of a mutilated problem (Bardeen 60, Bogoliubov 60, 60a).[6b] However, the mutilation of the Hamiltonian is much too severe. The reduced interaction is not even gauge-invariant, gives zero for quantities which are clearly not zero (the Hartree-Fock terms), and completely misses an entire branch of the excitation spectrum [the "collective" branch of Anderson (58) and Bogoliubov (58)]. Rather, one must look for the justification of the reduced interaction in terms of the wave function (2.2); *the model assumption is an assumption on the wave function, not on the interaction.* It is easy to see, starting from the wave function (2.2) and the *full* interaction, that the "reduced" matrix elements of the interaction are the

---

[6a] The experimental evidence for the BCS theory is reviewed by Pippard (58) and Ginsberg (62b). For reviews of the theory of superconductivity from the BCS point of view, see Bardeen (58, 61, 62a, 63) and Cooper (59, 60).

[6b] A voluminous literature has developed concerning the degree of exactitude of the BCS approximation to an exact solution of the reduced Hamiltonian, much of it concerned with the strong coupling limit for which the exact solution can be found explicitly. There is general agreement that the ground state energy is accurate to terms of relative order $1/N$ (Wada 58, 59; Mühlschlegel 59a; Katz 60; Thouless 60a; Mattis 61; Mitter 61; Galasiewicz 61; Yamazaki 61; Mittelstaedt 61, 62; Haag 62; Lüders 62a; Richardson 63), but there exist corrections of the relevant order (of order one) to *differences* between energy levels, in particular to the energy gap (Mühlschlegel 59a); the approximate wave function (2.2) also fails to reproduce certain finer details of the true solution, for example the fluctuations in the particle density (Mittelstaedt 61, 62; Lüders 62a). Nonetheless, considered as a trial function for the reduced Hamiltonian, the BCS Ansatz (2.2) is undoubtedly very successful. The difficulty lies in finding a justification for the reduced Hamiltonian.

only ones of qualitative importance; the Hartree-Fock terms are obtained correctly, but (as we have seen) they make little qualitative difference.

Of course, this rests on the assumption of an equivalent electron-electron interaction, to replace the Fröhlich coupling. Phonons are not included in the BCS theory explicitly. However, it seems clear that some sort of equivalent electron-electron interaction can be found to replace the Fröhlich term, and once this is accepted, the use of the wave function (2.2) is sufficient to obtain all the essential results of the theory at zero temperature.

But, how are we to justify this basic wave function? Within the confines of the BCS approach, no clear explanation exists for "... the deep mystery in the theory of superconductivity, the pairing condition" (Cooper 62). It is unusual to find, five years after the original publication of a theory, an admission by one of the authors that the basic assumption of the theory is a deep mystery to him.

If one goes outside the BCS theory, however, the "mystery" is not very difficult to dispel. Clearly, the wave function (2.2) as such is inadmissible, since it does not correspond to any one definite number of particles. This was recognized by BCS, who suggest one should project out, from (2.2), the component with exactly N particles present. Let us do this![7]

Under the assumption $u_{\mathbf{k}} \neq 0$, all $\mathbf{k}$,[8] we may multiply and divide the operator in (2.2) by $\Pi_{\mathbf{k}}\, u_{\mathbf{k}}$, which is a non-zero c-number. We define $\chi_{\mathbf{k}}$ by

$$\chi_{\mathbf{k}} = v_{\mathbf{k}}/u_{\mathbf{k}} \tag{2.15}$$

and write (2.2) as

$$\Psi_{\mathrm{BCS}} = \prod_{\mathbf{k}'} u_{\mathbf{k}'} \prod_{\mathbf{k}} (1 + \chi_{\mathbf{k}} \beta_{\mathbf{k}}^{+}) \mid 0\rangle \tag{2.16}$$

Next, we expand the product over $\mathbf{k}$, noting that the operators $\beta_{\mathbf{k}}^{+}$ commute with each other. This gives

$$\prod_{\mathbf{k}} (1 + \chi_{\mathbf{k}} \beta_{\mathbf{k}}^{+}) = 1 + \sum_{\mathbf{k}} \chi_{\mathbf{k}} \beta_{\mathbf{k}}^{+} + \tfrac{1}{2} \sum_{\mathbf{k} \neq \mathbf{m}} \chi_{\mathbf{k}} \chi_{\mathbf{m}} \beta_{\mathbf{k}}^{+} \beta_{\mathbf{m}}^{+} + \cdots \tag{2.17}$$

The operator

$$\begin{aligned} W^{+} &= \sum_{\mathbf{k}} \chi_{\mathbf{k}} \beta_{\mathbf{k}}^{+} = \sum_{\mathbf{k}} \chi_{\mathbf{k}}\, a_{\mathbf{k},+}^{+}\, a_{-\mathbf{k},-}^{+} \\ &= \sum_{\mathbf{k}_1 s_1 \mathbf{k}_2 s_2} \chi(\mathbf{k}_1 s_1 \mathbf{k}_2 s_2)\, a_{\mathbf{k}_1 s_1}^{+}\, a_{\mathbf{k}_2 s_2}^{+} \end{aligned} \tag{2.18}$$

[7] The discussion here follows Blatt (60); the same result had been obtained earlier by F. J. Dyson, by a different method (private communication to M. R. Schafroth), and independently by Bayman (60).

[8] This assumption is not essential; if $u_{\mathbf{k}} = 0$ and thus $v_{\mathbf{k}} = 1$, exactly, for $|\mathbf{k}| < k_0$, then there is a filled Fermi sea, with maximum momentum k_0, below the pairs; see Chapter IV, Section 1.

is, except for a constant factor, equal to the "pair" creation operator b^+ of the quasi-chemical equilibrium theory, Eq. (1.9), provided only that the pair wave function satisfies the "simple pairing" condition

$$\chi(\mathbf{k}_1 s_1 \mathbf{k}_2 s_2) = 0 \qquad \text{unless} \qquad \mathbf{k}_1 = -\mathbf{k}_2 \qquad \text{and} \qquad s_1 = -s_2 \tag{2.19}$$

Furthermore, as a result of the operator identity (2.5), the restriction $\mathbf{k} \neq \mathbf{m}$ in (2.17) is unnecessary and may be dropped. Thus, we obtain the identity

$$\Psi_{\text{BCS}} = \text{const}\left[1 + W^+ + \frac{(W^+)^2}{2!} + \cdots\right] |\, 0\rangle \tag{2.20}$$

The pth term in this series produces a wave function with exactly $2p$ electrons. Hence projection of (2.20) onto the N-particle space means retention of only one term in the series, namely, the term of power $p = \frac{1}{2}N$.

Thus *the projected BCS wave function is identically equal to the Schafroth-condensed pair wave function (1.11)*. The "deep mystery" is just the Schafroth condensation of electron pairs, which has been established on quite general grounds (see Chapter III, Section 9); the "simple pairing" condition (2.19) for the pair wave function is also based on general invariance arguments, see (III, 9.12) and the discussion there. Thus, the BCS wave function is a useful approximate computation device, to circumvent the rather elaborate computations needed with the wave function (1.11) (see Appendix B); the physical meaning of the BCS and quasi-chemical equilibrium wave functions is the same, but whereas this meaning is manifest from the start with the particle-conserving wave function (1.11), it is somewhat obscured by the approximate wave function (2.2).

There still remains the problem of what to do when the simple pairing condition (2.19) fails. Since the derivation of this condition, in Chapter III, Section 9 is based on translation and rotation invariance, Eq. (2.19) fails when these symmetry properties fail. This happens with annoying frequency: As soon as we introduce impurities, or a space-dependent magnetic field, or (even more simply) boundaries corresponding to the surface of the superconducting specimen, translation invariance is lost, and with it the property (2.19). Furthermore, even in the absence of a real magnetic field, it is possible to introduce a purely formal vector potential $\mathbf{A}$ with curl $\mathbf{A} = 0$ (and thus no real effect), by means of a gauge transformation; the condition (2.19) is not even invariant under this gauge transformation!

Fortunately, Zumino (62) has shown that the "simple pairing" property is not tied to special invariance requirements, but is actually a universal property of all pair wave functions. We first generalize the concept of "simple pairing" as follows: Let $u_m(q)$ be some complete orthonormal set of one-particle wave functions, and let φ_{kl} be the expansion coefficients (1.3) of the pair wave function $\varphi(q_1, q_2)$ with respect to this particular set. We then make the *definition*: The wave function φ has the simple pairing property with respect to the set $u_k(q)$ if and only if there exists a one-to-one correspondence $k \rightarrow \hat{k}$ between the indices k, such that $\hat{\hat{k}} = k$ (the correspondence is reflexive) and that $\varphi_{kl} = 0$ unless $l = \hat{k}$, for all k and l. This definition generalizes (2.19); the special correspondence in (2.19) is of course $\hat{k} = \overline{(\mathbf{k}, s)} = (-\mathbf{k}, -s)$.

The Zumino (62) theorem now reads: Given any antisymmetric pair wave function $\varphi(q_1, q_2) = -\varphi(q_2, q_1)$, there exists at least one set $u_k(q)$ of single-particle basis functions, such that $\varphi(q_1, q_2)$ has the simple pairing property with respect to this set. This theorem has also been found, independently, by Bloch (62).

Of course, the set in question is not always as simple as the plane wave functions which suffice if there is complete translation invariance. But at least we know that such a set exists, and sometimes it is even easy to find. For example, the purely formal vector potential $\mathbf{A}$ mentioned before, resulting from a gauge transformation of the zero-field problem, can be handled by performing this same gauge transformation on all the basis functions (the plane waves).

The Zumino theorem suffices to establish the complete equivalence between the BCS and quasi-chemical equilibrium theories for the ground state of the system. When we come to excited states (non-zero temperatures) this equivalence is not complete, for several reasons. One reason is the presence, at non-zero temperatures, of unpaired particles, such as the two "extra" particles in (2.10). It is then possible that the set of single-particle wave functions necessary to lead to simple pairing for the pair wave function, differs from the set of single-particle wave functions which are appropriate to describe the unpaired particles. This does not happen if there is complete translation invariance, but might happen in the absence of such invariance. In that case, the Zumino transformation would run into trouble, and with it the simple pairing. Fortunately, the relevant expectation values of operators in the quasi-chemical equilibrium theory can be evaluated even in the absence of simple pairing, i.e., for quite arbitrary pair wave functions φ_{kl} (Blatt 60a).

Whether one uses the general formulas for arbitrary pair wave functions, or employs Zumino's theorem to transform the pair wave function into simple pairing form, is purely a matter of convenience for the ground state. There is nothing basic about the simple pairing; the basic property which explains the "deep mystery" is the Schafroth condensation of pairs, which ensures that all the electron pairs occupy the *same* pair wave function.[9]

3. The Bogoliubov Transformation Applied to the Fröhlich Model

The most serious shortcoming of the theories discussed so far is the purely schematic treatment of the electron-electron interaction. We know (See Chapter I, Section 10) that the electron-phonon interaction is of major importance for superconductivity. Yet phonons do not appear explicitly in either the BCS or the quasi-chemical equilibrium theories.[10] Both these theories replace the Fröhlich interaction by an "equivalent" electron-electron interaction; although it is highly likely that this is indeed possible, one needs to know just what this equivalent interaction operator is. Bardeen (57) uses the Ansatz (2.9), but this Ansatz is derived by a perturbation calculation starting from the Fermi sea state of the electrons, whereas one should start from the Schafroth-condensed pair state (1.11). With a different starting point for the perturbation treatment, there is no reason to expect equivalent results.

The first real attack on this problem was made by Bogoliubov (58, 58a), who developed an entirely new and extremely elegant technique for this purpose. Rather than describing this technique in its full generality immediately, we shall do so in several stages. In the present section, we shall describe the Bogoliubov transformation and the Bogoliubov condition of "cancellation of dangerous terms," within the framework of ordinary perturbation theory (no renormalization). In Section 4, we shall establish the relationship between the Bogoliubov transformation and the BCS wave function (2.2), following the work of Valatin (58) and Yosida (58). Section 5 is devoted to the renormalized Bogoliubov theory.

In order to keep things simple, we shall ignore all interactions between electrons other than the Fröhlich interaction. This is of course incorrect, since the Coulomb repulsion, at the very least, should be included. However, this can be done in a crude fashion, at the end.

[9] Without this condensation, there would be many pair wave functions represented in the ground state, and the Zumino theorem could no longer be applied.

[10] It is possible to generalize the quasi-chemical equilibrium theory by allowing for a phonon cloud around each electron pair. The relevant operators have been written down by McKenna and Blatt (McKenna 62); but the evaluation of expectation values in this generalized theory becomes exceedingly difficult.

The Hamiltonian of the system electrons + phonons is

$$H = \sum_k \epsilon_k a_k^+ a_k + \sum_q \hbar\omega_q b_q^+ b_q + H_{\text{int}} \tag{3.1}$$

where $\epsilon_k = \hbar^2 k^2/2M$ is the electron energy in state k; a_k^+ and a_k are creation and destruction operators for an electron in state k, which satisfy the anticommutation rules (1.7), (1.13), and (1.18).

The energy of a phonon with wave number q is denoted by $\hbar\omega_q$. The b_q^+ and b_q are phonon creation and annihilation operators, respectively. They satisfy the commutation rules (1.28). The operator $b_q^+ b_q$ has eigenvalues 0, 1, 2, 3, $\cdots$, and denotes the number of phonons present in phonon state q.

Suppose an electron with momentum $\mathbf{k} + \mathbf{q}$ interacts with the solid lattice, resulting in a phonon with momentum $\mathbf{q}$ and an electron with momentum $\mathbf{k}$. The operator producing this same effect is

$$b_{\mathbf{q}}^+ \, a_{\mathbf{k}}^+ \, a_{\mathbf{k}+\mathbf{q}}$$

We shall assume that the electron spin is unaffected in this process, so that an up-spin electron remains with up-spin, a down-spin electron remains with down-spin. The operator is therefore

$$b_{\mathbf{q}}^+ (a_{\mathbf{k},+}^+ a_{\mathbf{k}+\mathbf{q},+} + a_{\mathbf{k},-}^+ a_{\mathbf{k}+\mathbf{q},-})$$

Besides emission of a phonon with momentum $\mathbf{q}$, the same electron transition can occur by means of absorption of a phonon with momentum $-\mathbf{q}$. We can allow for this by replacing $b_{\mathbf{q}}^+$ by the sum $b_{\mathbf{q}}^+ + b_{-\mathbf{q}}$. Finally, there is of course a factor in front of this operator. To a first approximation, this factor may be assumed to depend only on the phonon momentum $\mathbf{q}$, not on $\mathbf{k}$. It is conventional to write this factor in such a way that the interaction Hamiltonian H_{int} in (3.1) becomes

$$H_{\text{int}} = \sum_{\mathbf{k},\mathbf{q}} M(q) \left(\frac{\hbar\omega_{\mathbf{q}}}{2V}\right)^{1/2} (b_{\mathbf{q}}^+ + b_{-\mathbf{q}}) (a_{\mathbf{k},+}^+ a_{\mathbf{k}+\mathbf{q},+} + a_{-\mathbf{k}-\mathbf{q},-}^+ a_{-\mathbf{k},-}) \tag{3.2}$$

The quantity $M(q)$ is a q-dependent matrix element (it may be set equal to a constant for a first orientation); the factor $V^{-1/2}$ ensures the correct volume dependence of all final quantities; the $(\hbar\omega_{\mathbf{q}})^{1/2}$ serves to make $M(q)$ more nearly independent of q. Finally, the use of negative vector subscripts for electrons with down-spin is permissible (all these are dummy subscripts under summation signs) and turns out to be convenient for application of the Bogoliubov transformation later on.

The sum over $\mathbf{q}$ is cut off at the upper end by the Debye cutoff, i.e., phonon wave numbers q cannot be larger than $q_{\max}$, of the order of an inverse lattice distance. Formally, we may allow for this by setting $M(q) = 0$ for $q > q_{\max}$. However, this is only a formal device; strictly speaking, there are no modes of vibration of the lattice, and hence no operators $b_{\mathbf{q}}$ or $b_{\mathbf{q}}^{+}$, for $q > q_{\max}$.

The normal "Fermi sea" state Φ_F consists of occupied electron states $k < k_F$, and empty electron states $k > k_F$. It is usual, and convenient, to make a transformation to "hole theory" by introducing operators $\alpha_{\mathbf{k}}$ as follows:

$$\begin{aligned} \alpha_k &= a_k^+ , \qquad \alpha_k^+ = a_k \qquad \text{for} \quad k < k_F \\ &= a_k , \qquad\quad\; = a_k^+ \qquad \text{for} \quad k > k_F \end{aligned} \tag{3.3}$$

This has the result that

$$\alpha_k \Phi_F = 0 \qquad \text{all} \quad k \tag{3.4}$$

i.e., the Fermi sea state is the "vacuum" for the α_k operators. From the commutation rules (1.18) we have

$$\begin{aligned} a_k^+ a_k &= 1 - \alpha_k^+ \alpha_k \qquad \text{for} \quad k < k_F \\ &= \alpha_k^+ \alpha_k \qquad\quad\; \text{for} \quad k > k_F \end{aligned} \tag{3.5}$$

Besides the Fermi sea state itself, we may consider neighboring states, for example, the state obtained by lifting an electron from state k_0 inside the Fermi sea, to a state k_1 outside the Fermi sea. The excitation energy involved is

$$\epsilon_{k_1} - \epsilon_{k_0} = (\epsilon_{k_1} - \epsilon_F) + (\epsilon_F - \epsilon_{k_0}) \tag{3.6}$$

Formally, we may consider this as a sum of two separate excitation energies: the excitation energy necessary to create a hole in state k_0, namely, $\epsilon_F - \epsilon_{k_0}$, and the excitation energy necessary to create a particle in state k_1, namely, $\epsilon_{k_1} - \epsilon_F$. The Fermi energy ϵ_F was inserted here in order to make both these energies vanish for $k = k_F$; ϵ_F cancels out of the final result, of course.

Formally, the result (3.6) can be obtained from the free electron Hamiltonian by using (3.5) and considering, not H itself, but $H - \mu N$, where $\mu = \epsilon_F$ is the chemical potential:

$$H_{0,\text{electrons}} - \mu N = \sum_k (\epsilon_k - \mu)\, a_k^+ a_k = \sum_{k<k_F} (\epsilon_k - \mu) + \sum_k | \epsilon_k - \mu |\, \alpha_k^+ \alpha_k \tag{3.7}$$

The sum over $k < k_F$ is a constant, which equals $E - \mu N$ in the Fermi sea state itself. The other sum can be interpreted physically by saying that the elementary excitation described by the operator $\alpha_k^+\alpha_k$ has an excitation energy equal to $|\epsilon_k - \mu|$, i.e., positive for holes $(k < k_F)$ as well as for electrons $(k > k_F)$.

To a first approximation, these "elementary excitations" may be considered independent of each other; i.e., to a given hole in state k_0, say, we can associate an excited electron in a large number of states k_1, and the combined excitation energy is made up linearly of the excitation energy of the electron and the hole [see Eq. (3.6)]. There is only one condition which we must obey: to conserve the actual number of electrons, it is necessary to excite equally many electrons and holes. To this extent, then, the elementary excitations are not quite independent.

In the presence of the electron-phonon interaction (3.2), the Fermi sea state is unstable. It is therefore not an adequate starting point for an approximation method, and the transformation (3.3) which is based on the Fermi sea state as a starting point, is no longer appropriate. Bogoliubov and Valatin have generalized (3.3) as follows[11]:

$$\begin{aligned} \alpha_{\mathbf{k}0} &= u_k\, a_{\mathbf{k},+} - v_k\, a_{-\mathbf{k},-}^+; & \alpha_{\mathbf{k}0}^+ &= u_k\, a_{\mathbf{k},+}^+ - v_k\, a_{-\mathbf{k},-} \\ \alpha_{\mathbf{k}1} &= u_k\, a_{-\mathbf{k},-} + v_k\, a_{\mathbf{k},+}^+; & \alpha_{\mathbf{k}1}^+ &= u_k\, a_{-\mathbf{k},-}^+ + v_k\, a_{\mathbf{k},+} \end{aligned} \tag{3.8}$$

Here u_k and v_k are real numbers, yet to be determined, which satisfy the condition

$$u_k^2 + v_k^2 = 1 \tag{3.9}$$

As a result of this condition, the transformation (3.8) preserves the commutation rules, i.e., we have (s can take the values 0 and 1)

$$\alpha_{\mathbf{k}s}\, \alpha_{\mathbf{k}'s'}^+ + \alpha_{\mathbf{k}'s'}^+\, \alpha_{\mathbf{k}s} = \delta_{\mathbf{k}\mathbf{k}'}\, \delta_{ss'}; \qquad \alpha_{\mathbf{k}s}\, \alpha_{\mathbf{k}'s'} + \alpha_{\mathbf{k}'s'}\, \alpha_{\mathbf{k}s} = 0 \tag{3.10}$$

Let us show, first, that the Bogoliubov transformation (3.8) reduces to (3.3), or rather, to an equivalent transformation, for certain special values of the u_k and v_k, namely,

$$\begin{aligned} u_k^0 &= 0, & v_k^0 &= 1 & &\text{for} \quad k < k_F \\ u_k^0 &= 1, & v_k^0 &= 0 & &\text{for} \quad k > k_F \end{aligned} \tag{3.11}$$

[11] The different treatment of electrons with up-spin and down-spin is related to the fact that the electron pairs consist of electrons with opposite momenta and spins, i.e., an electron in state $\mathbf{k}, +$ is paired up with an electron in state $-\mathbf{k}, -$. The precise relationship between the Bogoliubov transformation and the electron pairing will be discussed in Section 4. The use of the notation u_k, v_k in (2.2) as well as in (3.8) is done with malice aforethought.

Substitution of (3.11) into (3.8) yields

$$\alpha_{\mathbf{k}0} = -a^{+}_{-\mathbf{k},-}, \qquad \alpha_{\mathbf{k}1} = a^{+}_{\mathbf{k},+} \qquad \text{for} \quad k < k_F \tag{3.12a}$$

$$\alpha_{\mathbf{k}0} = a_{\mathbf{k},+}, \qquad \alpha_{\mathbf{k}1} = a_{-\mathbf{k},-} \qquad \text{for} \quad k > k_F \tag{3.12b}$$

Except for signs, which do not matter in this case, (3.12) is the same as (3.3); that is, with the choice u_k^0 , v_k^0 in (3.8), the $\alpha_{\mathbf{k}s}$ are electron creation operators (hole annihilation operators) below the Fermi surface, and electron annihilation operators above the Fermi surface.

In order to write the Hamiltonian (3.1), (3.2) in terms of the new operators, we require the inverse transformation to (3.8). A straightforward calculation gives

$$a_{\mathbf{k},+} = u_k\alpha_{\mathbf{k}0} + v_k\alpha^{+}_{\mathbf{k}1}, \qquad a^{+}_{\mathbf{k},+} = u_k\alpha^{+}_{\mathbf{k}0} + v_k\alpha_{\mathbf{k}1} \tag{3.13a}$$

$$a_{-\mathbf{k},-} = u_k\alpha_{\mathbf{k}1} - v_k\alpha^{+}_{\mathbf{k}0}, \qquad a^{+}_{-\mathbf{k},-} = u_k\alpha^{+}_{\mathbf{k}1} - v_k\alpha_{\mathbf{k}0} \tag{3.13b}$$

Since the Bogoliubov transformation combines electron creation and annihilation operators for states k and $-k$ [see Eq. (3.8)], the condition that the total number of particles, N, remain constant becomes quite complicated. It is an essential part of the Bogoliubov theory that this condition, instead of being imposed exactly, is imposed only on the average; i.e., we relax the condition that the wave function describe an exactly constant number of particles N, and replace it by the weaker condition that

$$(\Psi, N_{\mathrm{Op}}\, \Psi) = N \tag{3.14}$$

for the permissible wave functions. Here N_{Op} is the number operator (1.23) which we write in the form

$$N_{\mathrm{Op}} = \sum_k a_k^{+} a_k = \sum_{\mathbf{k}} (a^{+}_{\mathbf{k},+}\, a_{\mathbf{k},+} + a^{+}_{-\mathbf{k},-}\, a_{-\mathbf{k},-}) \tag{3.15}$$

This step has a number of consequences: First, the use of creation and annihilation operators in the Bogoliubov theory (as well as in the BCS theory) is not merely a convenience, but is a necessity; second, in the physical interpretation of the theory, some care must be exercised; third, however, the theory becomes very much simpler to use than the quasi-chemical equilibrium theory (in which the number of particles is conserved).

The relaxation of the condition of constant N in the Bogoliubov theory must be distinguished from the use of a generating function for functions of N in statistical mechanics, see for example (III, 4.5). In the grand canonical ensemble of statistical mechanics, we deal with an ensemble of systems, with a distribution of particle numbers N, systems with N particles being weighted by a factor z^N. However, each separate system within that assembly has its own definite number of particles N, and that number N remains constant in time[12]; that is, the averaging over different values of N is done *incoherently* in the grand canonical ensemble of statistical mechanics. By contrast, the averaging over N in the BCS and Bogoliubov theories is a *coherent* average: any one wave function Ψ contains components with different particle numbers N, and these components are coherent, i.e., their phases are related.

Of course, the actual metal has a constant number of electrons, and the replacement of the condition $N =$ constant by (3.14) is not a basic physical assumption, but merely a convenient mathematical artifice, which must be (and can be) justified afterward, by showing that the correction terms are small and unimportant.

We now handle the condition (3.14) in the usual way, by a Lagrange multiplier μ; that is, we work with $H - \mu N_{\text{Op}}$ rather than with H itself, and adjust the chemical potential μ in the end so as to satisfy (3.14). When we write down the operator $H - \mu N_{\text{Op}}$ and use (3.13), the result assumes the form

$$H - \mu N_{\text{Op}} = C + H_0 + H' + H'' \tag{3.16a}$$

$$C = 2 \sum_{\mathbf{k}} (\epsilon_k - \mu)\, v_k^2 = \text{constant} \tag{3.16b}$$

$$H_0 = \sum_{\mathbf{k}} (\epsilon_{\mathbf{k}} - \mu)(u_k^2 - v_k^2)(\alpha_{\mathbf{k}0}^+ \alpha_{\mathbf{k}0} + \alpha_{\mathbf{k}1}^+ \alpha_{\mathbf{k}1}) + \sum_{\mathbf{q}} \hbar\omega_q b_q^+ b_q \tag{3.16c}$$

$$\begin{aligned} H' = \sum_{\mathbf{k},\mathbf{q}} M(q) \left(\frac{\hbar\omega_q}{2V}\right)^{1/2} (b_q^+ + b_{-q}) \Big[&(u_k u_{k+q} - v_k v_{k+q}) \\ &\times (\alpha_{\mathbf{k}0}^+ \alpha_{\mathbf{k}+\mathbf{q},0} + \alpha_{\mathbf{k}+\mathbf{q},1}^+ \alpha_{\mathbf{k}1}) \\ &+ (u_k v_{k+q} + v_k u_{k+q})(\alpha_{\mathbf{k}0}^+ \alpha_{\mathbf{k}+\mathbf{q},1}^+ + \alpha_{\mathbf{k}1} \alpha_{\mathbf{k}+\mathbf{q},0}) \Big] \end{aligned} \tag{3.16d}$$

$$H'' = 2 \sum_{\mathbf{k}} (\epsilon_k - \mu)\, u_k v_k (\alpha_{\mathbf{k}0}^+ \alpha_{\mathbf{k}1}^+ + \alpha_{\mathbf{k}1} \alpha_{\mathbf{k}0}) \tag{3.16e}$$

The constant C, Eq. (3.16b), is the analog of the sum over $k < k_F$ in (3.7). The operator H_0 is in the right form for a superposition of

---

[12] Sometimes one considers "open" system, connected not only to a constant temperature bath but also to a supply of particles. However, in principle one should start from a larger, closed system, which includes the "particle bath," and consider the "open" system as a subsystem.

independent excitations, provided we define the *Bogoliubov ground state* Φ_B by the conditions

$$\alpha_{\mathbf{k}0}\,\Phi_B = \alpha_{\mathbf{k}1}\,\Phi_B = b_{\mathbf{q}}\,\Phi_B = 0 \qquad (\text{all } \mathbf{k}, \text{ all } \mathbf{q}) \tag{3.17}$$

This is the analog of (3.4) in the Bogoliubov theory.

The operators H' and H'' are to be treated by means of perturbation theory, and if this can be done, then the method is consistent. However, it is not obvious at first sight that H' and H'' are at all small perturbations. On the contrary, the condition that $H' + H''$ is a small perturbation turns out to be a highly non-trivial condition, which determines the transformation coefficients u_k and v_k in (3.8).

To see this, consider the effect of the perturbation H'' on the Bogoliubov ground state Φ_{B}. By general first-order perturbation theory (Schiff 49), the first-order perturbation in the wave function Φ_B can be written as a sum over states Φ_s, say, such that H'' has non-zero matrix elements between Φ_B and Φ_s. A look at (3.16e) shows that the states Φ_s in question are obtained from Φ_B by application of the operator $\alpha^+_{\mathbf{k}0}\alpha^+_{\mathbf{k}1}$, i.e., there are two excitations labeled by $\mathbf{k}$, one with $s = 0$, the other with $s = 1$. The matrix element is

$$\langle \mathbf{k}0, \mathbf{k}1 \mid H'' \mid B\rangle = 2(\epsilon_k - \mu)\, u_k v_k$$

and the energy denominator is, from (3.16c),

$$E_s^0 = \langle \mathbf{k}0, \mathbf{k}1 \mid H_0 \mid \mathbf{k}0, \mathbf{k}1\rangle = 2(\epsilon_k - \mu)\,(u_k^2 - v_k^2)$$

The first-order perturbation in the wave function, due to H'', is

$$\Phi^{(1)} = -\sum_s \frac{\langle s \mid H'' \mid B\rangle}{E_s^0}\,\Phi_s \tag{3.18}$$

and this becomes

$$\Phi^{(1)} = -\sum_{\mathbf{k}} \frac{u_k v_k}{u_k^2 - v_k^2}\,\Phi_{\mathbf{k}0,\mathbf{k}1} \tag{3.19}$$

The coefficient is infinitely big when $u_k = v_k$, showing that the perturbation (3.19) is not small.[13]

However, H'' is not the only perturbation operator acting on the Bogoliubov ground state. We also have H', Eq. (3.16d). This operator

[13] In the Fermi sea state, the product $u_k v_k = 0$ for all k, and $u_k^2 - v_k^2$ is never zero [see Eq. (3.11)]. The quantity (3.19) then vanishes identically, showing that the Fermi sea is one possible solution of the Bogoliubov equation. This is not the only solution, however.

cannot lead to states of type (3.19) in first order; the first-order states admixed by H' are of type $\Phi_{\mathbf{q};\mathbf{k},0;\mathbf{k}+\mathbf{q},1}$, i.e., a phonon in state $\mathbf{q}$, an $s=0$ excitation in state $\mathbf{k}$, and an $s=1$ excitation in state $\mathbf{k}+\mathbf{q}$. But in second-order perturbation theory, the operator H' can lead from this state to the state $\Phi_{\mathbf{k}0;\mathbf{k}1}$. The relevant matrix elements are, from (3.16d),

$$\langle \mathbf{q}; \mathbf{k}0; \mathbf{k}+\mathbf{q}, 1 \mid H' \mid B\rangle = M(q)\left(\frac{\hbar\omega_{-q}}{2V}\right)^{1/2}(u_k v_{k+q} + v_k u_{k+q})$$

$$\langle \mathbf{k}0; \mathbf{k}1 \mid H' \mid \mathbf{q}; \mathbf{k}0; \mathbf{k}+\mathbf{q}, 1\rangle = M(-q)\left(\frac{\hbar\omega_{-q}}{2V}\right)^{1/2}(u_{k+q}u_k - v_{k+q}v_k)$$

The general formula for the wave function in second order perturbation theory[14] reduces here to

$$\Phi^{(2)} = \sum_{s,t} \frac{\langle s \mid H' \mid t\rangle \langle t \mid H' \mid B\rangle}{E_s^0 E_t^0} \tag{3.20}$$

In order to write the energy denominators more conveniently, we introduce the notation

$$\hat{\epsilon}_k = (\epsilon_k - \mu)(u_k^2 - v_k^2) \tag{3.21}$$

This is the energy of an "elementary excitation" in the Bogoliubov theory.[15] [With the Fermi sea choice (3.11), we have $\hat{\epsilon}_k = |\epsilon_k - \mu|$, as expected.] With this notation, the energy denominators are

$$E_s^0 = 2\hat{\epsilon}_k, \qquad E_t^0 = \hat{\epsilon}_k + \hat{\epsilon}_{k+q} + \hbar\omega_q$$

There are actually two different intermediate states t for the same state $\Phi_s = \Phi_{k0;\,k1}$. One of them has been listed; the other is of type $t' = (-\mathbf{q}; \mathbf{k}+\mathbf{q}, 0; \mathbf{k}, 1)$. Their contributions are equal and add. The result for $\Phi^{(2)}$ is then

$$\Phi^{(2)} = \sum_{\mathbf{k},\mathbf{q}} M^2(q)\frac{\hbar\omega_q}{V}\frac{(u_k u_{k+q} - v_k v_{k+q})(u_k v_{k+q} + v_k u_{k+q})}{2\hat{\epsilon}_{\mathbf{k}}(\hat{\epsilon}_{\mathbf{k}} + \hat{\epsilon}_{\mathbf{k}+\mathbf{q}} + \hbar\omega_{\mathbf{q}})}\Phi_{\mathbf{k}0,\mathbf{k}1}$$
$$+ \text{ terms involving states with 2 phonons present} \tag{3.22}$$

[14] Schiff (49), formula (25.14) on p. 152; note that $H'_{mm} = H'_{km} = 0$ here, i.e., the operator H' has no matrix elements connecting the states Φ_B and $\Phi_{\mathbf{k}0,\mathbf{k}1}$ with each other or with themselves.

[15] In order to keep the discussion simple, we have not used the renormalization technique of quantum field theory to determine $\hat{\epsilon}_k$, but have used straightforward perturbation theory throughout. The penalty is that we have missed the energy gap which actually exists in the Bogoliubov theory in its complete form. Expression (3.21) is only approximate, and is qualitatively wrong for $u_k = v_k$. The renormalized Bogoliubov theory is given in Section 5.

The *Bogoliubov compensation condition* now demands that the terms of Eq. (3.22) which we have written down explicitly, exactly cancel out the "dangerous" terms (3.19), for each value of **k**. If this condition is satisfied, the Bogoliubov ground state Φ_B is stable against small perturbations induced by the combination of H' and H'', and can be used as a starting state for a perturbation treatment.[16] The compensation condition is (we use $\mathbf{k}' = \mathbf{k} + \mathbf{q}$)

$$(\epsilon_k - \mu)\, u_k v_k = \frac{1}{2V} \sum_{\mathbf{k}'} \frac{M^2(\mathbf{k}' - \mathbf{k})\, \hbar\omega_{\mathbf{k}'-\mathbf{k}} (u_k u_{k'} - v_k v_{k'})\,(u_k v_{k'} + v_k u_{k'})}{\hat{\epsilon}_{\mathbf{k}} + \hat{\epsilon}_{\mathbf{k}'} + \hbar\omega_{\mathbf{k}'-\mathbf{k}}} \tag{3.23}$$

In order to simplify this rather horrible equation, we use the identity

$$(u_k u_{k'} - v_k v_{k'})\,(u_k v_{k'} + v_k u_{k'}) = (u_k^2 - v_k^2)\, u_{k'} v_{k'} + (u_{k'}^2 - v_{k'}^2)\, u_k v_k$$

and the definitions

$$L_k = -\frac{1}{4V} \sum_{k'} \frac{M^2(\mathbf{k}' - \mathbf{k})\, \hbar\omega_{\mathbf{k}'-\mathbf{k}}}{\hat{\epsilon}_{\mathbf{k}} + \hat{\epsilon}_{\mathbf{k}'} + \hbar\omega_{\mathbf{k}'-\mathbf{k}}} (u_{k'}^2 - v_{k'}^2) \tag{3.24}$$

and

$$\Delta_k = \frac{1}{V} \sum_{k'} \frac{M^2(\mathbf{k}' - \mathbf{k})\, \hbar\omega_{\mathbf{k}'-\mathbf{k}}}{\hat{\epsilon}_{\mathbf{k}} + \hat{\epsilon}_{\mathbf{k}'} + \hbar\omega_{\mathbf{k}'-\mathbf{k}}} u_{k'} v_{k'} \tag{3.25}$$

This leads to

$$(\epsilon_k - \mu + 2L_k)\, u_k v_k + \tfrac{1}{2} (v_k^2 - u_k^2)\, \Delta_k = 0 \tag{3.26}$$

We now make the identification, to be justified later on in Section 4,

$$\chi_k = \frac{v_k}{u_k} \tag{3.27}$$

and thus [see (IV, 1.7) and (IV, 1.10)]

$$v_k^2 = \frac{v_k^2}{u_k^2 + v_k^2} = g_k \tag{3.28}$$

$$u_k^2 = 1 - v_k^2 = 1 - g_k \tag{3.29}$$

$$u_k v_k = \frac{u_k v_k}{u_k^2 + v_k^2} = \psi_k \tag{3.30}$$

[16] We have written down the compensation condition to second order; if one wishes to go to a higher order of perturbation theory, the compensation condition must be modified suitably. This leads to minor alterations in the coefficients u_k, v_k, it is hoped of no importance.

If we divide (3.26) by u_k^2 and use (3.27), the equation becomes formally the same as (IV, 1.17). However, the equivalence is not perfect, because (3.24) and (3.25) are not, as they stand, equivalent to (IV, 1.15) and (IV, 1.16), no matter what we choose to be the matrix elements of v. The trouble comes from the energy denominators in (3.24) and (3.25), which involve $\hat{\epsilon}_k$, and thus, by (3.21), are themselves dependent on u_k and v_k. By contrast, the matrix elements of v which appear in (IV, 1.15) and (IV, 1.16), though arbitrary functions of their indices, are given functions, independent of χ_k, g_k, or ψ_k. Thus (3.26) is not strictly equivalent in structure to (IV, 1.17).

However, we may approximate to (3.24) and (3.25) in such a way that the approximate equations are equivalent to those of Chapter IV, Section 1. The sums over k' in (3.24) and (3.25) get contributions from a whole range of values of k'. Now, over most of that range, we expect u_k and v_k to be nearly equal to their "Fermi sea" values (3.11), and thus[17]

$$\hat{\epsilon}_k = (\epsilon_k - \mu)(u_k^2 - v_k^2) \cong | \epsilon_k - \epsilon_F | \tag{3.31}$$

When we make the approximation (3.31) in (3.24) and (3.25), the approximate equations have the same structure as the equations of Chapter IV. The solution obtained in Chapter IV, Section 2 is such that this approximation is permissible *a posteriori*, i.e., most of the contribution to the sums over k' in (3.24) and (3.25) turns out to come from regions of k' for which (3.31) is an excellent approximation.

The Bogoliubov compensation condition applied to the Fröhlich Hamiltonian therefore leads to the non-linear integral equation for Δ_k, Eq. (IV, 1.26), with the matrix elements of the "effective interaction" being (Tolmachev 58)

$$\langle k, -k \mid v_{\text{eff}} \mid k', -k' \rangle = -\frac{1}{V} \frac{M^2(\mathbf{k}-\mathbf{k}')\, \hbar\omega_{\mathbf{k}-\mathbf{k}'}}{| \epsilon_{\mathbf{k}} - \epsilon_F | + | \epsilon_{\mathbf{k}'} - \epsilon_F | + \hbar\omega_{\mathbf{k}-\mathbf{k}'}} \tag{3.32}$$

The most important aspect of (3.32) is its *sign*, indicating that the crucial terms of the Fröhlich interaction amount to an effective attraction between the electrons forming the pairs.[18]

---

[17] In the presence of an interaction, it is no longer true that $\mu = \hbar^2 k_F^2/2M$ where k_F is the usual Fermi wave number (determined from the condition that the Fermi sea, filled up to $k = k_F$, has the right number of particles in it). We continue to use the symbol ϵ_F for $\hbar^2 k_F^2/2M$. The distinction between μ and ϵ_F is beyond the accuracy of the crude approximations employed in this section. It requires the renormalization procedure for the energies ϵ_k of the elementary excitations, as well as for the phonon energies $\hbar\omega_q$, as carried out in Section 5. One result of this, more complicated, calculation is that $\hat{\epsilon}_k$ should be approximated here by $| \epsilon_k - \epsilon_F |$ rather than by $| \epsilon_k - \mu |$.

[18] We have not yet shown that the Bogoliubov transformation is equivalent to introducing Schafroth-condensed electron pairs; this will by done in Section 4, however.

The formula (3.32) is not free from doubt. It is, after all, based on a consideration of the low terms of a perturbation series in the phonon field. The *electron* wave function has been adjusted so that this series is at least not divergent; but there remains a question whether the lowest few *phonon* terms suffice. Physically, the use of low-order perturbation theory corresponds to a picture in which only one phonon at a time crosses between the two electrons making up the pair. If there are many phonons "crossing over" at any one time, then higher terms in the series would be required.

Quite apart from these questions, there remains the problem of including the Coulomb repulsion between electrons. This problem is far from trivial. The difficulty arises from the collective effects (plasma waves, Debye screening) discussed in Chapter I, Section 9 and the interaction of these effects with the electron-phonon matrix elements $M(q)$. As a crude, preliminary device, we may assume a simple screened Coulomb potential

$$v(r_{ij}) = \frac{e^2}{r_{ij}} \exp(-\gamma r_{ij}) \tag{3.33}$$

where γ is the inverse of the Debye screening distance, of the order of 10^8 cm^{-1}. When we add the relevant matrix element of (3.33) to (3.32), we obtain the crude "effective interaction" (we use the notation $\mathbf{q} = \mathbf{k} - \mathbf{k}'$)

$$\langle k, -k \mid v_{\text{eff}} \mid k', -k' \rangle = \frac{1}{V} \left\{ \frac{4\pi e^2}{q^2 + \gamma^2} - \frac{M^2(q)\, \hbar\omega_q}{\mid \epsilon_{\mathbf{k}} - \epsilon_F \mid + \mid \epsilon_{\mathbf{k}'} - \epsilon_F \mid + \hbar\omega_q} \right\} \tag{3.34}$$

As expected, the Coulomb repulsion term has positive sign, opposite to the sign of the term contributed by the Fröhlich interaction. Thus, a weak Fröhlich interaction (such as occurs in good ordinary conductors like copper and silver) could possibly be overcompensated by the Coulomb repulsion, so that no attractively coupled pairs can be formed. A strong Fröhlich interaction[19] is clearly favorable for superconductivity.

4. The Relationship between the Bogoliubov and BCS Ground States

So far, the Bogoliubov transformation (3.8) has appeared as a purely formal device for modifying the zero-order Hamiltonian, before per-

---

[19] I.e., a large $M^2(q)$ in Eq. (3.34).

forming perturbation theory on the electron-phonon coupling. In this section, we establish the physical meaning of this transformation, by investigating in more detail the nature of the Bogoliubov ground state Φ_B defined by (3.17). This was first done by Valatin (58), quite independently of Bogoliubov; here we shall follow the later work of Yosida (58).

We remarked in Section 1 that the anticommutation rules (1.7), (1.13), and (1.18) define the operators involved completely, up to a similarity transformation (1.19). Since the Bogoliubov operators α_k and α_k^+ satisfy the same anticommutation rules, such a correspondence must exist for them. Yosida (58) starts by exhibiting this similarity transformation explicitly. He writes:

$$\alpha_{\mathbf{k}0} = \exp(iS)\, a_{\mathbf{k},+} \exp(-iS); \qquad \alpha_{\mathbf{k}0}^+ = \exp(iS)\, a_{\mathbf{k},+}^+ \exp(-iS) \tag{4.1a}$$

$$\alpha_{\mathbf{k}1} = \exp(iS)\, a_{-\mathbf{k},-} \exp(-iS); \qquad \alpha_{\mathbf{k}1}^+ = \exp(iS)\, a_{-\mathbf{k},-}^+ \exp(-iS) \tag{4.1b}$$

We define $\beta_{\mathbf{k}}^+$ as before, (2.1), and we define an angle $\theta_{\mathbf{k}}$ by

$$u_{\mathbf{k}} = \cos\theta_{\mathbf{k}}, \qquad v_{\mathbf{k}} = \sin\theta_{\mathbf{k}} \tag{4.2}$$

Yosida then asserts that (4.1) is equivalent to (3.8) provided S is the operator:

$$S = -i \sum_{\mathbf{k}} \theta_{\mathbf{k}}(\beta_{\mathbf{k}}^+ - \beta_{\mathbf{k}}) \tag{4.3}$$

In order to prove that (4.1) and (4.3) are equivalent to the Bogoliubov transformation (3.8), we introduce the operators $s_{\mathbf{k}}$ by

$$s_{\mathbf{k}} = i(\beta_{\mathbf{k}} - \beta_{\mathbf{k}}^+) = i(a_{-\mathbf{k},-}a_{\mathbf{k},+} - a_{\mathbf{k},+}^+a_{-\mathbf{k},-}^+) \tag{4.4}$$

so that

$$S = \sum_{\mathbf{k}} \theta_{\mathbf{k}} s_{\mathbf{k}} \tag{4.5}$$

It follows from the commutation rules (2.3) and (2.7) that

$$[s_{\mathbf{k}}, s_{\mathbf{k}'}] = 0 \qquad \text{all} \quad \mathbf{k}, \mathbf{k}' \tag{4.6}$$

Thus, the first equation (4.1), for example, can be reduced to a form involving operators with indices $\mathbf{k}$ only:

$$\alpha_{\mathbf{k}0} = \exp(iS)\, a_{\mathbf{k},+} \exp(-iS) = \exp(i\theta_{\mathbf{k}} s_{\mathbf{k}})\, a_{\mathbf{k},+} \exp(-i\theta_{\mathbf{k}} s_{\mathbf{k}}) \tag{4.7}$$

Let us differentiate this equation with respect to $\theta_{\mathbf{k}}$. We also employ the commutator result (which is a direct consequence of the basic Fermi-Dirac anticommutation rules for the $a_{\mathbf{k},s}$ and $a_{\mathbf{k},s}^+$)

$$[s_{\mathbf{k}}, a_{\mathbf{k},+}]_- = ia_{-\mathbf{k},-}^+ \tag{4.8}$$

This gives

$$\begin{aligned}\frac{d\alpha_{\mathbf{k}0}}{d\theta_{\mathbf{k}}} &= i \exp(i\theta_{\mathbf{k}} s_{\mathbf{k}}) [s_{\mathbf{k}}, a_{\mathbf{k},+}] \exp(-i\theta_{\mathbf{k}} s_{\mathbf{k}}) \\ &= -\exp(i\theta_{\mathbf{k}} s_{\mathbf{k}})\, a^+_{-\mathbf{k},-} \exp(-i\theta_{\mathbf{k}} s_{\mathbf{k}}) = -\alpha^+_{\mathbf{k}1}\end{aligned} \tag{4.9}$$

A completely analogous derivation leads to the differential equation

$$\frac{d\alpha^+_{\mathbf{k}1}}{d\theta_{\mathbf{k}}} = \alpha_{\mathbf{k}0} \tag{4.10}$$

The coupled equations (4.9) and (4.10) must be solved subject to the initial conditions, at $\theta_{\mathbf{k}} = 0$,

$$\alpha_{\mathbf{k}0}(0) = a_{\mathbf{k},+}, \qquad \alpha^+_{\mathbf{k}1}(0) = a^+_{-\mathbf{k},-} \tag{4.11}$$

These initial conditions are trivial consequences of the second form of Eq. (4.7), and its analog for $\alpha^+_{\mathbf{k}1}$. The differential equations (4.9), (4.10) and initial conditions (4.11) define a unique solution, which is easily verified to be

$$\alpha_{\mathbf{k}0} = \cos(\theta_{\mathbf{k}})\, a_{\mathbf{k},+} - \sin(\theta_{\mathbf{k}})\, a^+_{-\mathbf{k},-} \tag{4.12a}$$

$$\alpha^+_{\mathbf{k}1} = \sin(\theta_{\mathbf{k}})\, a_{\mathbf{k},+} + \cos(\theta_{\mathbf{k}})\, a^+_{-\mathbf{k},-} \tag{4.12b}$$

With the identification (4.2), Eqs. (4.12) are in full agreement with the Bogoliubov transformation (3.8); the other two equations in (3.8) can be obtained by the same procedure.

In what follows, we shall also need an explicit expression for the operator $\exp(iS)$. Since the individual terms in the sum (4.5) commute with each other, we have the immediate result

$$\exp(iS) = \exp\Big(\sum_{\mathbf{k}} i\theta_{\mathbf{k}} s_{\mathbf{k}}\Big) = \prod_{\mathbf{k}} \exp(i\theta_{\mathbf{k}} s_{\mathbf{k}}) \tag{4.13}$$

Let us drop the index **k** henceforth, to save writing; when we compute $\exp(i\theta s)$ by expansion in a power series, there are tremendous simplifications, due to the fact that the square of s is a projection operator:

$$s^2 = P = \beta^+\beta + \beta\beta^+ = \begin{cases} 1 & \text{if states } (\mathbf{k},+) \text{ and } (-\mathbf{k},-) \text{ are either} \\ & \text{both empty or both occupied} \\ 0 & \text{otherwise} \end{cases} \tag{4.14}$$

Clearly, we have the projection operator identity

$$P^2 = P \tag{4.15}$$

and, since s is zero whenever P is zero, we also have

$$s = sP = Ps \tag{4.16}$$

and

$$s^3 = sP = s \tag{4.17}$$

We now use these relations to write

$$\begin{aligned}\exp(i\theta s) &= 1 + i\theta s/1! + (i\theta)^2 P/2! + (i\theta)^3 s/3! + \cdots \\ &= 1 - P + \cos(\theta) P + i \sin(\theta) s\end{aligned} \tag{4.18}$$

and thus, finally,

$$\exp(iS) = \prod_{\mathbf{k}} (1 - P_{\mathbf{k}} + u_{\mathbf{k}} P_{\mathbf{k}} + i v_{\mathbf{k}} s_{\mathbf{k}}) \tag{4.19}$$

The ground state of the system in the Bogoliubov theory is defined by the conditions (3.17). We assert that these conditions are satisfied by the state

$$\Phi_B = \exp(iS) \mid 0\rangle \tag{4.20}$$

where $\mid 0>$ is the conventional vacuum state (no electrons present, no phonons present). Equation (4.20) clearly satisfies the last condition (3.17), for phonons; as far as the electrons are concerned, we use (4.1) to write

$$\alpha_{\mathbf{k}0} \Phi_B = \exp(iS)\, a_{\mathbf{k},+} \exp(-iS) \exp(iS) \mid 0\rangle = \exp(iS)\, a_{\mathbf{k},+} \mid 0\rangle = 0$$

which is the first condition (3.17); the second condition follows the same way.

An explicit form for the operator $\exp(iS)$ has been derived above, Eq. (4.19). This expression simplifies considerably when applied to the vacuum state, since each $P_{\mathbf{k}}$ gives 1 on the vacuum state [see Eq. (4.14)], and $s_{\mathbf{k}}$, Eq. (4.4), is equivalent to $-i\beta_{\mathbf{k}}^{+}$. Thus we have

$$\Phi_B = \exp(iS) \mid 0\rangle = \prod_{\mathbf{k}} (1 - 1 + u_{\mathbf{k}} + v_{\mathbf{k}} \beta_{\mathbf{k}}^{+}) \mid 0\rangle = \Psi_{\mathrm{BCS}} \tag{4.21}$$

where the last equality follows from (2.2). *The ground states of the Bogoliubov and BCS theories are identical, provided u_k and v_k are the same.*

The condition for choosing u_k and v_k is at first sight entirely different in the BCS and Bogoliubov theories. The BCS u_k and v_k are chosen so as to get the minimum expectation value for the reduced Hamiltonian, whereas the Bogoliubov u_k and v_k are determined by the condition of "compensation of dangerous graphs," Eq. (3.23). However, Yosida (58) shows that the compensation condition of Bogoliubov ensures, automatically, that the Bogoliubov ground state leads to a minimum energy (the energy being computed by using second-order perturbation theory on the phonon coupling).

The BCS and Bogoliubov ground states have the same mathematical form, and both of them reduce to the ground state of the quasi-chemical equilibrium theory if one projects out the terms corresponding to some one definite number N of particles.[20]

Compared to the quasi-chemical equilibrium theory, both the BCS and Bogoliubov theories are much more elegant and easy to use, but harder to motivate and less general. Compared to each other, the BCS theory differs from the Bogoliubov theory in a number of ways. The two main differences are:

(1) The Bogoliubov theory, being based on a canonical transformation applied to the full Hamiltonian, not only includes the electron-phonon interaction explicitly, but also tells us clearly which terms are kept and which discarded, and allows an estimate of the contribution of the discarded terms.

(2) If all terms are retained, the Bogoliubov theory is gauge-invariant; the BCS reduced interaction is violently gauge-dependent.

5. Renormalization of the Bogoliubov Theory. Upper Limit on ρ

In Section 3, we deliberately oversimplified the Bogoliubov theory in order to emphasize the essentials. As a result, we failed to obtain the energy gap, as well as the upper limit on the coupling constant ρ, Eq. (IV, 2.8). In this section, we carry out the renormalization of the theory of Section 3, so as to obtain this additional information.

In one respect, however, we shall continue to oversimplify: the interaction will be taken purely as the electron-phonon interaction of Fröhlich, and the Coulomb repulsion between electrons will be ignored. The complications introduced by the Coulomb interaction are so enormous, and the theory becomes so involved, that it is not suitable for this book.

In the renormalized theory, there are two essential changes compared to Section 3:

(i) Not only are the electron operators transformed according to (3.8), but a similar transformation is applied to the phonon operators.

(ii) The energies of the elementary excitations, which appear in the

[20] The physical equivalence between the Bogoliubov transformation and the idea of a Bose-Einstein-like condensation of electron pairs was recognized immediately and stated clearly by Bogoliubov (58).

"zero-order" Hamiltonian (3.16c), are taken to be adjustable parameters which must be determined self-consistently at a later stage.

The transformation applied to the phonon operator is based on the following picture. A phonon may be "virtually absorbed," with the creation of a pair of Bogoliubov excitations[21]; these excitations may then recombine to restore the original phonon. The net effect of this process is a self-energy of the phonon, which should be included in the "zero-order" phonon energy. The transformation of the phonon operators is designed to replace "bare" phonons by "clothed" phonons.

Bogoliubov (58) writes this transformation as

$$b_{\mathbf{q}} = \lambda_q B_{\mathbf{q}} + \mu_q B^+_{-\mathbf{q}}, \qquad b^+_{\mathbf{q}} = \lambda_q B^+_{\mathbf{q}} + \mu_q B_{-\mathbf{q}} \tag{5.1}$$

where $B_{\mathbf{q}}$ and $B^+_{\mathbf{q}}$ are the destruction and creation operators, respectively, for the "clothed" phonons; and where λ_q and μ_q are real constants, yet to be determined, which satisfy the relation

$$\lambda_q^2 - \mu_q^2 = 1 \tag{5.2}$$

This relation ensures that the new operators $B_{\mathbf{q}}$, $B^+_{\mathbf{q}}$ satisfy the same commutation rules (1.28) as the "bare" phonon operators $b_{\mathbf{q}}$, $b^+_{\mathbf{q}}$.

The second change discussed above means that we specify the *form* of the "zero-order" Hamiltonian H_0, but leave the *coefficients* (renormalized energies) open for the time being. That is, we replace (3.16c) by

$$H_0 = \sum_{\mathbf{k}} \hat{\epsilon}_{\mathbf{k}} (\alpha^+_{\mathbf{k}0}\alpha_{\mathbf{k}0} + \alpha^+_{\mathbf{k}1}\alpha_{\mathbf{k}1}) + \sum_{\mathbf{q}} \hbar\hat{\omega}_{\mathbf{q}} B^+_{\mathbf{q}} B_{\mathbf{q}} \tag{5.3}$$

where $\hat{\epsilon}_k$ and $\hat{\omega}_q$ are adjustable parameters, to be determined self-consistently at a later stage.

When we carry out the combined transformations (3.13) and (5.1) on the Hamiltonian (3.1), (3.2), and separate out the contribution (5.3), we obtain an expression similar to (3.16), but more complicated in detail, namely,

$$H - \mu N = C + H_0 + H' + H'' \tag{5.4a}$$

$$C = 2\sum_{\mathbf{k}} (\epsilon_k - \mu)\, v_k^2 + \sum_{\mathbf{q}} \hbar\omega_q \mu_q^2 \tag{5.4b}$$

$$\begin{aligned} H' = \sum_{\mathbf{k},\mathbf{q}} M(\mathbf{q}) \left(\frac{\hbar\omega_q}{2V}\right)^{1/2} (\lambda_q + \mu_q)\,(B^+_{\mathbf{q}} + B_{-\mathbf{q}}) \\ \Big[(u_k u_{k+q} - v_k v_{k+q})\,(\alpha^+_{\mathbf{k}0}\alpha_{\mathbf{k}+\mathbf{q},0} + \alpha^+_{\mathbf{k}+\mathbf{q},1}\alpha_{\mathbf{k}1}) \\ + (u_k v_{k+q} + v_k u_{k+q})\,(\alpha^+_{\mathbf{k}0}\alpha^+_{\mathbf{k}+\mathbf{q},1} + \alpha^+_{\mathbf{k}1}\alpha^+_{\mathbf{k}+\mathbf{q},0})\Big] \end{aligned} \tag{5.4c}$$

[21] The operator for this process is the $b_{-\mathbf{q}}\alpha^+_{\mathbf{k}0}\alpha^+_{\mathbf{k}+\mathbf{q},1}$ which appears in (3.16d); the Hermitean conjugate of this operator restores the phonon.

$$H'' = 2 \sum_{\mathbf{k}} (\epsilon_k - \mu) u_k v_k (\alpha_{\mathbf{k}0}^+ \alpha_{\mathbf{k}1}^+ + \alpha_{\mathbf{k}1} \alpha_{\mathbf{k}0})$$

$$+ \sum_{\mathbf{q}} \hbar \omega_q \lambda_q \mu_q (B_{\mathbf{q}}^+ B_{-\mathbf{q}}^+ + B_{-\mathbf{q}} B_{\mathbf{q}})$$

$$+ \sum_{\mathbf{k}} [(\epsilon_k - \mu)(u_k - v_k) - \hat{\epsilon}_k] (\alpha_{\mathbf{k}0}^+ \alpha_{\mathbf{k}0} + \alpha_{\mathbf{k}1}^+ \alpha_{\mathbf{k}1})$$

$$+ \sum_{\mathbf{q}} [\hbar \omega_q (\lambda_q^2 + \mu_q^2) - \hbar \hat{\omega}_q] B_{\mathbf{q}}^+ B_{\mathbf{q}} \qquad (5.4\text{d})$$

The last two sums in (5.4d) involve subtraction of the "zero-order" Hamiltonian (5.3); the other expressions in (5.4) are generally similar to corresponding expressions in (3.16).

The Bogoliubov ground state is defined by conditions analogous to (3.17), namely,

$$\alpha_{\mathbf{k}0} \Phi_B = \alpha_{\mathbf{k}1} \Phi_B = B_{\mathbf{q}} \Phi_B = 0 \qquad (\text{all } \mathbf{k}, \text{ all } \mathbf{q}) \qquad (5.5)$$

The only difference between (3.17) and (5.5) is the replacement of the "bare" phonon operators $b_{\mathbf{q}}$ by the "clothed" phonon operators $B_{\mathbf{q}}$.

We then impose two types of conditions:

(i) The "compensation" conditions: The "effective" interaction operator must not lead from state Φ_B to either one of the states $\alpha_{\mathbf{k}0}^+ \alpha_{\mathbf{k}1}^+ \Phi_B$ and $B_{\mathbf{q}}^+ B_{-\mathbf{q}}^+ \Phi_B$.

(ii) The self-consistency conditions. The excitation energies of states $\alpha_{\mathbf{k}0}^+ \Phi_B$, $\alpha_{\mathbf{k}1}^+ \Phi_B$, and $B_{\mathbf{q}}^+ \Phi_B$ must be given (in lowest order of perturbation theory) by $\hat{\epsilon}_k$, $\hat{\epsilon}_k$, and $\hbar \hat{\omega}_q$, respectively. That is, they must be given by H_0 alone, and the "effective" interaction operator must yield zero excitation energies for these three types of states.

The effective interaction operator in either case arises from H'' in first-order perturbation theory, and H' in second-order perturbation theory (see the discussion in Section 3). Using an obvious symbolic notation, the effective operator for the compensation conditions is

$$W_1 = H'' - H' \frac{1}{H_0} H' \qquad (5.6)$$

so that the compensation conditions become

$$\langle \mathbf{k}0, \mathbf{k}1 \mid W_1 \mid B \rangle = \langle B \mid \alpha_{\mathbf{k}1} \alpha_{\mathbf{k}0} W_1 \mid B \rangle = 0 \qquad (5.7)$$

$$\langle \mathbf{q}, -\mathbf{q} \mid W_1 \mid B \rangle = \langle B \mid B_{-\mathbf{q}} B_{\mathbf{q}} W_1 \mid B \rangle = 0 \qquad (5.8)$$

We shall also need the expectation value of W_1 over the Bogoliubov ground state. The operator H'' has vanishing expectation value in that state, and the other term gives

$$\langle B \mid W_1 \mid B \rangle = -\left\langle B \left| H' \frac{1}{H_0} H' \right| B \right\rangle \tag{5.9}$$

Let us now consider the self-consistency condition for the state $\alpha_{\mathbf{k}0}^{+}\Phi_B$. The effective interaction operator for this state differs from (5.6) only through the energy denominator, which is $(H_0 - \hat{\epsilon}_k)$, rather than $(H_0 - 0) = H_0$ itself. Thus the effective energy of the state $\alpha_{\mathbf{k}0}^{+}\Phi_B$ in the same (lowest) order of perturbation theory is

$$\left\langle \mathbf{k}0 \left| H'' - H' \frac{1}{H_0 - \hat{\epsilon}_k} H' \right| \mathbf{k}0 \right\rangle = \left\langle B \left| \alpha_{\mathbf{k}0}(H'' - H' \frac{1}{H_0 - \hat{\epsilon}_k} H')\, \alpha_{\mathbf{k}0}^{+} \right| B \right\rangle \tag{5.10}$$

The self-consistency condition demands that the entire excitation energy of the state $\alpha_{\mathbf{k}0}^{+}\Phi_B$ be contributed by the zero-order Hamiltonian H_0, i.e., this excitation energy should be just $\hat{\epsilon}_k$. Consequently, the difference of (5.10) and (5.9) must vanish, and we obtain the condition

$$\langle B \mid \alpha_{\mathbf{k}0} H'' \alpha_{\mathbf{k}0} \mid B \rangle - \left\langle B \left| \alpha_{\mathbf{k}0} H' \frac{1}{H_0 - \hat{\epsilon}_k} H' \alpha_{\mathbf{k}0}^{+} \right| B \right\rangle + \left\langle B \left| H' \frac{1}{H_0} H' \right| B \right\rangle = 0 \tag{5.11}$$

As we shall see shortly, this is an implicit equation for $\hat{\epsilon}_k$.

By a precisely analogous argument, applied to the state $\alpha_{\mathbf{k}1}^{+}\Phi_B$, we get an equation which differs from (5.11) only in the replacement of $\alpha_{\mathbf{k}0}$ by $\alpha_{\mathbf{k}1}$. This equation turns out to give nothing new; this is not surprising, since a change of subscript from $s = 0$ to $s = 1$ in the Bogoliubov operators amounts to nothing more than flipping the spin of the excitation.

Finally, the self-consistency condition for the state $B_{\mathbf{q}}^{+}\Phi_B$, treated in the same way, gives rise to

$$\langle B \mid B_{\mathbf{q}} H'' B_{\mathbf{q}}^{+} \mid B \rangle - \left\langle B \left| B_{\mathbf{q}} H' \frac{1}{H_0 - \hbar\hat{\omega}_q} H' B_{\mathbf{q}}^{+} \right| B \right\rangle + \left\langle B \left| H' \frac{1}{H_0} H' \right| B \right\rangle = 0 \tag{5.12}$$

Equations (5.7), (5.8), (5.11), and (5.12) are the basic equations in the renormalized theory. All involve expectation values over the Bogoliubov ground state, and all these "vacuum" expectation values can be evaluated by standard methods, using the basic commutation and anticommutation rules. We merely write down the results. They are

$$(\epsilon_k - \mu)\, u_k v_k = \frac{1}{2V} \sum_{\mathbf{k}'} \frac{\hbar\omega_q M^2(q)\,(\lambda_q + \mu_q)^2}{\hat{\epsilon}_k + \hat{\epsilon}_{k'} + \hbar\,\hat{\omega}_q} (u_k u_{k'} - v_k v_{k'})\,(u_k v_{k'} + v_k u_{k'}) \tag{5.13}$$

where we have used the notation $\mathbf{q} = \mathbf{k} - \mathbf{k}'$; Eq. (5.13) is derived from (5.7). Equation (5.8) yields

$$\hbar\omega_q \lambda_q \mu_q = \frac{\hbar\omega_q M^2(q)\,(\lambda_q + \mu_q)^2}{2V} \sum_{\mathbf{k}-\mathbf{k}'=\mathbf{q}} \frac{(u_k v_{k'} + v_k u_{k'})^2}{\hat{\epsilon}_k + \hat{\epsilon}_{k'} + \hbar\,\hat{\omega}_q} \tag{5.14}$$

Eq. (5.11) gives rise to

$$\begin{aligned}&(\epsilon_k - \mu)\,(u_k^2 - v_k^2) - \hat{\epsilon}_k \\ &= \frac{1}{2V} \sum_{\mathbf{k}'} \hbar\omega_q M^2(q)\,(\lambda_q + \mu_q)^2 \left\{ \frac{(u_k u_{k'} - v_k v_{k'})^2}{\hat{\epsilon}_{k'} + \hbar\,\hat{\omega}_q - \hat{\epsilon}_k} - \frac{(u_k v_{k'} + v_k u_{k'})^2}{\hat{\epsilon}_{k'} + \hbar\,\hat{\omega}_q + \hat{\epsilon}_k} \right\}\end{aligned} \tag{5.15}$$

and, finally, Eq. (5.12) becomes

$$\begin{aligned}\hbar\omega_q(\lambda_q^2 + \mu_q^2) - \hbar\,\hat{\omega}_q &= \frac{\hbar\omega_q M^2(q)\,(\lambda_q + \mu_q)^2}{2V} \sum_{\mathbf{k}-\mathbf{k}'=\mathbf{q}} (u_k v_{k'} + v_k u_{k'})^2 \\ &\quad \times \left\{ \frac{1}{\hat{\epsilon}_k + \hat{\epsilon}_{k'} + \hbar\,\hat{\omega}_q} + \frac{1}{\hat{\epsilon}_k + \hat{\epsilon}_{k'} - \hbar\,\hat{\omega}_q} \right\}\end{aligned} \tag{5.16}$$

The "compensation" condition (5.13) is very similar to (3.23) in the simpler, unrenormalized theory. Equation (3.23) can be obtained from (5.13) by setting $\lambda_q = 1$, $\mu_q = 0$, and ignoring the difference between ω_q and $\hat{\omega}_q$.

The reduction of the set of equations (5.13)—(5.16) to manageable form is a formidable task, which would take too much space here. The two essential results are the limitation on the size of ρ, and the existence of the energy gap. Since we have already shown the latter using the BCS theory, we shall here restrict ourselves to the former. Fortunately, the upper limit on ρ can be deduced from just one of the above equations, namely, (5.14).

We introduce the notation

$$S(q) = \frac{M^2(q)}{2V} \sum_{\mathbf{k}-\mathbf{k}'=\mathbf{q}} \frac{(u_k v_{k'} + v_k u_{k'})^2}{\hat{\epsilon}_k + \hat{\epsilon}_{k'} + \hbar\,\hat{\omega}_q} \tag{5.17}$$

so that (5.14) becomes

$$\lambda_q \mu_q = (\lambda_q + \mu_q)^2\, S(q) \tag{5.18}$$

The quantities λ_q and μ_q are related to each other by (5.2). We introduce the quantity $z = (\lambda_q + \mu_q)^2$ and write, using (5.2),

$$4\lambda_q\mu_q = (\lambda_q + \mu_q)^2 - (\lambda_q - \mu_q)^2 = z - \frac{1}{z}$$

When we substitute this into (5.18), we obtain a quadratic equation for z, with the unique solution (the negative root is impossible, from the definition of z)

$$z = (\lambda_q + \mu_q)^2 = \frac{1}{[1 - 4S(q)]^{1/2}} \tag{5.19}$$

This first result of the renormalized theory becomes meaningless if $S(q)$ exceeds 1/4. Thus, we have as a *necessary condition* on consistency of the theory

$$S(q) < \tfrac{1}{4} \qquad \text{all } q \tag{5.20}$$

It turns out this is the most stringent condition on the coupling constant, i.e., if (5.20) is satisfied, the other equations (5.13)—(5.16) can be made consistent; in particular, the same condition (5.20) ensures that the renormalized frequency $\hat{\omega}_q$, as determined from (5.16), is real. An imaginary frequency would indicate breakdown of the assumed lattice structure of the solid.

We now evaluate (5.17) approximately. Since the renormalized excitation energies $\hat{\epsilon}_k$ and $\hbar\hat{\omega}_q$ are positive quantities, (5.17) is a sum of positive terms, and it is thus permissible to make fairly sweeping approximations. In particular, we are allowed to replace u_k and v_k by their "Fermi sea" values u_k^0 and v_k^0, Eq. (3.11). This is permissible because most of the contribution comes from energies which differ from the Fermi energy by more than Δ. This procedure yields

$$S(q) \cong \frac{M^2(q)}{V} \sum_{\mathbf{k}-\mathbf{k}'=\mathbf{q}} \frac{u_k^0 v_{k'}^0}{\hat{\epsilon}_k + \hat{\epsilon}_{k'} + \hbar\hat{\omega}_q} \tag{5.21}$$

Here we have used the fact that $v_k^0 u_{k'}^0$ vanishes whenever $u_k^0 v_{k'}^0$ is nonzero, and that both quantities assume the values 0 and 1 only.

Next, we simplify still further: since the main contribution to (5.21) comes from energies well away from the Fermi surface, we can ignore the phonon energy $\hbar\hat{\omega}_q$ in the energy denominator, and can replace $\hat{\epsilon}_k$ by $|\epsilon_k - \epsilon_F|$. In (5.21), k is always outside, and k' always inside, the Fermi sphere; the ϵ_F drops out (as it must), and we obtain

$$S(q) \cong \frac{M^2(q)}{V} \sum_{\substack{k>k_F \\ k'<k_F \\ \mathbf{k}-\mathbf{k}'=\mathbf{q}}} \frac{1}{\epsilon_k - \epsilon_{k'}} \tag{5.22}$$

We now evaluate the sum (5.22). We replace the sum by an integral in the usual fashion. In the integral over the three-dimensional **k**-space, we introduce elliptic coordinates as follows:

$$x = \frac{|\mathbf{k}| + |\mathbf{k}'|}{q}, \qquad y = \frac{|\mathbf{k}| - |\mathbf{k}'|}{q} \tag{5.23}$$

The third coordinate is the polar angle φ about the direction of the vector $\mathbf{q}$ as an axis. The volume element for this coordinate system is

$$d^3k = (\tfrac{1}{2}q)^3 (x^2 - y^2)\, dx\, dy\, d\varphi \tag{5.24}$$

With the notation

$$X = \frac{2k_F}{q} \tag{5.25}$$

the range of integration which corresponds to the restrictions on the sum (5.22) is given by

$$0 < y < 1, \qquad X - y < x < X + y \tag{5.26}$$

Finally, we note the identity

$$\epsilon_k - \epsilon_{k'} = \frac{\hbar^2}{2M} q^2 xy \tag{5.27}$$

We then obtain

$$\begin{aligned} S(q) &= \left(\frac{M(q)}{4\pi\hbar}\right)^2 qM \int_0^1 dy \int_{X-y}^{X+y} dx\, \frac{x^2 - y^2}{xy} \\ &= \frac{\rho_E}{4V} M^2(q) \left[1 + \frac{X^2 - 1}{2X} \ln \frac{X+1}{X-1}\right] = \frac{\rho_E}{4V} M^2(q)\, F(X) \end{aligned} \tag{5.28}$$

where ρ_E is the density of states at the Fermi surface, Eq. (IV, 2.6), and $F(X)$ is defined by the last equation in (5.28). The inequality (5.20) then becomes

$$\frac{\rho_E}{V} M^2(q) < \frac{1}{F(X)} = \frac{1}{F(2k_F/q)} \qquad \text{all } q \tag{5.29}$$

The function $F(X)$ is a very slowly varying function of its argument. We have $F(\infty) = 2$ and $F(2) = 1.82$; the value $X = 2$ corresponds to $q = k_F$ and is practically the smallest value which X can assume (the largest value which a phonon momentum q can assume); for $X < 2$ the region of integration is not given by (5.26), there being the additional condition $x > 1$, which is then no longer satisfied automatically.

The region of k-space in which both k and k' are close to the Fermi momentum k_F corresponds to $x = X$, $y = 0$, according to (5.23) and (5.25). We see from (5.28) that this gives a very small contribution to the integral for $S(q)$, thereby justifying *a posteriori* our various approximations.

The Bogoliubov parameter ρ, Eq. (IV, 2.8), is a measure of the average strength of the effective interaction, the average being taken over the region which makes the dominant contribution to the integral (IV, 1.26) for Δ_k. We shall now show that, with the interaction of the Bogoliubov form (3.32), the strength parameter ρ is equal to an appropriate average of the very quantity which is limited by (5.29).

To see this, we note that the main contributing region for the sum (IV, 1.26) is $|\xi_k| = |\epsilon_k - \epsilon_F| < \Delta$, i.e., the immediate neighborhood of the Fermi surface. Since Δ is much less than typical phonon energies, by the exponential factor $\exp(-1/\rho)$, we can ignore both $|\epsilon_k - \epsilon_F|$ and $|\epsilon_{k'} - \epsilon_F|$ in (3.32), compared to $\hbar\omega_{\mathbf{q}} = \hbar\omega_{\mathbf{k}-\mathbf{k}'}$. [This is exactly the opposite approximation to the one used for the evaluation of $S(q)$.] We then obtain, with $\mathbf{q} = \mathbf{k} - \mathbf{k}'$

$$\langle k, -k \mid v_{\text{eff}} \mid k', -k' \rangle \cong -\frac{M^2(q)}{V} \tag{5.30}$$

The parameter ρ is a suitable average of (5.30), multiplied by the density of states ρ_E. Thus, it is a suitable average of the quantity on the left side of the inequality (5.29).

Since $F(X)$ is such a slowly varying function, the detailed nature of the average does not matter much; replacing $F(X)$ by 2 throughout, we then obtain the condition

$$\rho < \tfrac{1}{2} \tag{5.31}$$

This is the desired result, which is in excellent agreement with experiment.

Let us make some comments on this derivation. First of all, the limitation (5.29) can be derived, and was in fact derived much earlier by Kohn (51) and Wentzel (51), by starting from the Fermi sea state rather than the electron-pair ground state of the Bogoliubov theory. However, such a derivation would not be conclusive for superconductors. The Bogoliubov derivation of (5.31) starts from the superconducting ground state, and is therefore free from this objection.

A more subtle point concerns the quantity $M^2(q)$ which appears in (5.30). This should really be replaced by the square of the "renormalized" electron-phonon matrix element. Bogoliubov (58) shows that the equations of this section lead naturally to the definition

$$\hat{M}^2(q) = \frac{M^2(q)}{1 - 4S(q)} \tag{5.32}$$

It is this quantity, rather than the "bare" $M^2(q)$, which enters into (5.30). If the renormalization correction $1 - 4S(q)$ is close to 1, this makes little difference. However, the upper limit (5.31) just corresponds to the case where $1 - 4S(q)$ vanishes for some value of q (for $q = 0$). It is not clear at this time just how the estimate (5.31) is affected by this.[22]

Next, consider the nature of the "suitable average" involved in the calculation of the strength parameter ρ. We must average the effective interaction (5.30) over all *directions* of the vectors **k** and **k**′, both their magnitudes being equal to k_F. Since (5.30) depends only on $\mathbf{q} = \mathbf{k} - \mathbf{k}'$, we can drop the average over directions of **k**, assuming that **k** always points in the z-direction. The magnitude of the vector **q** is then given by

$$q = 2k_F \sin(\theta/2) \tag{5.33}$$

where θ is the angle between **k** and **k**′. We thus have

$$\rho = \frac{\rho_E}{V} \frac{1}{2} \int_0^\pi \hat{M}^2(q) \sin\theta \, d\theta \tag{5.34}$$

where q is determined as a function of θ by (5.33).

The matrix element $\hat{M}^2(q)$ is not known very accurately; but all we need to know here is that it is a slowly varying function of q. Under these conditions, the main contribution to the integral (5.34) comes from values of q of the order of magnitude k_F, i.e., *the main contributing phonons have exceedingly high wave numbers.* The Debye approximation is really not adequate, therefore, in the theory of superconductivity, and one should consider the spectrum of lattice waves in much more detail. Accurately, the Einstein approximation could well be more accurate.

Qualitatively, the importance of phonons with high q is related to the fact that the Bogoliubov theory stops with terms corresponding to the exchange of *one* phonon at a time, between the two electrons making up the pair. If processes involving many phonons at a time are important, then lower phonon momenta could become significant.

Still, it is unlikely that really low values of q, of the order of the ones measured in determinations of the velocity of sound, matter for the theory of superconductivity. Fortunately, the same high values of q which matter for superconductivity, also matter for ordinary conductivity in the normal state of the metal. There, small deflections of the electron momentum

[22] The situation here is complicated by the nature of renormalization theory. Once the renormalization factors become really significant, one should not work with "bare" quantities at all. Presumably, one obtains a better approximation by using the renormalized $\hat{M}(q)$ in other equations as well; no detailed investigation is available in this region.

do not contribute to the resistance, since a small deflection does not alter the current significantly. Thus, the matrix element $\hat{M}(q)$ which appears in the theory of superconductivity, can be estimated sensibly by looking at experimental measurements of the resistivity of the metal in its normal state at room temperature.[23] We thus recover Fröhlich's rule: high normal resistivity favors superconductivity.

The most serious deficiency of the theory of this section is the complete omission of the Coulomb repulsion. The upper limit (5.29) is concerned with the phonons only. But the effective electron-electron interaction should not be taken from (3.32) [which gives rise to (5.30)] but rather from (3.34). Thus the parameter ρ is the average of a quantity other than the quantity on the left of the inequality (5.29).

However, the *direction* of this effect is such as to *strengthen* the inequality (5.31): The Coulomb correction works in such a way as to *decrease* ρ, and this is all we need to ensure the continued validity of (5.31). Actually, it is more surprising that metals exist in which ρ is not much below $\frac{1}{2}$, than that there are no metals with ρ in excess of $\frac{1}{2}$.

This is as far as we shall carry the renormalized Bogoliubov theory in this book. The other main result of Eqs. (5.13)–(5.16) is the energy gap in the excitation spectrum $\hat{\epsilon}_k$. Bogoliubov (58) shows, by a series of transformations and approximations, that $\hat{\epsilon}_k$ is given by

$$\hat{\epsilon}_k \cong [(\hat{\epsilon}_{n,k})^2 + (\Delta_k)^2]^{1/2} \tag{5.35}$$

where $\hat{\epsilon}_{n,k}$ is the renormalized excitation energy in the normal (Fermi sea) state. $\hat{\epsilon}_{n,k}$ differs from $|\epsilon_k - \epsilon_F|$ only by an insignificant factor [see IV, 1.22) and the footnote there.] Since we have already obtained (5.35) by a simpler method [see Eq. (2.14)], we shall not carry through these transformations here, but refer the reader to Bogoliubov's original work, and also to Moskalenko (58) and Zubarev (58a, 60).

6. The Criterion for Superconductivity

The zero-temperature theory which has been our concern so far cannot predict the transition temperature for onset of superconductivity. This, however, is not necessary to decide whether a given metal becomes superconducting at all. We can answer this latter question, in principle,

[23] The normal resistance at *low* temperatures is determined by impurity scattering, not by electron-phonon scattering.

by comparing the ground state energy computed with the condensed pair wave function (1.11), with the ground state energy obtained from a Fermi sea wave function. If the pairing wave function gives lower energy per particle, then the metal becomes superconducting at some, sufficiently low, temperature.

The first criterion for superconductivity, like so many other firsts in this field, is due to Fröhlich (50). In terms of the "average" ρ of Eq. (5.34), Fröhlich's criterion is $\rho > 1$. Kohn (51) and Wentzel (51) quickly called attention to the fact that such a large ρ is inconsistent with stability of the assumed lattice structure, i.e., condition (5.31); and modern data bear out this criticism. Nevertheless, Fröhlich's criterion represents a contribution of first-rate importance: Fröhlich was the first to point out the importance of the quantity ρ, for this purpose; and he was the first to call attention to the strong correlation between superconductivity and poor normal conductivity.

In order to obtain a quantitative criterion from the present microscopic theory, we need two things, neither of which we possess in fact:

(1) We must know the matrix elements $\langle k, -k \mid v_{\text{eff}} \mid k', -k' \rangle$ in the integral equation (IV, 1.26).

(2) we must be able to solve this integral equation.

In the absence of a complete mathematical theory of non-linear integral equations of the type (IV, 1.26), rather crude methods have been employed in the past.

The BCS criterion for superconductivity (Bardeen 57) is derived as follows: first, the effective interaction is taken as a superposition of a screened Coulomb repulsion and the Bardeen-Pines (Bardeen 55a) effective electron-electron interaction due to electron-phonon coupling, i.e., a superposition of (3.33) and (2.9). Then, we determine the average reduced interaction for two electrons k and k' both on the Fermi surface, i.e., $|\mathbf{k}| = |\mathbf{k}'| = k_F$. This average reduced interaction is defined by

$$v_{\text{av}} = \int \frac{d\Omega}{4\pi} \int \frac{d\Omega'}{4\pi} \langle k, -k \mid v_{\text{eff}} \mid k', -k' \rangle, \qquad |\mathbf{k}| = |\mathbf{k}'| = k_F \tag{6.1}$$

The BCS criterion for superconductivity is

$$v_{\text{av}} < 0 \tag{6.2}$$

That is, the average reduced interaction must be attractive. A calculation based on this criterion has been carried out by Pines (58). The results are in disagreement with experiment: for example, Pines finds $\rho = 0.074$ for tin compared to an experimental value $\rho \simeq 0.3$; the corresponding critical field at zero temperature (see Chapter IV, Section 3), would be $H_c \cong 0.01$ gauss, compared to the experimental value of 306 gauss.

The reason for this discrepancy is not difficult to find. The "average interaction" (6.1) must not be used if strong repulsions exist. An extreme example of this is provided by adding an infinite repulsive core, with an exceedingly small core radius r_0, say of the order of 10^{-10} cm. Such a repulsive interaction should not matter at all, physically. Yet, such an interaction would play havoc with the BCS criterion. If the repulsive potential in the core region has strength $+ V_0$, its contribution to the reduced interaction is

$$\langle \mathbf{k}, s, -\mathbf{k}, -s \mid v_{\text{core}} \mid \mathbf{k}', s', -\mathbf{k}', -s' \rangle = \delta_{ss'} V_0 \frac{\frac{4}{3}\pi r_0^3}{V} \quad \text{for} \quad r_0 \mid \mathbf{k} - \mathbf{k}' \mid \ll 1 \tag{6.3}$$

If we go to the limit $V_0 \to + \infty$, the expression (6.3) becomes infinitely large, and so does (6.1).

This situation is quite well-known in ordinary quantum mechanics: if there is a strongly repulsive potential, in a certain region of configuration space, the wave function tends to avoid that region; thus a straightforward average of the potential always overestimates the repulsive regions, and underestimates the attractive regions.

The non-linear integral equation (IV, 1.26) is much more complicated than an ordinary Schrödinger equation; but there is reason to expect that some of this behavior still persists. The Coulomb repulsion between electrons is very strong indeed at close distances of approach, and therefore behaves qualitatively like a repulsive core. The BCS criterion ignores this effect, and thus predicts a much more "repulsive" value of ρ than one would get from a correct solution of the integral equation. This is the main reason for the failure of Pines' calculation.

Another early comparison between theory and experiment (Vonsovskii 58a), based on Bogoliubov's work, was rather more successful.

Let us now discuss the present situation, both as regards the effective interaction and the solution of the integral equation.

There are many methods of arriving at an effective interaction; we have presented only one of them here, namely, the original method of Bogoliubov (58). Since we are concerned primarily with matters of principle and basic assumptions of the theory, rather than with mathematical methods, we shall only mention the other techniques. The *random phase approximation*, pioneered by Bohm (53) for plasma waves, has been adapted to superconductivity theory and suitably generalized by Anderson (58, 58a), Bogoliubov (58), and Rickayzen (59).

The *Tamm-Dancoff approximation* of quantum field theory can be used in modified form in the theory of superconductivity (Yamazaki 61, Mitter 61, Nigam 61).

In quantum field theory, the method of *Green's functions* has been developed quite highly. This method has been extended to statistical mechanics by Matsubara (55), Galitskii (58, 58a), Klein (58), Martin and Schwinger (Martin 59), Iwamoto (60), Wiser (63), and others. The method is exceedingly general and powerful, but not without troubles of its own. Besides the possibility of spurious solutions of the equations (Balian 62), there is the general difficulty that it is necessary to make a number of assumptions about the behavior of the various types of Green's functions, before one can arrive at significant results. These assumptions are by no means always obvious, and least so in the early papers on the use of Green's functions in the theory of superconductivity (Gorkov 58, Migdal 59, Nambu 60, Eliashberg 60, Kadanoff 61, Betbeder-Matibet 61, Maleev 62). The assumptions are discussed in more detail by Sawicki (61), Klein (62), and, in particularly lucid fashion, by Maschke (63).

Although straightforward perturbation treatments fail in superconductivity theory,[24] more elaborate treatments, involving selective summation of certain classes of Feynman diagrams, are not excluded. Such *graph-theoretical methods* are based on the expansion of the partition function first given by Matsubara (55), and refined by Bloch (59, 59a), Luttinger (60), and Balian (61, 61a). Selective summation of terms has been performed by Blum (63) and Katz (63). The latter obtains results reminiscent of the quasi-chemical equilibrium theory, in the sense of an approximate Bose distribution and the existence of a Bose-Einstein-like condensation phenomenon. The rather complicated graphs which must be retained and summed, make this approach quite difficult.

In spite of this profusion of methods, or perhaps because of it, the correct form for the effective interaction is by no means certain. The electron-phonon coupling leads to (3.32) in the Bogoliubov theory. Using the Green's function technique, Eliashberg (60) arrives at a rather different kernel. This discrepancy has been discussed by Liu (62); according to Liu, the Eliashberg effective interaction can be written in the form ($\mathbf{q} = \mathbf{k} - \mathbf{k}'$)

$$\langle k, -k \mid v_{\text{eff}} \mid k', -k' \rangle = -\frac{1}{V} \hat{M}^2(q)\, \hbar\hat{\omega}_q \frac{1}{2}\left[\frac{1}{\hat{\epsilon}_k + \hat{\epsilon}_{k'} + \hbar\hat{\omega}_q} + \frac{1}{\hat{\epsilon}_k - \hat{\epsilon}_{k'} + \hbar\hat{\omega}_q}\right] \tag{6.4}$$

[24] Indeed, the transition temperature can be found from the requirement that the "ladder diagrams" of the usual graph theory lead to a divergent sum (Thouless 60, Emery 60, Wild 60, Kenworthy 60a, Fukushima 62, Tolmachev 62).

where the $\hat{M}(q)$, $\hat{\epsilon}_k$, and $\hat{\omega}_q$ are renormalized quantities. The renormalized Bogoliubov theory leads to

$$\langle k, -k \mid v_{\text{eff}} \mid k', -k' \rangle = -\frac{1}{V} \hat{M}^2(q)\, \hbar\hat{\omega}_q \frac{1}{\hat{\epsilon}_k + \hat{\epsilon}_{k'} + \hbar\hat{\omega}_q} \tag{6.5}$$

Liu (62) shows that the difference between (6.4) and (6.5) arises from different interpretations of the "compensation" condition.

The essential difference between (6.4) and (6.5) is the occurrence of zero energy denominators in (6.4). Although the Eliashberg calculation uses the entire formalism of field-theoretic renormalization methods, Liu shows that the result (6.4) can be obtained much more simply by what amounts merely to a second-order perturbation calculation.

On purely theoretical grounds, it is not easy to decide between (6.4) and (6.5): Green's function theory seems to lead more naturally to (6.4) whereas a straightforward variational approach (minimizing the energy in second-order perturbation theory on the phonons, with respect to variation of the parameters u_k and v_k) is equivalent to the original Boboliubov theory, i.e., yields (6.5) (Yosida 58, Liu 62). For further discussion of the electron-phonon contribution, see Morel (58), Chun-Sian (59), and Buikov (60).

The other main component of the effective interaction is the Coulomb repulsion between electrons. Some, e.g., Morel and Anderson (Morel 62), start from a conventional screened Coulomb repulsion, i.e., from (3.33). Others, in particular Bogoliubov (58), insist that one must start from an *un*screeened Coulomb repulsion, and *derive* the form which the screening takes. Chester (61) strongly favors this latter view, but adds the rider that the starting point should be the Coulomb interaction between *all* particles in the solid (ions as well as electrons), and that the derived result should contain not only the screening of the electron-electron Coulomb force, but also the Fröhlich coupling itself. We agree with Chester in principle, but no such calculation has been carried through, and it seems exceedingly difficult to do so.

Fortunately, in spite of the quite different starting points of Bogoliubov (58) and Morel (62), their final results are extremely similar. This is due to the fact that Bogoliubov, after an exceedingly complicated development, arrives at an "effective" Coulomb interaction which is just the ordinary screened interaction (3.33). Thus, the Bogoliubov (58) calculation indicates that, contrary to the doubts expressed by Chester, the screening of the Coulomb field in the superconducting state does not differ much from the screening in the normal state. Bogoliubov's calculation is not a rigorous proof, since he employs the random phase approximation as an intermediate step. However, the result is entirely reasonable on physical grounds: the screening of the electron-electron Coulomb repulsion by the "correlation hole" around any one electron takes place in distances of the order of 10^{-8} to 10^{-7} cm. By contrast, the electron-electron correlations characteristic of the superconducting state (studied in Chapter IV, Section 4) extend out to electron-electron distances of the order of 10^{-4} cm, and are extremely small compared to the zero-order, Fermi-gas correlations wherever the latter do not vanish. Thus, as far as electron-electron screening of the Coulomb force is concerned, the differences between the normal and superconducting states should be negligible, and this is exactly what Bogoliubov finds. Bogoliubov's conclusions are confirmed by Green's function calculations of Shirkov (59, 60), Tewordt (60), and Tolmachev (61).

The upshot of all this discussion is that we are almost exactly back where we started, with formula (3.34) of the simplest version of the Bogoliubov theory. The replacement of renormalized by unrenormalized quantities, and the other approximations in (3.34), makes no qualitative difference to the final solution. However, in the phonon term of (3.34), the Bogoliubov result (6.5) may have to be replaced by the Eliashberg result (6.4).

This leaves us with the problem of solving the non-linear integral equation (IV, 1.26) with such a complicated kernel. Surprisingly enough, it was not until 1962 that anyone approached this problem the only way in which it can be settled eventually, namely by purely numerical methods. The first numerical calculations (Swihart 62, 62a, Culler 62) showed immediately that the crude approximations used in Chapter IV do indeed reproduce the qualitative behavior of the solution of the integral equation, in the region of main interest (the neighborhood of the Fermi surface). The quantity Δ_k does have an energy dependence, of course, but the energy scale is given by the Debye cutoff energy $\hbar\omega_c$ rather than by Δ itself. The repulsive Coulomb interaction makes Δ_k turn negative away from the Fermi surface, at values of $|\epsilon_k - \epsilon_F|$ of the order of $\hbar\omega_c$.

A more extensive calculation by Swihart (63) permits an experimental decision between the Bogoliubov and Eliashberg forms of the matrix elements. By making a detailed comparison with measured curves of critical field vs. temperature, Swihart concludes that Δ_k must *increase* with $|\epsilon_k - \epsilon_F|$ initially, for very small values of $|\epsilon_k - \epsilon_F|$. The Eliashberg interaction (6.4) leads to a solution with that property, whereas the Bogoliubov interaction fails to do so.

Green's function methods seem particularly powerful for "strong-coupling" superconductors such as lead, where the simple predictions break down (Decker 58). Detailed measurements on lead by Rowell (63) have been analyzed in considerable detail by Schrieffer (63), using numerical solutions of the Green's function equations appropriate to a realistic phonon frequency spectrum for lead. The agreement between theory and experiment is strikingly good.

At the present time, the numerical solutions are not extensive enough to permit an over-all discussion of the criterion for superconductivity, i.e., not enough elements have yet been calculated numerically.

Analytic approximations to the solution of the integral equation have been given by Bogoliubov (58) and Morel and Anderson (Morel 62). Their results agree closely with each other. A detailed comparison with experiment was made by Morel (62). They get rather good agreement between the "theoretical" and "experimental" values of the parameter ρ; in no case is the discrepancy bad enough to raise real doubts, and in many cases the two values are in nearly perfect agreement. Surprisingly enough, they find that *all* metals should become superconducting, even the alkali metals! This appears like a *reductio ad absurdum* of the theory, at first sight. However, the theoretical values of ρ for the alkali metals are all around 0.1 or lower. The resulting critical fields, even at the absolute zero, would be so small that even a tiny admixture of magnetic impurities would spoil the transition. Furthermore, the transition temperatures are predicted to be in the range below 10^{-3} °K, which has not yet been investigated experimentally with materials of superhigh purity. Thus, rather than a contradiction, we have a challenging theoretical prediction for future experiments.

Morel and Anderson are nonetheless pessimistic in their outlook. Their main "bad" point is the dependence of the energy gap Δ on the isotopic mass of the ions; Δ is proportional to $M^{-1/2}$ if the Fröhlich interaction is the only interaction (Swihart 59), but the presence of the Coulomb interaction spoils this simple result. Morel and Anderson find exponents n in $\Delta \sim M^{-n}$ significantly lower than $n = \frac{1}{2}$. *If* the transition temperature T_c is directly proportional to the energy gap Δ at temperature $T = 0$, then there is an actual discrepancy between theory and experiment for some of the substances, since T_c follows an M^{-n} law with n very close to $\frac{1}{2}$. However, there may be other influences at work in determining the transition temperature, and thus the pessimism of Morel and Anderson appears to us to be somewhat premature.

Taken all in all, the theoretical and experimental results at zero temperature are in quite good agreement for simple metals of non-transition type. In some cases, particularly in lead (Schrieffer 63, Rowell 63), there is striking detailed agreement. But some discrepancies remain (Ziman 62).

The situation is much less clear for transition metals and for alloys.

The evidence is summarized by Matthias (63, 63a) and we shall discuss it in Chapter X, Section 3. Matthias' conclusion is that the electron-phonon mechanism of Fröhlich is *not* the main interaction responsible for superconductivity in transition metals. Other mechanisms have been suggested and discussed theoretically (Vonsovskii 58, 61; Moskalenko 59; Suhl 59a; Suffczynski 62; Kondo 63), but more work is needed.

The difficulties of predicting on purely theoretical grounds whether a system of Fermions will become superfluid are exemplified strikingly by the case of liquid He^3, the rare isotope which has become available for experimentation only fairly recently. In a highly interesting paper, Brueckner, Soda, Anderson, and Morel (Brueckner 60) suggested that liquid He^3 might undergo a Schafroth condensation at sufficiently low temperatures. The $l = 0$ pairs characteristic of electrons in superconducting metals would not be favored in liquid He^3, because of the strong repulsion between He^3 atoms at close distances; but condensation might occur into pair states with $l \neq 0$, particularly into a state with $l = 2$. The suggestion was elaborated by Emery (60), Anderson (61), and Gorkov (61). These papers were in disagreement and a theoretical discussion ensued (Hone 62; Klein 62a, 62b; Balian 62, 63), resulting in substantial agreement among the theorists concerning the right method of calculating the transition temperature. The only trouble appears to be that there is no experimental evidence of any transition at or near the predicted temperatures (A. C. Anderson 61, Strongin 62).

In conclusion, we may summarize by saying that theory and experiment are in satisfactory agreement in those cases where theory has been developed in close correspondence to, and following on, detailed experimental evidence. But theory is still not in a good enough state to venture out into new territory, without stumbling quite badly.

VI

Excitation Spectrum and Thermodynamics. The Theory of Bogoliubov, Zubarev, and Tserkovnikov

1. The Excitation Spectrum Near the Ground State

The renormalized version of the Bogoliubov theory, discussed in Chapter V, Section 5, leads not only to a determination of the ground state energy and wave function, but contains within itself a complete description of the low-lying excited levels of the system.

To obtain this, we need merely neglect the parts H' and H'' of the transformed Hamiltonian (V; 5.3, 5.4). The approximate Hamiltonian obtained in this way is

$$H \simeq \mu N + U + H_0 \tag{1.1}$$

where μ is the chemical potential in the limit of zero temperature, N is the number of electrons, U is the c-number constant (V, 5.4b), and H_0 is given by

$$H_0 = \sum_{\mathbf{k}} \hat{\epsilon}_{\mathbf{k}} (\overset{+}{\alpha}_{\mathbf{k}0}\alpha_{\mathbf{k}0} + \overset{+}{\alpha}_{\mathbf{k}1}\alpha_{\mathbf{k}1}) + \sum_{\mathbf{q}} \hbar\hat{\omega}_{\mathbf{q}} B_{\mathbf{q}}^{+} B \tag{1.2}$$

To a first approximation, we may put

$$\hat{\epsilon}_{\mathbf{k}} = [(\epsilon_{\mathbf{k}} - \epsilon_F)^2 + (\Delta_{\mathbf{k}})^2]^{1/2} \tag{1.3}$$

$$\hbar\hat{\omega}_{\mathbf{q}} = \hbar v_s \mid \mathbf{q} \mid \tag{1.4}$$

where v_s is the actual (renormalized) sound velocity in the metal; and we may ignore the dependence of $\Delta_{\mathbf{k}}$ on $\mathbf{k}$, so that $\Delta_{\mathbf{k}} \simeq \Delta$, a constant.

The Bogoliubov ground state $| B >$ is given by (V, 4.21), and represents a Schafroth condensate of electron pairs. The ground state energy is equal to $\mu N + U$ in the approximation (1.1); to this, we should really add the contribution (V, 5.9) from the perturbation terms $H' + H''$; since this contribution is just another c-number constant, it has no effect on the excitation spectrum.[1]

The spectrum of excitations is given by the Hamiltonian H_0. There are excitation of Fermi-Dirac type, with excitation energies $\hat{\epsilon}_{\mathbf{k}}$, and phonon excitations, of Bose-Einstein type, with excitations energies $\hbar\hat{\omega}_{\mathbf{q}}$. To the accuracy of the approximation (1.1), these excitations are *independent* of each other. The interaction between excitations is contained in the parts H' and H'' of the complete Bogoliubov Hamiltonian (V, 5.4).

The phonon excitations are not significantly different in the normal and superconducting states of the metal [we recall that the renormalization factor for phonons, (V, 5.17), could be approximated quite accurately by (V, 5.21), its value in the normal state].

The Fermi-type excitations, however, are significantly different, since their energies $\hat{\epsilon}_k$, (1.3), are all larger than the "energy gap" Δ. We shall now investigate the nature of the excited states concerned, by use of the Yosida (58) transformation discussed in Chapter V, Section 4. If $| B >$ represents the Bogoliubov ground state, then the state with one "Bogolon" of type $\mathbf{k}$, 0 is given by

$$\alpha_{\mathbf{k}0}^{+} | B > = \exp(iS)\, a_{\mathbf{k},+}^{+} \exp(-iS) | B > = \exp(iS)\, a_{\mathbf{k},+}^{+} | 0 > \tag{1.5}$$

where $| 0 >$ is the ordinary vacuum state; we have used (V, 4.1) and (V, 4.20). We now use (V, 4.19) together with the relations:

$$P_{\mathbf{k}} a_{\mathbf{k},+}^{+} = 0, \qquad \beta_{\mathbf{k}}^{+} a_{\mathbf{k},+}^{+} = 0 \tag{1.6}$$

These relations are direct consequences of the anticommutation rules (V, 1.7) and (V, 1.18). Substitution of (V, 4.19) and (1.6) into (1.5) yields

$$\alpha_{\mathbf{k}0}^{+} \Big| B > = \Big\{ \prod_{l \neq \mathbf{k}} (u_l + v_l \beta_l^{+}) \Big\} a_{\mathbf{k},+}^{+} \Big| 0 > \tag{1.7}$$

The Bogoliubov wave functions do not conserve the particle number, and thus we must expect all particle numbers within any one wave function. Nonetheless, (1.7) is *not* a permissible excited state of the system whose ground state is (V, 4.21). Since the operators β_l^{+} create pairs of

[1] Conditions (V,5.11) and (V,5.12) ensure that, to first order, the terms $H' + H''$ do not alter the excitation spectrum obtained from (1.1).

particles, the total number of particles in (V, 4.21) is *even*, whereas the total number of particles in (1.7) is *odd*.

We get permissible excited states by combining *two* (or, more generally, an even number of) Bogolons; for example,

$$\overset{+}{\alpha}_{\mathbf{k}0}\overset{+}{\alpha}_{l1}\Big| B> = \Big\{ \prod_{\mathbf{m}\neq\mathbf{k},l} (u_{\mathbf{m}} + v_{\mathbf{m}}\overset{+}{\beta}_{\mathbf{m}}) \Big\} \overset{+}{a}_{\mathbf{k},+}\overset{+}{a}_{-l,-}\Big| 0> \tag{1.8}$$

This is just the excited state (V, 2.10) used for the discussion of the energy gap in the BCS theory. Thus, not only do the ground states of the BCS and Bogoliubov theories coincide, but so do the excited states, to the approximation (1.1).

According to the Hamiltonian (1.1)–(1.2), the excitation energy of state (1.8) is $\hat{\epsilon}_k + \hat{\epsilon}_l$, which is larger than 2Δ. Thus, the minimum energy for exciting states of this type is *twice* the "energy gap" parameter Δ.

The need to produce even numbers of Bogolons should not be surprising; this is exactly the same as in the hole theory for the normal state of the metal. An excitation of the normal metal is produced by moving an electron from inside the Fermi sea to a state outside the Fermi sea. Thus, we produce one particle (outside the Fermi sea) and one hole (inside the Fermi sea), and these are counted as two excitations in the hole theory Hamiltonian (V, 3.7). Furthermore, it is not unreasonable to count in this fashion, since the electrons and holes are largely independent of each other. Conservation of particles demands that there should at all times be equally many electrons and holes; but each electron can range over an infinite set of k-values, and so can each hole. Statistically speaking, they are independent excitations.

The low-lying excitations in the Bogoliubov theory are therefore represented as "configurations" of occupation numbers. There are two types of occupation numbers, namely,

$$f_{\mathbf{k}s} = \overset{+}{\alpha}_{\mathbf{k}s}\alpha_{\mathbf{k}s} = 0, 1 \quad \text{only} \tag{1.9}$$

(the number of Bogolons of type $\mathbf{k}$, s), and

$$\nu_{\mathbf{q}} = B_{\mathbf{q}}^{+}B_{\mathbf{q}} = 0, 1, 2, 3, \cdots \tag{1.10}$$

(the number of phonons of type $\mathbf{q}$). The energy of the configuration

$$\alpha = \{f_{\mathbf{k}s};\, \nu_{\mathbf{q}}\} \tag{1.11}$$

is given by

$$E_\alpha = \sum_{\mathbf{k},s} \hat{\epsilon}_{\mathbf{k}} f_{\mathbf{k}s} + \sum_{\mathbf{q}} \hbar\hat{\omega}_{\mathbf{q}}\nu_{\mathbf{q}} \tag{1.12}$$

We note that the total number of Bogolons, i.e., the sum of all the $f_{\mathbf{k}s}$, is *not* conserved, and has no direct physical meaning; the same is true of the total number of phonons, the sum of all the $\nu_{\mathbf{q}}$. The ground state of the system, $\alpha = 0$, is the configuration (1.11) in which all numbers $f_{\mathbf{k}s}$ and $\nu_{\mathbf{q}}$ vanish.

So far, we have ignored the perturbation terms $H' + H''$ in the Bogoliubov Hamiltonian. These perturbation terms induce "configuration interactions" between the excited states (1.11). It is conceivable that, as a result of such configuration interactions, some energy levels might decrease sharply, thereby producing a low-lying "collective branch" of the excitation spectrum. By using the random phase approximation, Bogoliubov (58) established that such collective excitations do exist. The collective excitations found by Bogoliubov have Bose-Einstein character, i.e., there is no upper limit on the number n_q of collective excitations of type q. If we ignore the Coulomb interaction between electrons, then the excitation energy of the collective excitation with wave number q is given by

$$\eta_q = \frac{\hbar v_F \mid \mathbf{q} \mid}{\sqrt{3}} \tag{1.13}$$

where $v_F = \hbar k_F/M$ is the velocity of an electron at the Fermi surface. These states are density waves in the electron pair gas.

At low temperatures, the lowest possible excitation energies are the most important ones. Since (1.13) has no energy gap, this collective branch might be expected to be very significant near zero temperature. However, the spectrum (1.13) has been obtained by ignoring the Coulomb interactions between electrons, and is therefore unrealistic. As soon as the Coulomb repulsions are included in the calculation, the picture changes drastically. Density waves now become plasma waves, and the excitation energy of a plasma wave of low momentum q is nearly independent of q, and equal to

$$\eta_q \cong \hbar\omega_P \tag{1.14}$$

where ω_P is the plasma frequency (I, 9.5). A number of calculations (Anderson 58, 58a, Yosida 59, Rickayzen 59, Belyaev 60, Galitsky 60, Schrieffer 60, Foldy 61, Bardasis 61) all give this result. The plasma energy (1.14) is so enormous that excitations of this type have no influence whatever on the thermodynamic properties of a superconductor. We shall, however, return to these collective states in Chapter VIII, in the discussion of electrodynamic properties.

This particular collective branch of the spectrum having been effectively excluded, there remains the question whether other collective states exist, particularly states with excitation energies η_q lower than the energy gap 2Δ. There is every reason to expect such states to occur, and this is confirmed by a calculation (Bardasis 61) based on the random phase approximation. However, for reasons which will become apparent later, these states do not show up significantly in the specific heat.[1a]

2. The Specific Heat at Low Temperatures

The energy level spectrum (1.12), together with the assumption that the excitation numbers $f_{\mathbf{k}s}$ and $\nu_{\mathbf{q}}$ are statistically independent of each other, suffices to obtain the internal energy and specific heat of the system in the limit of low temperatures. The phonons form a Bose-Einstein gas; since the number of phonons is not conserved (phonons can be created and destroyed), the chemical potential of the phonons must be set equal to zero, so that the average number of phonons of type q is given by the Planck law

$$\bar{\nu}_{\mathbf{q}} = \frac{1}{\exp(\beta\hbar\hat{\omega}_{\mathbf{q}}) - 1} \tag{2.1}$$

This is the special case of the Bose-Einstein distribution law (II, 1.13) corresponding to $\mu = 0$. The internal energy contributed by the phonon gas is then, according to (1.12),

$$\bar{E}_{\mathrm{ph}} = \sum_{\mathbf{q}} \hbar\hat{\omega}_{\mathbf{q}}\bar{\nu}_{\mathbf{q}} \tag{2.2}$$

The Bogolons form a Fermi-Dirac gas, so that the appropriate distribution law is (II, 1.12). Since Bogolons can be created and destroyed freely, their chemical potential must also be set equal to zero; hence the average number of Bogolons of type $\mathbf{k}$, s is given by

$$\bar{f}_{\mathbf{k}s} = \frac{1}{\exp(\beta\hat{\epsilon}_{\mathbf{k}}) + 1} \tag{2.3}$$

The average energy contributed by the Bogolon gas is therefore

$$\bar{E}_{\mathrm{Bg}} = \sum_{\mathbf{k}} \sum_{s=0,1} \hat{\epsilon}_{\mathbf{k}}\bar{f}_{\mathbf{k}s} \tag{2.4}$$

[1a] The treatment of collective excitations is quite complex (Valatin 58a, Vaks 61, Maschke 63), and it is by no means certain that all the significant collective modes have been found. In particular, superconductors might exhibit "zeroth sound" (Gottfried 60) and/or "second sound" (Velibekov 62, 62a, 62b).

The total internal energy of the superconductor at temperature $T = 1/\mathfrak{k}\beta$ is then, in this approximation,

$$\bar{E}(T) = \bar{E}_{\mathrm{ph}} + \bar{E}_{\mathrm{Bg}} \tag{2.5}$$

and the specific heat c_V at constant volume is the derivative of $\bar{E}(T)$ with respect to T.

From here on, it is merely a matter of straightforward evaluation of the sums (2.2) and (2.4). The phonon internal energy (2.2) is well-known; we replace the sum over q by an integral, and supply a factor 3 because there are three phonon modes (one longitudinal wave, two transverse waves) for every wave vector $\mathbf{q}$. Using the Debye approximation (1.4), which is valid for low q, we obtain from (2.2)

$$\bar{E}_{\mathrm{ph}} = 3\frac{V}{(2\pi)^3}\int_0^{q_m} 4\pi q^2 \frac{\hbar v_s q}{\exp(\beta\hbar v_s q) - 1}\, dq \tag{2.6}$$

The upper limit q_m is the Debye cutoff (see Chapter I, Section 10). For the present calculation, which is restricted to low temperatures, we may extend the integration to infinity. This gives

$$\bar{E}_{\mathrm{ph}} \cong V\frac{(\mathfrak{k}T)^4}{(\hbar v_s)^3} I \tag{2.7}$$

where I is a non-dimensional integral

$$I = \frac{3}{2\pi^2}\int_0^{\infty} \frac{x^3}{e^x - 1}\, dx = \frac{3}{2\pi^2}\frac{\pi^4}{15} = \frac{\pi^2}{10} = 0.987 \tag{2.8}$$

If we differentiate (2.7) with respect to T, we obtain the low-temperature form of the Debye specific heat

$$c_{V,\mathrm{ph}} = kV\frac{\pi^2}{10}\left(\frac{\mathfrak{k}T}{\hbar v_s}\right)^3 \tag{2.9}$$

This is the well-known T^3 law of Debye. We get a more conventional form for the coefficient by introducing the number of ions, N_i, the Debye cutoff wave number q_m, and the Debye temperature Θ by

$$q_m = \frac{3N_i}{2\pi^2 V}, \qquad \mathfrak{k}\Theta = \hbar v_s q_m \tag{2.10}$$

In terms of these quantities, we obtain the non-dimensional formula

$$\frac{c_{V,\mathrm{ph}}}{N_i \mathfrak{k}} = \frac{12\pi^4}{5}\left(\frac{T}{\Theta}\right)^3 \cong 234\left(\frac{T}{\Theta}\right)^3 \tag{2.11}$$

Before going on to the contribution of the Bogolon gas, let us note that (2.9) allows us to eliminate a collective excitation branch with a spectrum similar to (1.13), as a serious contributor to the specific heat. The spectrum (1.13) is exactly like a Debye phonon spectrum, except that the velocity of sound, v_s, has been replaced by $v_F/\sqrt{3}$. Since v_s enters into (2.9) as $(v_s)^{-3}$, the specific heat contributed by an excitation branch of type (1.13) would be smaller by a factor $(\sqrt{3}\, v_s/v_F)^3$, which is of the order of 10^{-7} to 10^{-8}. With an identical temperature dependence, proportional to T^3, such a contribution could never be separated experimentally from the Debye specific heat (2.11).

This shows that energies alone are not a sufficient criterion in statistical mechanics; the density of states must be taken into account. Low-lying states with a very small density of states are indetectable by purely thermal measurements. This same argument can be used to show that the other "collective excitations" found by Bardasis (61) also make negligible contributions to the specific heat.

We now turn to the Bogolon contribution (2.4). We replace the sum over k by an integral, and supply a factor 2 for the two possible values of s. This leads to

$$\bar{E}_{\mathrm{Bg}} = 2 \frac{V}{(2\pi)^3} \int_0^\infty \frac{\hat{\epsilon}_k}{\exp(\beta\hat{\epsilon}_k) + 1} 4\pi k^2\, dk \tag{2.12}$$

This is a fairly complicated integral, and we shall evaluate it only in the low-temperature limit where we can make the assumption

$$\mathfrak{k}T \ll \Delta \tag{2.13}$$

Since the excitation energies $\hat{\epsilon}_k$ are all larger than Δ, the exponential $\exp(\beta\hat{\epsilon}_k)$ is much larger than unity, for all k, and we can neglect the "1" in the denominator of (2.12). Furthermore, the main contributions to the integral come from values of k close to the Fermi momentum k_F, since $\hat{\epsilon}_k$ is lowest there. We may therefore replace one of the factors k in the integrand of (2.12) by its Fermi value k_F; we use the identity

$$k\, dk = (M/\hbar^2)\, d(\epsilon_k - \epsilon_F) = (M/\hbar^2)\, dx$$

and we note that the lower limit of integration on x can be extended to minus infinity without significant error; the integrand is an even function of x, and we obtain

$$\bar{E}_{\mathrm{Bg}} \cong \frac{2Mk_FV}{(\pi\hbar)^2} \int_0^\infty (x^2 + \Delta^2)^{1/2} \exp[-\beta(x^2 + \Delta^2)^{1/2}]\, dx \tag{2.14}$$

We use the transformation

$$x = \Delta \sinh u, \qquad dx = \Delta \cosh u\, du$$

and the formula (Watson 48, p. 181)

$$K_n(z) = \int_0^\infty \cosh(nu) \exp(-z \cosh u)\, du \tag{2.15}$$

The identity

$$(\cosh u)^2 = \tfrac{1}{2} + \tfrac{1}{2} \cosh(2u)$$

then suffices to bring (2.14) into a form where (2.15) can be used.

In order to obtain an intensive expression, we divide both sides of (2.14) by the number of electrons, N. We note that the number density N/V is related to the Fermi momentum k_F by $N/V = k_F^3/3\pi^2$. Combining the results so far, we obtain

$$\frac{\bar{E}_{\mathrm{Bg}}}{N} \cong \frac{3\Delta^2}{2\epsilon_F}[K_0(\beta\Delta) + K_2(\beta\Delta)] \qquad \text{for} \quad \beta\Delta \ll 1 \tag{2.16}$$

The electronic contribution to the specific heat is the derivative of Eq. (2.16) with respect to the temperature. We use the formulas (Watson 48, p. 79)

$$K_0' = -K_1, \qquad K_2' = -\tfrac{1}{2}(K_1 + K_3)$$

to obtain

$$\frac{c_{V,es}}{N\mathfrak{k}} \cong \frac{3\Delta}{4\epsilon_F}\left(\frac{\Delta}{\mathfrak{k}T}\right)^2 [3K_1(\beta\Delta) + K_3(\beta\Delta)] \qquad \text{for} \quad \beta\Delta \ll 1 \tag{2.17}$$

In order to obtain a quick insight into the behavior of this contribution at very low temperatures, we note that the asymptotic form, (IV, 4.20) for $K_0(z)$ is the same for all $K_n(z)$. Substitution of this asymptotic form into (2.17) yields

$$\frac{c_{V,es}}{N\mathfrak{k}} \cong 3\left(\frac{\pi}{2}\right)^{1/2} \frac{\Delta}{\epsilon_F}\left(\frac{\Delta}{\mathfrak{k}T}\right)^{3/2} \exp\left(-\frac{\Delta}{\mathfrak{k}T}\right) \qquad \text{for} \quad \beta\Delta \ll 1 \tag{2.18}$$

Thus the electronic specific heat in the superconducting state vanishes exponentially in the limit of very low temperatures. Such a behavior is of course expected as soon as there is an energy gap.

More interesting than the exponential behavior itself is the coefficient in the exponential. The "activation energy" in (2.18) is not equal to the energy gap 2Δ, but rather to half that value! This is due to the fact that an excitation by the minimum excitation energy 2Δ corresponds to the creation of two independent Bogolons; it is the same factor 2 which occurs in the theory of chemical equilibrium between binary molecules and their dissociation products [see (III, 2.15) and (III, 2.17)].[2]

Being exponentially small, the electronic specific heat in the superconducting state is quite negligible compared to its value in the normal state (Kittel 53)

$$c_{V,en} = N\mathfrak{k}\frac{\pi^2\mathfrak{k}T}{2\epsilon_F} = \gamma T \tag{2.19}$$

This normal value can be measured experimentally even below the superconducting transition temperature, by placing the material in a magnetic field larger than the critical field, so that the metal remains normal down to the absolute zero. It is useful to observe that an experimental measurement of γ in (2.19) gives a way of determining the density of states ρ_E directly. Comparison of (2.19) and (IV, 2.6) yields

$$\rho_E = \frac{3\gamma}{2\pi^2\mathfrak{k}^2} \tag{2.20}$$

where $\mathfrak{k}$ is Boltzmann's constant; both sides of (2.20) are proportional to the volume V of the specimen.

A useful property of the specific heats (2.17) and (2.19) is that their ratio is a function of $\beta\Delta = \Delta/\mathfrak{k}T$ only; no other parameters enter:

$$\frac{c_{V,es}}{c_{V,en}} = \frac{3}{2\pi^2}X^3[3K_1(X) + K_3(X)] \equiv G(X) \qquad \text{where} \quad X = \frac{\Delta}{\mathfrak{k}T} \tag{2.21}$$

Thus this ratio allows us to determine Δ directly.

The formulas of this section are valid only in the limit of low temperatures. Hence we shall defer the discussion of how to analyze experimental specific heat data until the end of Section 5. As an example of the sort of agreement one obtains from the simple low-temperature theory, however, let us consider the data of Phillips (59) for aluminium. The transition temperature of Al is $T_c = 1.16°$K, and Phillips' data extend from about 0.2°K to temperatures higher than T_c. We are interested in the low end of this range, and shall fit Δ to the data between 0.3 and 0.5°K.[3] The result is $\Delta/\mathfrak{k} = 1.94°$K, and the fit between theory and experi-

[2] In this derivation, we have ignored the condition that the total number of Bologons must be even. This condition has no visible effect on the specific heat, i.e., the effect is of lower order in N.

[3] There is reason to doubt the accuracy of the data between 0.2 and 0.3°K. In that region, the *normal* state specific heat measured by Phillips deviates from the expected behavior, i.e., $c_{V,n}/T$ deviates from straight-line behavior as a function of T^2.

ment is illustrated in Fig. 6.1, where we have plotted the ratio of the superconducting to the normal electronic specific heat, Eq. (2.21). The points are derived from Phillips' data. We emphasize that this is a

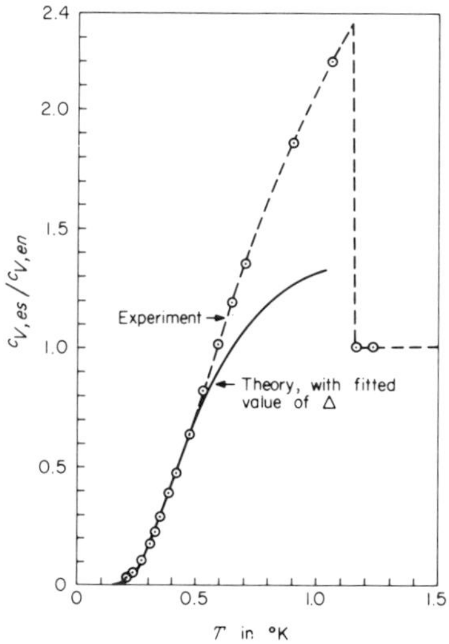

FIG. 6.1. *Ordinate:* Ratio of electronic specific heats in the superconducting and normal states of aluminum. *Abscissa:* Temperature. The theoretical curve is based on the low-temperature approximation, and fitted to the experimental points between 0.3 and 0.5°K. Deviations from the low-temperature theory occur already well below the superconducting transition temperature, 1.16°K.

one-parameter fit. The low-temperature deviations are probably experimental in origin,[3] but the high-temperature deviations are due to the deficiencies of the simple theory.

As a check on the theory, we may determine an independent value of the energy-gap parameter Δ from the measured critical field $H_c(0)$ at zero temperature, using the theoretical relations (IV, 3.1), (IV, 3.11), and (2.20). The measured critical field is $H_c(0) = 103.0$ gauss; the coefficient γ in the normal-state specific heat is $\gamma = 1.35 \times 10^4$ erg/mole-degree. Substitution yields $\Delta/\mathfrak{k} = 2.03$°K, in reasonable agreement with the value from the specific heat measurements.

3. The Theory of Bogoliubov, Zubarev, and Tserkovnikov

The simple theory of Section 2 is obviously insufficient to describe the thermodynamic transition between the superconducting and normal states of the metal. Furthermore, Fig. 6.1 shows that the insufficiency is apparent already well below the actual transition temperature.

To develop a theory of a thermodynamic transition is an exceedingly difficult task. Bitter experience with approximation methods in the (purely classical) Ising model has shown that none of the approximations gave anything like the exact result (Kramers 41, 41a; Onsager 44; Kaufman 49, 49a).[3a] Similarly, measurements of the specific heat in the neighborhood of a transition without latent heat must be exceedingly accurate: for many years, it was generally accepted that the specific heat of liquid helium near the lambda point has a simple discontinuity, whereas actually it has a logarithmic singularity on both sides of the transition temperature (Buckingham 61).

In the theory of superconductivity, this problem was attacked and solved in the path-breaking work of Bogoliubov, Zubarev, and Tserkovnikov (Bogoliubov 57, 60a), called BZT henceforth. By starting from the Hamiltonian in the "reduced" form (V, 2.9a), BZT managed to obtain an *exact* solution for the free energy by means of a completely novel method.[4]

Quite apart from its importance for the theory of superconductivity, the BZT theory is a real *tour de force*—one of the very few exact theories in statistical mechanics.

BZT start from the reduced Hamiltonian in the form (V, 2.9a). They assume that the effective electron-electron coupling is temperature-independent. The relevant matrix elements are written in the form

$$\langle \mathbf{k}, +, -\mathbf{k}, - \mid v_{\text{eff}} \mid \mathbf{k}', +, -\mathbf{k}', - \rangle = -\frac{J_{\mathbf{k}\mathbf{k}'}}{V} \tag{3.1}$$

[3a] In spite of an extensive subsequent development of the exact treatment, no approximate treatment exists which is valid near the transition point (Houtappel 50; Husimi 50; Newell 50; Wannier 50; Syozi 51; Utiyama 51; Yamamoto 51; Kac 52; Kano 53; Potts 55; Fisher 59; Domb 60, 61; Hurst 60, 63; Sykes 61; Meijer 62; Rushbrooke 62, 62a; Green 62, 63; Pais 63; Fisher 63; Glauber 63; Kasteleyn 63; Montroll 63).

[4] Simultaneously and independently, Bardeen, Cooper, and Schrieffer (Bardeen 57) developed a treatment based on a variational approach. Although a variational calculation in statistical mechanics, just as much as in ordinary quantum mechanics, gives in general only an upper bound on the energy, comparison with the exact BZT solution shows that the BCS treatment also leads to the exact result, in the end. We present the BZT method here; a variational approach is used in Chapter VII to investigate the meaning of the assumption of the reduced Hamiltonian in the thermodynamic theory.

where $J_{\mathbf{kk}'} = J_{\mathbf{k}'\mathbf{k}}$ is real. With the volume V appearing explicitly in (3.1), the quantities $J_{\mathbf{kk}'}$ are volume-independent. Furthermore, the minus sign in (3.1) means that $J_{\mathbf{kk}'}$ is positive for an attractive interaction.

In this notation, the reduced Hamiltonian becomes

$$H_{\text{red}} - \mu N = \sum_{\mathbf{k},s} (\epsilon_{\mathbf{k}} - \mu)\, a_{\mathbf{k}s}^{+} a_{\mathbf{k}s} - \frac{1}{V} \sum_{\mathbf{k},\mathbf{k}'} J_{\mathbf{kk}'} \beta_{\mathbf{k}}^{+} \beta_{\mathbf{k}'} \tag{3.2}$$

where, just as before,

$$\beta_{\mathbf{k}}^{+} = a_{\mathbf{k},+}^{+} a_{-\mathbf{k},-}^{+}, \qquad \beta_{\mathbf{k}} = a_{-\mathbf{k},-} a_{\mathbf{k},+} \tag{3.3}$$

The technique of BZT consists in making a *displacement* of these $\beta_{\mathbf{k}}$ operators; that is, we introduce a set of real constants $w_{\mathbf{k}}$ and new operators $B_{\mathbf{k}}$ by

$$\beta_{\mathbf{k}} = B_{\mathbf{k}} + w_{\mathbf{k}}\,, \qquad \beta_{\mathbf{k}}^{+} = B_{\mathbf{k}}^{+} + w_{\mathbf{k}} \tag{3.4}$$

When (3.4) is substituted into (3.2), we obtain terms independent of the B operators, terms linear in the B-operators, and finally a term quadratic in the B-operators. If this last term is ignored, the resulting Hamiltonian allows an exact solution. The idea of the BZT method is to adjust the constants $w_{\mathbf{k}}$ in such a way that the correction terms become utterly negligible, at the given temperature T.

We define the quantity $\Delta_{\mathbf{k}}$ by

$$\Delta_{\mathbf{k}} = \frac{1}{V} \sum_{\mathbf{k}} J_{\mathbf{kk}'} w_{\mathbf{k}'} \tag{3.5}$$

and we then obtain from (3.4) and (3.2)[5]

$$H_{\text{red}} - \mu N = H_0 + H' = \Big(C + \sum_{\mathbf{k}} H_{\mathbf{k}}\Big) + H' \tag{3.6a}$$

where

$$C = -\sum_{\mathbf{k}} w_{\mathbf{k}} \Delta_{\mathbf{k}} = \text{constant} \tag{3.6b}$$

$$H_{\mathbf{k}} = (\epsilon_{\mathbf{k}} - \mu)\,(a_{\mathbf{k},+}^{+} a_{\mathbf{k},+} + a_{-\mathbf{k},-}^{+} a_{-\mathbf{k},-}) - \Delta_{\mathbf{k}}(B_{\mathbf{k}}^{+} + B_{\mathbf{k}}) \tag{3.6c}$$

$$H' = -\frac{1}{V} \sum_{\mathbf{k},\mathbf{k}'} J_{\mathbf{kk}'} B_{\mathbf{k}}^{+} B_{\mathbf{k}'} \tag{3.6d}$$

[5] We emphasize that H_0 in (3.6a) is *not* an approximation to H_{red}, but rather to the operator $H_{\text{red}} - \mu N_{\text{Op}}$.

Although the full operator (3.6a) commutes with the operator for the number of particles, the separate operators H_0 and H' do not commute with N. This causes no trouble if the final result can be shown (as indeed it can be) to be exact in the limit of infinite volume V.

In Chapter III, Section 4, we introduced the grand canonical potential $\Omega(N, T, \mu)$ [see Eqs. (III, 4.5)–(III, 4,9)]. The same quantity in quantum statistical mechanics was introduced in Chapter III, Section 5; by combining Eqs. (III, 4.5) and (III, 5.1), we obtain

$$\exp [-\beta\Omega(V, T, \mu)] = \sum_N \exp (\beta\mu N) \operatorname{Trace}' \exp (-\beta H_N) \tag{3.7}$$

where the prime on the Trace indicates restriction to quantum states of the correct symmetry (antisymmetric for electrons). If we write the Hamiltonian H_N in terms of creation and destruction operators, for example (3.2), the number N does not appear explicitly, and furthermore the restriction to states of correct symmetry becomes automatic. We thus obtain the neater form

$$\exp [-\beta\Omega(V, T, \mu)] = \operatorname{Trace} \exp [-\beta(H - \mu N_{\mathrm{Op}})] \tag{3.8}$$

where now the trace indicates a sum over all possible configurations of occupation numbers $n_{\mathbf{k}s} = a^+_{\mathbf{k}s} a_{\mathbf{k}s}$, without worrying about keeping the sum of all the $n_{\mathbf{k}s}$ constant. If we transpose the c-number $\exp(-\beta\Omega)$ to the other side of (3.8), we find that the *operator*

$$U = \exp [\beta(\Omega + \mu N_{\mathrm{Op}} - H)] \tag{3.9}$$

has unit trace

$$\operatorname{Trace} U = 1 \tag{3.10}$$

Since N_{Op} and H are Hermitean operators and commute with each other, the eigenfunctions of U can be chosen to be also eigenfunctions of N_{Op} and H. If $\psi_{N\alpha}$ is such an eigenfunction, with eigenvalue N for N_{Op} and eigenvalue $E_{N\alpha}$ for $H = H_N$, the corresponding eigenvalue of the operator U is real and given by

$$P_{N\alpha} = \exp [\beta(\Omega + \mu N - E_{N\alpha})] \tag{3.11}$$

and (3.10) becomes

$$\sum_N \sum_\alpha P_{N\alpha} = 1 \tag{3.12}$$

The numbers $P_{N\alpha}$ are real, positive, and add up to unity. Each $P_{N\alpha}$ can therefore be interpreted as the probability of finding, in the "grand canonical ensemble," a member system with exactly N particles, in the quantum state $\psi_{N\alpha}$.

If Q is an arbitrary operator, its thermodynamic expectation value is the average

$$\langle Q \rangle = \sum_N \sum_\alpha \langle N\alpha \mid Q \mid N\alpha \rangle P_{N\alpha} = \text{Trace}\,(QU) \tag{3.13}$$

The operator U is called the "density matrix" or "statistical operator."

For the problem under discussion here, the exact statistical operator is given by (3.9), where Ω is determined by (3.8). Instead of this, we wish to use the approximation[6]

$$U_0 = \exp\,[\beta(\Omega_0 - H_0)] \tag{3.14}$$

where $\Omega_0(V, T, \mu)$ is our approximation for the grand canonical potential, determined by the condition that the trace of U_0 must equal unity, i.e.,

$$\exp\,[-\beta\Omega_0(V, T, \mu)] = \text{Trace}\exp\,(-\beta H_0) \tag{3.15}$$

The form of U_0, Eq. (3.14) simplifies greatly when we notice that the different operators $H_{\mathbf{k}}$, Eq. (3.6c), all commute with each other. Putting c-number constants out in front, we obtain from (3.6) and (3.14)

$$U_0 = \exp\,[\beta(\Omega_0 - C)] \prod_{\mathbf{k}} \exp\,(-\beta H_{\mathbf{k}}) \tag{3.16}$$

In many respects, the approximate statistical operator will turn out to be equivalent to the exact statistical operator (3.9). However, there are exceptions, one of which is of particular importance for the theory: let us consider the statistical expectation value of the operator $\beta_{\mathbf{k}}$, i.e.,

$$\langle \beta_{\mathbf{k}} \rangle = \text{Trace}\,(\beta_{\mathbf{k}} U) \tag{3.17}$$

Since $\beta_{\mathbf{k}}$ decreases the total number of electrons by 2, all its diagonal matrix elements $\langle N\alpha \mid \beta_{\mathbf{k}} \mid N\alpha \rangle$ vanish identically, and thus, by (3.13), so does the trace.

However, the approximate expectation value

$$\langle \beta_{\mathbf{k}} \rangle_0 = \text{Trace}\,(\beta_{\mathbf{k}} U_0) \tag{3.18}$$

does *not* vanish, because the approximate statistical operator U_0 has non-zero matrix elements connecting states with different particle numbers N.

[6] We recall that H_0 is an approximation to $H - \mu N$, not to H.

We now use this property in order to make the correction term H in (3.6a) unimportant. Since H', (3.6d), is quadratic in the $B_{\mathbf{k}}$, it is desirable to make the operators $B_{\mathbf{k}}$ as "small" as possible; we shall insist that the statistical expectation value of each $B_{\mathbf{k}}$ vanish over the approximate statistical ensemble (3.16), i.e.,

$$\langle B_{\mathbf{k}} \rangle_0 = \langle \beta_{\mathbf{k}} \rangle_0 - w_{\mathbf{k}} = 0 \tag{3.19a}$$

or,

$$w_{\mathbf{k}} = \langle \beta_{\mathbf{k}} \rangle_0 \tag{3.19b}$$

This is the BZT condition which determines the constants $w_{\mathbf{k}}$. With this condition satisfied, BZT show that *the approximate grand canonical potential (3.15) differs from the exact value (3.8) by terms which become completely negligible in the limit of infinite volume V; i.e., for purposes of statistical mechanics, the condition (3.19) transforms the approximation (3.15) into an exact result.*

The completely rigorous proof of this assertion is quite complicated (Bogoliubov 60a). However, a heuristic proof can be given (Bogoliubov 57) by using the perturbation expansion for statistical mechanics.

This perturbation expansion differs from the ordinary quantum-mechanical perturbation theory, since we are not interested in perturbations of individual energy levels, but rather in an expansion, in powers of H', of the trace

$$\text{Trace } G(H_0 + H')$$

where G is some arbitrary function. $G(x) = \exp(-\beta x)$ in statistical mechanics. This perturbation expansion, in its most general form, was derived by Schafroth in his first paper on the theory of superconductivity (Schafroth 51). For an arbitrary function $G(x)$, Schafroth's form of the expansion is still the only known form. For the special case of an exponential function, however, alternative forms are possible (Matsubara 55, Goldberger 52). We define the operator $H'(t)$ by

$$H'(0) = H' \tag{3.20a}$$

$$\frac{dH'(t)}{dt} = [H_0 \, , H'(t)]_- \tag{3.20b}$$

A formal solution of this differential equation is

$$H'(t) = \exp(tH_0) \, H' \exp(-tH_0) \tag{3.20c}$$

This solution is only formal, since the operator $\exp(tH_0)$ usually does not exist for positive values of t. However, the final expansion, (3.22), can be put into such a form that the quantities t which appear in such exponents are always negative.[7]

Furthermore, we define the average of an arbitrary operator over the unperturbed system by

$$\langle Q \rangle_0 = \frac{\text{Trace}\,[Q \exp(-\beta H_0)]}{\text{Trace} \exp(-\beta H_0)} = \text{Trace}\, Q U_0 \tag{3.21}$$

Then the perturbation expansion reads

$$\frac{\text{Trace} \exp[-\beta(H_0 + H')]}{\text{Trace} \exp(-\beta H_0)} = 1 - \int_0^\beta dt_1 \langle H'(t_1) \rangle_0 + \int_0^\beta dt_1 \int_0^{t_1} dt_2 \langle H'(t_1) H'(t_2) \rangle_0 - \int_0^\beta dt_1 \int_0^{t_1} dt_2 \int_0^{t_2} dt_3 \langle H'(t_1) H'(t_2) H'(t_3) \rangle_0 + \dots \tag{3.22}$$

Unlike the perturbation expansion for individual energy levels, there is no difficulty whatever with degenerate levels in (3.22).

Let us now look at the first non-trivial term on the right of (3.22). When we substitute (3.6d) into (3.21), we obtain

$$H'(t) = -\frac{1}{V} \sum_{\mathbf{k},\mathbf{k}'} J_{\mathbf{k}\mathbf{k}'} B_{\mathbf{k}}^+(t) B_{\mathbf{k}'}(t) \tag{3.23}$$

where[8]

$$B_{\mathbf{k}}(t) = \exp(tH_0) B_{\mathbf{k}} \exp(-tH_0), \qquad B_{\mathbf{k}}^+(t) = \exp(tH_0) B_{\mathbf{k}}^+ \exp(-tH_0) \tag{3.24}$$

The first integral on the right of (3.22) thus becomes

$$-\int_0^\beta dt \langle H'(t) \rangle_0 = +\frac{1}{V} \int_0^\beta dt \sum_{\mathbf{k},\mathbf{k}'} J_{\mathbf{k}\mathbf{k}'} \text{Trace}\,[B_{\mathbf{k}}^+(t) B_{\mathbf{k}'}(t) U_0] \tag{3.25}$$

---

[7] As an example, consider the third term on the right of (3.22). Using the definition (3.21), the numerator becomes

$$\int_0^\beta dt_1 \int_0^{t_1} dt_2 \,\text{Trace}\,\{H'(t_1) H'(t_2) \exp(-\beta H_0)\}$$
$$= \int_0^\beta dt_1 \int_0^{t_1} dt_2 \,\text{Trace}\,\{\exp(-\beta H_0) H'(t_1) H'(t_2)\}$$
$$= \int_0^\beta dt_1 \int_0^{t_1} dt_2 \,\text{Trace}\,\{\exp[-(\beta - t_1) H_0] H' \exp[-(t_1 - t_2) H_0] H' \exp(-t_2 H_0)\}$$

Considering the limits of integration, we see that only negative exponentials appear in the last form. A similar reduction is clearly possible for the general term of (3.22).

[8] Note that $B_{\mathbf{k}}^+(t)$ defined by (3.24) is *not* the Hermitean conjugate of $B_{\mathbf{k}}(t)$, except for $t = 0$. This makes no difference subsequently.

At this stage, we use the form (3.16) of U_0, in which it is apparent that U_0 is a product of separate mutually commuting factors, one for each value of $\mathbf{k}$. If $\mathbf{k} \neq \mathbf{k}'$, we thus obtain

$$\langle B_{\mathbf{k}}^{+}(t)\, B_{\mathbf{k}'}(t)\rangle_0 = \langle B_{\mathbf{k}}^{+}(t)\rangle_0 \langle B_{\mathbf{k}'}(t)\rangle_0 \qquad \text{for} \quad \mathbf{k} \neq \mathbf{k}' \tag{3.26}$$

We now show that the second factor vanishes by (3.19), for all values of t. To see this, we use the fact that Trace (AB) = Trace (BA), and write

$$\langle B_{\mathbf{k}'}(t)\rangle_0 = \frac{\text{Trace}\,[\exp{(tH_0)}\, B_{\mathbf{k}'} \exp{(-tH_0)} \exp{(-\beta H_0)}]}{\text{Trace}\,\exp{(-\beta H_0)}}$$

$$= \frac{\text{Trace}\,[B_{\mathbf{k}'} \exp{(-\beta H_0)}]}{\text{Trace}\,\exp{(-\beta H_0)}} = \langle B_{\mathbf{k}'}\rangle_0$$

Thus $\langle B_{\mathbf{k}'}(t)\rangle_0$ is actually independent of t, and condition (3.19) ensures that it vanishes not merely for $t = 0$, but for all values of t.

Returning to (3.25), we now see that the only terms which survive are terms with $\mathbf{k} = \mathbf{k}'$. Thus there is only a *single* sum over $\mathbf{k}$-space in (3.25); this gives only *one* factor V, which cancels the $1/V$ in front. Thus, finally, the correction term (3.25) is independent of volume.

An extension of this argument shows that *all* the terms of (3.22) are independent of the volume V. If the series converges,[9] its limit is therefore also independent of V. Thus we have

$$\frac{\text{Trace}\,\exp{[-\beta(H_0 + H')]}}{\text{Trace}\,\exp{(-\beta H_0)}} = \exp{[\beta(\Omega_0 - \Omega)]} = O(1) \tag{3.27}$$

where $O(1)$ means "of order 1," i.e., independent of the volume V. Since both Ω_0 and Ω are extensive quantities, proportional to V, a difference $\Omega - \Omega_0$ of order 1 is utterly negligible in statistical mechanics.

From this point on, it is purely a matter of calculation; we must evaluate the average $\langle \beta_{\mathbf{k}}\rangle_0$ explicitly, set it equal to $w_{\mathbf{k}}$, and ensure that Eq. (3.5) for $\Delta_{\mathbf{k}}$ is consistent. Since $\Delta_{\mathbf{k}}$ appears in H_0 [see Eqs. (3.6b) and (3.6c)], this leads to a non-trivial condition, which is the desired temperature-dependent integral equation.

We proceed to diagonalize $H_0 = C + \Sigma_{\mathbf{k}} H_{\mathbf{k}}$. C is just a constant, and the operators $H_{\mathbf{k}}$ for different $\mathbf{k}$ commute with each other. We may therefore diagonalize each $H_{\mathbf{k}}$ separately.

---

[9] This is the point at which the proof becomes heuristic. The rigorous proof of Bogoliubov (60a) makes no use of the expansion (3.22).

For more information concerning perturbation expansions in statistical mechanics, see Kirkwood 33, 34; Husimi 40; Bogoliubov 46, 47a; Vlasov 50; Green 51, 52a; Siegert 52; Saenz 55; Oppenheim 57; Mühlschlegel 62; Uhlenbeck 62.

Although $H_{\mathbf{k}}$ looks at first sight like a very difficult operator to handle, it is actually a simple matter to diagonalize $H_{\mathbf{k}}$ explicitly. The essential point to notice is that $H_{\mathbf{k}}$ is a *quadratic form* in the Fermi operators $a_{\mathbf{k},+}$ and $a_{-\mathbf{k},-}$ (and their Hermitean conjugates); it can therefore be diagonalized by a transformation to normal modes. The transformation in question is exactly the Bogoliubov transformation Eqs. (V, 3.13). Substitution of (V, 3.13) into the definition of $B_{\mathbf{k}}$, (3.3) and (3.4), yields

$$B_{\mathbf{k}} = u_{\mathbf{k}}^2 \alpha_{\mathbf{k}1} \alpha_{\mathbf{k}0} - v_{\mathbf{k}}^2 \alpha_{\mathbf{k}0}^+ \alpha_{\mathbf{k}1}^+ - u_{\mathbf{k}} v_{\mathbf{k}} (\alpha_{\mathbf{k}0}^+ \alpha_{\mathbf{k}0} + \alpha_{\mathbf{k}1}^+ \alpha_{\mathbf{k}1}) + (u_{\mathbf{k}} v_{\mathbf{k}} - w_{\mathbf{k}}) \tag{3.28}$$

The last term on the right of (3.28) is a c-number; the expression for $B_{\mathbf{k}}^+$ is obtained from (3.28) by Hermitean conjugation. When these expressions are substituted into the definition of $H_{\mathbf{k}}$, Eq. (3.6c), we obtain after collecting terms

$$\begin{aligned} H_{\mathbf{k}} = {} & [(\epsilon_{\mathbf{k}} - \mu)(u_{\mathbf{k}}^2 - v_{\mathbf{k}}^2) + 2\Delta_{\mathbf{k}} u_{\mathbf{k}} v_{\mathbf{k}}] (\alpha_{\mathbf{k}0}^+ \alpha_{\mathbf{k}0} + \alpha_{\mathbf{k}1}^+ \alpha_{\mathbf{k}1}) \\ & + 2(\epsilon_{\mathbf{k}} - \mu) v_{\mathbf{k}}^2 - 2\Delta_{\mathbf{k}} (u_{\mathbf{k}} v_{\mathbf{k}} - w_{\mathbf{k}}) \\ & + [2(\epsilon_{\mathbf{k}} - \mu) u_{\mathbf{k}} v_{\mathbf{k}} - \Delta_{\mathbf{k}} (u_{\mathbf{k}}^2 - v_{\mathbf{k}}^2)] (\alpha_{\mathbf{k}1} \alpha_{\mathbf{k}0} + \alpha_{\mathbf{k}0}^+ \alpha_{\mathbf{k}1}^+) \end{aligned} \tag{3.29}$$

If we interpret $\alpha_{\mathbf{k}s}^+ \alpha_{\mathbf{k}s} = f_{\mathbf{k}s}$ as the number operator for Bogolons of type **k**, s, then the first line of (3.29) gives the effective excitation energy due to the excitations of type (**k**, 0) and (**k**, 1). The middle line of (3.29) is a c-number, which need not worry us. Thus, we are all right provided the last line of (3.29) can be made to vanish. The condition for this is

$$2(\epsilon_{\mathbf{k}} - \mu) u_{\mathbf{k}} v_{\mathbf{k}} - (u_{\mathbf{k}}^2 - v_{\mathbf{k}}^2) \Delta_{\mathbf{k}} = 0 \tag{3.30}$$

Except for the absence of the term $L_{\mathbf{k}}$ (which vanishes identically with the reduced interaction), Eq. (3.30) is the same equation as (V, 3.26) in the zero-temperature Bogoliubov theory. We therefore obtain exactly the same solutions; introducing the non-dimensional quantity $y_{\mathbf{k}}$ by

$$y_{\mathbf{k}} = \frac{\epsilon_{\mathbf{k}} - \mu}{\Delta_{\mathbf{k}}} = \frac{\xi_{\mathbf{k}}}{\Delta_{\mathbf{k}}} \tag{3.31}$$

we have, in particular,

$$v_{\mathbf{k}}^2 = \frac{1}{2} \left[1 \pm \frac{y_{\mathbf{k}}}{(1 + y_{\mathbf{k}}^2)^{1/2}} \right] \tag{3.32}$$

and

$$u_{\mathbf{k}} v_{\mathbf{k}} = \pm \frac{1}{2(1 + y_{\mathbf{k}}^2)^{1/2}} \tag{3.33}$$

The upper sign should be chosen for a purely attractive interaction (all $J_{\mathbf{kk}'}$ positive). More generally, the signs in (3.32) and (3.33) are correlated, i.e., one should choose either the upper sign in both, or the lower sign in both.

By using (3.31)–(3.33) in (3.29), we obtain the following expression for the coefficient in the first line:

$$\hat{\epsilon}_{\mathbf{k}} = (\epsilon_{\mathbf{k}} - \mu)(u_{\mathbf{k}}^2 - v_{\mathbf{k}}^2) + 2\Delta_{\mathbf{k}} u_{\mathbf{k}} v_{\mathbf{k}} = \pm \Delta_{\mathbf{k}}(1 + y_{\mathbf{k}}^2)^{1/2} \tag{3.34}$$

where the sign is correlated with the signs in (3.32) and (3.33). Now, however, we can fix the sign more definitely, by requiring that the excitation energy $\hat{\epsilon}_{\mathbf{k}}$ of a Bogolon must not be negative.[10] Thus the sign convention is: choose the upper signs in all of (3.32), (3.33), and (3.34) if $\Delta_{\mathbf{k}}$ is positive; otherwise choose the lower sign. With this sign convention, we obtain

$$\hat{\epsilon}_{\mathbf{k}} = +[(\epsilon_{\mathbf{k}} - \mu)^2 + (\Delta_{\mathbf{k}})^2]^{1/2} \tag{3.35}$$

Substitution of these expressions into the full formula (3.29) then leads to the diagonal form of $H_{\mathbf{k}}$:

$$H_{\mathbf{k}} = \hat{\epsilon}_{\mathbf{k}}(\alpha_{\mathbf{k}0}^{+}\alpha_{\mathbf{k}0} + \alpha_{\mathbf{k}1}^{+}\alpha_{\mathbf{k}1}) + (\epsilon_{\mathbf{k}} - \mu - \hat{\epsilon}_{\mathbf{k}} + 2\Delta_{\mathbf{k}} w_{\mathbf{k}}) \tag{3.36}$$

and substitution of this into (3.6) gives the full diagonal form of the operator H_0:

$$H_0 = C' + \sum_{\mathbf{k},s} \hat{\epsilon}_{\mathbf{k}} \alpha_{\mathbf{k}s}^{+}\alpha_{\mathbf{k}s} = C' + \sum_{\mathbf{k}} W_{\mathbf{k}} \tag{3.37a}$$

$$C' = \sum_{\mathbf{k}} (\epsilon_{\mathbf{k}} - \mu - \hat{\epsilon}_{\mathbf{k}} + \Delta_{\mathbf{k}} w_{\mathbf{k}}) \tag{3.37b}$$

The operators $f_{\mathbf{k}s} = \alpha_{\mathbf{k}s}^{+}\alpha_{\mathbf{k}s}$ commute with each other, and each $f_{\mathbf{k}s}$ can take the values 0 and 1, only. An eigenstate of H_0 is specified by prescribing these occupation numbers, i.e., the eigenstates of H_0 are "configurations" $\{f_{\mathbf{k}s}\}$. The energy E_γ of the configuration $\gamma = \{f_{\mathbf{k}s}\}$ is

$$E_\gamma = C' + \sum_{\mathbf{k},s} \hat{\epsilon}_{\mathbf{k}} f_{\mathbf{k}s} \tag{3.38}$$

This completes the process of diagonalizing the operator H_0. We are now in a position to evaluate the statistical expectation value $\langle \beta_{\mathbf{k}} \rangle_0$, set it

[10] If this condition is violated, the number operator for Bogolons of type $(\mathbf{k}, s)$ would have to be identified with $\alpha_{\mathbf{k}s}\alpha_{\mathbf{k}s}^{+} = 1 - \alpha_{\mathbf{k}s}^{+}\alpha_{\mathbf{k}s}$. This is thus mainly a question of terminology. Statistical expectation values, which are sums over both eigenvalues (0 and 1) of either operator, are unaffected by this choice.

equal to $w_{\mathbf{k}}$, and thus write down the condition that $\Delta_{\mathbf{k}}$, (3.5), be a consistent equation.

Let $Q_{\mathbf{k}}$ be an arbitrary operator which, however, involves the index **k** only. Since all the operators $W_{\mathbf{k}}$ in (3.37a) commute with each other, we obtain quite generally

$$\langle Q_{\mathbf{k}}\rangle_0 = \frac{\text{Trace}\,[Q_{\mathbf{k}}\exp(-\beta H_0)]}{\text{Trace}\exp(-\beta H_0)} = \frac{\text{Trace}\,[Q_{\mathbf{k}}\exp(-\beta W_{\mathbf{k}})]}{\text{Trace}\exp(-\beta W_{\mathbf{k}})} \tag{3.39}$$

As a first application of this formula, let us evaluate the average number of Bogolons of type ($\mathbf{k}$, s). This is

$$\bar{f}_{\mathbf{k}s} = \langle \alpha^+_{\mathbf{k}s}\alpha_{\mathbf{k}s}\rangle_0 = \frac{\text{Trace}\,[\alpha^+_{\mathbf{k}s}\alpha_{\mathbf{k}s}\exp(-\beta W_{\mathbf{k}})]}{\text{Trace}\,[\exp(-\beta W_{\mathbf{k}})]} \tag{3.40}$$

The numerator of this expression is explicitly, for $s = 0$ (say),

$$\begin{aligned}\text{Trace}\,&[\alpha^+_{\mathbf{k}0}\alpha_{\mathbf{k}0}\exp(-\beta W_{\mathbf{k}})]\\ &= \sum_{f_{\mathbf{k}0}=0,1}\ \sum_{f_{\mathbf{k}1}=0,1} f_{\mathbf{k}0}\exp[-\beta\hat{\epsilon}_{\mathbf{k}}(f_{\mathbf{k}0}+f_{\mathbf{k}1})]\\ &= \exp(-\beta\hat{\epsilon}_{\mathbf{k}})\,[1+\exp(-\beta\hat{\epsilon}_{\mathbf{k}})]\end{aligned} \tag{3.41}$$

Evaluating the denominator by the same technique, Eq. (3.40) leads to

$$\bar{f}_{\mathbf{k}s} = \frac{1}{\exp(\beta\hat{\epsilon}_{\mathbf{k}})+1} \tag{3.42}$$

Thus the average number of Bogolons of type (**k**, *s*) *is given by the ordinary Fermi distribution law.*

Now let us put $Q_{\mathbf{k}} = \beta_{\mathbf{k}}$ in (3.39). We use the Bogoliubov transformation, (V, 3.13) to express $\beta_{\mathbf{k}}$ in terms of Bogoliubov operators. The result is

$$\beta_{\mathbf{k}} = u_{\mathbf{k}}v_{\mathbf{k}}(1-\alpha^+_{\mathbf{k}0}\alpha_{\mathbf{k}0}-\alpha^+_{\mathbf{k}1}\alpha_{\mathbf{k}1}) + u^2_{\mathbf{k}}\alpha_{\mathbf{k}1}\alpha_{\mathbf{k}0} - v^2_{\mathbf{k}}\alpha^+_{\mathbf{k}0}\alpha^+_{\mathbf{k}1} \tag{3.43}$$

Although the operator H_0 fails to conserve the number of electrons, the form (3.37) makes it clear that H_0 does conserve the number of Bogolons. The last two terms in (3.43) alter the number of Bogolons by -2 and $+2$, respectively. Therefore these terms give zero in the statistical expectation value (3.39). The remaining terms involve the average numbers (3.42) and thus

$$\langle\beta_{\mathbf{k}}\rangle_0 = u_{\mathbf{k}}v_{\mathbf{k}}(1-\bar{f}_{\mathbf{k}0}-\bar{f}_{\mathbf{k}1}) = u_{\mathbf{k}}v_{\mathbf{k}}\tanh(\tfrac{1}{2}\beta\hat{\epsilon}_{\mathbf{k}}) \tag{3.44}$$

According to (3.19b), this must be set equal to $w_{\mathbf{k}}$. Substitution of (3.33) and attention to the sign convention yields

$$w_{\mathbf{k}} = \pm\frac{\tanh(\tfrac{1}{2}\beta\hat{\epsilon}_{\mathbf{k}})}{2(1+y^2_{\mathbf{k}})^{1/2}} = +\frac{\Delta_{\mathbf{k}}}{2\hat{\epsilon}_{\mathbf{k}}}\tanh(\tfrac{1}{2}\beta\hat{\epsilon}_{\mathbf{k}}) \tag{3.45}$$

The $\hat{\epsilon}_k$ is defined as the positive square root [see Eq. (3.35)] and the sign of w_k agrees with the sign of Δ_k.[11]

When we substitute (3.45) into the definition (3.5), we obtain the *temperature-dependent integral equation*

$$\Delta_k = \frac{1}{V} \sum_{k'} \frac{J_{kk'} \Delta_{k'}}{2\hat{\epsilon}_{k'}} \tanh\left(\tfrac{1}{2} \beta \hat{\epsilon}_{k'}\right) \tag{3.46}$$

In this equation, $\hat{\epsilon}_k$ is defined by (3.35). Thus, for given temperature (given β) and given chemical potential μ, Eq. (3.46) is a non-linear integral equation for the unknown quantities $\Delta_k = \Delta_k(\beta, \mu)$.

In practice, we are given the number of particles N, not the chemical potential μ. We thus need an additional equation to relate μ and N. This equation is obtained by evaluating the expectation value of the number operator N_{Op}, over the approximate density matrix U_0, Eq. (3.16). We first transform the number operator to Bogoliubov form by using (V, 3.13). The result is

$$\begin{aligned} N_{\text{Op}} &= \sum_k [(a^+_{k,+} a_{k,+} + a^+_{-k,-} a_{-k,-}) \\ &= \sum_k [2v_k^2 + (u_k^2 - v_k^2)(\alpha^+_{k0}\alpha_{k0} + \alpha^+_{k1}\alpha_{k1}) + 2u_k v_k (\alpha^+_{k0}\alpha^+_{k1} + \alpha_{k1}\alpha_{k0})] \end{aligned} \tag{3.47}$$

We now use (3.39) to separate the contributions of different **k**-values. For each value of **k**, the last term in (3.47) fails to conserve the number of Bogolons and thus gives zero expectation value. The other terms reduce to results already calculated, and give

$$\begin{aligned} \langle N \rangle_0 &= \sum_k [2v_k^2 + (u_k^2 - v_k^2)(f_{k0} + f_{k1})] \\ &= \sum_k \left\{ 1 - \frac{\epsilon_k - \mu}{\hat{\epsilon}_k} + \frac{2(\epsilon_k - \mu)}{\hat{\epsilon}_k [\exp(\beta \hat{\epsilon}_k) + 1]} \right\} \end{aligned} \tag{3.48}$$

For any given μ, we can (in principle) solve (3.46) to get $\Delta_k = \Delta_k(\beta, \mu)$ and thence $\hat{\epsilon}_k = \hat{\epsilon}_k(\beta, \mu)$ from (3.35). Substitution of these results into (3.48) then yields $\langle N \rangle_0$, and the condition determining μ is that $\langle N \rangle_0$ must equal the actual particle number N.

In the limit of zero temperature, i.e., infinite β, the hyperbolic tangent becomes unity, and (3.46) reduces to the zero-temperature equation

[11] Actually, the sign of (3.45) is the same whichever sign is chosen for $\hat{\epsilon}_k$. This bears out our earlier remark that the sign convention is not essential for the final results.

(IV, 1.26).[12] Since we have seen that the BZT technique of dealing with the reduced Hamiltonian leads to an *exact* result for the free energy at all temperatures, we conclude that in particular *the zero-temperature results of Chapters IV and V are exact results for the reduced Hamiltonian.*

There still remains the question of the physical meaning of the assumption of the reduced Hamiltonian. We defer this question to Chapter VII, and devote the remainder of this chapter to the thermodynamic consequences of the results which we have obtained.

4. Solution of the Integral Equation. The Transition Temperature and the Isotope Effect

In order to obtain a solution of the non-linear integral equation (3.46), we make the same (unjustified) assumption of separability of the interaction matrix elements as we did for the zero-temperature case [Eq. (IV, 2.1)]. In our present notation, this assumption becomes

$$J_{\mathbf{k}\mathbf{k}'} = +\gamma_{\mathbf{k}}\gamma_{\mathbf{k}'} \tag{4.1}$$

where $\gamma_{\mathbf{k}}$ is volume-independent. The solution of Eq. (3.46) then has the form

$$\Delta_{\mathbf{k}} = S\gamma_{\mathbf{k}} \tag{4.2}$$

where S is a constant which must be determined from the equation

$$S = \frac{1}{V}\sum_{\mathbf{k}} \frac{\gamma_{\mathbf{k}}\Delta_{\mathbf{k}}}{2\hat{\epsilon}_{\mathbf{k}}} \tanh\left(\tfrac{1}{2}\beta\hat{\epsilon}_{\mathbf{k}}\right) \tag{4.3}$$

Substitution of (4.2) into (4.3) leads to

$$\frac{1}{V}\sum_{\mathbf{k}} \frac{\gamma_{\mathbf{k}}^2}{2\hat{\epsilon}_{\mathbf{k}}} \tanh\left(\tfrac{1}{2}\beta\hat{\epsilon}_{\mathbf{k}}\right) = 1 \tag{4.4}$$

where now $\hat{\epsilon}_{\mathbf{k}}$ is defined by

$$\hat{\epsilon}_{\mathbf{k}} = [(\epsilon_{\mathbf{k}} - \mu)^2 + (S\gamma_{\mathbf{k}})^2]^{1/2} \tag{4.5}$$

In the limit of zero temperature, $\beta \to \infty$, Eq. (4.4) reduces to (IV, 2.4), with

$$\xi_{\mathbf{k}} = \epsilon_{\mathbf{k}} - \mu \tag{4.6}$$

[12] The quantity $L_{\mathbf{k}}$, which appears in (IV,1.26) implicitly through (IV,1.18), vanishes for the reduced Hamiltonian. Observe also the change in notation, our equation (3.1).

We again use the supersimplified interaction (IV, 2.5); since the terms L_k of Chapter IV vanish identically for the reduced interaction, we now have

$$\mu = \epsilon_F = \frac{\hbar^2 k_F^2}{2M} \qquad \text{for} \quad T = 0 \tag{4.7}$$

where k_F is the usual Fermi momentum.

We shall assume for the moment that the chemical potential μ remains equal to its zero-temperature value (4.7) throughout the superconducting region, and we shall verify this assumption afterward. With this assumption, we have

$$\xi_k = \epsilon_k - \mu \cong \epsilon_k - \epsilon_F \qquad \text{all} \quad T < T_c \tag{4.8}$$

Furthermore, γ_k is by assumption constant for $|\xi_k| < \hbar\omega_c$, and zero outside that region. Thus $\Delta_k = S\gamma_k$ is also constant for $|\xi_k| < \hbar\omega_c$, and zero elsewhere. Replacing the sum over k by an integral over ξ_k, and noting that the density of states $\rho_E(\xi)$ is nearly constant over the small energy interval in question, Eq. (4.4) can be written in the form

$$\int_0^{\hbar\omega_c} \frac{\tanh\left[\frac{1}{2}\beta(\xi^2 + \Delta^2)^{1/2}\right]}{(\xi^2 + \Delta^2)^{1/2}}\, d\xi = \frac{1}{\rho} \tag{4.9}$$

where ρ is the non-dimensional coupling parameter used earlier, $\rho = \gamma^2 \rho_E / V$. In the limit of zero temperature (4.9) reduces to (IV, 2.7) as it should. We write this zero temperature limit in the form

$$\int_0^{\hbar\omega_c} (\xi^2 + \Delta_0^2)^{-1/2}\, d\xi = \frac{1}{\rho} \tag{4.10}$$

where the subscript on Δ_0 indicates the zero-temperature value of this quantity.

For any finite value of β, the hyperbolic tangent in the numerator of (4.9) is less than unity. Since the value of the integral as a whole must remain constant, we conclude that Δ *must be a decreasing function of temperature.*

The parameter Δ enters (4.9) only through Δ^2. Hence the smallest significant value of Δ is $\Delta = 0$. This happens, according to (4.9), at a temperature $T_c = 1/\mathfrak{k}\beta_c$ such that

$$\int_0^{\hbar\omega_c} \frac{\tanh\left(\frac{1}{2}\beta_c \xi\right)}{\xi}\, d\xi = \frac{1}{\rho} \tag{4.11}$$

For larger values of T (smaller values of β), Eq. (4.9) has no solution Δ at all. Thus T_c defined by (4.11) is the *critical temperature* of the superconductor.

For $T > T_c$, the basic equation (3.30) has only the "trivial" solution

$$\Delta_k = 0, \qquad u_k = u_k^0, \qquad v_k = v_k^0 \qquad \text{(trivial solution)} \tag{4.12}$$

With this solution, $u_k v_k = 0$ for all k, and (3.30) is satisfied trivially. The solution (4.12) describes the normal state of the metal, in the approximation of no interaction at all between the electrons, [see (V, 3.11)].

The complete absence of electron-electron interactions in the "normal" state in the theory of Bogoliubov, Zubarev, and Tserkovnikov is due to their initial starting point, the reduced interaction of BCS. The reduced interaction retains only a vanishingly small fraction of all the matrix elements of the true interaction. In the superconducting state, these specially selected matrix elements suffice to obtain a finite (proportional to the volume V) contribution to the free energy of the system. But in the normal state, keeping only the reduced interaction is equivalent to keeping no interaction at all.

In the normal state, the "Bogolon" excitations reduce to ordinary electrons for $k > k_F$, to holes in the Fermi sea for $k < k_F$. This was shown in Chapter V, Section 3 for the special values of u_k and v_k given by (4.12).

The temperature-dependent equation (4.9) has been discussed in detail in a very nice paper by Mühlschlegel (59), who also tabulates numerical values of several interesting functions. We reproduce Mühlschlegel's tables here, and refer the reader to Mühlschlegel for the derivations. Apart from the purely numerical results, there emerges an interesting relation between the critical temperature T_c [defined by Eq. (4.11)] and the energy gap parameter Δ_0 at zero temperature [defined by Eq. (4.10)]. This relation, first found by Bardeen (57), is

$$\mathfrak{k} T_c = 0.567 \Delta_0 \tag{4.13}$$

where the numerical coefficient is the ratio γ/π, with $\gamma = \exp(C) = 1.781...$ being Euler's constant.

The zero-temperature energy gap parameter Δ_0 is given by Eq. (IV, 2.9). This equation contains the non-dimensional coupling parameter ρ in the form $\exp(-1/\rho)$, and we pointed out that this exponential factor is very significant numerically, making the energy gap one or two orders of magnitude smaller than characteristic phonon energies. For the same reason, we now see that the superconducting transition temperature T_c is significantly smaller than a characteristic phonon temperature (the Debye temperature). In Section 5 of Chapter V we showed that the parameter ρ is always less than $\frac{1}{2}$, and thus $\exp(-1/\rho)$ is always less than $\exp(-2) = 0.136$. Application of the methods of Bogoliubov to the Fröhlich model at finite temperatures (Moskalenko 58, Zubarev 60) shows that the cutoff energy $\hbar\omega_c$ in (IV, 2.9) should be taken as

$$\hbar\omega_c = 0.303\,\mathfrak{k}\Theta$$

where Θ is the Debye temperature. Combining this equation with (IV, 2.9) and (4.13), we obtain

$$T_c = 0.344\,\Theta\exp(-1/\rho) \tag{4.13a}$$

Using $\rho < \frac{1}{2}$, we find that T_c is always less than 1/20th of the Debye temperature Θ, and usually considerably below this upper limit. The exponential factor $\exp(-1/\rho)$ therefore accounts for the fact that there are no room-temperature superconductors.

The relation (4.13) can be checked directly against experimental data. For example, using the data of Phillips (59) for aluminium, the transition temperature is $T_c = 1.16°\text{K}$. Thus (4.13) gives $\Delta_0/\mathfrak{k} = 1.764\,T_c = 2.05°\text{K}$. This value is in excellent agreement with the value $\Delta_0/\mathfrak{k} = 2.03°\text{K}$ deduced from the critical field $H_c(0)$ at zero temperature, and in reasonable agreement with the value $\Delta_0/\mathfrak{k} = 1.94°\text{K}$ deduced from a fit to the specific heat data at low temperatures. Considering the extreme crudity of the model used, this agreement is a remarkable success of the theory.

Formula (4.11) also has an important qualitative consequence: the *isotope effect* for the transition temperature T_c. The coupling constant ρ can be shown to be independent of the isotopic mass of the metallic ions, to a first approximation. On the other hand, the cutoff energy $\hbar\omega_c$ for the model interaction should be some definite fraction of the Debye cutoff energy of lattice vibrations. That is, ω_c is a cutoff frequency of an elastic vibration. Elastic vibration frequencies are of the form $(K/M_i)^{1/2}$ where K is related to the force constant, and M_i is the mass of the ion. The force constant is derived from electronic motions and is thus independent of the ionic mass. Hence, finally, $\hbar\omega_c$ is expected to be proportional to $(M_i)^{-1/2}$, and so is the transition temperature T_c by (4.11).

Although this derivation of the isotope effect is suggestive, it can hardly be called conclusive. Since the coupling constant ρ enters the final result in the form $\exp(-1/\rho)$ [see (4.13) and (IV, 2.9)] quite small effects of the ionic mass on ρ could alter the result significantly. Just such effects are produced by the repulsive Coulomb interaction between electrons, to which the scaling argument for the Fröhlich interaction fails to apply. This was first pointed out by Swihart (59, 59a), who gave an estimate of the correction. Morel and Anderson (Morel 62) give numerical values of the exponent n in a power law $\Delta_0 \sim (M_i)^{-n}$. The numerical values of n deviate from $n = \frac{1}{2}$ significantly more than the experimental values of n' in the power law $T_c \sim (M_i)^{-n'}$. Thus, if relation (4.13) still holds, there is a real disagreement between theory and experiment for the isotope effect. However, this pessimistic conclusion of Morel (62) is perhaps somewhat premature, in view of the absence, so far, of detailed numerical solutions of the full integral equation, and in view of the possibility of effects (especially near the transition temperature) of interaction matrix elements other than the reduced interaction.

We still have to verify that the assumption of constant chemical potential $\mu = \epsilon_F$ is consistent with the general requirement (3.48). The easiest way is to observe that for $T = T_c$, the BZT theory yields the same relation between μ and particle number as an ideal Fermi gas. For the latter, the relation is given by (III, 3.9), from which it is easily deduced that

$$\mu = \epsilon_F \left[1 - \frac{\pi^2}{12}\left(\frac{\mathfrak{k}T}{\epsilon_F}\right)^2 + \cdots\right] \qquad \text{normal Fermi gas} \tag{4.14}$$

For typical superconductors, the ratio $\mathfrak{k}T_c/\epsilon_F$ is of the order of 10^{-3}, so that the correction term (4.14) is of the order of one part in a million, which is completely negligible here. The overall variation of μ with temperature in the superconducting region is equal to that of the normal Fermi gas, so that our procedure here is justified.

5. The Specific Heat and the Critical Field

In order to obtain all the thermodynamic properties, it suffices to determine the grand canonical potential $\Omega(V, T, \mu)$, defined by (3.7) or (3.8). In the BZT theory, this is replaced by $\Omega_0(V, T, \mu)$, Eq. (3.15).

Let us therefore evaluate Ω_0 from (3.15) and (3.37). We obtain

$$\exp(-\beta\Omega_0) = \exp(-\beta C') \prod_{\mathbf{k}} \text{Trace} \exp(-\beta W_{\mathbf{k}}) \tag{5.1}$$

where C' and $W_{\mathbf{k}}$ are defined in (3.37). The trace in (5.1) is evaluated by the same technique we used to obtain Eq. (3.41), with the result

$$\text{Trace} \exp(-\beta W_{\mathbf{k}}) = [1 + \exp(-\beta\hat{\epsilon}_{\mathbf{k}})]^2 \tag{5.2}$$

Thus

$$\Omega(V, T, \mu) = \Omega_0(V, T, \mu) = C' - 2\mathfrak{k}T \sum_{\mathbf{k}} \ln[1 + \exp(-\beta\hat{\epsilon}_{\mathbf{k}})] \tag{5.3}$$

Finally, we use Eq. (3.37b) for C', and replace $w_{\mathbf{k}}$ in this equation by its actual value, Eq. (3.45). This gives the *final result*

$$\Omega(V, T, \mu) = \sum_{\mathbf{k}} \left\{ \epsilon_{\mathbf{k}} - \mu - \hat{\epsilon}_{\mathbf{k}} + \frac{\Delta_{\mathbf{k}}^2 \tanh(\frac{1}{2}\beta\hat{\epsilon}_{\mathbf{k}})}{2\hat{\epsilon}_{\mathbf{k}}} - 2\mathfrak{k}T \ln[1 + \exp(-\beta\hat{\epsilon}_{\mathbf{k}})] \right\} \tag{5.4}$$

From here on, it is only a matter of straightforward application of well-known thermodynamic formulas (see Appendix A). The evaluation of the thermodynamic quantities is discussed in detail by Mühlschlegel (59). We shall merely quote results here.

It is convenient to introduce the "reduced temperature" t by

$$t = T/T_c \tag{5.5}$$

For $T > T_c$, i.e., $t > 1$, the BZT theory reduces to the thermodynamics of the ideal Fermi-Dirac gas. In particular, the electronic specific heat is given by Eq. (2.19) in that region.

At $T = T_c$, i.e., at $t = 1$, the energy gap parameter Δ vanishes, as we have seen. In the neighborhood of the critical temperature, Δ approaches zero as follows:

$$\frac{\Delta}{\Delta_0} \cong 1.74(1 - t)^{1/2} \qquad \text{for} \quad t \to 1 \tag{5.6}$$

This infinite slope of $\Delta(T)$ vs. T at $T = T_c$ results in a jump of the electronic specific heat, i.e., we obtain a second-order transition at $T = T_c$.[13] The magnitude of the jump in the specific heat is given by

$$\frac{c_{V,\mathrm{es}}}{c_{V,\mathrm{en}}} = 2.43 \qquad \text{at} \quad T = T_c \tag{5.7}$$

In the immediate neighborhood of the transition temperature, the electronic specific heat in the superconducting state decreases linearly with temperature; the ratio of specific heats in the two states follows the law

$$\frac{c_{V,\mathrm{es}}}{c_{V,\mathrm{en}}} = 2.43 - 3.76(1 - t) \qquad \text{for} \quad t \to 1 \tag{5.8}$$

The other interesting region is the low-temperature domain, $T \ll T_c$ or $t \ll 1$. In that region, the energy gap parameter is nearly constant:

$$\frac{\Delta}{\Delta_0} = 1 - \left(\frac{2\pi\mathfrak{k}T}{\Delta_0}\right)^{1/2} \exp\left(-\frac{\Delta_0}{\mathfrak{k}T}\right) \qquad \text{for} \quad \mathfrak{k}T \ll \Delta_0 \tag{5.9}$$

[13] We emphasize once more that the BZT theory is an *exact* evaluation of the consequences of the reduced Hamiltonian. Thus the detailed nature of the transition is a mathematically valid deduction.

Therefore, we recover all the results of Section 2, in which the variation of the effective energy spectrum with temperature was ignored altogether.

In the intermediate region of temperatures, one must rely on the numerical results of Mühlschlegel (59). These are reproduced in Table 6.1.

TABLE 6.1

Various thermodynamic quantities as functions of the reduced temperature $t = T/T_c$

Column 2: $\Delta(T)/\Delta(0)$ Ratio of energy gap at temperature T to zero-temperature gap
Column 3: S/C_n Entropy at temperature T, divided by normal specific heat at $T = T_c$
Column 4: $-F/C_nT_c$ Negative free energy at T, divided by $T_cC_n(T_c)$
Column 5: $H_c^2/4\pi C_nT_c$ Square of the critical field $H_c(T)$, divided by $4\pi T_cC_n(T_c)$
Column 6: $1 - \Lambda_0/\Lambda_T$ Temperature dependence of London parameter $\Lambda(T)$.
Column 7: $C(T)/C_n(T_c)$ Electronic specific heat at T, divided by normal value at T_c.

This table is taken from Mühlschlegel (59).

$t = T/T_c$	Δ_T/Δ_0	S/C_n	$-F/C_nT_c$	$H_c^2/4\pi C_nT_c$	$1-\Lambda_0/\Lambda_T$	$C_s(T)/C_n(T_c)$
1.00	0.0000	1.0000	0.5000	0.0000	1.0000	2.4261
0.98	0.2436	0.9519	0.4805	0.0003	0.9601	2.3314
0.96	0.3416	0.9048	0.4619	0.0011	0.9206	2.2378
0.94	0.4148	0.8587	0.4443	0.0025	0.8814	2.1454
0.92	0.4749	0.8136	0.4276	0.0044	0.8425	2.0541
0.90	0.5263	0.7694	0.4117	0.0067	0.8041	1.9639
0.88	0.5715	0.7263	0.3968	0.0096	0.7660	1.8750
0.86	0.6117	0.6842	0.3827	0.0129	0.7283	1.7874
0.84	0.6480	0.6432	0.3694	0.0166	0.6911	1.7010
0.82	0.6810	0.6032	0.3569	0.0207	0.6544	1.6159
0.80	0.7110	0.5643	0.3453	0.0253	0.6182	1.5321
0.78	0.7386	0.5266	0.3344	0.0302	0.5826	1.4498
0.76	0.7640	0.4900	0.3242	0.0354	0.5475	1.3689
0.74	0.7874	0.4546	0.3148	0.0410	0.5131	1.2894
0.72	0.8089	0.4203	0.3060	0.0468	0.4793	1.2115
0.70	0.8288	0.3873	0.2979	0.0529	0.4463	1.1352
0.68	0.8471	0.3554	0.2905	0.0593	0.4140	1.0605
0.66	0.8640	0.3249	0.2837	0.0659	0.3825	0.9874
0.64	0.8796	0.2956	0.2775	0.0727	0.3518	0.9162
0.62	0.8939	0.2676	0.2719	0.0797	0.3221	0.8467
0.60	0.9070	0.2410	0.2668	0.0868	0.2933	0.7792
0.58	0.9190	0.2157	0.2622	0.0940	0.2656	0.7136
0.56	0.9299	0.1918	0.2582	0.1014	0.2389	0.6501
0.54	0.9399	0.1693	0.2545	0.1087	0.2133	0.5888
0.52	0.9488	0.1482	0.2514	0.1162	0.1890	0.5298
0.50	0.9569	0.1285	0.2486	0.1236	0.1660	0.4731
0.48	0.9641	0.1103	0.2462	0.1310	0.1442	0.4190
0.46	0.9704	0.0937	0.2442	0.1384	0.1239	0.3675
0.44	0.9760	0.0784	0.2425	0.1457	0.1055	0.3188
0.42	0.9809	0.0646	0.2410	0.1528	0.0878	0.2731
0.40	0.9850	0.0524	0.2399	0.1599	0.0721	0.2305
0.38	0.9885	0.0416	0.2389	0.1667	0.0580	0.1913
0.36	0.9915	0.0322	0.2382	0.1734	0.0456	0.1555

$t = T/T_c$	Δ_T/Δ_0	S/C_n	$-F/C_nT_c$	$H_c^2/4\pi C_nT_c$	$1-\Delta_0/\Delta_T$	$C_s(T)/C_n(T_c)$
0.34	0.9938	0.0243	0.2376	0.1798	0.0348	0.1233
0.32	0.9957	0.0177	0.2372	0.1860	0.0257	0.0950
0.30	0.9971	0.0124	0.2369	0.1919	0.0182	0.0706
0.28	0.9982	0.0082	0.2367	0.1975	0.0123	0.0502
0.26	0.9989	0.0051	0.2366	0.2028	0.0078	0.0338
0.24	0.9994	0.0030	0.2365	0.2077	0.0046	0.0212
0.22	0.9997	0.0016	0.2365	0.2123	0.0024	0.0121
0.20	0.9999	0.0007	0.2364	0.2164	0.0011	0.0061
0.18	1.0000	0.0003	0.2364	0.2202	0.0005	0.0027
0.16	1.0000	0.0001	0.2364	0.2236	0.0001	0.0009
0.14	1.0000	0.0000	0.2364	0.2266	0.0000	0.0002

TABLE 6.2

First entry: Ratio of superconducting to normal electronic specific heat, at temperature T

Second entry: Ratio of zero-temperature energy gap to $\mathfrak{k}T$, on the basis of the thermodynamic theory

$\frac{c_{V,es}}{c_{V,en}}$	$\frac{\Delta_0}{\mathfrak{k}T}$	$\frac{c_{V,es}}{c_{V,en}}$	$\frac{\Delta_0}{\mathfrak{k}T}$	$\frac{c_{V,es}}{c_{V,en}}$	$\frac{\Delta_0}{\mathfrak{k}T}$	$\frac{c_{V,es}}{c_{V,en}}$	$\frac{\Delta_0}{\mathfrak{k}T}$
0.001	17.771	0.038	8.538	0.32	5.3917	1.20	3.0880
0.002	12.701	0.040	8.467	0.34	5.2934	1.25	3.0115
0.003	12.077	0.042	8.397	0.36	5.2003	1.30	2.9379
0.004	11.642	0.044	8.333	0.38	5.1117	1.35	2.8668
0.005	11.319	0.046	8.271	0.40	5.0271	1.40	2.7982
0.006	11.062	0.048	8.212	0.42	4.9461	1.45	2.7318
0.007	10.848	0.050	8.155	0.44	4.8685	1.50	2.6677
0.008	10.636			0.46	4.7940	1.55	2.6055
0.009	10.478	0.06	7.900	0.48	4.7223	1.60	2.5454
0.010	10.338	0.07	7.682	0.50	4.6532	1.65	2.4870
		0.08	7.491			1.70	2.4304
0.012	10.094	0.09	7.322	0.55	4.4904	1.75	2.3755
0.014	9.888	0.10	7.169	0.60	4.3402	1.80	2.3221
0.016	9.710			0.65	4.2006	1.85	2.2703
0.018	9.552	0.12	6.9017	0.70	4.0701	1.90	2.2199
0.020	9.411	0.14	6.6730	0.75	3.9477	1.95	2.1709
0.022	9.282	0.16	6.4722	0.80	3.8322	2.00	2.1232
0.024	9.165	0.18	6.2935	0.85	3.7229	2.05	2.0768
0.026	9.056	0.20	6.1322	0.90	3.6192	2.10	2.0316
0.028	8.955	0.22	5.9848	0.95	3.5205	2.15	1.9876
0.030	8.863	0.24	5.8490	1.00	3.4264	2.20	1.9447
0.032	8.774	0.26	5.7231	1.05	3.3364	2.25	1.9029
0.034	8.691	0.28	5.6057	1.10	3.2502	2.30	1.8622
0.036	8.613	0.30	5.4955	1.15	3.1675	2.35	1.8225
						2.40	1.7837
						2.4261	1.7640

The independent variable in this table is the reduced temperature (5.5), which is known as soon as the transition temperature has been determined.

From the point of view of comparing theory and experiment on the specific heat, this Table is not in the most suitable form: the scaling of temperatures according to (5.5) relies on the correctness of the theory at the transition point, and therefore overemphasizes one temperature region. A better procedure is the following. We start from the observation that the electronic specific heat ratio is a function of $\Delta_0/\mathfrak{k}T$ only:

$$\frac{c_{V,\mathrm{es}}}{c_{V,\mathrm{en}}} = F(y), \qquad y = \frac{\Delta_0}{\mathfrak{k}T} \tag{5.10}$$

Let us invert this function to obtain

$$y = G(x), \qquad x = \frac{c_{V,\mathrm{es}}}{c_{V,\mathrm{en}}} \tag{5.11}$$

This inversion is possible since $F(y)$ is a monotonic function; it can be performed numerically, and the results are given in Table 6.2.

We may now plot the product $Ty = TG(x)$ vs. the temperature T.[14] If the experiment is in agreement with the theory, the resulting points should lie on a horizontal straight line, at a height equal to $\Delta_0/\mathfrak{k}$. Thus, this plot allows us to determine, at a glance, the extent of agreement between theory and experiment, and the best value of $\Delta_0/\mathfrak{k}$ from the specific heat data.

---

[14] It is necessary to subtract out the Debye specific heat in order to obtain the electronic contributions. In the normal state of the metal, one obtains a straight line by plotting c_V/T versus T^2; the slope and intercept of the straight line determine γ in (2.19) and the Debye temperature Θ in (2.11).

In the superconducting state, it is usually assumed that the Debye specific heat term is the same as in the normal state. This is what we expect from the Bogoliubov theory, i.e., the phonon term in (1.2) is substantially the same as in the normal metal. Measurements of Yaqub (62) on tin, using the Mössbauer effect, indicate that the Debye temperature is indeed unchanged. If one makes this assumption, the electronic specific heat contribution $c_{V,es}$ is the difference between the actual specific heat in the superconducting state, and the Debye specific heat in the normal state. At all except the very lowest temperatures, the correction is quite small, of the order of a few percent at most.

However, there exist superconductors [niobium, see Boorse (60) and Hirschfeld (60, 62), and indium, see Bryant (60)] for which the *total* specific heat in the superconducting state is less than the normal-state Debye term, at very low temperatures. Explanations by Alers (60), Daunt (61) and Ferrell (61), are disputed by Alers (61), Chandrasekhai (61), Kulik (62), and Prange (62a); an alternative explanation, involving a term proportional to $T^3 \ln T$ in the specific heat of the *normal* state, has been proposed by Eliashberg (62); more theoretical work is highly desirable.

Figure 6.2 contains such a plot for the data of Phillips (59) on aluminum, i.e., the same data which were compared with the low-temperature theory in Fig. 6.1. We see that the agreement is now much better,

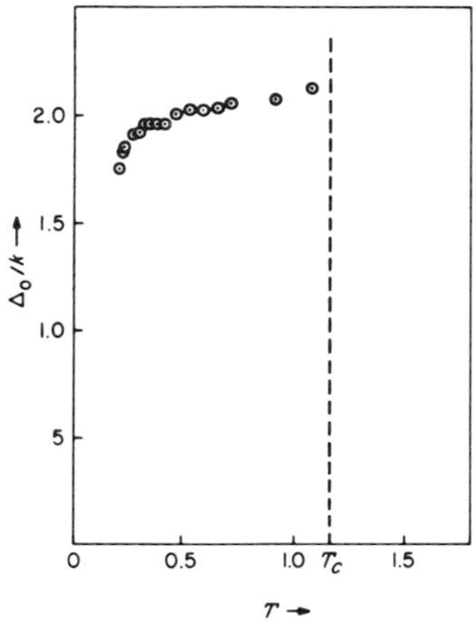

FIG. 6.2. The data of Fig. 6.1 on the specific heat ratio of aluminum are replotted here on the basis of the full thermodynamic theory. At any temperature T, the ratio $c_{V,es}/c_{V,en}$ has been used, in conjunction with Table 6.2, to deduce an "apparent" value of Δ_0, the energy gap parameter in the limit of zero temperature. Perfect agreement between theory and experiment would imply that all points lie on a horizontal straight line. The sharp break below 0.3°K is probably due to experimental errors. By way of comparison, the value of the energy gap deduced from the zero-temperature critical field of aluminum is $\Delta_0/\mathfrak{k} = 2.03$°K.

extending all the way up to the transition temperature. For temperatures above about 0.3°K, the agreement is really remarkably good; there is some evidence for a slight temperature dependence of $T\gamma$, but this is to be expected when one compares with such an oversimplified theory. The deviations below 0.3°K are probably experimental in origin.

The specific heat c_V has been measured in a number of superconductors (Phillips 58; Boorse 59; Martin 61, 61a; Bryant 61; Horwitz 62; Hein 63; Keesom 63). For most substances, there is reasonable, but not perfect, agreement with the thermodynamic theory. There is a pronounced discrepancy, however, in the low temperature electronic specific heat of lead (Keesom 63, Shiffman 63) which seems to follow a T^4 law rather than an exponential or a superposition of exponentials. Deviations from a simple exponential law at low temperatures have been attributed by Cooper (59) and Geilikman (61b) to anisotropy of the gap parameter $\Delta_{\mathbf{k}}$, i.e., to a dependence of $\Delta_{\mathbf{k}}$ on the direction of the vector $\mathbf{k}$. But this explanation has been questioned by Pokrovskii (61, 62) on theoretical grounds, and seems to be quite inadequate for the results of Keesom (63) on lead.

The other main thermodynamic quantity is the critical field $H_c(T)$. Its determination from thermodynamic data in zero magnetic field was discussed in Chapter I. In the neighborhood of the transition temperature, we obtain

$$\frac{H_c(T)}{H_c(0)} = 1.74(1 - t) \qquad \text{for} \quad t \to 1 \tag{5.12}$$

That is, the critical field goes to zero linearly with temperature. This is in full agreement with the thermodynamic relation of Rutgers (33), Eq. (I, 4.9).

Over the entire range of temperatures $0 \leqslant T \leqslant T_c$, the critical field turns out to conform quite closely (though not exactly) to the simple formula

$$\frac{H_c(T)}{H_c(0)} \cong 1 - t^2 \tag{5.13}$$

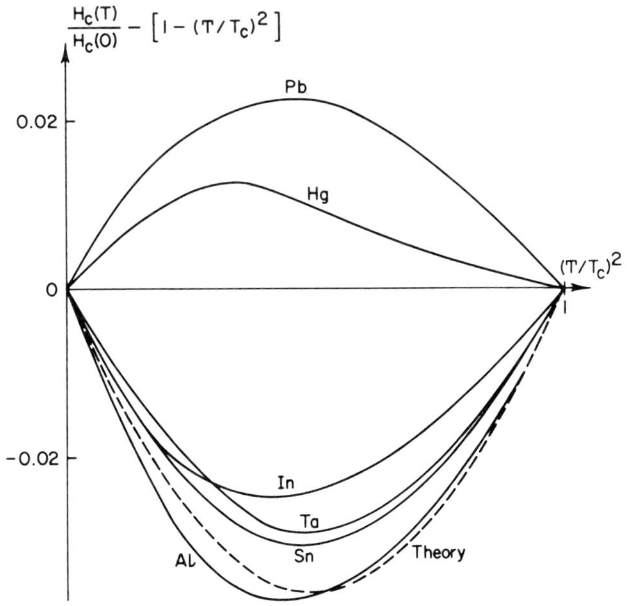

FIG. 6.3. The deviation of the critical field ratio $H_c(T)/H_c(0)$, from the simple approximation Eq. (5.13), vs. $t^2 = (T/T_c)^2$, for a number of substances. The curve based on the thermodynamic theory is also shown in this figure. There is good agreement for In, Ta, Sn, and Al, all of which are "weak-coupling" superconductors. Hg and Pb show marked deviations, as must be expected. This plot is taken from Mapother (62).

This formula was first suggested by Casimir and Gorter (Casimir 34) on the basis of a two-fluid model. Since the specific heat formula of the same model is rather different from that of the microscopic theory, we conclude that the critical field is not very sensitive to details of the model. Precision measurements are required to gain significant information (Shaw 60a).

By plotting the difference between the two sides of (5.13) vs. t^2, the small deviations from (5.13) are emphasized enormously. Such a plot, taken from Mapother (62), is shown as Fig. 6.3. We see that there is remarkably good agreement for "weak-coupling" superconductors such as In, Ta, Sn, and especially Al. On the other hand, mercury and lead are on the wrong side of the line. This is not surprising: both Hg and Pb have low Debye temperatures, and comparatively high transition temperatures. That is, the parameter $\exp(-1/\rho)$ is not very small, and the simple theory breaks down. Explicit numerical solution of the integral equation (3.46) is clearly desirable for these metals. Furthermore, both metals have complicated band structures, whereas the simple theory neglects all band structure effects.

In addition to temperature-dependence, there is of course a pressure-dependence or rather, since we are talking of solids here, a dependence on the stress tensor. The effect of pressure on the critical field is discussed by Rohrer (59), Gross (60), Hinrichs (61), Schirber (61), and Swenson (62). Squire (55) gives a review of the early work on pressure effects. For the influence of pressure on the transition temperature, see Jennings (58), Morel (59), Bowen (60), Garfinkel (61), Kan (61), Seraphim (61), and Schirber (62). A general theoretical discussion of stress tensor effects in superconductors, plus a summary of the experimental situation, is given by Seraphim and Marcus (Seraphim 62). Besides the effects already mentioned, there are also effects on elastic moduli (Alers 61, 62), on the speed of sound waves (Gibbons 59), on the thermal expansion coefficients (White 62, Andres 62), and there is a volume change at the transition from normal to superconducting behavior as the magnetic field passes through the critical value (Cody 58). The use of these thermodynamic data to construct a superconducting energy converter has been discussed by Chester (62).

For most of these effects, there is reasonable agreement between theory and experiment. The situation is less clear, however, in impure specimens (Anderson 59; Kenworthy 61, 62; Zuckermann 63) where mean free path effects are important (Lynton 62), and in specimens with a complex band structure (Clogston 61, Kondo 63); the isotope effect on the transition temperature is sometimes much smaller than expected on the simple theory (Devlin 60); we shall discuss these matters further, in Chapter X.

There are interactions between the variables characteristic of the superconducting state, for example, the energy gap Δ depends on the applied magnetic field (Douglass 61a), and so does the penetration depth (Douglass 61b). These effects are in substantial agreement with the non-linear theory of Ginzburg and Landau (see Section 5 of Chapter I).

The study of thermodynamic properties of superconductors, especially in the presence of an applied magnetic field, is complicated by the frequent appearance of irreversible phenomena. The basic work on the kinetics of the phase transition at $H = H_c$ was done long before the microscopic theory, by Lifshitz (50) and Pippard (50a), and a great deal of information is available by now (Faber 52, 53, 54, 55; Ittner 58; Keller 58; Näbauer 58, 61; Duijvestijn 59; Seraphim 59; Fink 59a; Meissner 59, 59a; Yaqub 60; Cohen 61; Swihart 62b). The discussion of these effects is beyond the scope of this book.

6. The Meaning of a Temperature-Dependent Hamiltonian

The BZT theory, giving an exact solution of a well-defined problem, does not require qualitative physical arguments for its validation. However, now that we have seen the results it leads to, we may well ask what is the physical meaning of an approximate Hamiltonian[15] H_0, which is an explicit function of *temperature*, of all things. The temperature is not a microscopic parameter, it cannot even be defined sensibly for a microscopic system; rather, it is a property of a macroscopic system, or, more precisely, of a whole assembly of macroscopic systems. Nevertheless, diagonalization of the operator H_0 [see Eq. (3.37)] leads to a temperature-dependent ground state energy (!) $C' = C'(T)$, and to an excitation spectrum described by temperature-dependent excitation energies $\hat{\epsilon}_k = \hat{\epsilon}_k(T)$. The approximation in which these temperature dependences are ignored leads to the results of Section 2, which are quite insufficient.

To understand this physically, let us consider the actual spectrum of energy levels of the N-particle system described by the true Hamiltonian H. This spectrum is exceedingly dense; the density of levels at energy E is related to the thermodynamic entropy $S(E, V)$ by the approximate relationship (see Appendix A):

$$\rho_E \cong \exp\left[\frac{S(E, V)}{\mathfrak{k}}\right] \tag{6.1}$$

[15] H_0 is actually an approximation to $H - \mu N$, rather than to H. This distinction is of no importance here.

The entropy is an extensive quantity, hence the exponent in (6.1) is itself proportional to the volume.

The *a priori* probability of finding a system of the ensemble in a quantum state with energy E is a decreasing exponential, $\exp(-\beta E)$. Multiplication by the exponentially increasing density of states (6.1) leads to an energy distribution with an exceedingly sharp maximum, for any given value of $\beta = 1/\mathfrak{k}T$. The position of this maximum is the thermodynamic "internal energy" $\bar{E}$. The spread of the distribution-in-energy about $\bar{E}$ is very small [15a]:

$$\overline{(E - \bar{E})^2} = \mathfrak{k}T^2 c_V \tag{6.2}$$

where c_V is the specific heat at constant volume, an extensive quantity of order V. Thus the root-mean-square fluctuation of E around $\bar{E}$ is of order $V^{1/2}$, negligible compared to $\bar{E}$ itself which is of order V.

We conclude that, at any given temperature T and corresponding internal energy $\bar{E}(T, V)$, we are by no means interested in the entire level spectrum of the Hamiltonian H, but rather only in a small region of excitation energies[16] in the immediate neighborhood of the mean energy $\bar{E}$. If we can find an approximate operator $H_0 = H_0(T)$ which fits the actual level spectrum *in this narrow region of energies*, we have all we need.

Qualitatively, we may think of $H_0(T)$ as a "tangent Hamiltonian," in the sense that a straight line tangent to some curve $y = f(x)$ is a good approximation to the true curve in the immediate neighborhood of the point in question. It is not at all surprising that the slope and intercept of the tangent depend upon the point at which the tangent is evaluated. Conversely, the tangent to a curve should not be used as an approximation outside of a narrow region around the tangent point. $H_0(T)$ is in no sense a good approximate Hamiltonian at temperatures $T' \neq T$. $H_0(T)$ represents the true energy level spectrum of H near the energy $\bar{E}(T)$, and nowhere else. Thus in particular, the ground state energy of $H_0(T)$ has no significance whatever, and must not be expected to lie close to the ground state energy of the true Hamiltonian H.

[15a] See Appendix A for a proof of this relation.

[16] The size of the region increases as $V^{1/2}$, but it is infinitesimally small compared to $\bar{E}$ itself.

7. Transport Phenomena Involving Bogolon Excitations

Quite apart from purely thermodynamic data (specific heat), one would expect that the temperature-dependent effective Hamiltonian can also be used to discuss non-equilibrium phenomena in which only a few Bogolons are involved. The simplest of these involve the scattering of one Bogolon from some state **k** to a state **k**′. Since the energy change in such an event is intensive, whereas the energy region over which the effective Hamiltonian may be used is proportional to $V^{1/2}$, the effective Hamiltonian should be quite adequate.

The theory in question was worked out by Bardeen, Cooper, and Schrieffer (Bardeen 57), and represents a major success of the BCS theory. Two main effects show up in a particularly striking manner:

(1) interference phenomena associated with the time-reversal behavior of the operator responsible for the transitions,

(2) the altered density of states which follows from the energy gap.

We shall discuss these in turn.

Let the operator responsible for the scattering events be denoted by H_i. For example, in the absorption of sound waves, the operator H_i is the scattering operator for particles, due to the time-dependent (with circular frequency ω) distortion of the crystal lattice. In the spin-lattice relaxation, the operator H_i describes the interaction between a typical nuclear spin, precessing with some frequency ω, and a conduction electron (hyperfine structure interaction). We assume that the interaction in question is capable of affecting one electron at a time, and that the one-electron term is the dominant one for the final result. The operator acting on the electron wave function is a typical one-particle operator (V, 1.22), which we write as

$$H_i = \sum_{\mathbf{k},s,\mathbf{k}',s'} \langle \mathbf{k}, s \mid H_i \mid \mathbf{k}', s' \rangle \, a^+_{\mathbf{k}s} a_{\mathbf{k}'s'} \tag{7.1}$$

For this discussion, it is important to retain the spin indices explicitly; we use the convention $s = +1$ for spin up, $s = -1$ for spin down (rather than $s = +\frac{1}{2}$ and $-\frac{1}{2}$, respectively).

The time-reversal behavior of the interaction H_i imposes certain restrictions on the matrix elements $\langle \mathbf{k}, s \mid H_i \mid \mathbf{k}', s' \rangle$ (Wigner 32). For electron wave functions, the time reversal operation is a combination of taking the complex conjugate (which reverses the momentum vector **k**) and multiplication by the Pauli spin matrix σ_y (which reverses the spin direction); denoting the combined operation by T, we have

$$T\psi_{\mathbf{k}s} = \sigma_y \psi_{\mathbf{k}s}^* = is\psi_{-\mathbf{k},-s} \tag{7.2}$$

where the factor $is = \pm i$ comes from the Pauli spin matrix σ_y.

The time-reversal operation on the operator H_i consists of taking the complex conjugate (*not* the Hermitean conjugate!) of H_i, and pre-multiplying as well as post-multiplying the result by σ_y. Usually, we deal with operators which have a simple behavior under this operation: they are either invariant, or else they change sign. We introduce a quantity $\zeta = \pm 1$, characteristic of the operator in question, and write

$$T(H_i) = \sigma_y H_i^* \sigma_y = \zeta H_i \tag{7.3}$$

Operators with $\zeta = +1$ are called "real"; ordinary potentials in x-space are of this type; since the lattice distortion due to a sound wave acts like an ordinary potential on each electron, the operator H_i for scattering of electrons by sound waves is real in this sense.[17] Operators with $\zeta = -1$ are called "imaginary"; all velocity and momentum operators are of this type (since velocities and momenta reverse direction if one reverses the sense of time). The hyperfine structure interaction responsible for the spin-lattice relaxation phenomenon is an example of such an imaginary operator.

Let us now derive a relation between matrix elements of "reciprocal" processes, that is, between the process $\mathbf{k}'s' \to \mathbf{k}s$ and the process $-\mathbf{k}-s \to -\mathbf{k}'-s'$. We use, first the Hermiticity of H_i, then Eq. (7.3), as follows:

$$\begin{aligned}
\langle \mathbf{k}s \mid H_i \mid \mathbf{k}'s' \rangle &= \int \psi_{\mathbf{k}s}^* H_i \psi_{\mathbf{k}'s'} \, d\tau \\
&= \int (H_i \psi_{\mathbf{k}s})^* \, \psi_{\mathbf{k}'s'} \, d\tau \\
&= \int \psi_{\mathbf{k}'s'} \, H_i^* \, \psi_{\mathbf{k}s}^* \, d\tau \\
&= \zeta \int \psi_{\mathbf{k}'s'} \sigma_y \, H_i \, \sigma_y \, \psi_{\mathbf{k}s}^* \, d\tau \\
&= \zeta \int (is' \, \psi_{-\mathbf{k}',-s'})^* \, H_i (is\psi_{-\mathbf{k}\,-s}) \, d\tau
\end{aligned}$$

so that, finally,

$$\langle \mathbf{k}, s \mid H_i \mid \mathbf{k}', s' \rangle = \zeta s s' \, \langle -\mathbf{k}', -s' \mid H_i \mid -\mathbf{k}, -s \rangle \tag{7.4}$$

---

[17] In this case, H_i is purely real and does not involve spins at all; thus H_i^* commutes with σ_y, and (7.3) with $\zeta = +1$ follows immediately from $(\sigma_y)^2 = 1$.

In normal metals, the only practical consequence of the relation (7.4) is obtained by taking its absolute square; the result is the usual reciprocity theorem (Onsager 31, 32, 32a; Blatt 52), stating that the transition probabilities for reciprocal processes are equal except for the density-of-state factor. This result is independent of ζ.

However, characteristic *interference* phenomena occur in the theory of superconductivity when (7.1) and (7.4) are combined. To the accuracy of the reduced Hamiltonian, the independent, elementary excitations of the system are associated with the Bogoliubov operators $\alpha_{\mathbf{k}0}$ and $\alpha_{\mathbf{k}1}$, rather than with the electron operators $a_{\mathbf{k}0}$. Thus, we must use the Bogoliubov transformation in the form (V, 3.13) to rewrite (7.1) in terms of Bogoliubov operators.

To keep things simple, let us consider an operator H_i which does not flip the electronic spin, i.e., for which the non-zero matrix elements have $s = s'$. We use the freedom to replace dummy indices by their negatives to write

$$\begin{aligned} H_i &= \sum_{\mathbf{k},\mathbf{k}'} \sum_{s=-1}^{+1} \langle \mathbf{k}s \mid H_i \mid \mathbf{k}'s \rangle \, a_{\mathbf{k}s}^{+} a_{\mathbf{k}'s} \\ &= \sum_{\mathbf{k},\mathbf{k}'} [\langle \mathbf{k}, + \mid H_i \mid \mathbf{k}', + \rangle \, a_{\mathbf{k}+}^{+} a_{\mathbf{k}'+} + \langle -\mathbf{k}', - \mid H_i \mid -\mathbf{k}, - \rangle \, a_{-\mathbf{k}',-}^{+} a_{-\mathbf{k},-}] \\ &= \sum_{\mathbf{k},\mathbf{k}'} \langle \mathbf{k}, + \mid H_i \mid \mathbf{k}', + \rangle \, (a_{\mathbf{k},+} a_{\mathbf{k}'+} + \zeta a_{-\mathbf{k}',-}^{+} a_{-\mathbf{k},-}) \\ &= \sum_{\mathbf{k},\mathbf{k}'} \langle \mathbf{k}, + \mid H_i \mid \mathbf{k}', + \rangle \, [(u_k u_{k'} - \zeta v_k v_{k'}) (\alpha_{\mathbf{k}0}^{+} \alpha_{\mathbf{k}'0} + \zeta \alpha_{\mathbf{k}'1}^{+} \alpha_{\mathbf{k}1}) \\ &\qquad + (u_k v_{k'} + \zeta v_k u_{k'}) (\alpha_{\mathbf{k}0}^{+} \alpha_{\mathbf{k}'1}^{+} + \zeta \alpha_{\mathbf{k}1} \alpha_{\mathbf{k}'0})] \end{aligned} \tag{7.5}$$

In the last expression (7.5), we have terms involving scattering of a Bogolon from state $\mathbf{k}'0$ to state $\mathbf{k}0$, and from state $\mathbf{k}1$ to state $\mathbf{k}'1$; we also have creation of a pair of Bogolons in states $\mathbf{k}0$ and $\mathbf{k}'1$, and destruction of a pair of Bogolons from states $\mathbf{k}'0$ and $\mathbf{k}1$.

In low frequency transitions, where the transition energy $\hbar\omega$ is much less than the energy gap parameter Δ, the Bogolon pair terms are unimportant. The essential terms are Bogolon scattering terms, and we see that the usual matrix element is multiplied by the "coherence factor"

$$F_{kk'} = u_k u_{k'} - \zeta v_k v_{k'} \tag{7.6}$$

The same coherence factor also enters into the effective matrix element for spin-flip processes.

The coherence factor distinguishes sharply between transition operators H_i which are even and odd under time reversal. This can be seen particularly clearly when the transition energy $\hbar\omega$ becomes so low that $u_k \simeq u_{k'}$ and $v_k \simeq v_{k'}$. In that case, the transition probability for "real" operators is depressed by the factor $(u_k^2 - v_k^2)^2$, which is zero for $k = k_F$, whereas the coherence factor for "imaginary" operators is equal to unity by (V, 3.9).

The other characteristic feature mentioned at the beginning of this section is the altered density of final states. The number of Bogolon states with energies $\hat{\epsilon}_k$ between $\hat{\epsilon}$ and $\hat{\epsilon} + d\hat{\epsilon}$ is given by

$$dN = 2\frac{V\,d^3k}{(2\pi)^3} = 2\frac{Vk^2}{2\pi^2}\frac{dk}{d\epsilon}\frac{d\epsilon}{d\hat{\epsilon}}\,d\hat{\epsilon} = \hat{\rho}_E\,d\hat{\epsilon}$$

Thus, using (3.35) to find $d\epsilon/d\hat{\epsilon}$ as a function of $\hat{\epsilon}$, we obtain

$$\hat{\rho}_E = \rho_E\frac{d\epsilon}{d\hat{\epsilon}} = \frac{\hat{\epsilon}}{[\hat{\epsilon}^2 - \Delta^2]^{1/2}}\,\rho_E \tag{7.7}$$

This expression applies only for $\hat{\epsilon} > \Delta$; for $\hat{\epsilon} < \Delta$, i.e., inside the gap, the density of states $\hat{\rho}_E$ vanishes altogether.

The ratio $\hat{\rho}_E/\rho_E$ exceeds unity, and becomes infinite at the gap edge. Qualitatively, this can be understood by observing that all the "normal" states which would otherwise lie in the gap region have been crowded out, and added to the density of states outside the gap.

For "imaginary" operators H_i, the coherence factor $F_{kk'}$ is close to unity at low frequencies, and the main effect on the transition probabilities is the altered density of states. We therefore expect an initial *enhancement* of transition probabilities, i.e., an increase of the transport phenomena of this type, at temperatures just below T_c. Eventually, at much lower temperatures, we expect a decrease due to the fact that the total number of available Bogolons [the sum of all the $\bar{f}_{ks}$, Eq. (3.42)] approaches zero at zero temperature.

For "real" operators, on the other hand, the coherence effect and density of states effect are in opposite directions, and a more detailed calculation (Bardeen 57), shows that they cancel each other almost precisely; in particular, if we compute the ratio of the total transition probabilities in the superconducting and normal states, this ratio is given entirely by the Bogolon occupation probability (3.42) right at the gap edge; if we denote the attenuation coefficients in the superconducting and normal states, and at frequency ω, by $A_s(\omega)$ and $A_n(\omega)$, respectively, then Bardeen (57) show that

$$\lim_{\omega\to 0}\frac{A_s(\omega)}{A_n(\omega)} = \frac{2}{\exp(\beta\Delta) + 1} \tag{7.8}$$

Since the energy gap parameter $\Delta(T)$ increases very rapidly as T drops below T_c, Eq. (7.8) predicts an exceedingly rapid drop in attenuation in the superconducting state for operators of "real" type.

These predictions are verified quite strikingly by experiment. The *attenuation of sound waves in superconductors* is due to scattering of phonons by Bogolons, and the corresponding operator is "real". The sharp drop of this attenuation with temperature below $T = T_c$ was observed already before the development of the microscopic theory, by Bömmel (54) and Mackinnon (55, 57, 59), and confirmed by the more recent experiments (Morse 57, 59, 59a, 59b, 62; Bezuglyi 59, 61; Horwitz 62; Mackintosh 61; Chopra 59, 59a; Kamigaki 61); more detailed theoretical work has been done by Migdal (58a), Kresin (59), Tsuneto (61), Pokrovskii (61a, 61b, 61c, 61d, 62b) and Privorotskii (62, 62a).[18]

By contrast, the *spin-lattice relaxation time in superconductors* is governed by the hyperfine structure interaction, which is an "imaginary" operator on the electron variables. The marked increase of the speed of relaxation (decrease of relaxation time) just below the transition temperature was first observed by Hebel and Slichter (Hebel 57, 59) and has been confirmed in many experiments afterward (Redfield 59, Hammond 60, Masuda 62, 62a, 62b). The experimental results do not, of course, reproduce the infinite peak predicted by (7.7) at $\hat{\epsilon} = \Delta$; but they show a very marked rise, and this rise can be fitted approximately by assigning a finite "width" to the Bogolon levels (Hebel 59a); the width in question is of the order of 10^{-6} ev, and is associated with anisotropy of Δ_k and with the effects of impurity scattering centers (Masuda 62, 62a, 62b; Anderson 59).[19]

A particularly neat verification of the increased density of states (7.7) is provided by *tunneling of electrons across a thin insulating layer separating a normal metal from a superconductor, or separating two superconductors*. The discovery of this phenomenon by Giaever (60, 60a, 61, 61a, 62) has been recounted in a delightful little article by Schmitt (61). The experiments can be analyzed quite well by ignoring all effects other than the changed density of states in the superconductor. By altering the voltage difference across the insulating layer, it is possible to juxtapose the Fermi level in the normal metal with an arbitrary region of the Bogolon energy spectrum in the superconductor. If, in particular, the Fermi level in the normal metal coincides with an energy inside the gap of the superconductor, no tunneling can occur; a sharp onset of tuneling is observed as the

---

[18] For the other superfluid system, liquid He II, the sound wave attentuation is discussed by Kawasaki (62), who give references to earlier work.

[19] See also Azbel (58).

Fermi level in the normal metal crosses the gap edge in the superconductor. The fact that this phenomenon seems to depend entirely on the density of states, i.e., the coherence factor $F_{kk'}$ does not enter, can be understood by observing (Bardeen 61b) that the rate of tunneling is determined by barrier penetration theory, and hence depends primarily on the exponentially decreasing wave functions in the insulating region; if the insulating region is predominantly "normal," i.e., if the electron pairing characteristic of the superconductor dies down quickly inside the insulating region, then the coherence factor $F_{kk'}$ should not enter the theoretical expression for the tunneling rate. There has been considerable further work, both experimental and theoretical, on the tunneling phenomenon (*experiment*: Nicol 60; Fisher 61; Minnigerode 61; Sherrill 61; Townsend 61, 62; Zavaritskii 61, 62; Chynoweth 62; Dayem 62; Dietrich 62; Rowell 62; Shapiro 62; Adler 63; *theory*: Franz 61; Harrison 61; Bardeen 62; Cohen 62; Josephson 62; Prange 62; Schrieffer 63, 63a; Tien 63).

A particularly detailed comparison of theory and experiment on tunneling has been made for lead. The data of Rowell (63) agree very well with Green's function calculations by Schrieffer (63); individual peaks in the phonon frequency spectrum of lead are visible in the tunneling curves. It seems paradoxical that just for this element there is a strong disagreement with theory in the low-temperature electronic specific heat (Keesom 63).

Parmenter (61) has suggested that tunneling may be used to maintain a nonequilibrium superconducting state at temperatures $T > T_c$, but the finite lifetime of Bogolon states (Ginsberg 62, Schrieffer 62, Tewordt 62, 62a) may prevent this possibility from being realized.

A particularly interesting aspect of tunneling is tunneling between two superconductors, rather than between a superconducting and a normal metal. Josephson (62) has predicted an anomalous tunneling current associated with direct transfer of electron pairs between the Fermi surfaces of the two superconducting metals. In such a case, the electron wave functions would retain their pairing character in spite of the insulating barrier region. The anomalous Josephson current is a direct current in the absence of an applied voltage across the barrier, the size of the current depending on the energy gap functions Δ_k in the two superconductors; if a DC voltage V is applied to the junction, the Josephson current becomes alternating (!), with frequency $2eV/h$, where h is Planck's constant. Probable observation of this Josephson anomalous tunneling current has been reported by Anderson and Rowell (Anderson 63).

A sharp threshold in the tunneling characteristic has been observed by Taylor (63), and has been associated with the onset of such "double particle tunneling" by Schrieffer (63a), who find reasonable agreement with a theoretical calculation.

In addition to these experiments, the Bogolon excitations can also be excited electromagnetically, in *infrared transmission and reflection experiments on thin films* (Glover 56, 57; Tinkham 58a, 58b, 59; Ginsberg 60, 62a; Richards 60; Tsuneto 60; Pokrovskii 62a). These experiments are in general agreement with the theory; i.e., the main absorption occurs for frequencies such that $\hbar\omega > 2\Delta$, the energy gap, and the temperature dependence of Δ checks (Biondi 59). But in some materials, such as lead and mercury, there is a "precursor absorption" for $\hbar\omega < 2\Delta$ (Ginsberg 59, 62a; Richards 60), which seems to be larger than one would expect on the basis of "collective excitations" (Maki 62); in other materials, such as tin, there is considerable structure just above the gap, as well as a strong dependence on impurity content (Richards 61, 62; Adkins 62). This latter dependence appears to be related to anisotropy of the energy gap parameter $\Delta_{\mathbf{k}}$ (i.e., $\Delta_{\mathbf{k}}$ depends on the orientation of the vector $\mathbf{k}$ with respect to the crystal lattice), which is observable in pure crystals but gets averaged out by multiple scattering on impurities in a "dirty superconductor" (Anderson 59, Tsuneto 62).

At microwave frequencies, the quantum energy is insufficient to excite "real" Bogolons, but their "virtual" effects can be seen (Mattis 58, Miller 60). Historically, study of the *surface impedance of superconducting metals at microwave frequencies* led to the coherence length concept of Pippard (Pippard 53, 54; Faber 55, 55a, 55b).

The London penetration depth is largely independent of impurity content, as would be expected from the simple Bose gas model, see Eq. (II, 5.27); this penetration depth is associated, qualitatively speaking, with the center-of-gravity motion of the electron pairs. The wave function for that motion is coherent all the way across the specimen, as a result of the Schafroth condensation. But the range of coherence of the "internal" part of the pair wave function (i.e., the expected distance between the two electrons making up the pair) turns out to be sensitive to the electron mean free path, i.e., sensitive to impurities. Qualitatively, one can understand this difference by observing that the individual momenta $\mathbf{k}_1$ and $\mathbf{k}_2$ of the two electrons are high, in the neighborhood of the Fermi momentum; whereas the center-of-gravity momentum $\mathbf{K} = \mathbf{k}_1 + \mathbf{k}_2$ vanishes for the usual pairing. Thus the center-of-gravity part of the pair wave function has macroscopic coherence range, and is insensitive to local disturbances such as scattering centers; whereas the internal part of the wave function has a microscopic range of coherence (see Chapter IV, Section 4), which is limited by mean free path effects in impure specimens. This latter range is the Pippard coherence length.

At temperatures close to the transition temperature T_c, the London penetration depth becomes very large, and the internal size of the pair need not be taken into account, to a first approximation. We are then within the range of validity of the Ginzburg-Landau theory (Chapter I, Section 5); the "wave function of the superelectron" corresponds to the center-of-gravity part of the wave function of the condensed electron pairs (see Chapter X, Section 2 for further discussion). Measurements of the surface impedance (Bezuglyi 58, Khaikin 58, Prozorova 58, Kaplan 59, Spiewack 59, Dresselhaus 60a, 60b) are in agreement with that theory (Abrikosov 58a, 59b).

Surface impedance measurements at microwave frequencies, performed on single crystal specimens, can yield information about the dependence of $\Delta_{\mathbf{k}}$ on the direction of $\mathbf{k}$ (Williams 62).

An interesting application of the threshold frequency for absorption of electromagnetic radiation, combined with the tunneling idea, has been suggested by Burstein (61): "photon-assisted" tunneling might be used for detection of quanta in the microwave and far infrared region; the low-frequency cutoff of the detector being readily adjustable by changing a bias voltage.

The Bogolon concept also enters the discussion of the radio-frequency Hall effect (Dresselhaus 60; Miller 61); the D.C. Hall effect vanishes in superconductors (Jaggi 59, Schindler 63). It is also required to understand the emission of electrons from a superconducting metal (Klein 61), and the capture of electrons injected into the metal into ground state pairs (Rothwarf 63). No pairing effect is predicted for Auger electron emission (Craig 63).

Although most differences between normal conductors and superconductors disappear at optical frequencies, Dresselhaus (62) suggests that a residual difference could be observed as a result of interband transitions. This effect has not yet been seen experimentally.

A very neat application of the Bogolon concept to transport properties has been suggested by Little (59, 61, 62a): the interaction between elementary electron excitations inside the metal, and phonons in liquid helium touching the surface of the metal, is different depending on whether the metal is in the normal or superconducting state. The normal electrons, without energy gap, permit such an interaction easily, whereas Bogolons are practically inert. One therefore expects a difference in the "*Kapitza resistance*" (Kapitza 41) against heat flow at a metal-liquid helium interface. This difference has been observed experimentally (Challis 61, 62; Gittleman 61, 62; Barnes 63).

Another transport property of interest if the *heat conductivity*. The theory has been given by Bardeen, Rickayzen, and Tewordt (Bardeen 59a), Kadanoff and Martin (Kadanoff 61), Greenwood (62), Tewordt (62, 62a, 63), Geilikman (58, 58a, 61), and Khalatnikov (59a). Unfortunately, the interpretation of the data (Hulm, 50; Laredo 55; Graham 58; Mendelssohn 58, 60a, 62; Zavaritskii 58, 60; Chaudhuri 59; Jones 60; Rowell 60; Guénault 61; Morris 61; Chang 62; Chaudhuri 62; Connolly 62; Satterthwaite 62; Toxen 62a; Dubeck 63) is complicated by the fact that there are a number of heat conduction mechanisms in the normal as well as in the superconducting states, and the separation between them is not easy. The data of Guénault (61) on tin, and of Jones (60) on indium, agree better with Kadanoff (61) and Tewordt (62, 62a, 63) than with Bardeen (59a).[20]

All these experiments are in satisfactory agreement with the effective Hamiltonian $H_0(T)$, and the appearance of the coherence factor (7.6), in particular, is a major contribution of the BCS theory. This coherence factor could not have been predicted, and was not in fact predicted, by any of the "two-fluid models" current before the microscopic theory.[21]

On the other hand, it must be realized that experiments sensitive to the excitation spectrum of a superconductor test properties of the "normal fluid," not properties of the "superfluid." Interesting though properties of the "normal fluid" undoubtedly are, the real "super"-properties are associated primarily with the coherent ground state, not with the excitations. We shall discuss the true "superfluid" properties (the Meissner effect and the persistent currents in superconducting rings) in Chapters VIII and IX, purely on the basis of the ground state wave function (IV, 1.1) characteristic of Schafroth condensed pairs, without having to consider excitations at all. First, however, we must devote our attention to some fundamental problems associated with the BZT theory, in particular, the meaning of the assumption of the reduced Hamiltonian.

---

[20] According to Geilikman (61a), the thermomagnetic effect is not influenced by the transition to the superconducting state.

[21] The concept of an effective Hamiltonian, describing independent quasi-particles, is only an approximation, of course. Actually, these Bogolons have only a finite lifetime (Migdal 58a; Harper 61; Pokrovskii 61a; Tewordt 62, 62a; Schrieffer 62; Ginsberg 62); and for sufficiently strong coupling, the approximation breaks down altogether (Schrieffer 63, Engelsberg 63).

VII

Thermodynamics in the Quasi-Chemical Equilibrium Theory

1. Introduction

The theory of Bogoliubov, Zubarev, and Tserkovnikov (Bogoliubov 57, 60a) gives an exact solution of the thermodynamics for the reduced Hamiltonian of Bardeen, Cooper, and Schrieffer (Bardeen 57). Furthermore the results are in good agreement with experimental measurements on the specific heat, the critical field, and the transition temperature.

Nevertheless, some questions remain, the most important being: *What is the physical meaning of the assumption of the reduced Hamiltonian?* In this chapter, we shall adress ourselves to this question.

In the limit of zero temperature, we already know the answer. The reduced Hamiltonian leads to a ground state wave function which is characteristic of Schafroth-condensed electron pairs. Since all the essential results of the theory can be obtained directly from that wave function, without omitting parts of the Hamiltonian, the physical meaning of the reduced Hamiltonian is just that it is one way of describing Schafroth condensation of pairs.

But the situation is not so simple at non-zero temperatures. Even an ideal Bose gas is not fully condensed for $T \neq 0$; thus we must expect to find "normal" electron pairs as well as "condensed" electron pairs, to say nothing of "unpaired" electrons. What is the relation, if any, between the "Bogolon" excitations and electron pairing? What types of correlations between electrons are described correctly by the reduced Hamiltonian, and hence by the BZT theory, and what types of correlations are omitted?

Ideally, one would like to have an exact solution for the full Hamiltonian, i.e., an explicit expression for the operator $\exp(-\beta H)$, or at least for the trace of this operator. Unfortunately, this is quite out of the question. Instead of this, we shall use a variational approach, similar to the Rayleigh-Ritz variation principle for the ground state energy from which we started in Chapter IV.[1]

In Chapter VI, Section 3, we introduced the "density matrix" or "statistical operator" $\mathscr{U}$ for the "grand canonical ensemble" [see Eq. (VI, 3.9)]. This operator has the same eigenfunctions $\Psi_{N\alpha}$ as the Hamiltonian H_N; the eigenvalues $P_{N\alpha}$ are given by (VI, 3.11), and can be interpreted as probabilities of finding a system of the ensemble to possess exactly N particles, and to be in the quantum state $\Psi_{N\alpha}$. The statistical expectation value of an arbitrary operator Q over this ensemble is given by (VI, 3.13); for example, the internal energy $\bar{E}$ is the statistical expectation value of the Hamiltonian operator:

$$\bar{E} = \text{Trace}\,(H\mathscr{U}) = \text{Trace}\,\{H \exp[\beta(\Omega + \mu N_{\text{Op}} - H)]\} \tag{1.1}$$

All operators in (1.1) are understood to be written in terms of creation and destruction operators, so that the total number of particles does not appear explicitly in (1.1); the operator N_{Op} is defined by (V, 1.23). H and N_{Op} commute with each other, and $\mathscr{U}$ commutes with both of them.[2]

[1] A variation principle was used by BCS to derive thermodynamic results. However, BCS started from the reduced Hamiltonian, for which BZT obtained an exact solution of the thermodynamics. We shall use the variation principle with the full Hamiltonian here.

Our discussion of variational principles in this chapter, and in Appendix A, is limited to our immediate needs. For further discussion, see Koppe (58), Mühlschlegel (60), Girardeau (62), and Mermin (63).

It is often convenient to use formulations of statistical mechanics in terms of reduced density matrices, i.e., projections of the full density matrix $\mathscr{U}$ onto the subspace of one, two, or some other small number of particles. If the interaction between particles is a sum of forces between pairs only (the usual assumption), then knowledge of the one-particle and two-particle reduced density matrices is in principle sufficient, even for a strongly interacting system (Meeron 58, 60; Van Leeuwen 59; Morita 60).

It is possible to find variational formulation of that theory, also, and such variation principles are of considerable usefulness (Yvon 35; Green 58, 60; Lee 59a, 60; Martin 59; Bloch 60; Luttinger 60; Balian 60, 60a, 61, 61a; Morita 61; De Dominicis 62, 63).

[2] The commutation law $[\mathscr{U}, N_{\text{Op}}]_- = 0$ is the mathematical statement of the law of conservation of particles in the theory. It implies that individual systems of the grand canonical ensemble have constant numbers of particles; the averaging over N is then done incoherently, and is merely a convenient mathematical device, not a physical assumption. All trial density matrices $\mathscr{U}$ in this chapter commute with N_{Op}. On the other hand, the operator $\mathscr{U}_0$ of the BZT theory, (VI,3.14), does not commute with N_{Op}. This shows itself by unusually large fluctuations in the local number density, with that model (Mittelstaedt 62, Bell 63).

The statistical average of the particle number N over the ensemble is

$$\bar{N} = \text{Trace}\,(N_{\text{Op}}\,\mathscr{U}) \tag{1.2}$$

We shall also need the entropy S. It is shown in Appendix A that the statistical mechanical definition of this quantity is

$$S = -\mathfrak{k}\,\text{Trace}\,(\mathscr{U}\ln\mathscr{U}) \tag{1.3}$$

where $\mathfrak{k}$ is Boltzmann's constant. Unlike (1.1), the entropy (1.3) is a functional of the density matrix itself only; it does not involve knowledge of the Hamiltonian of the system.

If we use the exact form of the operator $\mathscr{U}$, Eq. (VI, 3.9), and the unit trace property (VI, 3.10), we obtain

$$\text{Trace}\,(\mathscr{U}\ln\mathscr{U}) = \beta(\Omega - \bar{E} + \mu\bar{N}) \tag{1.4}$$

When we solve (1.4) for Ω and use (1.3), we obtain the well-known thermodynamic definition of the grand canonical potential Ω:

$$\Omega = \Omega(V, T, \mu) = \bar{E} - \mu\bar{N} - TS \tag{1.5}$$

In thermodynamics, the state of equilibrium is shown to lead to a minimum value of the free energy. If the volume V, the temperature T, and the chemical potential μ are considered to be the independent variables of the system, then Ω is the appropriate free energy, and Ω assumes its minimum value in thermodynamic equilibrium.

In Appendix A, it is proved that the same minimum principle holds also in statistical mechanics; that is

$$\begin{aligned}\Omega &= \text{Trace}\,(H\mathscr{U}) - \mu\,\text{Trace}\,(N_{\text{Op}}\mathscr{U}) + \mathfrak{k}T\,\text{Trace}\,(\mathscr{U}\ln\mathscr{U})\\ &= \text{minimum}\end{aligned} \tag{1.6a}$$

subject to

$$\text{Trace}\,\mathscr{U} = 1 \tag{1.6b}$$

$$\mathscr{U} = \text{Hermitean, non-negative} \tag{1.6c}$$

The side conditions ensure that the eigenvalues $P_{N\alpha}$ of the trial density matrix $\mathscr{U}$ are non-negative real numbers which add up to unity—necessary requirements for a set of probabilities.

The minimum principle (1.6) will serve us as an extension of the Rayleigh-Ritz principle of quantum mechanics to non-zero temperatures. For given volume V, temperature T, and chemical potential μ, the true density matrix $\mathscr{U}$, (VI, 3.9), is the unique solution of (1.6). If we replace the true $\mathscr{U}$ (which, needless to say, is difficult to find with the complete

Hamiltonian H) by some trial density matrix $\mathscr{U}_t$, Eq. (1.6a) gives a trial value Ω_t of the grand canonical potential, with the property

$$\Omega_t \geqslant \Omega \tag{1.7}$$

Thus, if the trial form $\mathscr{U}_t$ contains adjustable parameters, we can vary these parameters so as to get the lowest Ω_t. All this is in full analogy to the usual variation procedure in quantum mechanics; the main change is that the trial wave function has now been replaced by a trial density matrix.

2. The Trial Density Matrix of the Quasi-Chemical Equilibrium Theory

In this section, we describe and make plausible the particular form for the trial density matrix employed in the quasi-chemical equilibrium theory. Section 3 deals with the consequences of the non-negativity condition (1.6c); Section 4 describes the further (extreme) simplifications in the Ansatz which are required to obtain calculable results. The results themselves, and their relation to the BZT theory, are contained in Section 5.

Although the use of creation and destruction operators is most convenient in actual calculations, it is easier to discuss fundamentals by more pedestrian methods. Thus, the expressions in this section are written out in detail, for various numbers N of particles. The formulation in terms of creation and destruction operators is contained in Appendix B.

It is preferable to ignore the requirements of the Pauli principle in the beginning, i.e., work as if the particles were distinguishable, and to antisymmetrize all expressions later on. Thus we start with a "Boltzmann" system, without any statistical correlations; the correlations we introduce explicitly are all dynamical in nature. The statistical correlations are supplied afterwards by the antisymmetrization of all expressions.

Let H_N be the Hamiltonian of an N-particle Boltzmann system. Its statistical operator $\mathscr{U}_{B,N}$ is proportional to $\exp(-\beta H_N)$. *If* the system is an independent particle system, we have

$$H_N = \sum_{i=1}^{N} H(i) \qquad \text{(independent particles)} \tag{2.1}$$

where the operator $H(i)$ acts on the coordinates of particle number i, only. Thus $H(i)$ and $H(j)$ commute with each other, for all i and j, and the statistical operator becomes a matrix direct product of one-particle statistical operators

$$\begin{aligned}\langle k_1 k_2 \cdots k_N \mid \mathscr{U}_{B,N} \mid k_1' k_2' \cdots k_N' \rangle \\ &= C_N \langle k_1 k_2 \cdots k_N \mid \exp(-\beta H_N) \mid k_1' k_2' \cdots k_N' \rangle \\ &= C_N \langle k_1 \mid \hat{U} \mid k_1' \rangle \langle k_2 \mid \hat{U} \mid k_2' \rangle \cdots \langle k_N \mid \hat{U} \mid k_N' \rangle \qquad (2.2)\end{aligned}$$

where C_N is a normalization constant, and the one-particle operator $\hat{U}$ is defined by[3]

$$\langle k \mid \hat{U} \mid k' \rangle = \langle k \mid \exp[-\beta H(1)] \mid k' \rangle \qquad (2.3)$$

For an independent-particle system, (2.2) is an exact expression. It forms the basis for the theory developed in Chapter III, Section 7 [see Eq. (III, 7.3)]. The fact that the N-particle statistical matrix factors into a direct product of one-particle statistical matrices is the mathematical expression of the statistical independence of the particles. Of course, the subsequent antisymmetrization spoils this complete independence, but the correlations so introduced are "statistical" rather than "dynamical" correlations.

If the system is *not* an independent-particle system, i.e., if the Hamiltonian is not of the form (2.1), then the particles are correlated dynamically, and the density matrix $\mathscr{U}_{B,N}$ no longer factors as in (2.2); that is (2.2) is now incorrect no matter what we choose for the one-particle operator $\hat{U}$.

However, we may consider the *functional form* (2.2) as a *trial density matrix* for systems of interacting particles. The matrix elements $\langle k \mid \hat{U} \mid k' \rangle$ then take the place of the trial wave function in the ordinary variation principle of quantum mechanics, and we ask for the "best" matrix $\hat{U}$ in the sense of the variation principle (1.6).[4] This program has been carried out by Husimi (40). It leads to a natural extension of the Hartree-Fock theory to non-zero temperatures. No qualitative changes appear; in fact, most (but not all) of the results can be obtained

[3] This is substantially the operator U_1 of Chapter III, Section 7. We avoid the use of subscripts here, in order to simplify the appearance of the formulas. For the same reason, the two-particle dynamical correlation matrix U_2' of Chapter III is called $\hat{W}$ here.

[4] The functional form (2.2) applies to a Boltzmann system with given N. Before using (1.6), it is necessary to antisymmetrize (2.2), and to modify it for use with the grand canonical ensemble. We shall carry through this process for a more general functional form below; the corresponding formulas for the form (2.2) can be obtained by setting $\hat{W} = 0$ systematically in all the later formulas of this section.

by the simple recipe of using an ordinary Fermi distribution but replacing the single-particle energies ϵ_k by "renormalized" single-particle energies:

$$\hat{\epsilon}_k = \epsilon_k + \sum_{k'<k_F} [\langle kk' \mid v \mid kk' \rangle - \langle kk' \mid v \mid k'k \rangle] \tag{2.4}$$

where v is the operator for the particle-particle interaction potential. The additional term on the right of (2.4) is the average potential seen by a particle in state k, due to all the other particles in the Fermi sea.[5]

Of course, the detailed formula (2.3) applies only to a true independent-particle system. The "best" matrix $\hat{U}$ of the Husimi theory is *not* given by (2.3).

In the quasi-chemical equilibrium theory, it is our aim to extend the Ansatz (2.2) by including dynamical pair correlations.[6] We introduce an arbitrary pair correlation matrix $\langle k_1k_2 \mid \hat{W} \mid k_1'k_2' \rangle$ whose matrix elements will, later on, serve as additional parameters in the variational calculation.

Omitting the normalization factor C_N for the moment, the Husimi Ansatz for $N = 2$ is

$$\langle k_1k_2 \mid \mathscr{U}_{B,2} \mid k_1'k_2' \rangle = \langle k_1 \mid \hat{U} \mid k_1' \rangle \langle k_2 \mid \hat{U} \mid k_2' \rangle \qquad \text{(Husimi)} \tag{2.5}$$

In the quasi-chemical equilibrium theory, this is augmented to read

$$\langle k_1k_2 \mid \mathscr{U}_{B,2} \mid k_1'k_2' \rangle = \langle k_1 \mid \hat{U} \mid k_1' \rangle \langle k_2 \mid \hat{U} \mid k_2' \rangle + \langle k_1k_2 \mid \hat{W} \mid k_1'k_2' \rangle \qquad \text{(QCE)} \tag{2.6a}$$

This Ansatz is actually still quite general for $N = 2$; however, restrictions appear when we go to higher values of N. For example, the quasi-chemical equilibrium Ansatz for three particles is

$$\begin{aligned}\langle k_1k_2k_3 \mid \mathscr{U}_{B,3} \mid k_1'k_2'k_3' \rangle &= \langle k_1 \mid \hat{U} \mid k_1' \rangle \langle k_2 \mid \hat{U} \mid k_2' \rangle \langle k_3 \mid \hat{U} \mid k_3' \rangle \\ &+ \langle k_1k_2 \mid \hat{W} \mid k_1'k_2' \rangle \langle k_3 \mid \hat{U} \mid k_3' \rangle + \langle k_2k_3 \mid \hat{W} \mid k_2'k_3' \rangle \langle k_1 \mid \hat{U} \mid k_1' \rangle \\ &+ \langle k_1k_3 \mid \hat{W} \mid k_1'k_3' \rangle \langle k_2 \mid \hat{U} \mid k_2' \rangle\end{aligned} \tag{2.6b}$$

The first term on the right of (2.6b) is the Husimi (independent-particle)

[5] This procedure is slightly inconsistent, since the other particles occupy a Fermi distribution for non-zero temperature, rather than a zero-temperature Fermi sea state. The Husimi theory in its complete form allows for this effect, and thus leads to a non-linear integral equation. The form (2.4) corresponds to the first approximation (usually quite sufficient) to the solution of that integral equation.

[6] Statistical correlations of all orders, not just pair correlations, are included automatically through the later antisymmetrization.

term; the remaining terms allow for dynamical pair correlations. (2.6b) is not completely general because it omits any explicit three-particle correlation, i.e., there is no term in (2.6b) of type

$$\langle k_1k_2k_3 \mid \hat{U}_3 \mid k_1'k_2'k_3' \rangle$$

with an operator $\hat{U}_3$ which cannot be reduced to a product of a $\hat{W}$ and a $\hat{U}$.

For four particles, the trial density matrix of the quasi-chemical equilibrium theory has the form (the dots indicate terms which can be obtained by permutations):

$$\begin{aligned}\langle k_1k_2k_3k_4 \mid \mathscr{U}_{B,4} \mid k_1'k_2'k_3'k_4' \rangle \\ &= \langle k_1 \mid \hat{U} \mid k_1' \rangle \langle k_2 \mid \hat{U} \mid k_2' \rangle \langle k_3 \mid \hat{U} \mid k_3' \rangle \langle k_4 \mid \hat{U} \mid k_4' \rangle \\ &+ \langle k_1k_2 \mid \hat{W} \mid k_1'k_2' \rangle \langle k_3 \mid \hat{U} \mid k_3' \rangle \langle k_4 \mid \hat{U} \mid k_4' \rangle + \cdots \\ &+ \langle k_1k_2 \mid \hat{W} \mid k_1'k_2' \rangle \langle k_3k_4 \mid \hat{W} \mid k_3'k_4' \rangle + \cdots \end{aligned} \tag{2.6c}$$

We can classify the terms which appear in these expressions by the number N_1 of "singles," i.e., the number of factors $\hat{U}$, and the number N_2 of pair correlations, i.e., the number of factors $\hat{W}$. The first term on the right of (2.6c) has $N_1 = 4$, $N_2 = 0$; this is the "Husimi" or independent-particle term. The next term has $N_1 = 2$, $N_2 = 1$. When we allow for the possible permutations, there are six terms of this type altogether. Finally, the last line of (2.6c) corresponds to $N_1 = 0$, $N_2 = 2$, and when we count possible permutations, there are three such terms [namely, in obvious notation, the terms (12) (34), (13) (24), and (14) (23), respectively].

In general, the Ansatz for $\mathscr{U}_{B,N}$ contains all terms of type (N_1, N_2) such that

$$N_1 + 2N_2 = N \tag{2.7}$$

A simple count of possible permutations yields the result that the number of distinct terms of type (N_1, N_2) is equal to

$$g(N_1, N_2) = \frac{N!}{N_1!\, N_2!\, 2^{N_2}} \tag{2.8}$$

For example, for $N = 4$, we have $g(4, 0) = 1$, $g(2,1) = 6$, and $g(0,2) = 3$, in agreement with our counting above. For $N = 3$, we have $g(3,0) = 1$, and $g(1,1) = 3$, in agreement with the explicit formula (2.6b).

The expressions given so far need an additional condition to define $\hat{U}$ and $\hat{W}$ uniquely. Looking at (2.6a), for example, we might replace $\hat{U}$

by some other one-particle operator, and include the change so produced as part of the operator $\hat{W}$. This would be a completely spurious "correlation" between particles. In order to avoid this possibility, we impose an additional condition, which is based on the physical requirement that the correlation $\hat{W}$ should have a *finite range*. If we think of $\hat{W}$ for the moment as a correlation function in x-space, $\hat{W}(\mathbf{r}_1 - \mathbf{r}_2)$, the condition is that the integral of this function over all $\mathbf{r}_1$ and $\mathbf{r}_2$ is proportional to the first power of the volume V, rather than to V^2. The correlation function $\hat{W}(\mathbf{r}_1 - \mathbf{r}_2)$ is given by the diagonal elements of the correlation matrix $\hat{W}$, expressed in configuration space. The integral of $\hat{W}(\mathbf{r}_1 - \mathbf{r}_2)$ is therefore equal to the trace of the correlation matrix $\hat{W}$, and this trace is independent of the representation used. Our condition on the pair correlation matrix $\hat{W}$ therefore assumes the form

$$\lim_{V\to\infty} \left(\frac{\text{Trace } \hat{W}}{V}\right) = \text{finite} \tag{2.9}$$

The purely formal "correlation" one would obtain by altering the one-particle operator $\hat{W}$ violates condition (2.9), and is thus excluded. Except for trivial factors and the lack of antisymmetrization so far, the limit (2.9) is just the contribution of dynamical pair correlations to the second virial coefficient, i.e., the b_2' of Eq. (III, 5.17). Equation (2.9) is related directly to the requirement (III, 4.18) that the integrals I_n arising from n-particle correlations must involve the volume V to the first power only.

In the Boltzmann case, the sequence of formulas (2.6) defines the general density matrix $\mathscr{U}_{B,N}$ in terms of the basic operators $\hat{U}$ and $\hat{W}$. The next step is the antisymmetrization. Instead of all possible N-particle states

$$| k_1 k_2 \cdots k_N \rangle$$

we now restrict ourselves to the configurations, namely, the antisymmetric functions (Slater determinants)

$$| \{k_1 k_2 \cdots k_N\} \rangle = (N!)^{-1/2} \sum_P (-1)^P \, | k_{1P} k_{2P} \cdots k_{NP} \rangle \tag{2.10}$$

The factor $(N!)^{-1/2}$ normalizes (2.10) to unity; the sum extends over all $N!$ permutations P of the N indices k_i. The antisymmetric form (2.10) vanishes identically if any two indices coincide with each other.

The properly antisymmetrized trial density matrix $\mathscr{U}_N$ is obtained from the Boltzmann form $\mathscr{U}_{B,N}$ by using (2.10) with the result

$$\begin{aligned} &\langle \{k_1 k_2 \cdots k_N\} \,|\, \mathscr{U}_N \,|\, \{k_1' k_2' \cdots k_N'\} \rangle \\ &= \frac{1}{N!} \sum_{P,P'} (-1)^P (-1)^{P'} \langle k_{1P} k_{2P} \cdots k_{NP} \,|\, \mathscr{U}_{B,N} \,|\, k_{1P'}' k_{2P'}' \cdots k_{NP'}' \rangle \end{aligned} \tag{2.11}$$

In this expression, P is a permutation of the indices $k_1, k_2, \ldots, k_N$, and P' is a permutation of the indices $k_1', k_2', \cdots k_N'$. The sums extend over all $N!$ permutations P and all $N!$ permutations P', i.e., there are $(N!)^2$ terms in all. The expression (2.11) vanishes automatically whenever two indices k_i coincide, or two indices k_i' coincide (Pauli principle).

Finally, we introduce the grand canonical ensemble, by weighting each density matrix $\mathscr{U}_N$ by a factor $z^N = \exp(N\beta\mu)$. In order to satisfy the condition of unit trace, (1.6b), we supply a c-number factor C in front. Thus *the trial density matrix of the quasi-chemical equilibrium theory has the form*

$$\mathscr{U} = C \sum_{N=0}^{\infty} z^N \mathscr{U}_N = C \sum_{N=0}^{\infty} \exp(N\beta\mu)\, \mathscr{U}_N \tag{2.12}$$

where $\mathscr{U}_N$ *is defined by* (2.11) *and* (2.6). This is the most general density matrix operator which can be built up from uncorrelated particles and dynamical pair correlations. Statistical correlations are included to all orders, through the antisymmetrization in (2.11).

The adjustable parameters, for use with the variation principle (1.6), are:

(a) The constant C in front; this is fixed by condition (1.6b);
(b) the matrix elements $\langle k \mid \hat{U} \mid k' \rangle$;
(c) The matrix elements $\langle k_1 k_2 \mid \hat{W} \mid k_1' k_2' \rangle$.

By using the minimum principle (1.6), we can determine, for any given temperature $T = 1/\mathfrak{k}\beta$ and chemical potential μ, the "best" matrices $\hat{U}$ and $\hat{W}$. Having found the "best" $\hat{U}$ and $\hat{W}$, the corresponding value of Ω, Eq. (1.6a) is the self-consistent approximation to the grand canonical potential of the system, in the quasi-chemical equilibrium approximation.

The extreme complexity of the expressions encountered makes it impossible to carry through this complete program at present. We shall find it necessary to restrict the trial density matrix rather severely, by making additional assumptions about $\hat{U}$ and $\hat{W}$. The restrictions are of two kinds: (1) To ensure that $\mathscr{U}$ is non-negative [condition (1.6c)], and (2) To enable actual calculations to be performed. The first type of restriction is discussed in Section 3, the second type in Section 4. Connection with the BZT theory is then made in Section 5.

3. Repulsive Correlations and the "Lifetime Problem"

Before we can use the trial density matrix (2.12) in the variation principle (1.6a), we must make sure that it is a non-negative operator, i.e., we must satisfy condition (1.6c).

For arbitrary basic matrices $\hat{U}$ and $\hat{W}$, it is terribly difficult to decide whether (2.12) is non-negative. In general, $\hat{U}$ and $\hat{W}$ fail to commute, and with sums of arbitrary non-commuting operators non-negativity of the sum is an exceedingly delicate condition.[7]

However, for use in a minimum principle such as (1.6), it is not necessary to determine a complete (necessary and sufficient) condition for $\mathscr{U}$ to be nonnegative. We may instead restrict ourselves to a more limited class of trial operators, by imposing additional restrictions on the basic operators $\hat{U}$ and $\hat{W}$. We shall do so, and we shall then discuss this new restriction from the physical point of view.

Non-negative operators have the following useful properties:

(i) The sum of a number of non-negative operators is itself non-negative

(ii) The direct (Kronecker) product of non-negative operators is itself non-negative,

(iii) An operator which is non-negative over some Hilbert space is automatically non-negative over any linear subspace of that Hilbert space.

We use these properties to prove the *theorem*: If the basic operators $\hat{U}$ and $\hat{W}$ are non-negative, then $\mathscr{U}$, Eq. (2.12), is also non-negative.

The proof is straightforward. First, with our initial assumption, expressions of type (2.6a), (2.6b), (2.6c), $\cdots$ are sums of direct products of non-negative operators, and hence are themselves non-negative. Next, the antisymmetrization operation (2.11) means that we restrict ourselves to the subspace of completely antisymmetric wave functions; i.e., the operators $\mathscr{U}_N$ and $\mathscr{U}_{B,N}$ have the same effect when acting on antisymmetric wave functions, but differ for all other wave functions,

[7] To mention one example: If the operators A and B are both non-negative, and commute with each other, then the operator

$$T = \exp(-A) - \exp(-A - B)$$

is positive definite. But if A and B fail to commute, T can have negative eigenvalues even if both A and B are non-negative! The proof (Delves, private communication) consists in constructing an example, using 2 by 2 matrices for A and B. A possible set is: $A = 1 + \sigma_z$, $B = 100(1 + \sigma_x)$, where the σ's are the usual Pauli spin matrices.

in the sense that $\mathscr{U}_{B,N}$ leads to some finite result whereas application of the operator $\mathscr{U}_N$ yields identically zero. Hence the replacement of $\mathscr{U}_{B,N}$ by $\mathscr{U}_N$ replaces some (unwanted) eigenvalues by zero, and retains the other (desired) eigenvalues intact. Thus a non-negative $\mathscr{U}_{B,N}$ implies a non-negative $\mathscr{U}_N$. Finally, in the last step, Eq. (2.12), we construct a sum of non-negative operators, which is non-negative, thereby proving the theorem.

As far as the operator $\hat{U}$ is concerned, the condition $\hat{U} > 0$ is not only sufficient, but is even necessary. To see this, let us work in a representation in which $\hat{U}$ is diagonal (this implies no loss of generality), and let us denote the eigenvalues of $\hat{U}$ by $\hat{\mathrm{u}}_k$

$$\langle k \mid \hat{U} \mid k' \rangle = \hat{\mathrm{u}}_k \, \delta_{kk'} \tag{3.1}$$

Instead of considering all possible configurations (antisymmetric states), let us fix our attention on the vacuum state and on the one-particle states. For these special states, the operator $\hat{W}$ does not contribute at all (since it requires the presence of at least two particles). Thus $\mathscr{U}$ is a diagonal operator for these states. Let us assume that a particular eigenvalue $\hat{\mathrm{u}}_m$ of $\hat{U}$ is negative. Then the eigenvalue of $\mathscr{U}$ in the vacuum state has opposite sign to the eigenvalue of $\mathscr{U}$ in the one-particle state $\mid m \rangle$. One of these two eigenvalues is therefore negative, in contradiction to our initial assumption that $\mathscr{U}$ is a non-negative operator.[8]

For the operator $\hat{W}$, on the other hand, the restriction to non-negative $\hat{W}$ is a significant physical assumption, which we now proceed to discuss.

Positive eigenvalues of the dynamical correlation matrix $\hat{W}$ correspond to attractive dynamical pair correlations, negative eigenvalues to repulsive dynamical pair correlations. *The omission of negative eigenvalues of $\hat{W}$ means that we are omitting all repulsive dynamical pair correlations.*[9] In view of the fact that a strong repulsive force, the Coulomb repulsion between electrons, is a prominent feature of the actual system, this omission of repulsive dynamical pair correlation is highly suspect. Nonetheless, it is a feature of all calculations carried out to date, and we must understand the meaning behind this apparently senseless limitation.

---

[8] To be complete, one must also show that the two eigenvalues in question do not both vanish. This is easy, since the eigenvalue of $\mathscr{U}$ in the vacuum state is just the constant C in front of (2.12). If C is zero, the entire operator vanishes, contrary to the normalization condition (1.6b).

[9] Some repulsive correlations are retained, however, namely, the statistical correlations, which are repulsive in a Fermi-Dirac system.

The term "pair correlation" can be understood in two entirely different ways. These two meanings are exemplified by the following two "pair-correlated" functions for 4 particles:

$$F = f(r_{12})f(r_{34}) + f(r_{13})f(r_{24}) + f(r_{14})f(r_{23}) \tag{3.2}$$

$$G = f(r_{12})f(r_{13})f(r_{14})f(r_{23})f(r_{24})f(r_{34}) \tag{3.3}$$

In the first function, F, any one particle i appears in at most one pair correlation factor $f(r_{ij})$. This is the meaning which we have used so far. If we think of the correlated pairs as pseudo-molecules, this is quite reasonable; it corresponds to saturation of chemical bonds. The basic wave function (IV, 1.1), of the superconducting ground state exhibits correlations of this type.

On the other hand, correlations of type (3.3) are needed if there are *repulsive* forces between all pairs of particles, e.g., the Coulomb force between electrons. The fact that the Coulomb force prevents electrons 1 and 2 from approaching each other closely, does not mean that close approaches of electrons 1 and 3 may be tolerated. Thus a factor $f(r_{13})$ is needed as well as a factor $f(r_{12})$, i.e., we are led to the function (3.3). *Repulsive pair correlations are not at all analogous to pseudo-molecules.*

If we do, unreasonably, attempt to include repulsive correlations within the formalism of the quasi-chemical equilibrium Ansatz (2.6), (2.11), (2.12), i.e., if we permit the operator $\hat{W}$ to have negative eigenvalues, then we find that the trial density matrix $\mathscr{U}$ is no longer a non-negative operator, and is therefore unacceptable. This formal difficulty is not an accident, but is an indication of a real difficulty with the approximation in these circumstances.

In normal systems, the presence of repulsive correlations is well known to produce important effects, and an elaborate formalism has been developed by Brueckner and others (Brueckner 58, Bell 61, and references quoted there), to allow calculation of these effects. The Brueckner theory is a modified perturbation expansion, starting from the Fermi sea state. In the theory of superconductivity, one will have to redo the Brueckner development, starting from the zero-order wave function, (IV, 1.1) rather than from the Fermi sea state. By this procedure, it should be possible to allow for the repulsive correlations in a meaningful fashion.

Another way of looking at the same problem has been suggested by Bardeen (61). As a result of the forces between electrons, the one-electron states $| \, k >$ may be thought of as "decaying states," rather than truly stable states. For example, an electron state well above the Fermi sea is unstable, because this electron may collide with other electrons, thereby sharing the excess energy. The "lifetime" of the one electron state becomes shorter, the farther its energy is removed from the Fermi surface. Similarly, a "hole" well inside the Fermi sea has a short lifetime. Bardeen (61) has suggested that the cutoff $\hbar\omega_c$ in the simplified interaction (IV, 2.5) may be determined by such lifetime effects, and that this cutoff will prevent the strong depression of the Coulomb repulsion term as calculated by Bogoliubov (58) and Morel (62). Since the calculation of Morel and Anderson (Morel 62) is actually in very much better agreement with experiment than the crude estimates of Pines (58), it is to be hoped that the lifetime effect is *not* as drastic as envisaged by Bardeen.

However, a cutoff of some sort seems desirable for quite another reason. In Chapter IV, Section 2, we pointed out that a theory with cutoff leads to a simple physical picture, in which most of the electrons sit in a filled single-particle Fermi sea, and there are only relatively few Schafroth-condensed electron pairs, made out of components close to the usual Fermi surface. This simple and intuitively acceptable picture is spoiled if the interaction has no natural cutoff, since then *all* electrons are contained in Schafroth-condensed electron pairs.

It is possible that a consistent calculation of the effects of repulsive correlations, based on an extension of the Brueckner theory, may restore the "chemical equilibrium" feature to the quasi-chemical equilibrium theory. No such calculation has been done as yet.

4. Reduction from a Three-Fluid to a Two-Fluid Model

The restriction to non-negative operators $\hat{W}$ does not lead to any appreciable formal simplification of (2.12). Since even the trace of the operator (2.12) defies explicit evaluation in the general case, some further drastic simplification of the Ansatz is required.

This is based on the idea of a Schafroth condensation of the attractive pair correlations. In the quasi-chemical equilibrium approximation, the total number of particles, N, is distributed among three different "fluids":

(i) the "single" particles, say N_1;
(ii) the normal, non-condensed pairs, containing $2N_{2n}$ particles;
(iii) the superfluid (condensed) pairs, containing $2N_{2s}$ particles.

We know that the independent-particle model (Husimi theory) is an excellent approximation above the transition point; we also know that the condensed pairs by themselves determine the behavior of the system in the limit of zero temperature (this was the basis of the work in Chapter IV).

It is therefore not completely unreasonable to *ignore the non-condensed pairs altogether*, as a first approximation. This eliminates the contribution (ii) above, and thus reduces the three-fluid model of the full quasi-chemical equilibrium theory, to a two-fluid model.

Mathematically, the reduction proceeds as follows: The operator $\hat{W}$ can always be written as a sum of factored terms:

$$\langle k_1 k_2 \mid \hat{W} \mid k_1' k_2' \rangle = \sum_\gamma \delta_\gamma w_\gamma(k_1, k_2)\, w_\gamma^*(k_1', k_2') \tag{4.1}$$

where $\delta_\gamma = \pm 1$ only. The number of positive and negative terms is invariant, but the actual functions $w_\gamma(k_1, k_2)$ are not unique (they can be made essentially unique by requiring mutual orthogonality, but this is not always the most convenient choice); the total number of terms in the sum (4.1) is called the rank of $\hat{W}$.

The restriction of Section 3, $\hat{W}$ non-negative, means that we omit all negative terms from (4.1).

The new restriction, to condensed pairs only, means that we retain *only one term* of (4.1). We saw this in Eq. (III, 9.6). It is this extreme and drastic simplification which underlies the microscopic theory of superconductivity!

Qualitatively speaking, each term of (4.1) with $\delta_\gamma = +1$ corresponds to a quantum state of the "quasi-molecule." In any real system, we therefore expect that the operator $\hat{W}$ has infinite rank [infinitely many terms in (4.1)]. The replacement of this operator by an operator of rank 1, i.e., by

$$\langle k_1 k_2 \mid \hat{W} \mid k_1' k_2' \rangle = w(k_1, k_2)\, w^*(k_1', k_2') \tag{4.2}$$

is therefore a very drastic assumption, which makes sense only if there exists an effective condensation of these quasi-molecules.

The fact that the "statistics of the quasi-molecule" enters the argument essentially, is brought out clearly by considering the possible effects of three-particle dynamical correlations. We may, of course, decompose the three-particle dynamical correlation matrix, call it $\hat{W}_3$, in a way analogous to (4.1):

$$\langle k_1 k_2 k_3 \mid \hat{W}_3 \mid k_1' k_2' k_3' \rangle = \sum_\varrho \delta_\varrho y_\varrho(k_1, k_2, k_3)\, y_\varrho^*(k_1', k_2', k_3') \tag{4.3}$$

However, if the particles are Fermions, so that the functions y_ρ are antisymmetric, it can be shown (Blatt 58) that each term of (4.3) appears at most once in any final result of the theory. This is the formal statement of the Pauli exclusion principle for 3-Fermion molecules in statistical mechanics.

As a result, a reduction of the infinite sum (4.3) to a single term, or even to any finite number of terms, does not yield a useful approximation. The same situation exists for dynamical correlations of all odd numbers of Fermions.

The theory based on retention of only a single term in (4.1) is given in Appendix B; some of the essential results have been stated in Chapter III, Section 9, (where U_2' is used for the matrix which we call $\hat{W}$ here). The function $w(k_1, k_2)$ in (4.2) is closely related to the "pair wave function" $\phi(k_1, k_2)$ [see (III, 9.6) and (III, 9.7)]. The results quoted in Chapter III, for this approximation, may be taken over directly. In particular, the Trace of the trial density matrix (2.12) is given by

$$\text{Trace}\,(\mathscr{U}) = C \exp\,[-\beta(\Omega^0 + \Omega')] = 1 \tag{4.4}$$

where Ω^0 is the contribution of the "singles," Eq. (III, 8.7), and Ω' is the contribution of the condensed pairs, Eq. (III, 9.17).

However, we now require more than just the trace of the trial density matrix (which is similar to the normalization integral in an ordinary variational calculation in quantum mechanics). We also need the traces (1.1), (1.2), and (1.3), all of which enter into (1.6). If the potential energy term in the Hamiltonian is a sum of interactions between pairs of electrons only, i.e., H is given by (IV, 1.3), then the traces (1.1) and (1.2) can be found explicitly (Blatt 60a). We quote the result here: with $g_k(x)$ defined by (III, 9.15) and $\bar{n}_k$ defined by (III, 8.9), the expectation value of the number operator is

$$\bar{N} = \text{Trace}\,(N_{\text{op}}\mathscr{U}) = \sum_k \nu_k = 2 \sum_{\mathbf{k}} \nu_k \tag{4.5a}$$

$$\nu_k = \bar{n}_k + (1 - \bar{n}_k)\, g_k \tag{4.5b}$$

The right-hand side of (4.5b) may be considered the sum of a contribution from single particles, and from condensed pairs, respectively. The parameter "x" in the definition of $g_k(x)$, (III, 9.15), is fixed eventually by requiring (4.5a) to equal N, the actual number of particles.

In order to write down the expectation value of the Hamiltonian, we introduce the definition

$$\psi_k = \frac{(2x)^{1/2}\,\phi_k}{1 + 2x\,|\,\phi_k\,|^2} \tag{4.6}$$

The same quantity played a role in the zero-temperature theory of Chapter IV [see (IV, 1.5) and (IV, 1.10)]. The expectation value of the

Hamiltonian (IV, 1.3) over the trial statistical matrix (2.12) is given by

$$\bar{E} = \text{Trace}\,(H\mathscr{U}) = \sum_k \epsilon_k \nu_k + \tfrac{1}{2}\sum_{k,k'} [\langle kk' \mid v \mid kk' \rangle - \langle kk' \mid v \mid k'k \rangle]\, \nu_k \nu_{k'}$$
$$+ \tfrac{1}{2}\sum_{k,k'} \langle k, -k \mid v \mid k', -k' \rangle\, (1 - \bar{n}_k)\,(1 - \bar{n}_{k'})\, \psi_k \psi_{k'} \tag{4.7}$$

This is very similar to the zero-temperature result (IV, 1.12), the changes resulting from the presence of unpaired particles.

The calculation of Trace ($\mathscr{U} \ln \mathscr{U}$) from (2.12) and (4.2) is quite complex; it has been carried out by Matsubara and Thompson (Thompson 63). The result is stated most easily in terms of quantities f_k defined by

$$f_k = \tfrac{1}{2}\{1 - [1 - 4\bar{n}_k(1 - \bar{n}_k)\,(1 - g_k)]^{1/2}\} \tag{4.8}$$

The result of Matsubara and Thompson is

$$S = -\,\mathfrak{k}\,\text{Trace}\,(\mathscr{U} \ln \mathscr{U})$$
$$= -\,\mathfrak{k} \sum_k [f_k \ln f_k + (1 - f_k) \ln (1 - f_k)] \tag{4.9}$$

We emphasize that (4.9) is not an assumption, but rather a mathematical consequence of (2.12) and (4.2); and furthermore, that (4.9) depends *only* on the trial density matrix $\mathscr{U}$, not on the assumed Hamiltonian.

Of course, the actual entropy does depend on the Hamiltonian; but this dependence is obtained only indirectly, through minimization of (1.6). For this minimization, the variational parameters of the theory are:

(a) the constant C in front of (2.12); this is fixed by (4.4);

(b) the constant x in (III, 9.15) and (4.6); this is fixed by requiring $\bar{N} = N$;

(c) the components ϕ_k of the pair wave function;

(d) the quantities $\bar{n}_k$ (average number of single particles in state k), which are related to the matrix $\langle k \mid \hat{U} \mid k' \rangle$ by (3.1), (III, 8.5), and (III, 8.9).

At this stage, we have all the results necessary to write down (1.5), or (1.6a). The side conditions (1.6b) and (1.6c) are satisfied, and we can proceed to minimize Ω.

5. Relationship to the BZT Theory

The theory developed so far differs from the thermodynamic theory of Bogoliubov, Zubarev, and Tserkovnikov in its entire approach; The BZT theory is an exact evaluation of the consequences of an assumed Hamiltonian (the reduced Hamiltonian), which represents a very drastic simplification of any reasonable Hamiltonian. The theory of this chapter starts from a reasonable Hamiltonian,[10] but makes drastic assumptions concerning the statistical matrix of the system.

Nonetheless, the two theories yield nearly identical results, as we shall now show. For this purpose, it is convenient to replace the variational parameters ϕ_k by new parameters u_k and v_k defined by

$$v_k^2 = \frac{\bar{n}_k + (1 - \bar{n}_k)\, g_k - f_k}{1 - 2f_k} = \frac{\nu_k - f_k}{1 - 2f_k} \tag{5.1}$$

$$u_k v_k = \frac{1 - \bar{n}_k}{1 - 2f_k}\, \psi_k \tag{5.2}$$

Using (5.1) and (5.2), it is possible to show that

$$u_k^2 + v_k^2 = 1 \tag{5.3}$$

The new variational parameters are u_k, v_k, and f_k; they replace the parameters $\bar{n}_k$ and $\chi_k = (2x)^{1/2}\phi_k$. The new parameters are defined in terms of the old ones through (4.8), (5.1), and (5.2). The inverse relations are

$$\bar{n}_k = \frac{f_k(1 - f_k)}{f_k v_k^2 + (1 - f_k)\, u_k^2} \tag{5.4}$$

and (5.2) solved for ψ_k. We also note that ν_k, Eq. (4.5b), is given in terms of the new parameters by

$$\nu_k = f_k + (1 - 2f_k)\, v_k^2 \tag{5.5}$$

[10] Our assumed Hamiltonian is reasonable, in the sense that all matrix elements of the assumed electron-electron interaction are retained. It is nevertheless insufficient from a quantitative point of view, because there is no explicit interaction between electrons and phonons. However, the close similarity between the zero temperature results with and without explicit electron-phonon interaction (see Chapters IV and V) makes it likely that the, much more difficult, calculation with phonons included will not give anything essentially different. This expectation is confirmed by variational calculations of Moskalenko (58) and Kvasnikov (58, 58a, 58b).

We now introduce the new parameters into (4.5a) [i.e., (5.5) replaces (4.5b)] and into (4.7); the entropy, Eq. (4.9), is already expressed in terms of the new parameters. It is furthermore convenient to introduce the definitions

$$L_k = \tfrac{1}{2} \sum_{k'} [\langle kk' \mid v \mid kk' \rangle - \langle kk' \mid v \mid k'k \rangle] \, \nu_{k'} \tag{5.6}$$

and

$$\Delta_k = - \sum_{k'} \langle k, -k \mid v \mid k', -k' \rangle \, (1 - 2f_{k'}) \, u_{k'} v_{k'} \tag{5.7}$$

These definitions are the finite-temperature generalizations of (IV, 1.15) and (IV, 1.16).[11]

Combining all the expressions and definitions so far, the quantity to be minimized according to (1.5) and (1.6) is

$$\Omega = \sum_k \{(\epsilon_k - \mu + L_k) [f_k + (1 - 2f_k) \, v_k^2] - \tfrac{1}{2} \Delta_k (1 - 2f_k) \, u_k v_k + \mathfrak{k}T[f_k \ln f_k + (1 - f_k) \ln (1 - f_k)]\} \tag{5.8}$$

The independent parameters are v_k and f_k; u_k is given in terms of v_k by (5.3). Differentiation of (5.8) with respect to v_k yields an equation not involving f_k at all. We introduce the quantity ξ_k exactly as before:

$$\xi_k = \epsilon_k - \mu + 2L_k \tag{5.9}$$

We then obtain the variational results

$$u_k v_k = \frac{\Delta_k}{2(\xi_k^2 + \Delta_k^2)^{1/2}} \tag{5.10}$$

and

$$v_k^2 = \frac{1}{2} \left[1 - \frac{\xi_k}{(\xi_k^2 + \Delta_k^2)^{1/2}} \right] \tag{5.11}$$

where the positive square root is implied throughout.

Next, we differentiate (5.8) with respect to f_k, and use (5.10) and (5.11) to simplify the resulting expression. This procedure yields

$$f_k = \frac{1}{\exp [\beta(\xi_k^2 + \Delta_k^2)^{1/2}] + 1} \tag{5.12}$$

Finally, we substitute (5.10) and (5.12) into the defining Eq. (5.7) to obtain an integral equation for Δ_k, namely,

$$\Delta_k = - \sum_{k'} \langle k, -k \mid v \mid k', -k' \rangle \frac{\Delta_{k'} \tanh [\frac{1}{2} \beta(\xi_{k'}^2 + \Delta_{k'}^2)^{1/2}]}{2(\xi_{k'}^2 + \Delta_{k'}^2)^{1/2}} \tag{5.13}$$

[11] For the sake of simplicity, we have assumed that the, always possible, antisymmetrization (V,1.27) of the interaction matrix elements has already been carried out.

We now note that these results of the variational approach are identical to the results of the BZT theory, if we use the reduced interaction.[12] In particular, (5.13) reduces to the BZT integral equation (VI, 3.46) if we ignore the energy renormalization term L_k in (5.9). With the reduced interaction, (5.6) vanishes identically, since the interaction matrix elements appearing in (5.6) are thrown away; the reduced interaction consists of the matrix elements appearing in (5.7), and *only* those matrix elements.

Since the BZT theory is an *exact* (in the limit of infinite volume) evaluation of the consequences of the reduced Hamiltonian, we conclude that the trial density matrix (2.12) and (4.2) is also exact in that case. This conclusion is not rigorous, since we have shown equivalence of the final results (traces) only, not equivalence of the density matrices. In fact, for finite numbers of particles, Eq. (2.12) is certainly *not* equivalent to the density matrix (VI, 3.16): the latter fails to commute with the number operator, whereas (2.12) does commute with N_{op}. Nevertheless, there is certainly a physical equivalence in spite of the mathematical differences.

Qualitatively, it is easy to see that the reduced interaction leads to a pair correlation matrix of separable form, (4.2): the operator for the reduced interaction is zero for all pairs of particles with $k_1 \neq -k_2$, and therefore fails to yield dynamical correlations of those particle pairs. The only pairs which are correlated at all are pairs with $k_1 = -k_2$. In this way, the infinity of terms in (4.1) is reduced to the single term (4.2), with $w(k_1, k_2) = 0$ unless $k_1 = -k_2$.

We are now in a position to answer the question posed at the beginning of this chapter: What is the physical meaning of the assumption of the reduced Hamiltonian?

All the essential results derivable from the reduced Hamiltonian are also derivable from the trial density matrix (2.12)–(4.2). The differences between the two treatments eventually appear only in energy renormalization terms L_k, which make no qualitative difference in the superconducting state. In the normal state, these terms are exactly what is expected from the independent-particle calculation of Husimi (40); to this extent the variational calculation based on (2.12)–(4.2) is superior to the calculation based on the reduced interaction.

It appears reasonable to suppose, therefore, that *the physical meaning of the reduced Hamiltonian is precisely the assumption of this trial density*

[12] The fact that, for the reduced interaction, the variational calculation yields the same results as the exact theory of BZT, follows from the work of Bardeen, Cooper, and Schrieffer (Bardeen 57).

matrix for the system. Considered as a Hamiltonian, the reduced Hamiltonian has no justification whatever; but considered as the equivalent of (2.12)–(4.2), the assumption appears quite tenable, even though far from perfect. In Chapter III, we gave direct physical arguments for restriction to dynamical pair correlations only [i.e., for (2.12)], and we gave both a mathematical argument and a physical plausibility argument for the assumption of a Schafroth condensation of these pair correlations, i.e., for (4.2).

Knowing the physical meaning of the theory developed so far, we are now in a position to discuss the limitations of that theory. These limitations arise from:

(i) the neglect of higher dynamical correlations;

(ii) the neglect of pair correlations corresponding to non-condensed pairs.

The most important higher dynamical correlations are repulsive correlations between *all* particles, caused by the repulsive Coulomb interaction. This effect not only gives rise to plasma waves (which appear neither in the quasi-chemical equilibrium theory nor in the BZT theory), but it is likely that the repulsive correlations also underlie the "lifetime effect" and "cutoff" which were discussed at the end of Section 3.

By comparison, the neglect of non-condensed pairs is unlikely to be very serious. In the limit of zero temperature, the non-condensed pairs are probably related to the "exciton" states of Anderson (58), Bogoliubov (58) and Bardasis (61). These states are known to be rather unimportant, at least for the thermodynamics. Above the transition point, the non-condensed pairs must give the first-order correction to the band theory of metals; they therefore include such phenomena as exciton states in normal metals, which are again not of major importance for metallic properties. Thus, the only region in which non-condensed pairs may turn out to play an essential role is the transition region itself, T near T_c. In this region, the energy gap approaches zero, and the extreme favoring of the condensed pair state over all other pair states disappears. Whether introduction of the non-condensed pairs alters the nature of the thermodynamic transition is not known at present. The only calculation available, that of Wentzel (60), gives a drastic change: the transition found by Wentzel is first-order (with a latent-heat), not second order as in the BZT theory. However, it is doubtful whether the perturbation expansion used by Wentzel is sufficiently accurate in the neighborhood of the transition, and more work is required.

VIII

The Meissner Effect

1. Introduction

Having discussed the thermodynamics of superconductivity, we now turn our attention to the spectacular electromagnetic properties of superconductors: the Meissner effect (this Chapter), and persistent currents (Chapter IX).

In Chapter II, we saw that a condensed ideal Bose-Einstein gas has a Meissner effect; to a very good approximation, the magnetic behavior of such a gas is described by the phenomenological London equation (I, 2.7) which we repeat here for convenience:

$$\text{curl}\,(\Lambda \mathbf{j}_s) = -\frac{\mathbf{B}}{c} \tag{1.1}$$

The material constant Λ in this equation has the value, for the Bose gas at zero temperature,

$$\frac{1}{\Lambda} = \frac{N}{V}\frac{e^2}{M} \qquad \text{(Bose gas)} \tag{1.2}$$

where N/V is the number of bosons per cm^3, e is the charge of the boson, and M its mass.

The ground state of the superconducting system is described by a wave function corresponding to a Schafroth condensate of electron pairs:

$$\Psi = (b^+)^{N/2} \mid 0\rangle \tag{1.3}$$

where b^+ is the operator which creates a pair of electrons in a certain wave function $\varphi(k, k')$:

$$b^+ = 2^{-1/2} \sum_{k,k'} \varphi(k, k')\, a_{k'}^+ a_k^+ \tag{1.4}$$

It appears likely, therefore, that the analogy with a gas of elementary bosons extends also to magnetic properties, i.e., we may expect to find a Meissner magnetic field expulsion for the Schafroth-condensed pair gas as well.

In this chapter we shall demonstrate that this is indeed the case. In order to avoid unnecessary complexity, we shall restrict ourselves to the simplest assumptions. These are:

(1) Zero temperature, i.e., the calculation is based on the ground state.

(2) The electron-electron interaction responsible for the pairing is described by an ordinary potential; this same assumption was made in Chapter IV, and we saw in Chapter V that the explicit introduction of the Fröhlich electron-phonon coupling did not produce any qualitative changes in the results.

(3) In the absence of the applied magnetic field, the system has full translation and rotation invariance; this allows us to use "simple pairing" for the pair wave function in the field-free system. This assumption will be discussed later on; it is adequate for the essentials of the Meissner effect (the exponential law of field expulsion), but inadequate for the determination of the law of field expulsion in the neighborhood of the metallic surface.

(4) We restrict ourselves to the linear response of the system to an applied magnetic field; since the equations are by assumption linear, we make a Fourier analysis and treat each Fourier component by itself; finally, we confine ourselves to Fourier components with very long wavelengths (small wave numbers q). These latter components determine the law of field expulsion for large depths of penetration into the specimen.

Before going into the calculation itself, let us summarize the essential steps. In the absence of an applied field, the calculation was done already under assumptions (1), (2), and (3) (see Chapter IV). What, then, are the changes introduced by the applied magnetic field?

There is *no* change in the structure of the ground state wave function (1.3): for sufficiently weak applied fields (below the critical field), the system retains its character of a Schafroth-condensed pair gas. The quantity which changes is the wave function $\varphi(k, k')$ of the electron pair. Thus, if $\varphi_0(k, k')$ is the electron pair wave function in the absence of the applied field, we must now put

$$\varphi(k, k') = \varphi_0(k, k') + \delta\varphi(k, k') \tag{1.5}$$

where $\delta\varphi$ is (for a first-order calculation) directly proportional to the strength of the applied field.

As a result of this change in the pair wave function φ, the over-all wave function (1.3) undergoes a corresponding change, and in particular the expectation value of the current operator no longer vanishes. We intend to calculate this expectation value, and relate it linearly to the applied field; the leading term in this linear relationship will turn out to have the London form (1.1).

It should be noted that the Meissner effect at zero temperature is purely a ground state property. Excited states of the system need not, and do not, enter the calculation. Of course, the "perturbed" ground state wave function (1.3) may be expanded as a sum of wave functions of the unperturbed (field-free) electron system. But this expansion, which we shall discuss in Section 6 of this chapter, is neither necessary nor even convenient for our purposes. The simplicity of the pairing wave function (1.3) is obscured if that wave function is written as an infinite sum.

The treatment based on (1.3), (1.4), and (1.5) is the most natural one from the point of view of the quasi-chemical equilibrium theory. It is, however, not the only way to obtain a gauge-invariant Meissner effect.

Bogoliubov (58d, 59) has succeeded in generalizing his transformation so as to preserve gauge invariance of the theory. The "Hartree-Bogoliubov" method has been discussed and developed further by Baranger (61, 62, 63), Valatin (61), and Weller (61).

It can be shown that this generalized Hartree-Bogoliubov method is closely related to the generalized pairing wave function approach used in this chapter (Weller 61, Baranger 62).

An alternative way to obtain gauge-invariant results is the use of Green's function methods. The basic papers are by Gorkov (58), Nambu (60), and Kadanoff (61). Further references will be given in Sections 6 and 7.

A third method, the use of the theory of "collective excitations," is much less satisfactory; we shall return to this point in Section 6.

2. Expectation Values over General Pairing Wave Functions

The presence of an applied magnetic field spoils the translation and rotation invariance, which were used in Chapter III, Section 9, to establish the "simple pairing" property (III, 9.12):

$$\varphi_0(\mathbf{k}_1, s_1; \mathbf{k}_2, s_2) = 0 \quad \text{unless} \quad \mathbf{k}_1 = -\mathbf{k}_2 \quad \text{and} \quad s_1 = -s_2 \tag{2.1}$$

We expect, and shall find, that the additional term $\delta\varphi$ in (1.5) does not satisfy this selection rule.[1]

The formulas for expectation values of operators over the wave function (1.3) were given in Eqs. (IV, 1.8) and (IV, 1.11). We now need generalizations of these formulas for a general pair wave function $\varphi(k_1, k_2)$ which does not satisfy (2.1). These formulas (Blatt 60a) are merely quoted here; an outline of their derivation is contained in Appendix B.

We start by defining an unnormalized pair wave function $\chi(k_1, k_2)$ through

$$\chi(k_1, k_2) = (2x)^{1/2} \varphi(k_1, k_2) \tag{2.2}$$

This equation is the generalization of (IV, 1.5); the parameter x is a c-number to be determined later. Next, we define a one-particle "density matrix" based on this wave function, namely,

$$\langle k \mid \rho \mid k' \rangle = \sum_{k''} \chi(k, k'') \, \chi^*(k', k'') \tag{2.3}$$

This is a Hermitean, non-negative operator in k-space. In the simple pairing case, it reduces to

$$\langle k \mid \rho \mid k' \rangle = \chi_k^2 \, \delta_{kk'} \qquad \text{(simple pairing)} \tag{2.4}$$

The generalization of g_k, Eq. (IV, 1.7), is the one-particle matrix

$$\langle k \mid g \mid k' \rangle = \left\langle k \left| \frac{\rho}{1 + \rho} \right| k' \right\rangle \tag{2.5}$$

With this notation, the expectation value of a general single-particle operator J, Eq. (V, 1.22), becomes

$$\langle J \rangle = \frac{(\Psi, J\Psi)}{(\Psi, \Psi)} = \sum_{k,k'} \langle k \mid j \mid k' \rangle \langle k' \mid g \mid k \rangle \equiv \mathrm{tr}_1(jg) \tag{2.6}$$

In particular, the expectation value of the number operator (V, 1.23) is

$$\langle N \rangle = \sum_k \langle k \mid g \mid k \rangle \tag{2.7}$$

[1] Zumino (62) and Bloch (62) show that the simple pairing property is restored by using a different set of single-particle basis functions. However, the problem of finding these new basis functions is by no means trivial, and no calculation along these lines has been performed as yet. We shall, in this Chapter, use the term "simple pairing" in its narrow sense only, as a pairing of ordinary states in k-space according to (2.1). In this narrow sense, simple pairing is destroyed by the application of the magnetic field.

The parameter x in (2.2) enters implicitly into (2.7), and the value of x is determined by the condition that $\langle N \rangle$ should equal the actual number of particles in the system. Equation (2.6) reduces to (IV, 1.8), Eq. (2.7) to (IV, 1.9), in the simple pairing case.

In order to obtain expectation values of two particle operators (V, 1.26), we need further definitions. The generalization of ψ_k, Eq. (IV, 1.10), is

$$\psi(k_1, k_2) = \sum_{k'} \left\langle k_1 \left| \frac{1}{1+\rho} \right| k' \right\rangle \chi(k', k_2) \tag{2.8}$$

The structure of this function is quite simple: if we expand $(1 + \rho)^{-1}$ in a power series and use (2.3), the result is

$$\begin{aligned} \psi(k_1, k_2) &= \chi(k_1, k_2) - \sum_{m_1 m_2} \chi(k_1 m_1)\, \chi^*(m_1 m_2)\, \chi(m_2 k_2) \\ &\quad + \sum_{m_1 m_2 m_3 m_4} \chi(k_1 m_1)\, \chi^*(m_1 m_2)\, \chi(m_2 m_3)\, \chi^*(m_3 m_4)\, \chi(m_4 k_2) - \cdots \end{aligned} \tag{2.9}$$

The explicit form (2.9) can be used to derive the identity

$$\langle k \mid g \mid k' \rangle = \sum_{k''} \chi(k, k'')\, \psi^*(k', k'') \tag{2.10}$$

A second quantity which we require is the two-particle operator with matrix elements:

$$\langle kl \mid p \mid k'l' \rangle = \langle k \mid g \mid k' \rangle \langle l \mid g \mid l' \rangle - \langle k \mid g \mid l' \rangle \langle l \mid g \mid k' \rangle \tag{2.11}$$

In the simple pairing case, this reduces to

$$\langle kl \mid p \mid k'l' \rangle = g_k g_l (\delta_{kk'} \delta_{ll'} - \delta_{kl'} \delta_{lk'}) \qquad \text{(simple pairing)} \tag{2.12}$$

We use the natural notations

$$(\psi, t\psi) = \sum_{klk'l'} \psi^*(k, l) \langle kl \mid t \mid k'l' \rangle \psi(k', l') \tag{2.13}$$

and

$$\mathrm{tr}_2(tp) = \sum_{klk'l'} \langle kl \mid t \mid k'l' \rangle \langle k'l' \mid p \mid kl \rangle \tag{2.14}$$

The generalization of (IV, 1.11) is then

$$\langle T \rangle = \frac{(\Psi, T\Psi)}{(\Psi, \Psi)} = \tfrac{1}{2} [(\psi, t\psi) + \mathrm{tr}_2(tp)] \tag{2.15}$$

Formulas (2.6) and (2.15) suffice to determine the expectation value of the Hamiltonian H_N, Eq. (IV, 1.3), over the generalized pairing wave function (1.3), even when the magnetic field terms are included in that Hamiltonian. The N-particle Hamiltonian in the presence of a magnetic field with vector potential $\mathbf{A}(\mathbf{x})$ is[2]

$$H_N = \sum_{i=1}^{N} \frac{[\mathbf{p}_i - e\mathbf{A}(\mathbf{x}_i)/c]^2}{2M} + \sum_{i<j=1}^{N} v(\mathbf{x}_i - \mathbf{x}_j) \tag{2.16}$$

The first sum in (2.16) is a one-particle operator, the second sum is a two-particle operator.

The idea of the subsequent calculation is to minimize the expectation value $\langle H(A)\rangle$ of the operator (2.16). The parameters for this variation are the components $\varphi(k_1, k_2)$ of the pair wave function. With the wave function determined in this fashion, we then evaluate the expectation value of the current density. Since the current density is a one-particle operator, and the Hamiltonian a sum of one-particle and two-particle operators, the expressions (2.6) and (2.15) are in principle sufficient for this calculation.

3. Gauge Invariance

It is highly desirable to establish gauge invariance of the calculation, before doing any calculating. Fortunately, this is quite easy here (Blatt 60b). The gauge transformation

$$\mathbf{A}'(\mathbf{x}) = \mathbf{A}(\mathbf{x}) + \operatorname{grad} f(\mathbf{x}) \tag{3.1}$$

on the vector potential induces the following transformation on the pair wave function:

$$\varphi'(x_1, x_2) = \exp\left\{\frac{ie}{\hbar c}[f(x_1) + f(x_2)]\right\} \varphi(x_1, x_2) \tag{3.2}$$

Because of the simple structure of (3.2) in x-space, it is preferable to use x-space rather than k-space to establish gauge invariance.

In x-space, the definition of the operator ρ, Eq. (2.3), assumes the form (the integration over $d\tau''$ includes a sum over spin indices)

$$\langle x \mid \rho \mid x'\rangle = \int d\tau''\, \chi(x, x'')\, \chi^*(x', x'') \tag{3.3}$$

[2] This equation is obtained from (IV,1.3) by the usual replacement of the momentum $\mathbf{p}$ by $\mathbf{p} - e\mathbf{A}/c$. In this process, magnetic effects due to spins have been ignored. The Meissner effect is due to orbital magnetism, not spin magnetism. However, spin effects do enter into certain experiments, most notably the Knight shift.

The gauge transform of the operator ρ is therefore

$$\langle x \mid \rho' \mid x' \rangle = \exp \left\{ \frac{ie}{\hbar c} [f(x) - f(x')] \right\} \langle x \mid \rho \mid x' \rangle \tag{3.4}$$

This can also be written as a similarity transformation

$$\rho' = \exp(iS)\, \rho \exp(-iS) \tag{3.5}$$

with S diagonal in x-space,

$$S = \frac{e}{\hbar c} f(x) \tag{3.6}$$

The same S-transformation applies to the operator g, Eq. (2.5),

$$g' = \exp(iS)\, g \exp(-iS) \tag{3.7}$$

Let us now consider the one-particle part of the Hamiltonian (2.16), i.e., the first sum in (2.16). We introduce the notation

$$\mathbf{D} = \nabla - \frac{ie}{\hbar c} \mathbf{A}(x) \tag{3.8}$$

so that the one-particle part of H_N is

$$J = -\sum_{i=1}^{N} \frac{\hbar^2 \mathbf{D}_i^2}{2M} \tag{3.9}$$

Using (2.6) and (2.10), the expectation value of this operator can be written in the form

$$\mathrm{tr}_1(jg) = \frac{\hbar^2}{2M} \int d\tau \int d\tau'' \, [\mathbf{D}\psi(x, x'')]^* \cdot [\mathbf{D}\chi(x, x'')] \tag{3.10}$$

In (3.10), the operator $\mathbf{D}$ operates on the first coordinate, x, in both factors.

When we make a gauge transformation, the operator $\mathbf{D}$ changes because $\mathbf{A}$ appears explicitly in (3.8), and the wave functions ψ and χ change according to the same rule as (3.2). Denoting the gauge-transformed expressions by primes, a straightforward calculation yields the identities

$$\mathbf{D}'\psi' = \exp \left[\frac{ie}{\hbar c} f(x'') \right] \mathbf{D}\psi \tag{3.11a}$$

$$\mathbf{D}'\chi' = \exp \left[\frac{ie}{\hbar c} f(x'') \right] \mathbf{D}\chi \tag{3.11b}$$

Substitution of (3.11) into (3.10) yields the gauge identity

$$\mathrm{tr}_1(j'g') = \mathrm{tr}_1(jg) \tag{3.12}$$

The calculation is even simpler for the two-particle operators. The gauge transform of the two-particle operator p is given in x-space by

$$\langle x_1 x_2 \mid p' \mid x_1' x_2' \rangle = \exp\left\{ \frac{ie}{\hbar c} [f(x_1) + f(x_2) - f(x_1') - f(x_2')] \right\} \langle x_1 x_2 \mid p \mid x_1' x_2' \rangle \tag{3.13}$$

Thus, in particular, the diagonal elements of p are invariant under the gauge transformation. With an ordinary potential $v(x_i - x_j)$, however, only the diagonal elements of p enter into

$$\mathrm{tr}_2(vp) = \int d\tau_1 \int d\tau_2 \, v(x_1 - x_2) \langle x_1 x_2 \mid p \mid x_1 x_2 \rangle \tag{3.14}$$

This expression is therefore immediately gauge-invariant.

Since the gauge transformation on the wave functions, including $\psi(x_1, x_2)$, is given by (3.2), the square of a wave function is gauge-invariant in x-space. This suffices to establish gauge invariance of the term

$$(\psi, v\psi) = \int d\tau_1 \int d\tau_2 \, v(x_1 - x_2) \mid \psi(x_1, x_2) \mid^2 \tag{3.15}$$

Using (3.12), (3.14), and (3.15), we obtain from (2.6) and (2.15) the gauge identity

$$(\Psi', H'\Psi') = (\Psi, H\Psi) \tag{3.16}$$

The expectation value of the Hamiltonian is explicitly gauge-invariant. Since all subsequent expressions are obtained by minimizing this expectation value with respect to certain adjustable parameters, all subsequent expressions are automatically gauge-invariant.

The proof which we have just given depends on the fact that the electron-electron interaction is taken as a simple potential $v(x_1 - x_2)$, diagonal in configuration space. This is not an accident of our approximations, but is in the nature of the problem. In general, unless very special care is taken, a Hamiltonian with momentum-dependent interactions gives gauge-dependent results, even if one calculates *exactly*. In such a case, no approximation can be expected to give gauge-invariant results (Schafroth 58).

If electrons are the only particles considered explicitly, the Hamiltonian we have used is the most general one for which gauge-invariant results can be expected. In fact, however, electrons are only part of the picture in superconductivity. The electron-phonon interaction should really be included. In principle, one should start from ordinary Coulomb interactions involving lattice ions and electrons, and go from this to the electron-phonon picture. In the original picture, the interactions are all ordinary potentials, diagonal in configuration space, and gauge invariance is immediate. It would then be necessary to make sure that the reductions necessary to go to the electron-phonon picture are carried out in such a way that gauge invariance is preserved. This procedure has not yet been carried through.

However, a much simpler procedure is possible by using a bit of physical reasoning. The formal requirement of gauge invariance is a mathematical expression of the law of conservation of charge, i.e., of the continuity equation

$$\operatorname{div} \mathbf{j} + \frac{\partial \rho}{\partial t} = 0$$

In metals, the ions move much more slowly than the electrons. It is therefore a good first approximation to neglect the contribution of the ionic motions to the electrical current, compared to the contribution of electronic motions. Since phonons are a way of describing ionic motions, the phonons do not contribute to the electrical current density in this approximation. We obtain a gauge-invariant formulation, in this approximation, by considering all phonon operators as unaffected by the gauge transformation on the vector potential. In this simplified model, the gauge transformation operates on electron variables but not on phonon variables.

Even in this approximation, the actual calculations are so difficult that no full calculation of the Meissner effect along the present lines has been carried through as yet. Starting from the wave function of McKenna (62) rather than the pure electron wave function (1.3), gauge invariance is easy enough to establish; but the expressions for the expectation values of the relevant operators are exceedingly complicated for further work.

A quicker and more formal method is required, however, to establish this gauge invariance. We could have used this quicker method also for the pure electron case, but have preferred to keep all formulas explicit. Let us explain the quick method for the pure electron case.

We write the gauge transformation as a similarity transformation (3.5), but this time on the operators in second-quantization formalism. The operator b^+ creating a pair in the pair state $\varphi(x_1, x_2)$ is

$$b^+ = 2^{-1/2} \int d^3x_1 \int d^3x_2 \, \varphi(x_1, x_2) \, a^+(x_2) \, a^+(x_1)$$

where $a^+(x)$ is the x-space analog of the creation operator a_k^+ [this x-space operator is often denoted by $\Psi^+(x)$].

The S-transformation corresponding to the gauge transformation (3.1) is mediated by the operator

$$S = \frac{e}{\hbar c} \int d^3x \, f(x) \, a^+(x) \, a(x)$$

With this choice of S, it is possible to show that $\exp(iS) \, b^+ \exp(-iS)$ is an operator of precisely the same *form* as b^+ itself; the only change is that $\varphi(x_1, x_2)$ has been replaced by its gauge transform [see Eq. (3.2)]. Since the basic approximation of the theory, namely the trial function (1.3) and (1.4), makes no assumption about the specific pair function $\varphi(x_1, x_2)$, it follows immediately that the approximation procedure preserves gauge invariance of the formalism as a whole.

Put in this way, the argument goes through equally readily in the presence of phonons, since all phonon variables are considered invariant under the gauge transformation anyway.

The simplicity of the proof of gauge invariance in x-space representation may obscure the complexity of the gauge transformation in other representations, in particular, in k-space. As one example, consider the following gauge function $f(x)$ in (3.1):

$$f(x) = \frac{A_{||}(q)}{q} \left(\frac{2}{V}\right)^{1/2} \sin(\mathbf{q} \cdot \mathbf{x}) \tag{3.17}$$

which adds a single Fourier component of "longitudinal" vector potential

$$\operatorname{grad} f(x) = \frac{\mathbf{q} A_{||}(q)}{q} \left(\frac{2}{V}\right)^{1/2} \cos(\mathbf{q} \cdot \mathbf{x}) \tag{3.18}$$

Let $\varphi(k_1, k_2)$ be the pair wave function in k-space, and let the constant γ be defined by

$$\gamma = \left(\frac{2}{V}\right)^{1/2} \frac{e A_{||}(q)}{2\hbar c q} \tag{3.19}$$

Then the gauge transform φ' of the pair wave function is given by the double infinite sum

$$\varphi'(k_1, k_2) = \sum_{n,n'=-\infty}^{\infty} J_n(2\gamma) \, J_{n'}(2\gamma) \, \varphi(k_1 - nq, k_2 - n'q) \tag{3.20}$$

where $J_n(z)$ is the Bessel function of order n. What we have shown above is that the wave function Ψ, Eqs. (1.3) and (1.4), constructed from *this* φ' instead of the original φ, gives the same expectation value for the transformed Hamiltonian as the original wave function (1.3) gives for the original Hamiltonian. It is apparent that a direct proof of this statement within the k-space formalism would be lengthy.

Equation (3.20) also shows that the "simple pairing" condition (2.1) is inconsistent with gauge invariance. If the original wave function $\varphi(k_1, k_2)$ obeys (2.1), its gauge transform φ' fails to obey (2.1).

Since the "reduced interaction" leads automatically to a pair wave function with simple pairing, *the reduced interaction is by its very nature gauge-dependent.* If we had restricted ourselves to the "reduced interaction" matrix elements of v, instead of retaining the whole operator v as we actually did, the gauge identity (3.16) would not have been valid.

Whereas the difference between the true and the reduced interaction is, in the final event, of only minor importance for the thermodynamic theory, this difference is absolutely crucial for a theory of the Meissner effect. The full interaction is required to obtain gauge invariance, which is a necessary condition for an acceptable theory.

4. Variational Equation for the Pair Wave Function

In Chapter IV, Section 1 we used the condition $(\Psi, H\Psi) =$ minimum, to derive an integral equation whose solution determines the pair wave function $\varphi(k_1, k_2)$. The work of that chapter depended on the simple pairing condition (2.1), and we must now develop a generalization to the case of arbitrary pair wave functions φ.

The detailed calculation (Blatt 60b) is straightforward in principle but rather lenghty. We therefore confine ourselves to stating the results.

As before, we handle the condition of given N by means of a Lagrange multiplier, i.e., we minimize the expectation value of the operator $H - \mu N_{\text{op}}$. Since the particle number is conserved at all stages, this is merely a mathematical convenience.

Several auxiliary quantities enter, which are generalizations of corresponding quantities in Chapter IV. The generalization of L_k, Eq. (IV, 1.15), is the matrix[3]

$$\langle k \mid L \mid k' \rangle = \tfrac{1}{2} \sum_{l,l'} (\langle k'l' \mid v \mid kl \rangle - \langle k'l' \mid v \mid lk \rangle) \langle l' \mid g \mid l \rangle \tag{4.1}$$

[3] In writing this matrix, we assume that the (always possible) antisymmetrization of interaction matrix elements according to (V,1.27) has been carried out.

The generalization of the "energy gap parameter" Δ_k, Eq. (IV, 1.16), is a quantity with the structure of a wave function, namely,

$$\Delta(k, l) = -\sum_{k',l'} \langle kl \mid v \mid k'l' \rangle \, \psi(k', l') \tag{4.2}$$

When we vary the pair wave function $\varphi(k, k')$ and ask for the minimum expectation value of $H - \mu N$, the result is the integral equation

$$\sum_{k''} (\langle k \mid j \mid k'' \rangle - \mu \delta_{kk''} + 2 \langle k \mid L \mid k'' \rangle) \, \chi(k'', k')$$
$$- \tfrac{1}{2} \Delta(k, k') + \tfrac{1}{2} \sum_{k'',l''} \Delta^*(k'', l'') \, \chi(k'', k) \, \chi(l'', k') = 0 \tag{4.3}$$

In this equation, $\langle k \mid j \mid k'' \rangle$ is the matrix element of the one-particle operator in (3.9); this is the term containing the vector potential of the applied magnetic field.

In deriving (4.3), use has been made of the antisymmetry of all wave functions, i.e.,

$$\chi(k_2, k_1) = -\chi(k_1, k_2) \tag{4.4a}$$

$$\psi(k_2, k_1) = -\psi(k_1, k_2) \tag{4.4b}$$

$$\Delta(k_2, k_1) = -\Delta(k_1, k_2) \tag{4.4c}$$

Let us first determine the form which (4.3) takes if there is no applied field and simple pairing. In that case, all of χ, ψ, and Δ vanish unless $k_2 = -k_1$; the matrix $\langle k \mid g \mid k' \rangle$ becomes diagonal; thus the double sum in (4.1) reduces to a single sum, which vanishes (from conservation of momentum and over-all spin within the interaction v) unless $k = k'$ as well. The matrix elements of J are given by

$$\langle k \mid j \mid k' \rangle = \frac{\hbar^2 k^2}{2M} \delta_{kk'} = \epsilon_k \, \delta_{kk'} \qquad \text{(no applied field)} \tag{4.5}$$

Equation (4.3) remains a meaningful equation only when $k' = -k$, and in that case it yields [using the antisymmetries (4.4)]

$$(\epsilon_k - \mu + 2L_k) \, \chi(k, -k) - \tfrac{1}{2} \Delta(k, -k) + \tfrac{1}{2} \Delta^*(k, -k) \, [\chi(k, -k)]^2 = 0 \tag{4.6}$$

Since all quantities are real in this case, anyway, (4.6) is precisely the earlier equation (IV, 1.17) from which the Bogoliubov-BCS integral equation (IV, 1.26) follows in a few steps.

However, (4.3) is very much more general than (IV, 1.17). It allows, in principle, to determine the effect of an arbitrarily large applied field with equally arbitrary variation in space. Equation (4.3) also includes, in principle, the effects of finite boundaries of the specimen (these enter into the choice of the one-particle functions $|k\rangle$ in terms of which all matrices and wave functions are written). A self-consistent treatment of the Coulomb repulsion between the electrons can be obtained by including the corresponding term into the single-particle operator j.

No calculation including this degree of generality has been carried out as yet. In the absence of such a calculation, the details of the field penetration law in the neighborhood of the metallic surface (within a distance of the order of a Pippard length) remain uncertain (see also Section 7). However, it is reasonable to assume that a much simpler calculation, ignoring the "collective" part of the Coulomb repulsions as well as all surface effects, suffices to establish the *essence* of the Meissner effect, which is the exponential law of field expulsion at large depths of penetration. The surface effects should not matter if the distance from the surface is much larger than the internal size of the pair (the Pippard length). As far as collective Coulomb effects[4] are concerned, their main effect is to keep the wave function flat in the interior of the specimen; thus, apart from surface effects, the Coulomb repulsions lead to a ground state wave function of the same form as one obtains from periodic boundary conditions.

5. The Linear Magnetic Response

When a magnetic field is applied to the specimen, the vector potential appears in the kinetic energy part of the Hamiltonian [see Eqs. (2.16) or (3.8) and (3.9)]. London has shown (see Chapter I) that the essence of the Meissner effect can be deduced from a *linear* relation between the induced current and the applied magnetic field, namely, the London equation (1.1). We therefore restrict ourselves to calculating the linear magnetic response of the system. In particular, we decompose the applied field in a Fourier series, and consider each term of that series separately.

A single term of the Fourier series for the vector potential has the form

$$\mathbf{A}(\mathbf{x}) = \mathbf{A}(\mathbf{q}) \left(\frac{2}{V}\right)^{1/2} \cos(\mathbf{q} \cdot \mathbf{x}) \tag{5.1}$$

[4] The Coulomb potential, suitably screened, can of course be included in the arbitrary electron-electron interaction $v(x_i - x_j)$ of the theory, and to this extent Coulomb effects are included. However, the basic approximation does not include true collective effects, which are responsible for the flattening of the wave function for any one pair.

where $\mathbf{A}(\mathbf{q})$ is a constant vector. It is useful to decompose that vector into two components, one perpendicular to $\mathbf{q}$, the other parallel to $\mathbf{q}$:

$$\mathbf{A}(\mathbf{q}) = \mathbf{A}_{\perp}(\mathbf{q}) + \mathbf{A}_{\parallel}(\mathbf{q}) \tag{5.2}$$

Only the perpendicular component gives rise to a real magnetic field $\mathbf{B} = \text{curl}\,\mathbf{A}$: the parallel component is merely a gauge transformation, with gauge function $f(x)$ given by (3.17).

If we assume full translation and rotation invariance (see Section 7 for a discussion of that assumption), the induced current must have the same structure as (5.1), i.e., we expect (the brackets denote quantum mechanical averages)

$$\langle \mathbf{j}(\mathbf{x}) \rangle = \langle \mathbf{j}(\mathbf{q}) \rangle \left(\frac{2}{V}\right)^{1/2} \cos(\mathbf{q} \cdot \mathbf{x}) \tag{5.3}$$

We wish to find a linear relationship between $\langle \mathbf{j}(\mathbf{q}) \rangle$ and $\mathbf{A}_{\perp}(\mathbf{q})$. Gauge invariance requires that $\langle \mathbf{j}(\mathbf{q}) \rangle$ be independent of $\mathbf{A}_{\parallel}(\mathbf{q})$.

When we write the one-particle term of (2.16) in second quantization formalism, the resulting operator is

$$J = \sum_k \epsilon_k a_k^+ a_k - \frac{1}{c} \mathbf{A}(\mathbf{q}) \cdot \mathbf{j}(\mathbf{q}) \tag{5.4}$$

where $\epsilon_k = \hbar^2 k^2 / 2M$, and $\mathbf{j}(\mathbf{q})$ is the operator whose expectation value appears on the right-hand side of (5.3). The operator $\mathbf{j}(\mathbf{q})$ splits naturally into two parts, the second of which contains the vector $\mathbf{A}(\mathbf{q})$ explicitly:

$$\mathbf{j}(\mathbf{q}) = \mathbf{j}_1(\mathbf{q}) + \mathbf{j}_2(\mathbf{q}) \tag{5.5}$$

$$\mathbf{j}_1(\mathbf{q}) = \frac{e\hbar}{2M}\left(\frac{2}{V}\right)^{1/2} \sum_k \mathbf{k}(a_{k+\frac{1}{2}q}^+ a_{k-\frac{1}{2}q} + a_{k-\frac{1}{2}q}^+ a_{k+\frac{1}{2}q}) \tag{5.6}$$

$$\mathbf{j}_2(\mathbf{q}) = -\frac{e^2}{2Mc}\frac{1}{V} \mathbf{A}(\mathbf{q}) \sum_k (2a_k^+ a_k + a_k^+ a_{k+2q} + a_k^+ a_{k-2q}) \tag{5.7}$$

Corresponding to the splitting of the operator $\mathbf{j}(\mathbf{q})$ into these two parts, there is also a splitting of the expectation value:

$$\langle \mathbf{j}(\mathbf{q}) \rangle = \langle \mathbf{j}_1(\mathbf{q}) \rangle + \langle \mathbf{j}_2(\mathbf{q}) \rangle \tag{5.8}$$

Except for notation, the splitting involved here is the same as we had for the ideal Bose gas in Chapter II, Eq. (II, 4.9). The term $\mathbf{j}_1(\mathbf{q})$ arises from the "adjustment" of the ground state wave function to the applied vector potential; the term $\mathbf{j}_2(\mathbf{q})$ is the "London term," which gives a non-zero result even if the ground state wave function stays unchanged.

At this stage, we make use of the fact that $\mathbf{A}(\mathbf{q})$ is, by assumption, small; that is, we write the pair wave function $\varphi(k_1, k_2)$ in the form (1.5), in which $\varphi_0(k_1, k_2)$ is the field-free pair function which obeys the simple pairing condition (2.1), and $\delta\varphi$ is taken to be proportional to $\mathbf{A}(\mathbf{q})$. All higher order terms are dropped systematically. In particular, the contribution $-(1/c)\mathbf{A}(\mathbf{q}) \cdot \mathbf{j}_2(\mathbf{q})$ in (5.4) can be dropped, because it is quadratic in $\mathbf{A}(\mathbf{q})$. That is, we may replace the operator $\mathbf{j}(\mathbf{q})$ in (5.4) by $\mathbf{j}_1(\mathbf{q})$, to the accuracy of a linear theory. Of course, both $\mathbf{j}_1(\mathbf{q})$ and $\mathbf{j}_2(\mathbf{q})$ must be retained in the later evaluation of the expectation value of the current, from (5.8).[5]

When this is done, it turns out that the pair function perturbation $\delta\varphi(\mathbf{k}_1, s_1; \mathbf{k}_2, s_2)$ obeys the selection rule

$$\delta\varphi(\mathbf{k}_1, s_1; \mathbf{k}_2, s_2) = 0 \quad \text{unless} \quad \mathbf{k}_1 + \mathbf{k}_2 = \pm\,\mathbf{q} \quad \text{and} \quad s_1 + s_2 = 0 \tag{5.9}$$

We note that this selection rule differs from the simple pairing selection rule (2.1); the applied field introduces essentially different components into the pair wave function, even in the lowest order of a perturbation treatment.

When one introduces (1.5) into the general equation (4.3), and drops all terms of higher order in $\mathbf{A}(\mathbf{q})$, the result is by necessity a linear inhomogeneous integral equation for the unknown function $\delta\varphi$, with right hand side proportional to $\mathbf{A}(\mathbf{q})$. The detailed calculation is quite lengthy (Blatt 60b) and will not be reproduced here. The integral equation simplifies somewhat in the limit of very small q, which is of interest for field penetration deep into the specimen (small q means large x). In this limit, it is possible to solve the integral equation at least formally, i.e., to determine the structure of the unknown function $\delta\varphi(\mathbf{k}_1, s_1; \mathbf{k}_2, s_2)$ in considerable detail.

When the result is used to find the expectation value of the current from (2.6),

$$\langle \mathbf{j}(\mathbf{q}) \rangle = \mathrm{tr}_1[\mathbf{j}(\mathbf{q})\, g] \tag{5.10}$$

one finds the following, to terms linear in $\mathbf{A}(\mathbf{q})$:

$$\langle \mathbf{j}_1(\mathbf{q}) \rangle = \frac{Ne^2}{VMc} \mathbf{A}_{\parallel}(q) + cKq^2\mathbf{A}_{\perp}(q) + O(q^4) \tag{5.11}$$

$$\langle \mathbf{j}_2(\mathbf{q}) \rangle = -\frac{Ne^2}{VMc} \mathbf{A}(\mathbf{q}) \tag{5.12}$$

[5] From the point of view of a variational calculation, one should retain the full Hamiltonian, including the term in A^2, and deduce the current from the relation $j = -c(\partial E/\partial A)$, where E is the expectation value of the Hamiltonian. This expectation value has the expansion $E = E_0 + E_2A^2 + \cdots$, i.e., the term linear in A vanishes; this must be so to avoid non-zero currents in a field-free system in its ground state. It is possible to show that our procedure leads to the same final result.

In (5.11), K is a constant, independent of q, which can be evaluated in principle from the solution of a certain linear integral equation. We shall not require the actual value of K.

Equation (5.12) is the "London term" in the current. It is not by itself gauge-invariant, since it is proportional to the full $\mathbf{A}(\mathbf{q})$, rather than to the perpendicular part $\mathbf{A}_{\perp}(\mathbf{q})$ only. However, when one adds (5.11) and (5.12) to obtain the complete current, the result is gauge-invariant (as it must be), namely,

$$\langle \mathbf{j}(\mathbf{q}) \rangle = -\frac{Ne^2}{VMc}\mathbf{A}_{\perp}(\mathbf{q}) + cKq^2\mathbf{A}_{\perp}(\mathbf{q}) + O(q^4) \tag{5.13}$$

Not only is (5.13) gauge-invariant, i.e., $\mathbf{A}_{\parallel}(\mathbf{q})$ has dropped out, but the first term of (5.13) corresponds precisely to the London equation (1.1) with Λ given by (1.2). It is worth noting that the e^2/M which appears here is computed from the charge and mass of a single electron, not of an electron pair; however, the factor of $\mathbf{A}_{\perp}(\mathbf{q})$ in (5.13) can also be written as $-(\frac{1}{2}N)(2e)^2/V(2M)c$, showing that $\frac{1}{2}N$ pairs of charge $2e$ and mass $2M$ lead to the same result as N singles of charge e and mass M. This London contribution arises from the center-of-mass motion of the electron pairs.

The second term on the right of (5.13) describes the magnetic susceptibility arising from internal deformation of the electron pairs, i.e., it is analogous to the magnetic susceptibility of ordinary diatomic molecules. The fact that this is just an ordinary magnetic susceptibility can be seen most easily by transforming back to x-space. The quantity $q^2\mathbf{A}_{\perp}(\mathbf{q})$ is the Fourier transform of curl curl $\mathbf{A}$ = curl $\mathbf{B}$; if we write the current in terms of a density of magnetization, $\mathbf{j} = c$ curl $\mathbf{M}$, then the second term of (5.13) corresponds to the equation $\mathbf{M} = K\mathbf{B}$, i.e., ordinary magnetic behavior.

For purposes of the field penetration law at large distances from the surface, we need (5.13) in the limit $q \to 0$. In this limit, the first (London) term obviously dominates, and gives rise to the exponential fall-off of the field characteristic of the Meissner-Ochsenfeld effect.

We note that the essential thing for the Meissner effect is *not* the term (5.12) in the current (which is present in any material), nor even the cancellation of $\mathbf{A}_{\parallel}(\mathbf{q})$ between (5.11) and (5.12) (which is also true in all materials, by gauge invariance). Rather, the essential feature is the sharp distinction, in (5.11), between the response of the system to the parallel and perpendicular parts of the vector potential, respectively. The pair wave function φ is exceedingly "flexible" in the presence of a longitudinal vector potential; in fact, in that case, it has the full gauge flexibility implied by the gauge transformation (3.2). The same pair wave function is "stiff" when exposed to a perpendicular vector potential $\mathbf{A}_\perp(\mathbf{q})$, so stiff that the relevant term in (5.11) is proportional to q^2 for small q. It is this, rather peculiar, combination of flexibility and stiffness of the wave function, which is needed to establish the Meissner effect in a satisfactory (gauge-invariant) fashion.

6. Intensive and Extensive Perturbation Theory

The perturbation treatment of Section 5 is based on treating the perturbation $\delta\varphi$ of the pair wave function as a small quantity. Since $\delta\varphi$ is proportional to the strength of the applied field, this is a reasonable approximation at first sight, and has turned out to be a reasonable approximation *a posteriori* as well.

However, in many-body systems, a perturbation is not necessarily small merely because it is proportional to a small external parameter. In fact, one and the same physical perturbation may be "large" or "small," depending on the way it is treated. In order to illustrate this point, let us study the perturbation which the external field induces in the wave function Ψ of the N-body system.

The perturbed pair wave function (1.5) generates a change in the pair creation operator (1.4), i.e.,

$$b^+ = b_0^+ + \delta b^+ \tag{6.1}$$

where

$$\delta b^+ = 2^{-1/2} \sum_{k,k'} \delta\varphi(k, k')\, a_{k'}^+ a_k^+ \tag{6.2}$$

When we substitute (6.1) into the definition of the pairing wave function Ψ, Eq. (1.3), and note that the two operators on the right of (6.1) commute with each other, we obtain a perturbation expansion of the N-body wave function Ψ:

$$\Psi = \Psi_0 + \delta\Psi \tag{6.3}$$

with

$$\delta\Psi = \left[\frac{\frac{1}{2}N}{1!}(b_0^+)^{\frac{1}{2}N-1}\,\delta b^+ + \frac{\frac{1}{2}N(\frac{1}{2}N-1)}{2!}(b_0^+)^{\frac{1}{2}N-2}\,(\delta b^+)^2 + \cdots\right] | 0\rangle \tag{6.4}$$

This expression for the perturbation of the many-body wave function shows that *the perturbation function $\delta\Psi$ is never small:* with $\delta\varphi$, and hence

δb^+, proportional to the perturbing field $A(q)$, even the first term in (6.4) is proportional to $NA(q)$, where N is of the order of 10^{23}. The next term is proportional to $[NA(q)]^2$, etc., so that the series (6.4) is violently divergent for all physically reasonable applied fields.

The difference between the perturbation expansion (1.5) of the elementary pair wave function (which converges rapidly) and the perturbation expansion (6.3) of the N-particle wave function (which diverges horribly) is the volume-dependence of the perturbation term. The perturbation term $\delta\varphi$ is volume-independent, whereas the perturbation term $\delta\Psi$ increases violently with increasing volume (increasing N). We shall call the first kind of perturbation expansion an "intensive perturbation theory," the other an "extensive perturbation theory." The latter is valid, if at all, only by accident.

This distinction is well known in many body theory. It was stressed particularly by Wigner (34a, 38) in connection with the theory of normal metals. The systematic transformation (Kubo 62) of perturbation theory to reexpress everything in terms of "irreducible diagrams," so characteristic of the Brueckner theory (Brueckner 58, Bell 61) among others, is a transformation from extensive to intensive perturbation theory.

This transformation is by no means always possible. There are characteristic many-body effects which arise from the interactions of large groups of particles, and such effects cannot be handled in this fashion. One prominent example is the plasma wave, in which the smallest "unit" consists of the number of particles contained in a volume of the order of λ^3, where λ is the plasma wavelength. Since the plasma wavelength can be as large as we please, no intensive perturbation theory is possible.

This difficulty has been circumvented (although perhaps not altogether solved) by the theory of "collective excitations"; in its more mathematical form (Gell-Mann 57), this theory amounts to a selective summation of certain contributions to a perturbation expansion. One selects the "leading term" in each order of the perturbation expansion, and then performs a formal summation of all these leading terms.

As we have seen from the preceding sections, these complications are not required for the theory of the Meissner effect. A purely intensive perturbation treatment, based on (1.5), is feasible and leads to the desired result in a gauge-invariant fashion. The basic simplification making this treatment possible is the Schafroth condensation of electron pairs, which gives the straightforward and simple relation (1.3) between the pair wave function φ and the system wave function Ψ.

On the other hand, if one starts from conventional perturbation theory, one is led naturally to (6.3) and (6.4), and the theory runs into considerable difficulties. For one thing, the leading term of (6.4) taken by itself is not gauge-invariant. Gauge-invariance requires the full expression for $\delta\Psi$, and is therefore beyond the range of a straightforward perturbation calculation. Of course, one may consider the reduction of (6.3) and (6.4) to the simple form (1.3) as a special case of selective summation of higher-order terms, and in this sense "gauge invariance is restored by including collective excitations." One could hardly maintain, however, that this procedure illuminates the physics of the situation.

The "energy gap argument" for the Meissner effect is based on the extensive perturbation expansion (6.3). The argument, very simply, is that in lowest order perturbation theory, $\delta\Psi$ has the form

$$\delta\Psi = \sum_{\alpha} \frac{H'_{\alpha 0}}{E_0 - E_\alpha} \Psi_\alpha^{(0)} \tag{6.5}$$

where the $\Psi_\alpha^{(0)}$ and E_α are the unperturbed wave functions and energies. A straightforward calculation shows that the matrix elements $H'_{\alpha 0}$ of the perturbing Hamiltonian vanish as q goes to zero, whereas the energy denominators $E_\alpha - E_0$ are bounded from below by the energy gap. Thus, the perturbation term $\delta\Psi$ vanishes in the limit of small q, i.e., the energy gap makes the wave function "stiff"—and this ensures the Meissner effect.

This argument is invalid: One cannot restrict oneself to the lowest order term in *this* perturbation series, and therefore arguments based on properties of the lowest order term have no significance whatever. The obvious lack of gauge-invariance is an immediate indication of trouble. To restore gauge invariance, one must "include collective excitations," which means including an infinite sequence of higher-order perturbations—thereby losing the original argument. The wave function Ψ cannot be "inflexible" and gauge invariant at the same time. The "inflexibility" suggested by the energy gap argument means that $\langle j_1(q)\rangle$ would have to be always small compared to $\langle j_2(q)\rangle$; whereas in fact a strong cancellation between (5.11) and (5.12) is required for the gauge-invariant final result (5.13). The true "inflexibility" of the wave function is much more subtle than envisaged by the simple-minded energy gap argument.

In view of the fact that the energy gap argument, based on perturbation theory, nevertheless gives the right result, many people consider these objections mere quibbling. After all, plausibility arguments which give right answers are very common in theoretical physics, and full mathematical rigor is an unreasonable demand.

However, the Meissner effect is an exception to the general rule that one need not worry about validity of the argument, as long as the answer is right. *In the Meissner effect, the right answer is obtained by almost any* **incorrect** *calculation*! The current operator (5.5) is composed of two terms, (5.6) and (5.7); there is an enormous amount of cancellation between these two terms, in all materials. If, by any mistake in the calculation, this cancellation is destroyed, this shows up in two ways: (1) lack of gauge-invariance, (2) a spurious Meissner effect in London gauge.

It is worthwhile to remember that London obtained the correct value (1.2) of the London constant Λ from exactly such an invalid calculation, back in the 1930's, before there was any microscopic theory at all, and without either the concept of an energy gap or the concept of a condensation of electron pairs.

In this case, then, validity of the calculation is not an unreasonable request, and the success of an invalid calculation is no indication that there is some deeper truth hidden in it. From the physical point of view, the energy gap argument makes sense only in a negative way: Excitations subject to an energy gap are hard to produce, and make the system "stiff"; thus, excitations of that sort neither carry a supercurrent, nor do they stop a supercurrent if a supercurrent is produced in some other way. Usually, an energy gap is associated with insulators, i.e., most energy gap systems carry no current at all. Only rarely, in systems which possess some other, quite separate, mechanism for current transport, is an energy gap present at the same time as superfluid behavior. The two are related only in the sense that the excitations subject to the energy gap do not destroy the supercurrent; the energy gap is associated with supercurrents about as much as French rentiers are associated with the success of French avant-garde painting—the rentiers do not interfere with the painters.

There is an enormous literature on the Meissner effect, with more calculations than investigators. The energy gap argument (Welker 38) has been discussed at great length. Buckingham (57) called attention to identities which must exist between $\langle j_1(q)\rangle$ and $\langle j_2(q)\rangle$; Bardeen (57a) countered this with the suggestion that "collective excitations" are needed to preserve gauge invariance. The early calculations (Bardeen 57, Rickayzen 58, Bogoliubov 58) were gauge-dependent. The first gauge-invariant calculation was made by Wentzel (58) and Gupta (59, 61), but ironically enough this was the one calculation which gave an incorrect result; the trouble arose from a perturbation expansion in the electron-phonon coupling constant. A variational approach suggested by Blatt, Matsubara and May (Blatt 59) was used in an explicit calculation by May and Schafroth (May 59).

A method very close in spirit to the one used in this chapter has been developed by Bogoliubov (58d, 59). It is fully gauge-invariant, and has since been shown to be substantially equivalent mathematically (Weller 61, Baranger 62).

The "energy gap plus collective excitations" approach has given rise to much discussion and many calculations (Anderson 58, 58a; Pines 58a; Wentzel 58b, 59; Blatt 58a; Fukuda 59; Kuper 59; Nakajima 59, 59a; Rickayzen 59, 59a; Sewell 59; Wada 59; Yosida 59; Taylor 61; Valatin 61; Mori 62).

The use of a BCS type pairing of cylindrical, rather than plane, waves (Oliphant 60) cannot be considered adequate.

The method of Green's functions permits fully gauge-invariant calculations, but it is somewhat harder to motivate the assumptions which must be made about the behavior of the various Green's functions (Nambu 60, Kadanoff 61, Ambegaokar 61).

A problem related to the magnetic response is the response of a superconductor to an *electric* field (the dielectric constant). This has been discussed by a number of authors (Anderson 58, 58a; Rickayzen 59, 59a; Prange 62a; Nishiyama 62).

7. The Knight Shift, and Other Problems

The characteristic frequency for flipping of a nuclear spin depends upon the Zeeman splitting of the nuclear ground state in the magnetic field H_1 at the position of the nucleus. This magnetic field in general differs slightly from the magnetic field H_0 applied to the outside of the specimen. The difference between H_1 and H_0 depends upon the chemical surroundings of the nucleus; in particular, this "chemical shift," first discovered by Knight (55), is different in insulators and in metals. The spins of the conduction electrons in metals interact with the nuclear spins via the Fermi hyperfine structure interaction; in normal metals, then, the "Knight shift" provides a measure of the spin paramagnetism (Pauli paramagnetism) of the conduction electron gas.[6]

The measurement of the Knight shift in superconductors is made difficult by the Meissner effect. In order to have nearly the same "applied" magnetic field throughout the specimen, the specimen must be much

[6] The volume magnetic susceptibility of a normal metal also contains the Landau orbital diamagnetism (which goes over into the Meissner effect in superconductors).

thinner than a London penetration depth. The actual specimens are made by evaporating, in succession, layers of superconducting material and insulating layers, thereby building up a specimen with enough metallic volume to carry out measurements, and yet permitting substantially complete magnetic field penetration. The preparation of such specimens is exceedingly difficult.

The first experiments (Knight 56, Reif 57) on superconductors gave result in qualitative disagreement with each other. Knight, Androes, and Hammond obtained a frequency shift, as compared to the same nucleus imbedded in a salt, consistent with normal Pauli paramagnetism above the transition temperature, and decreasing rapidly to zero in the superconducting temperature range. Reif (57) observed a small decrease of the frequency shift with decreasing temperature $T < T_c$, but more than half the frequency shift still existed at the lowest temperature employed by him (about $T_c/5$).

The subsequent developments were all too typical of the general history of the theory of superconductivity. First of all, Yosida (58a) calculated the theoretical electron spin susceptibility on the BCS theory, and found that it goes to zero at zero temperature. This is of course to be expected: the electron pairs in the superconducting ground state are made up of electrons with opposite spins; thus, electron pairs must be broken up before any electron spin susceptibility can be observed, i.e., the spin susceptibility is a "normal fluid" phenomenon. Yosida's result agreed quite well with the measurement of Knight (56), but disgreed with that of Reif (57); at that stage, theoretical opinion was nearly unanimous that there must be something wrong with Reif's measurement, and nothing further need be said. Only Heine and Pippard (Heine 58) took the discrepancy seriously.

However, at the Cambridge Superconductivity Conference, in June 1959, Knight reported new experiments, which were in complete agreement with those of Reif. Later work (Androes 59, 61; Noer 61) confirmed this.

Within two months of the Conference, no less than four separate letters appeared in *Physical Review Letters* (Schrieffer 59, Martin 59a, Ferrel 59, Anderson 59a), with three distinct theories of a non-zero electron spin paramagnetism in superconducting specimens of small size. Abrikosov and Gorkov (Abrikosov 61, 62) independently suggested one of these explanations, as well. Still different theories have been advanced subsequently (Galasiewicz 60, Fisher 60, Ginzburg 61, Suzuki 61, Cooper 62, Balian 63).

The latest experimental work, however, suggests strongly that the observed Knight shift is not at all related to spin paramagnetism of the conduction electrons! Clogston, Gossard, Jaccarino, and Yafet (Clogston 62) note that the superconductors in question contain unsaturated *d*-shell electrons, which give rise to a non-zero, temperature-independent *orbital* paramagnetism (Kubo 56, Masuda 57). By a series of measurements, they show that this purely orbital effect can account for the entire frequency shift remaining in the limit of zero temperature, so that the true electron spin paramagnetism is zero within the accuracy of the experiments.

Clearly, further experimental and theoretical work is required before we can be sure that the explanation of Clogston (62) is sufficient. But, it is certain that, once more in the history of superconductivity, the theorists have come off second best.

Nor is the Knight shift the only unsolved problem in the theory of the magnetic response of a superconductor. The main phenomenon, the Meissner effect, is understood in essence, but a number of problems remain.

The first of these is inclusion of the Fröhlich interaction into the theory. We used an ordinary x-space interaction potential $v(x_i - x_j)$, and this assumption entered significantly. A calculation similar to the present one would probably start from the wave function of McKenna (62) which includes the phonons explicitly. There is no trouble with gauge invariance, but the reduction of the necessary expectation values to sufficiently simple form has not been possible so far. An alternative approach, using Green's functions, has been made by Nambu (60) and by Kadanoff (61).

The second problem relates to the field penetration law in the neighborhood of the metallic surface. In that region, it is not possible to start from plane wave functions (i.e., to assume translation invariance of the theory). The internal size of the electron pair (the Pippard length) is of order 10^{-4} cm, much larger than the London penetration depth (of order 10^{-5} to 10^{-6} cm). Thus the field decays to quite small values within distances smaller than the pair size; coupling between center-of-gravity and internal motion of the pair, produced by reflection from the metallic surface, cannot be neglected.

Furthermore, the pilot calculation of Schafroth (55a) for the ideal Bose gas shows that the Coulomb repulsion between the particles must be included into the theory as soon as we impose realistic boundary conditions. No such calculation has yet been done for the electron pair gas.

In order to obtain practically useful results concerning the penetration law in the neighborhood of the surface, Bardeen (57) and Mattis (58) performed a perturbation expansion in London gauge, assuming full translation invariance. This yields an explicit expression for the kernel $K(q)$ in the general linear relationship (I, 3.8) between the Fourier components of the magnetization $M(q)$ and of the magnetic field $B(q)$. The kernel obtained in that fashion is qualitatively very similar to the one suggested by Pippard on purely experimental grounds (see Chapter I). The most characteristic difference between such a "non-local" theory, and the London phenomenological theory, is the occurrence of a sign reversal of the magnetic field inside the specimen, whereas the London theory yields a simple exponential decay law. This sign reversal has been observed experimentally (Sommerhalder 61, 61a; Drangeid 62).[6a]

Since the perturbation treatment in London gauge, though unjustified fundamentally, does actually give a correct result for $K(q)$ in the limit of very small q, it is highly likely that the perturbation treatment also gives the correct function $K(q)$ for larger q. Unfortunately, this function is not really what is needed. The trouble comes from the initial assumption of full translation invariance. The x-space transform of the relationship $M(q) = K(q)B(q)$ is

$$\mathbf{M}(\mathbf{x}) = \int K(\mathbf{x} - \mathbf{x}')\, \mathbf{B}(\mathbf{x}')\, d^3x' \tag{7.1}$$

where the function K is a function of the vector displacement $\mathbf{x} - \mathbf{x}'$ only. Thus, at a typical point $\mathbf{x}$ in the neighborhood of the surface of the specimen, (7.1) would lead to contributions from points $\mathbf{x}'$ outside the superconductor altogether.

It is therefore necessary to look for a more general relation,

$$\mathbf{M}(\mathbf{x}) = \int K(\mathbf{x}, \mathbf{x}')\, \mathbf{B}(\mathbf{x}')\, d^3x' \tag{7.2}$$

in which the kernel K is a function of the positions $\mathbf{x}$ and $\mathbf{x}'$, separately, and vanishes when either or both are outside the specimen. It is clear that such a kernel can not be obtained from a treatment which starts with the assumption of full translation invariance.

To obtain practical results, Bardeen (57) and Mattis (58) attempt to construct an acceptable kernel $K(\mathbf{x}, \mathbf{x}')$ from their calculated $K(\mathbf{x} - \mathbf{x}')$, by using boundary conditions corresponding to either specular reflection or diffuse scattering of the superelectrons. Quite apart from the difficulty

[6a] For an early indication of non-London behavior of the penetration law, see Schawlow (58a) and Peter (58).

A very direct nuclear correlation method (Lewis 60) suffices to establish the Meissner field expulsion, but is not accurate enough to measure the detailed penetration law.

of deciding which of these alternative boundary conditions to employ, the physical basis of such a procedure is highly doubtful: it makes sense only if the internal size of the superelectrons is small compared to the distances over which the field varies rapidly, whereas in fact the opposite is the case. The internal size of the superelectrons is the Pippard distance, which gives a measure of the size of the correlated electron pairs; this correlation distance is some 10 to 100 times larger than the London penetration depth. Thus the superelectron cannot be thought of as a unit which is scattered by the surface, either in specular or diffuse fashion. The coupling between the internal motion and the center-of-gravity motion of the superelectron is of the essence of this problem.

In view of these difficulties of principle, it is surprising that the calculations give reasonable agreement with experimental measurements in a number of instances, including such detailed features as the dependence of the penetration depth on the density of impurity scattering centers (Mattis 58, 62; Miller 59).

Algebraic and Green's function method have been developed for this and related problems by a number of authors: Edwards (58), Weiss (58), Abrikosov (58), Khalatnikov (59), Miller (60), Anderson (59), Engelsberg (62), Langer (62), Hagenow (62), Stephen (62), Tsuneto (62).

The calculation of this chapter is based on the ground state, i.e., refers to the limit of zero temperature. Extension of that theory to non-zero temperatures is possible, but has not been done as yet. The perturbation-theoretic calculation (Bardeen 57, Mattis 58) has been extended to non-zero temperatures, and yields a prediction for the temperature-dependence of the penetration depth, among other things. The agreement with experiment is satisfactory but not perfect (Schawlow 59, Sarachik 60, McLean 62). Since the difficulties concerning the treatment of surface effects persist at non-zero temperatures, the interpretation of the experiments is not completely clear at present. Only in the immediate neighborhood of the transition temperature is it possible to be more definite. In that region, the penetration depth becomes large, and eventually exceeds the Pippard length. We may then neglect the effects arising from the internal structure of the pairs, and obtain a simple London exponential penetration law.

Another direction in which the theory should be extended is to larger fields, i.e., nonlinear terms in the field strength. The general variational equation (4.3) is sufficient in principle for such effects in the limit of zero temperature, but the actual solution of that equation is difficult. Non-linear effects include the dependence of the energy gap and the penetration depth on magnetic field strength (Gupta 61; Douglass 61, 61a, 61b; Mathur 62; Nambu 62), the surface energy between a normal and a superconducting region, and the critical field. In the immediate neighborhood of the transition temperature, where the internal structure of the pair can be ignored, one might expect the microscopic theory to reduce to the phenomenological theory of Landau and Ginzburg (see Chapter I, Section 5); a Green's function calculation by Gor'kov (59, 59a) gives just this result, provided one sets the "charge of the superelectron" equal to $2e$, rather than e. An analysis of the data by Ginzburg (59) indicates that this gives just a good, or even somewhat better, agreement with experiment.

This short survey of unsolved problems is by no means exhaustive; there is much more to be done. Yet, we must not underestimate the positive achievement of the existing theory of the Meissner effect: After all, the problem of highest fundamental interest, for many decades, has been a theoretical understanding, on a sound basis, of the well-nigh complete expulsion of the magnetic field from regions deep inside the specimen. This phenomenon is now completely understood, in terms of a Schafroth condensation of electron pairs: (1) through the analogy with the ideal Bose gas, and (2) through a direct calculation for the electron pair gas, with strong similarities between the two calculations, as well as characteristic differences where differences are to be expected [the second and higher terms on the right of (5.13)]. Thus, although much indeed remains to be filled in, the crucial problem can be considered solved.

IX

Persistent Currents

1. Introduction

So far, we have been able to present at least partial answers to the questions asked at the end of Chapter I. The existence of a thermodynamic transition, the qualitative importance of the Fröhlich interaction vis-à-vis the Coulomb interaction, the order-of-magnitude of the transition temperature, the nature of the ordering of electronic motions in a superconductor, the meaning of the Pippard coherence length, and the origin of the Meissner effect—all these questions may be considered solved, at least in principle.

Now, however, we come to a question which is still open. Ironically enough, this very question, concerning the lifetime of persistent currents, is actually the first question from the historical point of view. The incredibly high electrical conductivity at low temperatures was the first property of superconductors found experimentally by Onnes (11), and it is the property responsible for the name "superconductivity." Yet, in spite of all the development of superconductivity theory so far, the mechanism of electrical conduction is the least understood of the basic properties of superconductors (Brillouin 58).

From the theoretical point of view, this is not quite as surprising as it might appear at first. So far, we have been considering *equilibrium* properties (specific heat, Meissner effect), and furthermore properties which are not "structure-sensitive," i.e., which do not depend essentially on the crystal structure and/or the presence of impurities in the lattice. In fact, we have ignored crystal structure and deviations from a perfect lattice, in all the work so far.[1]

By contrast, the electrical conduction is a *non-equilibrium* phenomenon, and we know that in the normal state of the metal, the electrical conductivity at low temperatures is most definitely sensitive to deviations from a perfect lattice. As Peierls (55) emphasizes strongly, a perfect crystal lattice has exact quantum-mechanical eigenfunctions of the Hamiltonian with non-zero electrical current. Once such a wave function is contained in the ensemble, there is no way it can die out—i.e., a perfect lattice gives rise to infinite conductivity.

The actual conductivity of normal metals is limited by two main mechanisms: (1) deviations from a perfect crystal lattice due to thermal motion of the ions, and (2) deviations from a perfect crystal lattice caused by randomly located impurities, vacancies, or interstitial atoms, all of which we shall consider under the one heading of "lattice defects." At room temperature, the thermal motions of the ions are the dominant mechanism, but at low temperatures (of the order of typical superconducting transition temperatures) the second effect takes over.

It is important for the second effect that the lattice imperfections be randomly located. For suppose the lattice defects were themselves arranged in a "superlattice": we would then have a composite solid of perfect lattice structure (with a much bigger lattice constant, of course), and Peierls' argument for infinite conductivity would again apply.

All these remarks apply to normal metals. When we now turn to superconductors, the first requirement is that *any theory of electric current transport in superconductors must start from a model which is sufficiently realistic to lead to a finite electrical conductivity in the normal state*. It is an easy exercise to establish "superconductivity" in a model which has no resistance mechanism to start with; but such an exercise does not prove much. This requirement should be obvious. Yet, it is ignored in the literature all too often.

The second remark is equally obvious, and equally often overlooked: If it is true, as we believe it to be true, that superconductivity involves wave functions correlated over the whole specimen, then there is no *a priori* reason to believe that different lattice defects act independently of one another. On the contrary, the correlated wave function is expected to "see" the coherent effect of all the lattice defects acting in concert, and

[1] The electron-phonon interaction involved in the Fröhlich theory is predominantly with the zero-point phonons; the zero-point vibrations of the lattice are a consequence of the uncertainty principle, and the state in question is the nearest one can come to a perfect lattice within a quantum-mechanical approach.

this coherence must be included in any realistic theory. The coherence is a strictly wave mechanical effect, not obtainable from classical theories such as the Boltzmann transport theory.

As a third point, let us consider two different experimental situations for electrical conduction measurements:

(1) A section of superconducting material is included as part of a circuit connected to a source of emf, such as a battery. We measure the potential drop across the superconducting section.

(2) A superconducting wire is bent around and joined so as to form a closed ring. A current is induced in this ring through magnetic induction. We observe the magnetic field produced by the ring current so generated, and study the decay (with time) of that ring current.

From a purely classical point of view, these two situations are closely related to each other. One "material constant," the electrical conductivity, suffices to describe both phenomena. In superconductivity, on the other hand, these two situations are utterly different, and it is idle to expect to find a single theoretical treatment to encompass both of them. The most striking experimental verification of this distinction is the observation of *flux quantization* in situation (2) (Deaver 61, Doll 61). Clearly, the flux through the hole in a superconducting ring exists only if there is a superconducting ring. In situation (1), this flux has no meaning, and cannot be quantized.

From the theoretical point of view, the second situation is infinitely easier to handle than the first. If the superconducting material forms only part of the electrical circuit, the theory must allow for electrons entering and leaving the superconductor at the joining surfaces. Unlike the junction between two normal metals, the qualitative character of the electron wave function changes as we pass across such a junction. On one side, we have essentially independent electrons; on the other side, in the superconducting material, we have correlated electron pairs. The construction of wave functions which effect a smooth joining between these two limits is not an easy task, and thus no simple theory can be expected. As a consequence, the main theoretical work so far has been done on persistent currents in superconducting rings.

The phenomenological London equations (I, 2.7), (I, 6.1) are *linear* equations. Thus, we may restrict ourselves in the first instance to a linear microscopic theory, just as we have done in the Meissner effect. However, unlike the Meissner effect, the London equations do not contain the full essence of the observed phenomenon, for two reasons: (1) the flux quantization is not a consequence of the London equations [even though flux quantization was first suggested by London (50)], and (2) the "critical current" for destruction of superconductivity is not necessarily determined purely thermodynamically, in the way that the critical magnetic field is related to the specific heat. We leave open, at this stage, whether a strictly linear theory suffices to establish flux quantization; it clearly does not suffice for a study of critical currents.

Considerable theoretical work meeting the requirements which we have discussed has been done on the *Bose-Einstein gas model.* In view of the similarity between this model and actual superconductors, it is likely that the main *qualitative* points can be discussed on the basis of the Bose-Einstein model. However, since the model differs from actual superconductors in important respects (especially the absence of an energy gap), results from this model do not have any *quantitative* significance for actual superconductors.

In Section 2, we shall describe the quantum theory of electrical resistance, following the work of Kubo (57) and Kohn and Luttinger (Kohn 57a, Luttinger 58). Our formulation (Blatt 61) is somewhat different from theirs, in order to be adaptable to the peculiar features of the Bose-Einstein condensed state. In Section 3, we show that the division of the current into "normal current" and "super-current" follows naturally from the Bose-Einstein model in the condensation region. Sections 4, 5, and 6 contain preparatory work; Section 7 is devoted to flux quantization, Section 8 to the building up of a ring current in a changing external flux; Section 9, based on the work of Bloch and Rorschach (Bloch 62), is devoted to the "critical current" for destruction of superconductivity; finally, in Section 10 we describe the present situation concerning the ultimate lifetime of these persistent currents.

2. Quantum Theory of Normal Conductivity

We are interested in the electrical current $J(t)$ at time t, due to a given external electromagnetic field acting on the specimen. The first basic idea of the Kubo (57) theory is the assumption that this external field is zero until some given time (which we shall take as $t = 0$ here, but which Kubo takes as $t = -T$, with T approaching infinity), at which time the field-free system is in full thermodynamic equilibrium. Thereafter, the system develops according to the time-dependent Hamiltonian

$$H(t) = H_0 + H'(t) \tag{2.1}$$

where H_0 is the Hamiltonian in the absence of the external field, and $H'(t)$ contains the influence of that field.

Let E_α and Φ_α be the eigenvalues and eigenfunctions of the field-free Hamiltonian H_0, i.e.,

$$H_0\Phi_\alpha = E_\alpha\Phi_\alpha \tag{2.2}$$

Then, at time $t = 0$, the systems of the ensemble are distributed over the states α with probabilities

$$P_\alpha = \exp[\beta(F - E_\alpha)] \tag{2.3a}$$

$$\sum_\alpha P_\alpha = 1 \tag{2.3b}$$

It should be noted that it is here, and *only* here, that the initial temperature $T = 1/\mathfrak{k}\beta$ enters the Kubo theory. The statistical mechanics appears at this point; any subsequent "statistical relaxation" is ignored.

The switching on, in some given manner, of the external field now causes a time-development of the wave functions, according to

$$H(t)\,\Psi_\alpha(t) = i\hbar\,\frac{\partial\Psi_\alpha}{\partial t} \tag{2.4a}$$

$$\lim_{t\to 0}\Psi_\alpha(t) = \Phi_\alpha \tag{2.4b}$$

Unlike the wave functions $\Psi_\alpha(t)$, the probabilities P_α are time-independent: if a system is in state $\Phi_\alpha = \Psi_\alpha(0)$ at time $t = 0$, that same system will be found in state $\Psi_\alpha(t)$ at time t.

Let J_{Op} be the operator for the electrical current, or other measured property of the system. The ensemble average of the current, i.e., the observed current, at time t is given by

$$J(t) = \sum_\alpha P_\alpha\,(\Psi_\alpha(t), J_{\text{Op}}\Psi_\alpha(t)) = \text{Tr}\,[J_{\text{Op}}\mathscr{U}(t)] \tag{2.5}$$

Here $\mathscr{U}(t)$ is the time-dependent density matrix of the system; it is a non-negative operator with eigenfunctions $\Psi_\alpha(t)$ and corresponding eigenvalues P_α, the latter being independent of time.

If J_{Op} is the operator for the electrical current, it is odd under time-reversal (Wigner 32); if the field-free Hamiltonian H_0 is invariant under time-reversal (the usual case), then the expectation values $(\Phi_\alpha, J_{\text{Op}}\Phi_\alpha)$ all vanish, and we have from (2.4b) and (2.5)

$$J(0) = \sum_\alpha P_\alpha\,(\Phi_\alpha, J_{\text{Op}}\Phi_\alpha) = \text{Tr}\,[J_{\text{Op}}\mathscr{U}(0)] = 0 \tag{2.6}$$

The second basic idea of the Kubo theory is the restriction to a *linear* (or, more generally, low order) response of the system to the applied field. That is, instead of solving (2.4a) exactly, we solve it in a perturbation expansion with respect to $H'(t)$; since $H'(t)$ is in general proportional to the applied field, this is a perturbation expansion in the applied field. The first non-trivial term is proportional to the field strength, and this is the term which gives the linear response of the system.

The consistent working-out of that procedure leads to an explicit formula for the kernel of this linear relation between $J(t)$ and the applied fields $\mathscr{E}(t'), \mathscr{H}(t')$ at times $0 < t' < t$. The kernel is the trace of a definite, well-defined operator over the $t = 0$, equilibrium density matrix $\mathscr{U}(0)$. To this extent, then, the calculation of non-equilibrium linear response functions has been reduced to the purely mathematical problem of evaluating certain given traces; this is a problem of the same complexity as the one encountered in equilibrium statistical mechanics, where the free energy also appears as a trace of some given operator. We shall not use that formula in the remainder of this chapter, since we shall find it easier to work from (2.5) directly.[1a]

The traces which appear in Kubo's work are quite difficult to evaluate explicitly. Working independently, Kohn and Luttinger (Kohn 57a, Luttinger 58) developed an equivalent approach, and succeeded in carrying through an explicit calculation for the following *model system*: A set of independent particles is placed in a ring-shaped volume; the particles do not interact with each other, but each particle is subject to the influence of the external field as well as the influence of a set of randomly located scattering centers. Each of these is represented by a static potential $v(\mathbf{r}_i - \mathbf{r}_s)$, where $\mathbf{r}_i$ is the position of electron number i, and $\mathbf{r}_s$ is the position of scattering center number s. Thus, the full Hamiltonian for the model is

$$H(t) = \sum_{i=1}^{N} H_i(t) \tag{2.7a}$$

$$H_i(t) = \frac{1}{2M}\left[\mathbf{p}_i - \frac{e}{c}\mathbf{A}(\mathbf{r}_i, t)\right]^2 - e\varphi(\mathbf{r}_i, t) + \sum_{s=1}^{S} v(\mathbf{r}_i - \mathbf{r}_s) \tag{2.7b}$$

Here $\mathbf{A}(\mathbf{r}, t)$ and $\varphi(\mathbf{r}, t)$ are the vector and scalar potentials, respectively, of the applied field. In order to avoid the need to consider end effects, the volume is taken to have the shape of a ring, and the current is generated by changing the magnetic flux through the hole in the ring.

[1a] For transport theory in normal systems, see Bogoliubov (47a); Casimir (45); H. S. Green (52b); M. Green (55, 58); de Groot (51, 59, 62, 63); Henin (61); Kubo (56a, 57); Machlup (53); Mazur (59, 61); Meixner (41, 42, 43, 43a, 59, 63); Mori (62a); Onsager (53); Pfennig (59); Prigogine (47, 49, 62); Resibois (59, 63); Ranninger (63); Rice (60, 61).

When we set the external fields equal to zero, we obtain the Hamiltonian H_0 in (2.1). This Hamiltonian has the form

$$H_0 = \sum_{i=1}^{N} H_i(0) \tag{2.8a}$$

$$H_i(0) = \frac{p_i^2}{2M} + \sum_{s=1}^{S} v(\mathbf{r}_i - \mathbf{r}_s) \tag{2.8b}$$

This expression illustrates the practical difficulty encountered in this theory: The "zero-order" Hamiltonian is still quite complicated, since it contains the influence of all the scattering centers. We must not leave out the sum over s in (2.8b), for if we did, there would be no mechanism in the model to limit the current.

Thus, we must qualify our earlier statement that the Kubo approach leads to problems of the same order of complexity as the problem of equilibrium statistical mechanics. For the same model, the statement is true. But the study of non-equilibrium situations forces us to use models of greatly increased complexity. If the equilibrium free energy were all that is wanted, we could omit the scattering centers in (2.8b), and include them afterward by a simple perturbation treatment. By contrast, no *simple* perturbation treatment of the influence of scattering centers can succeed in the non-equilibrium case. We may restrict ourselves to terms linear in the applied field; but we must not restrict ourselves to terms linear in the scattering potentials $v(\mathbf{r}_i - \mathbf{r}_s)$.

The failure of a simple perturbation treatment here is similar to the situation in the theory of radiation damping (Weisskopf 30). After the external field has been removed, the system should decay back to equilibrium, i.e., we expect a time dependence of type $\exp(-\gamma t)$ The usual perturbation expansion in powers of v, however, leads naturally to a power series in the time t, also; the term proportional to v^n gives a time dependence proportional to t^n. Such a power series expansion is quite useless for functions like $\exp(-\gamma t)$.

Kohn and Luttinger therefore developed an alternative expansion method, closely related to the expansions of the theory of radiation damping. The theory is written in such a way that the exponential dependence $\exp(-\gamma t)$ is natural, and the expansion procedure amounts to a perturbation expansion for γ. In their first paper (Kohn 57a) the expansion is in powers of the scattering potential v. However, this is not the best expansion parameter here. Rather, even for quite strong scattering potentials, we should expect to obtain simple results if the scattering centers are far enough apart, on the average, so that they do not influence each other's action. The basic expansion parameter should therefore be proportional to the volume density of scattering centers and nearly independent of the strength of v. In their second paper (Luttinger 58), Kohn and Luttinger have developed this point of view into a complete calculation. In this calculation, it is essential that the scattering centers are located randomly; we know this *a priori*, from Peierls' "superlattice" argument, and it is equally evident *a posteriori*, from the explicit calculation. Kohn and Luttinger perform an average over the distribution of positions $\mathbf{r}_s$ of the scattering centers, and simple results emerge only after this average.

What is *not* evident *a priori*, but will be of great subsequent interest to us, is that the average over random locations of scattering centers is not by itself sufficient for the theory. In their first paper, Kohn and Luttinger estimate the fluctuations of the important matrix elements around their average values, and find that the fluctuations are so large that the average values have no meaning. It is necessary to perform a second averaging operation in order to obtain meaningful average matrix elements. This second averaging operation is an average over a small region of k-space, i.e., over a set of neighboring single-particle levels. Since the observed current finally appears as a sum of contributions from all the single-particle states, this second averaging is permitted.

The calculation of Kohn and Luttinger is quite complex, but the final result is gratifyingly simple: in the lowest order of their expansion (i.e., for widely separated scattering centers) their theory reduces to the classical Boltzmann transport equation! The only quantum-mechanical feature of the final result is the differential scattering cross section for scattering of a particle of energy ϵ by a scattering center; this cross section must now be calculated quantum mechanically rather than classically.

In higher order, Kohn and Luttinger find deviations from the classical Boltzmann transport equation; these terms are small if the average distance between scattering centers is much larger than the most important de Broglie wavelength, which is the wavelength of an electron at the top of the Fermi distribution.[2] Thus, the explicit calculation of Kohn and Luttinger bears out the argument of Landau that the criterion of validity of the Boltzmann transport equation for a Fermi system does not involve the temperature [see Peierls (55, Section 6.8), for a discussion of that argument].

[2] There is of course also the trivial condition that the average distance between scattering centers must be large compared to the range of action of the scattering potential function $v(\mathbf{r}_i - \mathbf{r}_s)$; otherwise we could not speak of separate scattering centers at all.

The calculation of Kohn and Luttinger does not depend strongly on the quantum statistics of the particles which carry the current. Most of the calculation is done for a "Boltzmann" system, i.e., ignoring symmetry requirements on the wave functions. In their Appendix F, Kohn and Luttinger show explicitly that the same calculation goes through for Fermi-Dirac and Bose-Einstein statistics. The Boltzmann transport equation is unchanged; we merely have a different zero-order number distribution, namely the Fermi-Dirac or Bose-Einstein distribution, rather than the Boltzmann distribution.

3. The Effect of Bose-Einstein Condensation

Although Kohn and Luttinger show explicitly that the statistics of the particles are unimportant, their argument breaks down in spectacular fashion for a Bose-Einstein system below the Bose-Einstein condensation point. This can be seen in a number of ways; most directly, perhaps, by noting that their theoretical treatment depends upon an averaging over neighboring states in k-space. In the presence of Bose-Einstein condensation, one quantum state (the single-particle ground state) has a macroscopic occupation number, and this state must not be averaged in with the other quantum states. Thus we may expect, and will find, that the ordinary resistivity found by Kohn and Luttinger does not apply to the temperature region in which the Bose gas is condensed.

The model system described by the Hamiltonian (2.7) is an independent particle system, and thus the system wave functions $\Psi_\alpha(t)$, Eq. (2.4), are built up from single-particle wave functions $\psi_k(r, t)$, each of which satisfies the single-particle Schrödinger equation

$$H_1(t)\,\psi_k(r_1, t) = i\hbar\,\frac{\partial\psi_k(r_1, t)}{\partial t} \tag{3.1}$$

The index α denotes a "configuration" of occupation numbers:

$$\alpha = \{n_0, n_1, n_2, \cdots, n_k, \cdots\} \tag{3.2}$$

The system wave function $\Psi_\alpha(t)$ is a symmetrized sum of products, each product containing n_0 factors ψ_0, n_1 factors ψ_1, and so on. In this description, the occupation numbers n_k are *independent of time*. The entire time dependence of $\Psi_\alpha(t)$ is generated by the time dependence of the single-particle functions $\psi_k(r_i, t)$ according to (3.1).

At time $t = 0$, before application of the external field, the system is assumed to be in thermal equilibrium. The initial energy E_α of the configuration (3.2) is given by

$$E_\alpha = \sum_k n_k \epsilon_k \tag{3.3}$$

where

$$H_1(0)\, \varphi_k(r) = \epsilon_k\, \varphi_k(r) \tag{3.4}$$

and

$$\lim_{t\to 0} \psi_k(r, t) = \varphi_k(r) \tag{3.5}$$

The probability P_α of finding a system of the ensemble in state α is given by (2.3), and this probability is also *independent of time.*

Let us now compute the current $J(t)$. The quantum-mechanical operator for the current is a sum of single-particle operators:

$$J_{\text{Op}} = \sum_{i=1}^{N} j_{\text{Op}}(i) \tag{3.6}$$

The quantum-mechanical expectation value of such an operator over a symmetrized product wave function can be expressed quite generally in terms of single-particle matrix elements

$$(\Psi_\alpha(t), J_{\text{Op}}\, \Psi_\alpha(t)) = \sum_k n_k^{(\alpha)} \int \psi_k^*(r, t)\, j_{\text{Op}}\, \psi_k(r, t)\, d\tau$$

$$= \sum_k n_k^{(\alpha)} j_{kk}(t) \tag{3.7}$$

Here we have added a superscript α to $n_k^{(\alpha)}$ to remind us of the particular configuration α. We note that only the *diagonal* matrix elements enter (3.7); there are no cross terms between different single-particle states $\psi_k(r, t)$.

The observed current $J(t)$ is taken to be the statistical average of (3.7) over all configurations α, with probabilities P_α; i.e.,

$$J(t) = \sum_\alpha P_\alpha \sum_k n_k^{(\alpha)} j_{kk}(t) \tag{3.8}$$

By interchanging the order of the two sums, we are led naturally to the quantity

$$\bar{n}_k = \sum_\alpha P_\alpha\, n_k^{(\alpha)} \tag{3.9}$$

This is the average occupation number of single-particle state number k. At time $t = 0$, $\bar{n}_k$ is given by the ordinary Bose-Einstein distribution

$$\bar{n}_k = \frac{1}{\exp\left[\beta(\epsilon_k - \mu)\right] - 1} \tag{3.10}$$

Since P_α and $n_k^{(\alpha)}$ are time-independent by construction, $\bar{n}_k$ *is independent of time also, and retains the value* (3.10) *at all later times.*

We now combine (3.8) and (3.9), and note that $\bar{n}_0$ is of order volume in the Bose condensation region . We therefore obtain

$$J(t) = \sum_k \bar{n}_k j_{kk}(t) = \bar{n}_0 j_{00}(t) + \sum_{k \neq 0} \bar{n}_k j_{kk}(t) \tag{3.11}$$

For the model under discussion, this result is *exact.*

In the Bose condensation region, the two contributions on the right of (3.11) have the same order of magnitude; the sum over $k \neq 0$ is the "normal current," to which the calculation of Kohn and Luttinger applies without significant changes; the single term $k = 0$ is the "supercurrent," which needs a quite separate discussion.

At sufficiently low (but still intensive) temperatures, the supercurrent term dominates (3.11) completely, since then $\bar{n}_0$ is nearly equal to N, the total number of particles. We shall restrict ourselves to this situation in what follows. Since $\bar{n}_0$ is independent of time, the essence of the problem lies in the calculation of the matrix element $j_{00}(t)$:

$$j_{00}(t) = \int \psi_0^*(r, t)\, j_{\text{op}}\, \psi_0(r, t)\, d\tau \tag{3.12}$$

where $\psi_0(r, t)$ is defined by (3.1) and (3.5), with $k = 0$. It is this single matrix element (3.12), rather than some average of matrix elements, which determines the persistent current in this model.

We note that the single-particle Hamiltonian in (3.1) still contains the effects of all the scattering centers [see (2.7b)]. Since $\psi_0(r, t)$ reduces to the ground state solution $\varphi_0(r)$ of (3.4) in the limit $t = 0$, and this solution has a de Broglie wavelength comparable to the size of the entire specimen, *the scattering centers affect the supercurrent term in a coherent fashion.*

In physical optics, the coherent effect of a large number of scattering is not a large amount of scattering, but rather an "index of refraction," i.e., a changed relation between frequency and wavelength of the light wave. This is true no matter whether the scattering centers are located at random (refraction of light in a dilute gas) or in lattice positions. The residual scattering (Rayleigh scattering) is proportional to the fluctuation of the number of scattering centers in a volume of the order of λ^3, λ being the wavelength of the light. If we apply this analogy to our present problem, the wavelength is as large as the specimen, the volume is the total volume of the specimen, and the fluctuation in question vanishes identically; thus, we expect no residual scattering at all, i.e., an infinite lifetime for the supercurrent.

This is the qualitative explanation of superconductivity; it arises from a Bose-Einstein-like condensation into a state of macroscopic de Broglie wavelength.

Just as in the Meissner effect, the important term is the ground state contribution. The excited states give rise to the "normal" current, which is of no special interest.

Since the excited states are unimportant anyway, it makes little difference whether they are separated from the ground state by an energy gap. In the present model, there is no energy gap, yet we obtain a persistent current and flux quantization.

Bardeen (59, 62b) has developed a two-fluid model for superconductors in analogy to the Landau (41) theory of liquid helium.[2a] In both these theories, the energy gap is essential to the discussion. As far as superconductivity is concerned, we are convinced that this is basically incorrect; the energy gap argument for persistent currents suffers from the same defect as the energy gap argument for the Meissner effect: both depend upon an invalid application of perturbation theory (see the discussion in Chapter VIII, Section 6).

The analogy between liquid helium and superconductivity is not necessarily close enough to draw conclusions. The nature of the thermodynamic transition, in particular, is so profoundly different (logarithmic peak vs. simple discontinuity in the specific heat) that liquid helium must be considered a problem in its own right.

4. The Applied Field

So far, the discussion has been quite general; it has not been necessary to specify in detail the nature of the perturbing Hamiltonian $H'(t)$ in (2.1). Now, however, we need to be more explicit.

In the London phenomenology, the London equations are *linear*. If we assume, for the moment, that a theory linear in the applied field is also meaningful in the microscopic approach, we may perform a

[2a] For further discussion, see Fukuda (63), Gross (63), Martin (63).

Fourier analysis of the applied field in space (not in time, though: the applied field is by assumption identically zero for $t < 0$), and consider each Fourier component by itself. Except for Section 9, which is concerned with the critical current for destruction of supercurrent flow, this will be our approach.

In the ring current experiment, the supercurrent is produced by magnetic induction. The flux through the hole in the ring is associated with the vector potential $\mathbf{A}(\mathbf{r}, t)$, not with the scalar potential $\varphi(\mathbf{r}, t)$. In a purely linear theory, the effects of $\mathbf{A}$ and φ superpose, and we may therefore set $\varphi = 0$ in what follows.[3]

In Chapter I, Section 6, we defined a vector field $\mathbf{v}_0(\mathbf{r})$ corresponding to "streamline flow" around the ring; the defining equations are (I, 6.12). This field $\mathbf{v}_0$ has vanishing curl and vanishing divergence, and $\mathbf{v}_0$ is parallel to the surface. In a ring-shaped volume, there is one and only one such vector field. In a simply connected volume, there is no such solution at all.

The current density $\mathbf{j}(\mathbf{r}, t)$ can be written as a sum of two terms:

$$\mathbf{j}(\mathbf{r}, t) = j_0(t)\, \mathbf{v}_0(\mathbf{r}) + \mathbf{j}_1(\mathbf{r}, t) \tag{4.1}$$

where[4]

$$\int \mathbf{v}_0(\mathbf{r}) \cdot \mathbf{j}_1(\mathbf{r}, t)\, dV = 0 \tag{4.2}$$

and

$$j_0(t) = \int \mathbf{v}_0(\mathbf{r}) \cdot \mathbf{j}(\mathbf{r}, t)\, dV \tag{4.3}$$

In Chapter I, Section 6, we pointed out that the part $\mathbf{j}_1(\mathbf{r}, t)$ of the current density makes no contribution at all to the total ring current; the entire ring current is contained in the first term on the right of (4.1).

It is natural, therefore, to make the same decomposition for the vector potential $\mathbf{A}(\mathbf{r}, t)$:

$$\mathbf{A}(\mathbf{r}, t) = A_0(t)\, \mathbf{v}_0(\mathbf{r}) + \mathbf{A}_1(\mathbf{r}, t) \tag{4.4}$$

$$\int \mathbf{v}_0(\mathbf{r}) \cdot \mathbf{A}_1(\mathbf{r}, t) dV = 0 \tag{4.5}$$

$$A_0(t) = \int \mathbf{v}_0(\mathbf{r}) \cdot \mathbf{A}(\mathbf{r}, t)\, dV \tag{4.6}$$

We then look for a linear relation between $j_0(t)$ and $A_0(t)$.[5]

[3] This is the scalar potential φ of the externally applied field. The argument cannot be used to exclude a self-consistent scalar potential φ due to the electrical charges of the particles themselves. To the extent that this effect is ignored, the theory is incomplete.

[4] The integration is over the volume of the ring.

[5] Within the confines of a linear theory, $j_0(t)$ may depend on other Fourier components of $\mathbf{A}$ as well; by confining ourselves to the linear relation between $j_0(t)$ and $A_0(t)$, we assume implicitly that the dependence, if any, of $j_0(t)$ on these other Fourier components does not make an essential difference. In the phenomenological theory of London, $j_0(t)$ depends only on $A_0(t)$.

Before doing so, however, we must first show that the component $A_0(t)$ of the vector potential is a meaningful physical quantity, and elucidate its meaning. We start by proving that $A_0(t)$ is gauge-invariant.

If we make a gauge transformation (VIII, 3.1), the change in $A_0(t)$, (4.6), is given by the integral

$$A_0'(t) - A_0(t) = \int \mathbf{v}_0(\mathbf{r}) \cdot \operatorname{grad} f \, dV$$

$$= -\int f \operatorname{div} \mathbf{v}_0 \, dV + \int f \, \mathbf{v}_0 \cdot \mathbf{n} \, dS \tag{4.7}$$

where the surface integral extends over the surface of the ring. Both contributions vanish identically, the volume integral because $\operatorname{div} \mathbf{v}_0 = 0$, the surface integral because $\mathbf{v}_0$ is parallel to the surface. *Thus $A_0(t)$ is gauge-invariant.*

Equation (4.4) is valid only within the volume occupied by the ring. Outside that volume, $\mathbf{A}(\mathbf{r}, t)$ cannot be decomposed in this way. Rather, $\mathbf{A}(\mathbf{r}, t)$ outside the ring must join continuously with $\mathbf{A}(\mathbf{r}, t)$ at the surface of the ring, as given by (4.4).

We shall restrict ourselves to the first term of (4.4), i.e., we assume that the applied vector potential is given by

$$\mathbf{A}(\mathbf{r}, t) = A_0(t) \, \mathbf{v}_0(\mathbf{r}) \qquad \text{(within the ring)} \tag{4.8}$$

We have seen that $A_0(t)$ is physically meaningful; what then does it mean? Since curl $\mathbf{v}_0$ vanishes by construction, the vector potential (4.8) gives zero magnetic field within the material of the ring! However, the magnetic field is non-zero outside the ring.

Let us calculate the flux F through the hole in the ring. Let C be a closed curve within the material of the ring, encircling the hole exactly once. Let S be a surface with curve C as its boundary. Then the flux through the hole in the ring is given by

$$F(t) = \int_S \mathbf{B} \cdot \mathbf{n} \, dS = \oint_C \mathbf{A} \cdot d\mathbf{s} = A_0(t) \oint_C \mathbf{v}_0(\mathbf{r}) \cdot d\mathbf{s}$$

Since curl $\mathbf{v}_0 = 0$, the last line integral is independent of the precise nature of the path C; it depends only upon the topology of that path; i.e., the quantity

$$F_0 = \oint_C \mathbf{v}_0(\mathbf{r}) \cdot d\mathbf{s} \tag{4.9}$$

is the same for all paths C which encircle the hole exactly one. We thus obtain the relation

$$F(t) = F_0 A_0(t) \tag{4.10}$$

Hence the portion (4.8) of the vector potential describes the magnetic flux through the hole in the ring, not any magnetic field within the material of the ring.

It may appear surprising that magnetic effects can arise from the vector potential (4.8): the particles within the ring are, after all, in a region free of magnetic field. This paradox has been discussed extensively by Bohm and others (Aharanov 59, 61; Peshkin 61; Tassie 61). Since this component of the vector potential is crucial for superconducting ring currents, the persistent currents in superconducting rings may be considered to be a particularly striking instance of the Bohm-Aharanov effect.

5. The Pseudo-Gauge Transformation

Since curl $\mathbf{v}_0 = 0$, the vector field $\mathbf{v}_0$ may be represented locally as the gradient of a scalar function f

$$\mathbf{v}_0(\mathbf{r}) = \operatorname{grad} f(\mathbf{r}) \tag{5.1}$$

This representation suffers from the trouble that $f(\mathbf{r})$ is not a single-valued function. Rather, if we go once around the ring, the value of f increases by the amount F_0, Eq. (4.9). Nevertheless, some useful results can be derived from consideration of (5.1) (Byers 61; Blatt 61, 61a; Brenig 61; Peshkin 62; Yang 62).

Let us consider a general, interacting system of particles in the ring, subject to the vector potential (4.8), with $A_0(t) = A_0$ independent of time. The Hamiltonian is

$$H(A_0) = \sum_{i=1}^{N} \frac{[\mathbf{p}_i - (eA_0/c)\,\mathbf{v}_0(\mathbf{r}_i)]^2}{2M} + V(\mathbf{r}_1, \mathbf{r}_2, \cdots, \mathbf{r}_N) \tag{5.2}$$

The potential function $V(\mathbf{r}_1, \cdots, \mathbf{r}_N)$ is quite arbitrary; it contains interactions of the particles with scattering centers, with each other, and with an external scalar potential. The only requirement is that V be a real function of its arguments.

Let Φ_α be an eigenfunction of $H(A_0)$, with eigenvalue $E_\alpha(A_0)$:

$$H(A_0)\,\Phi_\alpha = E_\alpha(A_0)\,\Phi_\alpha \tag{5.3}$$

Let us consider the quantity $S(\mathbf{r}_1, \mathbf{r}_2, \cdots, \mathbf{r}_N)$ defined by

$$S = \frac{eA_0'}{\hbar c} \sum_{i=1}^{N} f(\mathbf{r}_i) \tag{5.4}$$

and let us construct the new function

$$\Phi'_\alpha(\mathbf{r}_1, \mathbf{r}_2, \cdots, \mathbf{r}_N) = \exp(iS)\, \Phi_\alpha(\mathbf{r}_1, \mathbf{r}_2, \cdots, \mathbf{r}_N) \tag{5.5}$$

Although this transformation looks very similar to a gauge transformation for the vector potential $\mathbf{A}' = A'_0 \operatorname{grad} f$, this resemblance is misleading. We have seen in Section 4 that the coefficient A_0 is gauge-invariant and has immediate physical meaning. We shall call (5.4)–(5.5) a "pseudo-gauge transformation."

In general, the new wave function Φ'_α fails to be single-valued and is therefore not acceptable. If we take any one particle once around the ring, Φ'_α must remain the same (Peshkin 61, Tassie 61, Merzbacher 62). However, for certain special values of the constant A'_0, the exponential factor $\exp(iS)$ is unity in this operation. Inspection of (5.4) and (5.5) gives the condition

$$\frac{eA'_0}{\hbar c} F_0 = 2\pi n \qquad (n = \text{integral}) \tag{5.6}$$

Using the relation (4.10) for the flux F, we write the condition (5.6) in the form

$$F' = F_0 A'_0 = nF_1 \tag{5.7}$$

where

$$F_1 = hc/e \tag{5.8}$$

is the "flux quantum" of London (50). Thus the pseudo-gauge transformation is permissible if and only if the associated change in flux through the hole in the ring is an integral multiple of the flux quantum.

In view of the identity

$$H(A_0 + A'_0) \exp(iS) = \exp(iS)\, H(A_0) \tag{5.9}$$

we see that

$$H(A_0 + A'_0)\, \Phi'_\alpha = E_\alpha(A_0)\, \Phi'_\alpha \tag{5.10}$$

That is, the wave function Φ'_α is an eigenfunction of the Hamiltonian for the field with strength $A_0 + A'_0$, with the *same* eigenvalue as before.[6]

Since this is true for every single quantum state α, we conclude that *the entire eigenvalue spectrum is a periodic function of the flux through the hole in the ring, with period F_1*, (5.7). This is the first theorem of Byers and Yang (Byers 61).

---

[6] Considered as a differential equation, (5.10) holds for arbitrary A'_0. However, unless (5.7) is satisfied, Φ_α is multiple-valued and therefore not an acceptable wave function.

Their second theorem follows from taking the complex conjugate of (5.2). Since the potential function V is by assumption real, we obtain immediately

$$[H(A_0)]^* = H(-A_0) \tag{5.11}$$

By taking the complex conjugate of (5.3), we see that Φ_α^* is an eigenfunction of $H(-A_0)$ with unchanged eigenvalue $E_\alpha(-A_0) = E_\alpha(A_0)$. Hence, *the eigenvalue spectrum is an even function of the flux through the hole in the ring.*

Thus, if an eigenvalue $E_\alpha(A_0)$ has a minimum at zero flux, $A_0 = 0$, it has associated minima at flux values $F = F_0 A_0 = nF_1$, and maxima halfway in between. Conversely, if $E_\alpha(A_0)$ has a maximum at zero flux, it has associated maxima at flux values $F = nF_1$, and minima halfway in between. Since $E_\alpha(A_0)$ is an even function of A_0, these are the only two possibilities.

These theorems are completely general—and this is just the trouble! The theorems apply to a normal metal just as much as to a superconductor; yet flux quantization is not observed in normal metals, and in superconductors the observed flux quantum has half the London value (5.8), i.e., corresponds to a "superelectron" of charge $e^* = 2e$.

In a normal metal, the quantum states $E_\alpha(A_0)$ which enter the statistical distribution have alternate minima and maxima for $A_0 = 0$ (and associated flux values); when one performs the statistical averaging necessary to obtain the free energy, no dependence on A_0 remains in the leading term (Byers 61). Thus theorem 3 of Byers and Yang, that the free energy is an even, periodic function of the flux, with period F_1, is satisfied trivially.

But in a Bose-Einstein gas, in the condensation region, the situation is quite different. At sufficiently low (but still intensive) temperatures T, the vast majority of all particles occupies the single-particle ground state, and the free energy is given by

$$F(A_0) \cong N\,\epsilon_0(A_0) \tag{5.12}$$

where ϵ_0 is the single-particle ground state energy. The quantum-mechanical dependence of this energy on the flux through the hole in the ring (the Bohm-Aharanov effect) is then observable macroscopically. The flux quantum (5.8) must be computed with the boson charge for e.

As Onsager first realized,[7] the Schafroth (54) idea that the "super-

[7] This consideration was not published by Onsager until 1961, after the flux quantization experiments; but Onsager had made this point privately immediately after Schafroth's first publication, well before the more detailed microscopic theories were developed.

electrons" are to be identified with condensed electron pairs, results immediately in a flux quantum value

$$F_1 = \frac{hc}{e^*} = \frac{hc}{2e} \tag{5.13}$$

When this value was found experimentally (Deaver 61, Doll 61), Onsager (61) called attention to this relationship, and to the fact that a flux quantum of this size is not a fundamental property of electromagnetic fields, but rather a specific material property of metals in the superconducting state.

6. Single-Particle States in a Ring, with and without Scattering Centers

According to (5.12), we are interested in the low-lying single-particle energy levels in our ring-shaped volume, for given flux through the hole in the ring (given A_0).

In order to keep things simple, we consider a ring of precisely cylindrical shape. The cylinder has inner radius a, outer radius b, mean radius $R = (ab)^{1/2}$ and height L. For a volume of this shape, the vector function $\mathbf{v}_0(\mathbf{r})$ has field lines which are circles around the symmetry axis. Using cylindrical coordinates r, ϕ, z, and denoting a unit vector in the direction of increasing angle ϕ by $\mathbf{e}_\phi$ we have

$$\mathbf{v}_0(\mathbf{r}) = \frac{K}{r}\,\mathbf{e}_\phi \tag{6.1}$$

where K is a constant; the normalization condition (I, 6.12d) yields

$$K = [2\pi L \ln (b/a)]^{-1/2} \tag{6.2}$$

It is convenient to introduce a symbol for the ratio of the flux F to the flux quantum F_1, (5.13):

$$f = \frac{F}{F_1} = \frac{e^* K A_0}{\hbar c} \tag{6.3}$$

where e^* is the charge of the boson (we shall drop the asterisk henceforth).

The single-particle Hamiltonian then assumes the form

$$H = T + \sum_{s=1}^{S} v(\mathbf{r} - \mathbf{r}_s) \tag{6.4a}$$

crossings no longer occur. The level spectrum then assumes the form shown schematically in Fig. 9.2. The range of flux values (f-values) over which the two figures differ from each other, and the vertical gaps

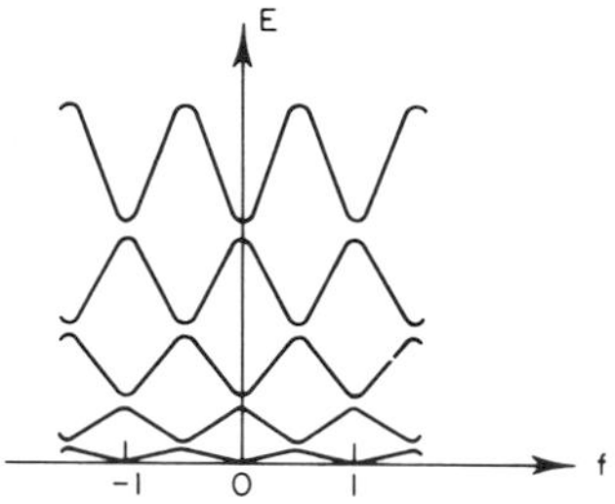

FIG. 9.2. Energy levels of a single particle in a cylindrical ring, with randomly located scattering centers within the volume of the ring. The energy levels are still periodic and even functions of the flux, but levels do not cross each other any more, since the scattering centers have destroyed the cylindrical symmetry.

between the upper and lower energies at each former crossing point, depend on the number density S/V of the scattering centers. For reasonable number densities, and random location of the scattering centers, the gaps are exceedingly narrow, and Eq. (6.9) is adequate for almost all f-values, the exceptions being very tiny neighborhoods of integral and half-integral values of f.[9] The periodicity of the energy vs. flux curves is related to an observed periodicity of the transition temperature T_c as a function of flux (Little 62).

It is interesting also to compute the ring current associated with these eigenstates. This is not necessarily equal to the actual ring current (which latter is generated in a time-dependent fashion, see Sections 7 and 8).

For a pure quantum state we have

$$j_0 = -c \frac{\partial E}{\partial A_0} \tag{6.10}$$

For a Bose-condensed state, $E = N\epsilon$; furthermore, when we relate the component j_0 of the current density Fourier expansion (4.1) to the

[9] Detailed estimates are given in Blatt (61); these are overestimates, since the self-consistent Coulomb effect decreases the energy gaps in Fig. 9.2. But even these over-estimates are exceedingly small numbers. The assumption of random locations of scattering centers enters at this point in the calculation; since the scattering centers affect a wave function which is coherent all the way around the ring, the scattering centers act coherently here, quite differently from the case of a normal metal.

actual ring current I, and relate A_0 to the flux F through (4.10), we obtain

$$I = -\frac{e}{2\pi\hbar} N \frac{\partial\epsilon}{\partial f} \tag{6.11}$$

Thus the ring current obtained from a pure eigenstate of the time-independent Hamiltonian is directly proportional to the slope of the curves in Figs. 9.1 and 9.2. Looking at Fig. 9.2, which includes the effects of scattering centers, we see that the ring current vanishes, for all quantum states, when f is integral or half-integral, i.e., when the flux F through the hole in the ring equals an integral or half-integral of flux quanta F_1. In particular, there is no current at all for zero flux; this is in accordance with a famous theorem by Felix Bloch: A system of particles, in thermal equilibrium and not subject to external fields, has vanishing current density everywhere.[10]

In Fig. 9.3 we show schematically the ring current I associated with

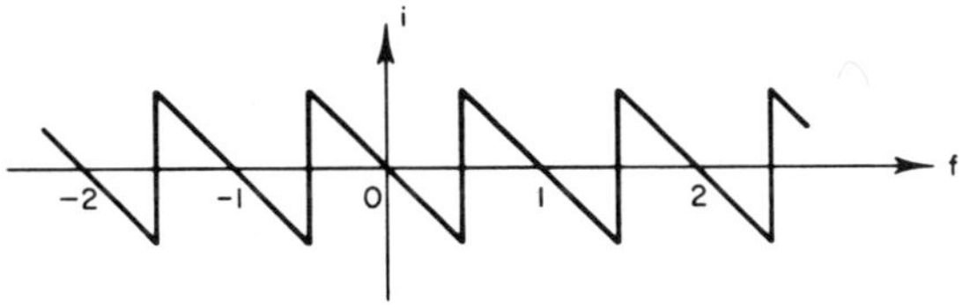

FIG. 9.3. The current associated with a single particle in the ground state of Fig. 9.2, as a function of the flux through the hole in the ring. The direction of the current is such that the flux produced by this current tends to bring the total flux through the hole closer to an integral number of flux quanta.

the ground state of the condensed Bose system, i.e., with the lowest level in Fig. 9.2. For values of f not too close to half-integers, the approximate equation (6.9) is adequate, provided we use for "m" the integer

$$m_0(f) = \text{integer closest to } f \tag{6.12}$$

We obtain from (6.9) and (6.11)

$$I = -N \frac{e\hbar}{2\pi MR^2} [f - m_0(f)] \qquad \text{(ground state current)} \tag{6.13}$$

Equation (6.13) breaks down in the immediate neighborhood of half-integral values of f; in these narrow regions the actual current I changes suddenly from a negative value (for f slightly less than half-integral) to an equal positive value (for f slightly more than half-integral); this is shown in Fig. 9.3.

If f remains within $\pm \frac{1}{2}$ of some integral value, the current I, Eq.

[10] For a number of years, this theorem was considered to imply the Pauli corollary: "All theories of superconductivity are wrong."

(6.13), decreases algebraically as the flux $F = fF_1$ increases; in this sense the current is "diamagnetic." However, this diamagnetism is highly peculiar. There is, after all, no magnetic field at all within the material of the ring. The current is related to the flux through the hole in the ring, i.e., the entire current I is a Bohm-Aharanov effect. Furthermore, unlike normal diamagnetic currents, I does not have a direction such as to bring the flux closer to zero, but rather the direction of I is such as to bring the flux closer to the nearest integral number of flux quanta.

The ground state current (6.13) is limited in magnitude, since the difference $f - m_0(f)$ cannot exceed $\frac{1}{2}$ in absolute value:

$$|I| \leqslant I_{\text{max}} = \frac{Ne\hbar}{4\pi MR^2} \tag{6.14}$$

Let us estimate the order of magnitude of this limiting current. We use two electronic charges for e, two electronic masses for M, a radius R of the order of 1 cm, and a total particle number of order $N \sim 10^{22}$ (this corresponds to a volume of order 1 cm^3 for reasonable particle densities). We then obtain a limiting current $I_{\text{max}} \sim 10^{12}$ esu, which is of the order of 100 amp. Thus even a slight departure of the flux from an integral number of flux quanta results in truly macroscopic currents around the ring.

This argument, incidentally, disposes completely of the idea of the "displaced Fermi sphere" as an explanation of supercurrents. According to the displaced Fermi sphere idea, the current-carrying state is one in which the "pairing" of electron levels is not between states k_1 , k_2 with $\mathbf{k}_1 + \mathbf{k}_2 = 0$, but rather with $\mathbf{k}_1 + \mathbf{k}_2 = \mathbf{q}$, where $\mathbf{q}$ is some small wave vector. This amounts to a sideway shift of the entire distribution of electrons in k-space. As P. W. Anderson first pointed out, the minimum value of q is of order $\hbar/L$ where L is the linear dimension involved, i.e., the circumference of the ring. The corresponding "quantum" of current is twice the maximum current (6.14). The insultingly macroscopic size of this "quantum" of current makes the displaced Fermi sphere idea untenable.[11]

---

[11] In normal metals, any one electron carries a quantized current, but the number of electrons involved is an arbitrarily small fraction of the total number N. In practice, this fraction is exceedingly small. By contrast, in the displaced Fermi sphere state, *all* electrons participate in current transport.

Unfortunately, the displaced Fermi sphere lies at the basis of a number of calculations (Rogers 60; Suzuki 60; Wentzel 61; Parikh 62; Maki 63, 63a).

For superconductors of very small size, and hence small total number of particles N, the upper limit (6.14) falls below the critical current; under those conditions, the displaced Fermi sphere may perhaps serve as a first approximation, though there are other difficulties associated with small specimens (see Chapter X, Section 4). The displaced Fermi sphere treatment of persistent currents in small specimens is reviewed by Bardeen (62b).

Returning to the discussion of Fig. 9.3, we see that for rings of ordinary macroscopic dimensions, the ground state current can be large enough to explain the observed supercurrents.

Fortunately, Garwin,[12] and Mercereau and Hunt (Mercereau 62), have performed ring current experiments with rings of much smaller dimensions. Garwin produces his rings be evaporating a thin, ring-shaped film of metal onto a neutral backing. The diameter of the ring is of order 10^{-1} cm, and the thickness is of order 10^{-6} cm. Thus the volume of a Garwin ring is of order 10^{-8} cm^3; with the same number density as before, we obtain the estimate $N \sim 10^{14}$. Allowing for the smaller mean radius R, $I_{\max}$ is of the order of 10^{-4} amp, well below the actual currents which Garwin is able to induce. We conclude that *ring current experiments cannot be explained completely by consideration of the quantum mechanical ground state of a time-independent Hamiltonian.* The process of inducing the current must be considered in detail.

If all particles are in the same single-particle state, but in a state other than the ground state, then higher currents are possible. In Fig. 9.4 we show the current I, Eq. (6.11), appropriate for the first

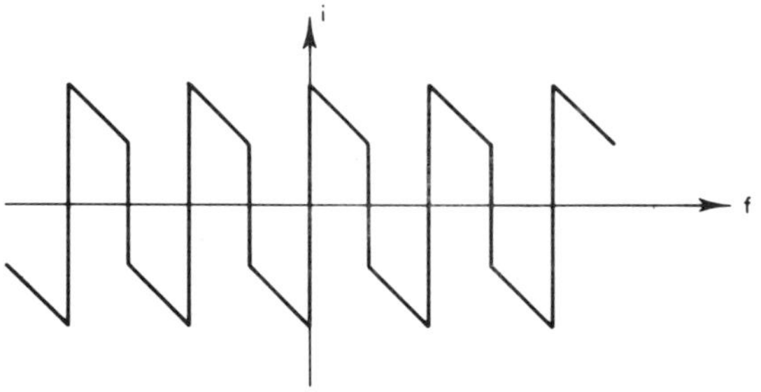

FIG. 9.4. The current associated with a single particle in the first excited state of Fig. 9.2, as a function of the flux through the hole in the ring. The direction of the current is the same as in Fig. 9.3, but now there are sharp breaks at integral numbers of flux quanta, as well as at half-integral numbers.

excited state $\epsilon_1(f)$ of Fig. 9.2. This current is zero for integral and half-integral values of f, as expected, but behaves in quite peculiar fashion for other values of f. For these other values, the current I ranges in magnitude from $I_{\max}$, Eq. (6.14), to twice that value; the direction of the current alters suddenly for every integral and half-integral value of f; the main behavior of the current is still "diamagnetic," in the sense that an increase of flux causes a change in current so as to oppose this flux increase.

[12] R. Garwin, private communication.

The current vs. flux curves for higher quantum states look similar to Fig. 9.4, with progressively higher absolute values of the current.

7. Flux Quantization upon Cooling

Let us follow along the experimental procedure used for establishing ring currents in superconductors, and let us see what our model predicts at each stage of that procedure. The procedure consists of three parts:

(1) The ring is placed, at room temperature, between the poles of a magnet, and we wait until thermal equilibrium is achieved. If we neglect the normal magnetic susceptibility of the metal, the flux through the hole in the ring is due entirely to the externally applied field. We shall call this flux F_e.

(2) The ring is then cooled to a very low temperature, well below the transition temperature T_c. We shall assume that the final temperature is low enough so that the Bose gas is substantially completely condensed, i.e., the occupation number $\bar{n}_0(T)$ at the final temperature will be taken equal to the total number of particles, N.

(3) The external magnetic field is then removed, for example, by pulling the magnet away from the ring.

In this section we shall be concerned with the first two of these three steps; the third step will be considered in Section 8.

Above the transition temperature, our Bose gas has a non-zero, normal resistivity, as shown explicitly by Kohn and Luttinger. Thus, whatever ring currents may be induced at the beginning of step 1 die down quickly; hence we may start our consideration from a state in which the gas carries no current.

As we cool down below the transition temperature, a temperature-dependent, macroscopic number of particles $\bar{n}_0(T)$ accumulate in one single-particle state. Clearly this state must be the state of lowest energy under the given conditions; i.e., we settle onto the lowest energy level in Fig. 9.2 for whatever actual flux F, or flux parameter $f = F/F_1$, passes through the hole in the ring (Brenig 61).

Initially, the relevant flux is entirely due to the external magnet, i.e., we start off with a flux parameter

$$f_e = F_e/F_1 \tag{7.1}$$

However, Fig. 9.3 shows that, unless f_e happens to be exactly equal to an integer, the particles in the ground state give rise to a ring current around the ring, and this current itself produces a flux, which we must take into account self-consistently.

Let $\mathscr{L}$ be the self-inductance of the ring; then the total flux F through the hole in the ring is

$$F = F_e + c\mathscr{L}I, \qquad f = F/F_1 \tag{7.2}$$

where I is found by modifying Eq. (6.13) suitably: we must replace N, the total number of particles, by $\bar{n}_0(T)$, the number of condensed particles at temperature T; this gives

$$I = -\bar{n}_0(T)\frac{e\hbar}{2\pi MR^2}[f - m_0(f)] \tag{7.3}$$

Equations (7.2) and (7.3) are simultaneous linear equations for f and I. The integer $m_0(f)$ is equal initially to

$$m_e = m_0(f_e) \tag{7.4}$$

Since the supercurrent (7.3) acts in such a direction as to make the total flux approach more closely to an integral number of flux quanta, this initial value of m_0 remains unchanged throughout the process, and preserves a "memory" for the initial, external flux $F_e = f_e F_1$.

When we eliminate I from (7.2) and (7.3), and use the notation (7.4), we obtain the result

$$f - m_e = \frac{f_e - m_e}{1 + Q} \tag{7.5a}$$

where

$$Q = \bar{n}_0(T)\frac{e\hbar c\mathscr{L}}{2\pi R^2 MF_1} = \bar{n}_0(T)\frac{e^2\mathscr{L}}{(2\pi R)^2 M} \tag{7.5b}$$

Thus the flux parameter f at temperature $T < T_c$ is closer to the integral value m_e than the original external flux parameter f_e, (7.1). *The act of cooling has brought about an approximate quantization of flux.*

It should be noted that the flux parameter f is not *exactly* equal to the integer m_e, i.e., the right-hand side of (7.5a) does not vanish identically. Thus flux quantization, even at this stage, is not perfect. But in practice the quantity Q is very large indeed, so large that the difference $f - m_e$ would be quite difficult to detect. The estimate of Q proceeds as follows: the self-inductance $\mathscr{L}$ of a long cylinder ($L \gg a$) is given by

$$\mathscr{L} = \frac{(2\pi a)^2}{c^2 L} \tag{7.6}$$

In the experiment of Deaver (61) the length L was of order 1 cm, the radius of order 10^{-3} cm, thus $L \gg a$. Inserting (7.6) into (7.5b) yields

$$Q = \bar{n}_0(T) \frac{e^2 a^2}{c^2 R^2 L M} \simeq 5 \times 10^{-13} \, \bar{n}_0(T) \tag{7.7}$$

where we have used two electronic charges for e, two electronic masses for M, and the approximate physical dimensions of the Deaver (61) experiment. The volume of the ring in that experiment was of order 10^{-6} cm^3, and with a number density of order 10^{22}, we get $N \sim 10^{16}$. Thus, at all superconducting temparatures T except in the immediate neighborhood of the transition temperature T_c, the quantity Q is quite large compared to unity; at low temperatures it is of the order of 5000. Equation (7.5a) then shows that the difference between f and the quantized (integral) value m_e is no more than one part in 10^4, which could not have been detected by this experiment.

Thus the Bose-Einstein model predicts flux quantization to within experimental accuracy as a consequence of simply cooling the ring, without moving either the ring or the magnet; and this conclusion is in full agreement with experiment.

It would be highly interesting to detect the small deviation $f - m_e$, which of course depends on the initial deviation $f_e - m_e$ before cooling. This has not been possible so far.

8. Build-up of the Ring Current by Electromagnetic Induction

Let us now discuss the third step of the normal experimental procedure, namely, the removal of the external magnetic field, for example, by pulling the external magnet away from the ring. We assume that this is done at a temperature T sufficiently low so that the Bose condensation is substantially complete.

This is definitely a time-dependent problem, and we are therefore concerned with the solution of the time-dependent Schrödinger equation (3.1). The time-dependence of the Hamiltonian is implicit: the Hamiltonian (6.4) depends explicitly upon the flux parameter f, and this parameter is time-dependent. Thus we have to solve the equation

$$H_1(f) \psi = i\hbar \frac{\partial \psi}{\partial t} \tag{8.1}$$

where the time-dependent flux parameter is given by [see (7.2)]

$$F(t) = c\mathscr{L}I(t) + F_e(t), \qquad f(t) = \frac{F(t)}{F_1} \tag{8.2}$$

Here the external flux $F_e(t)$ is the directly given quantity. The current $I(t)$ in the ring must be determined from the equation

$$I(t) = N(\psi(t), I_{\text{op}}\,\psi(t)) \tag{8.3}$$

Note that $I(t)$ is *not* given by Eq. (6.11), which refers to a purely time-independent situation.

The system of equations (8.1)–(8.3) must be solved subject to the initial condition

$$\lim_{t\to 0} \psi(t) = [\varphi_0]_{f=f(0)} \tag{8.4}$$

where φ_0 is the ground state wave function of the time-independent problem; the flux parameter $f = f(0)$ is determined from (7.5) with a temperature $T \simeq 0$, i.e., $\bar{n}_0(T) \simeq N$.

This system of equations is highly non-linear, since $I(t)$, Eq. (8.3), is quadratic in the wave function ψ, and $H_1(f)$ is quadratic in the flux parameter f which depends on $I(t)$. No explicit calculation has been carried out so far.

However, the essential results can be obtained by using general properties of such time-dependent problems (Schiff 49, section 31). There are two main regions, "adiabatic" and "sudden," depending upon the rapidity with which $f(t)$ changes with t. The time scale is determined by the quantum-mechanical frequencies $(\epsilon_k - \epsilon_{k'})/\hbar$, where the ϵ_k are the eigenvalues of the time-independent Hamiltonian for the given flux parameter f.

If the time necessary for f to change appreciably is long compared to all the quantum frequencies, we have a "slow" variation, and the "adiabatic" approximation is valid. In this approximation, the time-dependent function $\psi(t)$ is, at all times, equal to the ground state wave function $\varphi_0(f)$ for the flux value $f = f(t)$ appropriate at that time.

Conversely, if $f(t)$ changes rapidly compared to quantum frequencies, we may use the "sudden" approximation. In this approximation, the time-dependent wave function $\psi(t)$ is nearly independent of time and remains equal to the initial function $\psi(0)$, except for a phase factor.

Let us apply these concepts to the building-up of the ring current as the external flux $F_e(t)$ decreases to zero. The simplest case to consider is the one of macroscopic rings. We have seen, in Section 6, that the ring current associated with the time-independent ground state solution φ_0 can be as large as hundreds of amperes. Thus, there is no need for $f(t)$ to deviate appreciably from the integral value m_e, (7.4); quite small deviations make $I(t)$, (8.3), large enough to supply the necessary flux even in the absence of F_e. In this region, the "adiabatic" and "sudden" approximations give the same result! For this is the region in which the ground state wave function is almost precisely independent of f, anyway.[13]

We conclude that, in rings of normal macroscopic dimensions, the flux remains approximately quantized, and the system remains in the ground state, even after the external magnetic field is removed. The speed of removal of the external field is not an important factor.

The situation is quite different for "Garwin rings." We saw in Section 6 that the steady ring current associated with the ground state in such a ring is only of order 10^{-4} amp at the most. Since the experimental currents are larger, we conclude that the process of removing the external field must leave the system in an exited state; the final flux is at least several flux quanta removed from the initial, external value. Therefore, *in Garwin rings, the process of building up a ring current destroys the flux quantization.* This theoretical prediction has not yet been verified experimentally.[14]

Returning to the theoretical discussion, we see that the change of flux must be "sudden" rather than "adiabatic" in order to produce the desired result.[15]

---

[13] f enters only through the radial equation (6.7); if the ring is thin, i.e., $b - a \ll (ab)^{1/2}$, we may replace $(m - f)^2/r^2$ in that equation by $(m - f)^2/ab$, and we get a constant solution $R(r) =$ constant, provided that $k^2 = (m - f)^2/ab$. This is the approximation underlying the energy value (6.9). In this approximation, the wave function (6.5) is completely independent of f, except insofar as (for the ground state) m must be chosen to be the appropriate integer $m_0(f)$, nearest to f.

[14] If we assume "sudden" behavior of the system at all integral and half-integral values of the flux parameter f, then the final result is the same as if no scattering centers were present, and can therefore be obtained by much simpler theoretical considerations. If we ignore scattering centers, the system moves along the continuous curves of Fig. 9.1, right across all level intersections in that figure.

This model has been studied by Keller (61) and Lipkin (62), who find that the final flux reached, after removal of the external magnetic field, is such that the fluxoid of London, Eq. (I,6.5), remains quantized precisely; since the fluxoid differs slightly from the flux F, the flux F deviates from an integral number of flux quanta. The resulting final flux F, or flux parameter $f = F/F_1$, is the same as the value obtained from Eq. (7.5a) by setting $f_e = 0$.

The same expression has been derived in somewhat different fashion by Bardeen (61a) and Ginzburg (62).

For further work on flux quantization, see Bohr (62), Lüders (62), Maki (62a), Schwabl (62), and Weller (62, 62a).

[15] The idca that the wave function must remain unaltered, and thus the final wave function in this process is not the ground state wave function in the final flux, is quite in accordance with the conjectures of London (50). However, London sometimes used the unfortunate terminology "adiabatic" for this process. In the quantum-mechanical meaning of this word, the process must be the very opposite of "adiabatic," i.e., it must be "sudden."

The estimate of the relevant quantum frequencies is somewhat involved (Blatt 61); the important regions are $f \simeq$ integral and $f \simeq$ half-integral, where the randomly located scattering centers prevent the crossing of energy levels (see Figs. 9.1 and 9.2). However, the estimates are certainly consistent with a "sudden" process—the relevant quantum-mechanical time scale is of the order of decades! Furthermore, effects not included in these estimates (the self-consistent Coulomb repulsion between the particles, and the energy gap for actual superconductors) are bound to lengthen these time scales even more.

So far, we have shown how a large circulating ring current can be built up in the Bose-Einstein model, in spite of randomly located impurities which make the model a normal conductor above the Bose-Einstein condensation temperature. In the remaining two sections, we discuss the stability of these currents.

9. The Critical Current

So far, we have not enquired into the energetic stability of the current-carrying state. We have started from a Bose-condensed state at time $t = 0$, and have investigated the change of the wave function with time. In the strict independent-particle model, that is all we need to do: if all particles are in one and the same single-particle state at time $t = 0$, then they will also be in one quantum state at any later time. The independent-particle Hamiltonian leads to a change with time of that wave function, but there are no transitions to other states.

This is of course an idealization; interactions between particles exist, and in first approximation their effect is to produce transitions out of the Bose-condensed state, into states with partial Bose-condensation; this can happen in principle, and may be assumed to happen in practice, if energy can be liberated by taking a particle out of the highly occupied single-particle state, and placing it in some other single-particle state.

It can also happen in principle, but is *not* likely to happen within any reasonable time, if energy can be liberated by taking a *large number* of particles out of the condensed state, and putting them into some other state (or states). An extreme example of this is a transition involving all N particles, leading to a new Bose-condensed state with an entirely different wave function. If the ring carries a current initially, then there always exists such a transition with liberation of energy, namely to a final state with no trapped field and no current in the ring. This is the thermodynamic equilibrium state, i.e., the state of lowest free energy. But clearly, such a transition, involving cooperative action of all N particles, is a highly unlikely event. The mean time for such a transition is an increasing function of N, which is bound to be enormous for macroscopic values of N.

The energetic stability of ring currents, in the sense of transitions involving small numbers of particles, has been investigated in the Bose-Einstein model by Bloch and Rorschach (Bloch 62). This work is complementary to Blatt (61), in the sense that certain effects included in one are ignored in the other, and vice versa. Bloch (62) include non-linear effects in the field, i.e., they do not restrict themselves to a single Fourier component of the field distribution, and they include the Coulomb repulsion between the bosons self-consistently. On the other hand, they do not consider the effects of scattering centers, and they restrict themselves to time-independent situations. Since we have seen that ring currents can be built up by a time-variation of the applied magnetic field, and that this can be done in spite of the presence of scattering centers, the assumptions of Bloch (62) provide a reasonable starting point for a theory of critical currents.

The self-consistent treatment of Bloch (62) starts from the independent-particle Hamiltonian (2.7) with the assumption that $\mathbf{A}$ and φ are independent of time. One then finds the energy levels and single-particle wave functions from

$$H_1\psi_k = \epsilon_k\psi_k \tag{9.1}$$

The many-particle wave function Ψ_α is described by some configuration α, Eq. (3.2). Allowing for a uniform background of opposite charge to give overall electrical neutrality, the charge density at some point $\mathbf{r}$ in the specimen is

$$\rho(\mathbf{r}) = e\left\{\sum_k n_k \mid \psi_k(\mathbf{r}) \mid^2 - \frac{N}{V}\right\} \tag{9.2}$$

This charge density determines the self-consistent potential $\varphi(\mathbf{r})$ through the Poisson equation

$$\nabla^2\varphi = -4\pi\rho(\mathbf{r}) \tag{9.3}$$

This potential enters (9.1) through the Hamiltonian H_1, (2.7b). In the same way, the current density associated with the wave function Ψ_α is

$$\mathbf{j}(\mathbf{r}) = \frac{e}{M}\,\text{Real Part}\left\{\sum_k n_k\psi_k^*(\mathbf{r})\left[\frac{\hbar}{i}\nabla - \frac{e}{c}\mathbf{A}(\mathbf{r})\right]\psi_k(\mathbf{r})\right\} \tag{9.4}$$

This current determines the "internal" part $\mathbf{A}_i(\mathbf{r})$ of the vector potential through

$$\text{curl curl } \mathbf{A}_i(\mathbf{r}) = \frac{4\pi}{c} \mathbf{j}(\mathbf{r}) \tag{9.5}$$

The total vector potential, which appears in (9.4), is the sum of $\mathbf{A}_i$ and the externally applied vector potential $\mathbf{A}_e(\mathbf{r})$, which latter is assumed given:

$$\mathbf{A}(\mathbf{r}) = \mathbf{A}_i(\mathbf{r}) + \mathbf{A}_e(\mathbf{r}) \tag{9.6}$$

For given $\mathbf{A}_e(\mathbf{r})$ and given quantum configuration $\alpha = \{n_0, n_1, n_2, \cdots\}$, Eqs. (9.1)–(9.6) form a self-consistent set. The energy of the system as a whole is

$$E = \sum_k n_k \epsilon_k - \frac{eN}{V} \int \varphi \, d^3r + (8\pi)^{-1} \int | \text{curl } \mathbf{A}_i |^2 \, d^3r - (8\pi)^{-1} \int | \nabla\varphi |^2 \, d^3r \tag{9.7}$$

In this expression, the first term is the quantum mechanical energy of the bosons; the second term is the electrostatic potential energy of the uniform background charge, with its constant charge density $-eN/V$, in the electrostatic potential $\varphi(r)$; the third term is the magnetic field energy of the "internal" magnetic field set up by the particles; the last term is the negative of the electrostatic field energy associated with $\epsilon = -\text{grad } \varphi$; this energy must be subtracted since it not only appears implicitly, with positive sign, in the first two terms of (9.7), but in fact appears twice over.

The determination of energetic stability of the Bose-condensed state proceeds, in principle, as follows:

(1) We determine the self-consistent solution of lowest energy E for perfect condensation, i.e., for configurations of type $n_k = N$, some one k, $n_{k'} = 0$ for all $k' \neq k$.

(2) We then investigate the self-consistent solutions, and corresponding energies E, for all "neighboring" configurations; if all these other energies are higher, the Bose-condensed state is stable.

Bloch (62) show that this investigation can be greatly simplified. First of all, it is sufficient to consider configurations with only *one* particle moved out of state k, i.e., $n_k = N - 1$, $n_l = 1$, and $n_{k'} = 0$ for $k' \neq k, l$. Second, it is then not necessary to reinvestigate the self-consistent electric and magnetic field distributions; to sufficient accuracy, these may be taken to be the same as in the condensed state. The energy difference between the two configurations, under these assumptions, reduces to the contribution of the first term in (9.7), i.e.,

$$\Delta E = \epsilon_l - \epsilon_k \tag{9.8}$$

If this quantity is positive for all single-particle states $l \neq k$, the Bose-condensed state is stable energetically.

The detailed investigation of Bloch (62) is concerned with a ring-shaped volume consisting of the space between two cylinders, the inner one of radius a, the outer one of radius b. The magnetic field in the hole (for $r < a$) is denoted by B_1, the field outside the cylinder (for $r > b$) by B_2; both these fields are parallel to the cylinder axis. We now state the results of this calculation.

If $B_1 = B_2$, we do not have a net circulating ring current, but rather an ordinary Meissner effect. In that case, the stability criterion turns out to be

$$B_1 = B_2 < H_c = 4\pi \frac{N}{V} \mu_0 \tag{9.9}$$

where $\mu_0 = e\hbar/2Mc$ is the Bohr magneton for the bosons. This is precisely the critical field found by Schafroth, Eq. (II, 5.26). Coming from the other side (i.e., assuming field penetration and the absence of Bose-Einstein condensation), Schafroth (55a) was able to obtain a self-consistent solution for $B > H_c$. Although the calculation of Bloch (62) indicates only the onset of instability, without determination of the nature of the new stable state, it is clear that the new stable state is just the "normal" state investigated by Schafroth (55a). Thus, in this case, the stability criterion of Bloch (62) is not only necessary but also sufficient. Since the stability criterion is obtained by moving only *one* particle out of the condensed state, this gratifying result is by no means obvious.

If B_1 and B_2 differ from each other, there is a circulating current I in the ring. If the cylinder has a total height L, this current is given by

$$I = \frac{cL}{4\pi} (B_1 - B_2) \tag{9.10}$$

One might think that the ring current is limited by the condition that both B_1 and B_2 must have a magnitude less than H_c, Eq. (9.9), i.e., we might expect a critical current

$$I_c = \frac{cLH_c}{2\pi} \tag{9.11}$$

This result, it turns out, is true *only* if the ring is "thick" in the purely geometrical sense

$$b - a \gg a \qquad \text{(thick ring)} \tag{9.12}$$

It is *not* sufficient that the thickness $b - a$ exceed a London penetration depth. Rather, as soon as (9.12) is violated, we have a "thin" ring, and the stability condition becomes

$$\frac{H_c(B_1 - B_2)}{H_c^2 - B_1 B_2} < \frac{b^2 - a^2}{b^2 + a^2} \tag{9.13}$$

The case of most interest here is a "thin" ring, $b - a \ll a$, with no field at all on the outside, i.e., $B_2 = 0$. We then obtain from (9.13) the stability condition

$$B_1 < \frac{b - a}{a} H_c \tag{9.14}$$

Since $b - a \ll a$ by assumption, this means that the region of stability is reduced strongly compared to the expected result (9.11). For example, for $a \sim 10^{-3}$ cm and $b - a \sim 10^{-5}$ cm (an admittedly unfavourable case), the criterion (9.14) allows the trapping of at most 10^2 flux quanta in the hole.

If $B_1 = B_2$ initially, and the ring current is induced by removing the outside field, i.e., by decreasing B_2 to zero, then the results depends on whether B_1 is within the stable range (9.14). If B_1 exceeds the limit (9.14) initially, the removal of the external field B_2 leads to breakdown, at a value of B_2 determined by (9.13); as B_2 is decreased further, bosons shift in large numbers into successively lower quantum states,[16] until eventually we reach a self-consistent solution, still of Bose condensed type, with the inner field B_1 equal to its upper limit (9.14).

It is highly interesting that this process involves the breakdown of part of the flux through the hole in the cylinder, even though the cylinder thickness $b - a$ may be greatly in excess of the London penetration depth. The fact that this process follows from a microscopic investigation of a reasonable model, shows that one must take macroscopic arguments about "lack of penetration of field" with considerable scepticism. One simply can not reason directly from the Meissner field expulsion to the existence and stability of persistent currents.[16a]

Bloch (62) point out that the energy changes involved in their breakdown process are exceedingly small, and may thus take great a deal of time. No actual time estimate is given, however; nor would such a time estimate be useful, in view of the basic limitations of the Bose-Einstein model.

---

[16] The transition most favored energetically, and thus underlying the criterion (9.13), is a transition to the very next quantum state, from magnetic quantum number m to magnetic quantum number $m - 1$.

[16a] The Bloch (62) calculation is based on the Bose-Einstein gas model. The situation is not so clear for the electron pair gas (Nakajima 59b).

10. Unsolved Problems

Although considerable progress has been made with the Bose-Einstein model, the unsolved problems in the theory of persistent currents are numerous and perplexing.

The most obvious unsolved problem is the extension of the theoretical work to the electron pair gas. The analogy between the ideal Bose-Einstein gas and the electron pair gas in superconductors is helpful and suggestive, but it is incapable of giving quantitative results for detailed properties, such as relaxation times for the build-up or decay of persistent currents. The presence of an energy gap in superconductors, compared to its absence in the Bose-Einstein model, may have a considerable influence on all time estimates.

In the region where the internal size of the pairs is unimportant, quantitative results can be obtained from the theory of Landau and Ginzberg (see Chapter I, Section 5). Observation of lifetime effects, however, is most likely in thin film specimens, where the film thickness is smaller than the Pippard length. The Landau-Ginzberg theory replaces the actual pair wave function by a "wave function of the superelectron," depending on only one position variable. In the region of validity of that theory (close to the transition point), the position variable in question may be identified with the center-of-gravity of the electron pair. But in thin films, the coupling between center-of-gravity motion and internal motion of the electron pair cannot be ignored, and the Landau-Ginzberg theory must be used with considerable caution.

For discussion of the Landau-Ginzberg approach, see Ginzberg (59a) and Bardeen (62b).

Even in the Bose-Einstein model, there remain some serious unsolved problems. These are associated with the question of the lifetime of the non-equilibrium, current-carrying state. In any "normal" system, there is a natural tendency to return to thermodynamic equilibrium. If the system is isolated energetically (no heat transfer), the equilibrium state has the same energy as the initial state, but a higher entropy. Thus the return to equilibrium for an isolated system involves an increase in entropy, at constant energy.

The trouble with the treatment used so far is that it excludes such an increase of entropy! The time-development of the ensemble is given by Eq. (2.4), which determines the wave functions $\Psi_\alpha(t)$, together with the statement that the probabilities P_α are time-independent. If we define the entropy by (VII, 1.3), the entropy equals

$$S = -\mathfrak{k} \sum_{\alpha} P_{\alpha} \ln P_{\alpha} \tag{10.1}$$

and this quantity is strictly independent of time.

The usual way out of this dilemma, discussed for example by Tolman (38), is through the process of "coarse-graining." We assume that macroscopic experiments are inherently incapable of measuring enough detail about the system to make the entropy definition (10.1) meaningful. Rather, macroscopic experiments distinguish only between large groups of quantum states. We must then replace the "fine-grained" probabilities P_{α} by their respective group averages $\bar{P}_{\alpha}$. The corresponding "coarse-grained entropy" is

$$\bar{S} = -\mathfrak{k} \sum_{\alpha} \bar{P}_{\alpha} \ln \bar{P}_{\alpha} \tag{10.2}$$

Objections have been raised against this procedure (Blatt 59a), on the grounds that apparently "fine-grained" experiments do exist; i.e., the detailed information about the ensemble contained in (10.1) but not in (10.2) can be observed by a sufficiently elaborate experiment. One example is the spin-echo experiment of Hahn (50, 53), where a "spin-echo" is observed well *after* the system of spins has apparently returned to equilibrium, for ordinary experimental conditions.

There is every indication that persistent ring currents in superconductors are another example of "fine-grained" observations. The process of coarse-graining makes no sense if the main effect is produced by one coherent wave function; yet the very existence of persistent currents depends upon just such a coherent wave function. It is therefore possible[17] that the question of the lifetime of persistent currents requires an extension of the framework of statistical mechanics.

[17] This is the most interesting possibility, but it is not the only one. For example, the interactions between particles in the Bose model (or between different electron pairs in the quasi-chemical equilibrium model) may serve to deplete the occupation number of the condensed state, as time goes on. If the stability criterion of Bloch and Rorschach is satisfied, this cannot happen by means of transitions involving a few particles at a time; but it might still happen by way of transitions involving many particles simultaneously. The question here is the meaning of the "condensed state" in the presence of particle-particle interactions. The existence of a sharp thermodynamic transition, and the observation of "global" properties such as flux quantization, suggests strongly that something like an equivalent wave function, coherent throughout the specimen, has more than merely approximate meaning. This is the reason for our expectation that the fundamental difficulties which appear in the independent-particle Bose-Einstein model will also occur in the, still to be developed, theory of persistent currents in actual superconductors.

The complexity of the Hamiltonian H is not the issue here; no matter how complex the Hamiltonian is, the time-development of the density matrix is determined by a similarity transformation

$$\mathscr{U}(t) = \exp(-iHt/\hbar)\, \mathscr{U}(0) \exp(+iHt/\hbar) \tag{10.3}$$

and Trace ($\mathscr{U} \ln \mathscr{U}$) is invariant under similarity transformations. If coarse-graining is inappropriate, this trace is the entropy, and stays constant.

The situation is particularly clear if we imagine that we start from strictly zero temperature, i.e., from the ground state of the system. When we remove the external magnetic field, the wave function changes in time, and the final current-carrying state is not the ground state for the final flux (see Section 8). We therefore have excess energy, which would have to be converted to heat for approach to thermal equilibrium. However, as long as the system can be described by one single wave function, the entropy (10.1) is rigorously zero; it does not matter in the least what that wave function is, or how it changes in time. We can create entropy only by going from a pure quantum state to an *incoherent* mixture of quantum states, and this is precisely what the Hamiltonian equation of motion (2.4a) fails to permit.

Actually, we never start from the strict quantum-mechanical ground state of the system, since we are always at some non-zero temperature. But if the concept of a Bose-Einstein condensation is accepted, then one coherent wave function dominates the picture for all temperatures well below the transition temperature. The time-development of that coherent wave function is determined by some Hamiltonian, and that is just the trouble.

The difficulties connected with "Hotel Management in Ergodia" have been discussed by Trimmer (62), and (in more serious vein) by many others, e.g., Albertoni (60), E. Cohen (62); Dresden (62); de Groot (51, 59, 62, 63); van Hove (55, 57); Janner (62); Kubo (59, 63); Ludwig (63, 63a); Mazur (59, 60, 61); Meixner (59); Prigogine (60, 62); Prosperi (60); Verboven (63).

In the spin-echo experiment, the spin-echo cannot be observed after indefinitely long time intervals. The relaxation time is associated, not with the interactions within the system of nuclear spins, but rather with interactions betweeen the nuclear spins and the "lattice," i.e., the rest of the world. These latter interactions have an inherently random element, since we could predict them only by including the "outside" factors within the dynamical description of our system—i.e., we would have to let the whole Universe be our "system." With a truly stochastic element in the picture, the "neg-information" S, (10.1), can and does increase with time (Bergmann 55; Lebowitz 57, 57a, 62; Gross 56; Willis 62, 62a).

Returning to persistent ring currents, there is very little experimental evidence of a finite lifetime for such currents. Unsuccessful attempts to see a decay were made by Onnes (14), Grassman (36), Collins (57), Broom (61), Quinn (62), Mercereau (62), Hempstead (62), File (63). All but the last two used "soft" superconductors. A decay of the supercurrent has now been observed in "hard" superconductors (Kim 62). This decay phenomenon has been discussed theoretically by Anderson (62), and his treatment (see Chapter X, Section 2) makes it likely that the decay mechanism is characteristic of hard superconductors, and not related directly to the problems under discussion here.

De Feo (62) has discovered that a persistent ring current in a thin film decays if the material is subjected to alpha-particle bombardment. It is likely that this phenomenon can be explained in terms of a local heating of the film, which results in temporarily normal regions (Cabibbo 63).[18]

The further investigation of persistent currents is still a fascinating problem in both experimental and theoretical physics.

[18] Direct insertion of a normal metal layer (perpendicular to the current flow) into an otherwise superconducting ring naturally causes a time-decay (Fink 59); but the normal regions in the De Feo experiment do not extend all the way across a cross section of the ring.

X

Further Problems

1. Introduction

In the development of the theory of normal metals, the fundamental problem for many years was the justification of the concept of "free electrons," on which the whole Sommerfeld-Drude theory depended. On any classical picture, the lattice ions must scatter the conduction electrons, and the mean free path is of the order of one lattice distance, much too small to explain the observed high conductivity of metals. This fundamental problem was solved by Bloch's construction of electron wave functions in a periodic lattice; Bloch showed that a perfect lattice in which the topmost band is only partially occupied by electrons leads to infinite electrical conductivity.

Yet this work was only the start of the theory of normal metals, and much of that theory is concerned with the effects of deviations from lattice perfection. For example, the mechanical behavior of metals (hardness, plasticity, etc.) cannot be understood at all without reference to "dislocations" (Kittel 53, Read 53, Cottrell 53, Kröner 58, Eshelby 56, DeWit 60).

Similarly, the fundamental problem in the theory of superconducting metals has been to gain a qualitative understanding of the "coherent wave function" of London, and of the way in which that coherent wave function yields a Meissner effect and persistent ring currents. With the concept of a Schafroth condensation of electron pairs, we are well along the path toward a solution of that fundamental problem. This advance, however, only serves to throw into focus the much larger number of problems still requiring theoretical solution, or at least qualitative understanding. While the thermodynamic transition, the Meissner effect, and the persistent ring currents, are the "spectacular" problems in the theory of superconductivity, these properties are far from exhausting all the interesting (both theoretically and technologically) properties of superconducting substances.

In this last chapter, we shall give a survey of some of these other problems, and of the progress which has been made with them.

2. Hard Superconductors and the Intermediate State

So far, we have considered "ideal" superconductors, which show a perfectly sharp thermodynamic transition, a complete Meissner field expulsion, have a sharp critical field, etc. In practice, no such materials exist, but a number of substances (for example, tin, lead, and mercury) come rather close to this ideal behavior. These substances are often mechanically soft, and for that reason are referred to as "soft" superconductors.

There are two apparently distinct ways in which actual materials deviate from ideal behavior:

(1) If a specimen not of cylindrical shape (for example, a sphere) is exposed to an applied magnetic field of increasing strength, the magnetic field distribution at the surface of the specimen is non-uniform, i.e., the field at the surface is higher in some regions (the equator of the sphere) than in others (the poles). When the field reaches critical value over part, but not all, of the surface, a rather complicated process sets in, leading to the production of an "intermediate" state, in which there are alternating normal and superconducting domains. (Landau 37; London 36, 36a; Peierls 36; Gorter 34, 35).

(2) There exists a large group of materials, called "hard" superconductors, which fail to show "ideal" behavior even if the specimen is cylindrical with the applied magnetic field parallel to the cylinder axis (so that the field at the surface of the specimen is uniform). Even under those conditions, the magnetic field is not expelled uniformly; rather, the specimen retains "trapped flux" in narrow regions. Furthermore, superconducting (i.e., resistance-less) current transport continues until very much higher fields are applied, so that there are different critical fields for (partial) magnetic field expulsion and for supercurrent transport; often the latter field is more than 10 times the former. In the complete absence of an applied field, the thermodynamic transition in hard superconductors is spread out over a range of temperatures.

We shall discuss the present position regarding "hard" superconductors first, and then return to the intermediate state in soft superconductors. There has been considerable confusion in the literature between these two, rather different, deviations from ideal behavior.

Of course, the word "ideal" is equivocal here. From the strictly theoretical point of view, "ideal" is whatever theorists find easy to understand and calculate. From a practical point of view, however, the hard superconductors have exceedingly useful properties, which make them "ideal" for some very important applications. Their critical fields are much larger than those of soft superconductors, and so are their critical currents. Thus, hard superconductors can be used as wires in electromagnets to maintain fields of the order of 10^5 gauss, without any resistive loss (Autler 60; Bozorth 60a; Arp 61; Alekseevski 61; Betterton 61; Kropschot 61; Kunzler 61, 61a, 61b; LeBlanc 61; Aron 62; Devons 62; Furth 62; Saur 62, 62a; Swartz 62).

In principle, no power whatever is required to sustain a magnetic field; as Kolm[1] has put it so vividly: "sustaining magnetic fields is probably the only major type of operation that we perform with absolutely zero efficiency." Hard superconductors may make it necessary to put that statement into the past tense.

Quite apart from this practical interest, hard superconductors pose a number of interesting theoretical problems. As late as 1950, London (50) could say only: "Possibly dissolved gases or internal strains are the cause" of the altered behavior. We know a good deal more now, but the whole story is by no means clear yet.

A long time ago, Mendelssohn (35) suggested that hard superconductors are inhomogeneous internally, the material being permeated by an interconnected network of "superconducting filaments"; if the critical field of the filaments is larger than the critical field of the bulk material, then the filaments remain superconducting long after the external field has destroyed superconductivity in the remainder of the specimen. Since the filaments, by assumption, form an interconnected structure, they can carry a supercurrent; thus we have the phenomenon of supercurrent transport (vanishing electrical resistance) coupled with practically full penetration of the magnetic field into the specimen. The critical field for magnetic field expulsion is of the order of a few hundred to a few thousand gauss; when it is exceeded, the bulk of the material becomes normal, but the superconducting filaments are still capable of transporting current without Ohmic losses. The critical field for the onset of electrical

[1] H. H. Kolm, as quoted by Kunzler (61).

resistance is much higher; it corresponds to destruction of superconductivity in the "filaments."

The early experiments (Lasarew 44, Chotkevich 50, Grenier 55) were carried out on soft superconductors in polycrystalline form; Shaw and Mapother (Shaw 60) observed a small increase of critical field with increasing strain on a single crystalline sample of lead; they suggested that their results could be understood in terms of Mendelssohn's picture, with the "filaments" being the "dislocations" associated with the strain.[1a] Since no dislocation can terminate within the lattice, the dislocations must form a continuous network which can carry electrical current all the way across the specimen.

If this picture is right, a number of consequences follow from it: First, the critical *field* for onset of electrical resistance should be the critical field for destruction of superconductivity in the filaments, i.e., it should be independent of the number of filaments per unit cross-sectional area; the critical *current*, on the other hand, is clearly proportional to the number of filaments per cm^2, and this number can be altered by mechanical influences. By "working" the specimen, we increase the number of dislocations; by annealing the specimen at high temperatures, we decrease that number. These processes should therefore alter the critical current, but not the critical field. This is precisely what is observed (Cook 50, Berlincourt 59, Kunzler 61, Hauser 62, Hake 62a). Furthermore, if the current is carried throughout the volume of the specimen, by a network of filaments, the critical current should be proportional to the cross-sectional area; whereas in soft superconductors,

[1a] There is considerable supporting evidence for the importance of lattice dislocations. By direct observation of domain structure, Schawlow (59a) showed that the large-scale domains in hard superconductors are associated with local strains (i.e., with dislocations), rather than with intrinsic electronic properties of the material. Further evidence for large-scale structure (in Ta wires carrying supercurrents) was obtained by Baird (59).

Evidence of the influence of plastic deformation on superconducting bulk properties is the effect on the transition temperature (Minnigerode 59, Müller 62), and on the specific heat (Otter 62).

The superconducting properties of a Re single crystal are altered by deforming it mechanically (Hauser 62b). So are the properties of a superconducting film (Hauser 62c, 63).

Neutron irradiation produces lattice defects, and affects superconducting properties (Doulat 59, Chaudhuri 60).

The amount of trapped flux in a specimen subjected to an external field in excess of the critical field is affected by plastic deformation (Budnick 56, Lynton 58, Seraphim 59, Shaw 60, LeBlanc 61a). The detection of trapped flux is, however, not always free from difficulties (Jaggi 58, Serin 59).

For further evidence see Maxwell (63).

by an array of flux filaments threading the specimen. We emphasize that this mixed state of Abrikosov has no relation to the London "intermediate state" of superconductors with positive surface energy. The London intermediate state occurs only if the specimen is exposed to non-uniform magnetic fields at its surface. The Abrikosov mixed state is seen most clearly by considering a cylindrical specimen with applied magnetic field parallel to the cylinder axis.

The flux lines in Abrikosov's theory have an interesting physical interpretation in terms of the Landau-Ginzburg "wave function of the superelectron." This wave function must be single-valued, and it satisfies an equation closely analogous to the Schrödinger equation (see Chapter I, Section 5). Thus, excessively rapid variations of the phase of this wave function are excluded, since they would give rise to a very high kinetic energy. However, there is an exception to this rule. Suppose the wave function vanishes everywhere along a "nodal line" which goes through the entire specimen, from one surface to another; for the sake of illustration, we shall let the nodal line be along the z-axis, from $z = 0$ (the lower surface of the cylinder) to $z = L$ (the upper surface of the cylinder). A wave function proportional to $x + iy$ is then quite possible; it is single-valued, and although it has kinetic energy rather higher than a wave function without nodes, the additional kinetic energy is not excessive. The phase of such a wave function changes by 2π as we go once around the nodal line. Associated with this phase change is a circular current around the nodal line, and a magnetic flux parallel to the nodal line and concentrated in the neighborhood of that line. The precise distribution of the magnetic field about the nodal line, and the rate at which the absolute value of the wave function rises from zero (its value at the nodal line) to its value in the bulk medium (equal to the square root of the number density of superelectrons), adjust themselves so as to minimize the energy necessary to create such a line. Furthermore, different nodal lines tend to repel each other.

The lower critical field H_{cl} is determined by the condition that the magnetic energy gained by allowing one unit of flux within the superconducting material, just balances the energy required to create one such nodal line. The unit of flux in question is just the flux quantum of London, hc/e^*, for particles of charge e^*. In the light of present knowledge, e^*, the charge of the "superelectron," should be set equal to $2e$.

At the time the Landau-Ginzburg theory was developed, no clear microscopic meaning could be attached to the "wave function of the superelectron." However, by the use of Green's function methods and a series of sweeping approximations, Gor'kov (58, 59, 59a, 59b, 60) and

Abrikosov (58, 59, 59a, 60)[4a] have been able to *derive* the Landau-Ginzburg equations from the microscopic theory of superconductivity, in a certain limiting case.

In Chapter VIII, we remarked that the "energy gap parameter" Δ_k of the simple theory generalizes to a function of two variables, $\Delta(k, l)$, if there are external fields present. This function is defined by (VIII, 4.2). If we transform to x-space, we obtain a function $\Delta(\mathbf{r}, \mathbf{r}')$; since the electron-electron interaction appears explicitly in (VIII, 4.2), $\Delta(\mathbf{r}, \mathbf{r}')$ vanishes unless the points $\mathbf{r}$ and $\mathbf{r}'$ are within interaction range of each other. Thus $\Delta(\mathbf{r}, \mathbf{r}')$ is a much more "concentrated" function than $\psi(\mathbf{r}, \mathbf{r}')$, the Fourier transform of $\psi(k, k')$; the latter having a "range" corresponding to the Pippard coherence length.

If the range of the interaction is small compared to distances over which other quantities vary, the function $\Delta(\mathbf{r}, \mathbf{r}')$ may be replaced approximately by a function of one variable

$$\Delta(\mathbf{r}, \mathbf{r}') \cong \Delta(\mathbf{r})\, \delta(\mathbf{r} - \mathbf{r}') \tag{2.2}$$

Except for a numerical factor, this function $\Delta(\mathbf{r})$ turns out to be the Landau-Ginzburg "wave-function of the superelectron."

The Schafroth condensation of electron pairs shows itself in that all the "superelectrons" are described by one and the same wave function. The electron pair nature of the "superelectrons" is reflected directly in the superelectron charge $e^* = 2e$. It is remarkable that Ginzburg suggested a value of e^* in this neighborhood on purely phenomenological grounds already in 1955 (Ginzburg 55).

In addition to this value of the charge e^*, the microscopic derivation also leads to a specific prediction for the parameter κ, (2.1), in terms of microscopic parameters of the metal. κ turns out to be proportional to $(\lambda_L/\xi)^2$, where ξ is the coherence length of Pippard and λ_L is the London penetration depth. In soft superconductors, ξ is of the order of 10^{-4} cm, some 10 to 100 times larger than λ_L. Thus κ lies well below the Abrikosov value $2^{-1/2}$, and the Abrikosov flux lines are unstable. They become stable only in "dirty" superconductors, where the Pippard length ξ is limited by the average mean free path of an electron (Shapoval 61).

However, in these dirty, or "hard," superconductors, it is possible that the Abrikosov flux lines play an important role. If the high magnetic field regime of a hard superconductor is characterized by the presence of large numbers of these quantized flux lines, threading everywhere through the specimen, then we have here an alternative picture of the internal state of a hard superconductor (Gorter 62, 62a). It is important

[4a] For a more transparent derivation, see Rapoport (62).

to keep in mind the qualitative difference between this flux line picture, and the Mendelssohn "sponge" picture. In the flux line picture, the material remains in the superconducting state almost everywhere; the flux lines are associated with nodes of the wave function of the super-electrons. In the "sponge" picture of Mendelssohn, the material is mostly in the normal state, with superconducting "filaments" threading through the volume. In the flux line picture, dislocations are invoked to retard the motion of flux lines, i.e., to hinder the return to equilibrium. In the modern version of the "sponge" picture, the dislocation lines are themselves identified with the superconducting "filaments." In any one substance, these pictures are mutually exclusive. But it is possible that one picture may apply to certain hard superconductors, the other to others.

The experimental evidence in favor of the flux line picture is rather impressive.[4b] Particularly striking is the analysis of the specific heat of the hard superconductor V_3Ga when exposed to an enormous magnetic field (Morin 62, Goodman 62a); if the "sponge" picture is valid, most of the material should be in the normal state; the measurement implies, however, that some 88% of the material is in the superconducting state.[4c]

A second line of evidence comes from the study of magnetization curves and magnetic hysteresis effects (we recall that hysteresis is troublesome to explain on the "sponge" picture). In a number of materials, the magnetization curves agree nicely with the Abrikosov theory (Schubnikov 36; Berlincourt 61, 62; Jurisson 62; Kamper 62; Kinsel 62; Swartz 62a; Goedemoed 63).

Of particular interest is the fact that not all these materials are "dirty" superconductors. Reproducible results in agreement with Abrikosov have been obtained in single crystals of Mo_3Re (Blaugher 62a) and even in a pure element, niobium (Autler 62, Stromberg 62).

The thermal conductivity of an In-Bi alloy in a magnetic field shows clear evidence for the upper and lower critical fields of Abrikosov (Dubeck 63).

For further evidence in favor of the Abrikosov picture, see Berlincourt (63), Bon Mardion (62), Chandrasekhar (63), Chiou (63), Connell (63). Glover (63), Kinsel (63), and Livingston (63).

In Abrikosov's treatment, the material is considered homogeneous,

[4b] This is quite apart from the copious experimental evidence in favor of the Landau-Ginzburg theory in *soft* superconductors; see footnote 9c, p. 367.

[4c] This favors some sort of flux line picture, though not necessarily the detailed picture of Abrikosov. Alternative theories have been advanced (Goodman 61, Parmenter 62), but a more detailed discussion of the evidence (Goodman 62) shows better agreement with the Abrikosov theory.

with no hindrance to the motion of the flux lines. As Abrikosov himself observes, however, the flux lines are not actually free to move, since lattice imperfections of various types tend to "pin down" the flux lines. Even in carefully annealed specimens (Budnick 56, Lynton 58, Calverley 60) these barriers to flux line motion must be taken into account (Fleischer 62).

The most interesting phenomenon which has been interpreted in terms of hindered motion of these flux lines is the decay of persistent currents (Kim 62, 63, 63a; Hempstead 63). They work with a specimen in the shape of a hollow cylinder, with unequal magnetic fields inside and outside. The difference in magnetic field is associated with a persistent ring current in the cylinder. If there are flux lines inside the cylinder and if these flux lines "creep" from one surface of the specimen to the other, they tend to equalize the magnetic fields inside and outside, and therefore result in a time-decay of the persistent current. The observed time-decay[4d] is in agreement with a rough theory by Anderson (62), which includes order-of-magnitude estimates of the relevant constants, as well as a prediction for the functional form of the time-dependence of the persistent current. In Anderson's theory, the flux line creep is in a direction determined by the systematic Lorentz force $\mathbf{J} \times \mathbf{B}$, and is hindered by thermal barriers associated with lattice imperfections.

The observed time-decay of the ring current is incredibly slow: the difference between the inside and outside magnetic field decreases (after initial transient behavior) proportionately to $\ln(t)$; the highest observed rate of decrease is $dH/d(\ln t) = 10$ gauss per decade, from an initial field difference of 1000 gauss. If this rate of decay were to continue indefinitely, Kim *et al.* estimate that the persistent current in their sample would die out after 3×10^{99} years!

It is certainly interesting that a finite rate of decay of a supercurrent has been observed experimentally. From the fundamental point of view, however, this process of supercurrent decay is probably unrelated to the considerations of Section 10 of the preceding chapter.

Having discussed our present knowledge (or ignorance) of the physics of hard superconductors, let us turn our attention to the "intermediate state" of soft superconductors. That is, we now assume that the material is of sufficient purity, homogeneity, and sufficiently free of strain, to show close to "ideal" behavior in weak applied fields; and we enquire

---

[4d] Note, however, that this phenomenon does not occur in *all* hard superconductors (File 63).

Motion of flux has also been observed under irradiation by photons (Marchand 62). The explanation of this effect is not certain.

what occurs when a specimen of arbitrary shape is subjected to ever-increasing fields.

When the field reaches critical value on part, but not all, of the surface, that part turns normal. This cannot happen, however, in the simple way one would imagine first, namely, that an outer shell near the surface becomes fully normal, and the center remains fully superconducting. One should then have $H = H_c$ on the boundary, $H > H_c$ outside the boundary, and $H < H_c$ in the inner, superconducting region. This does not permit any curvature of the field lines toward the center of the sphere and is inconsistent with the magnetic field equations.

Instead, the whole specimen must split into alternate normal and superconducting "domains"; the normal domains carry a flux density H_c, so that there is equilibrium between neighboring domains. This domain structure was predicted theoretically by Landau (37, 43) and has been confirmed experimentally by Shalnikov (45) and Meshkovsky (47), in tin.

The nature of the domains depends sensitively on the "surface energy" at the interface between normal and superconducting regions. If there is no surface energy, the "normal" domains must branch out into finer parts near the surface of the specimen, so that, right at the surface, there would be no observable structure. Actually, domain structure right at the surface is shown beautifully by a number of experiments (Meshkovsky 47a; Schawlow 56, 58; Alers 57, 59; Balashova 57; DeSorbo 60, 60a, 62; Faber 58), and this possibility was anticipated theoretically by Landau (38, 43) as a consequence of a positive surface energy. Although various theories of the domain structure have been developed (Andrew 48; Landau 37, 47b; Lifshitz 51), all "non-branching" theories give the same order of magnitude for the linear size of the domains, in terms of the surface energy and the London depth: Let σ be the surface energy, and define a length L by the relation

$$\sigma = LH_c^2/8\pi \tag{2.3}$$

where H_c is the bulk critical field; then the size of the domains is of the order of the geometric mean of this length L and the London penetration depth. Comparison with experiments on soft superconductors leads to values of L of the order of 10^{-5} to 10^{-4} cm, which is comparable to the Pippard coherence length (Makei 58; Sharvin 59, 60; Davies 60; Batrakov 62).

In Section 5 of Chapter I, we mentioned that any *linear* relationship between field strength and magnetization leads to a negative sign for this surface energy. On the other hand, the non-linear Landau-Ginzburg theory, discussed there and also earlier in this section, gives a surface energy which can be either positive or negative, depending upon whether κ, Eq. (2.1), is less than or greater than $2^{-1/2}$. From a purely phenomenological point of view, the Landau-Ginzburg theory must treat κ as an adjustable parameter, which can be determined from the experimental data on the penetration depth and the critical field, from (2.1). However, the microscopic derivation of that theory relates κ to the ratio of the London penetration depth to the Pippard coherence length, and it becomes clear immediately that κ must be small in pure materials (soft superconductors), where the Pippard length greatly exceeds the London depth. The resulting, positive surface energy is in full agreement with experimental observations on the domain structure of the intermediate state in soft superconductors.

The non-linearities characteristic of the intermediate state can and do lead to quite complicated phenomena.

Although superconductors in the ideal, London regime are perfect diamagnets, they can assume paramagnetic behavior in the intermediate state (Steiner 37, 49; W. Meissner 51, 51a, 52, 53, 55; Mendelssohn 52; Teasdale 53; J. C. Thompson 54; H. Meissner 55, 56, 56a, 58a, 58b, 58c; Shibuya 55, 58, 58a).

Besides this absolute paramagnetic effect, there is also a "differential paramagnetic effect" in the intermediate state (Daunt 49; Smith 52; Steele 52, 53; Hein 57, 61).

The nature of the intermediate state in the presence of a current larger than critical can be exceedingly complex (Ballmoos 60); and the critical current itself can depend upon the direction of the applied field (Daunt 62).

The thermal conductivity of materials in the intermediate state shows anomalous behavior (Mendelssohn 50, 50a, 60a; Detwiler 52; Olsen 52; Webber 53, Renton 55; Zavaritskii 60). Both the phonon contribution (Cornish 53, Laredo 55a) and the electron contribution (Hulm 53, Strässler 63) to the thermal conductivity are altered by the presence of the domain structure.

3. Superconducting Alloys and Compounds

As Wigner[5] once said, there are two sorts of discrepancy between theory and experiment: (1) theory may predict something which is not

[5] E. P. Wigner, private communication, quoted from memory.

observed experimentally, and (2) Experimentalists may see regularities for which there is no theory. The first type is relatively harmless (contrary to what some philosophers of science maintain); one can usually find half a dozen reasons why the theory is not precisely applicable to that particular experiment, and one can point to a dozen effects tending in the right direction to decrease the discrepancy. It is the *second* type of discrepancy which is really serious! If experimentalists find simple regularities for which there is no theoretical explanation whatever, then the theory may well be wrong from the ground up.

Bernd Matthias, for years now, has been discovering just such regularities in the field of superconductivity. Since this is a book on the *theory* of superconductivity, and no theory exists for these effects, all we can do is to summarize the experimental situation, and avert our heads in shame. This we shall now do.

The early investigations of Matthias are summarized in a review article (Matthias 57), entitled "Superconductivity in the periodic system." When Matthias started his work, some 40 superconducting. substances were known, most of them being chemical elements. By studying regularities of the pattern of superconducting transition temperatures among the superconductors, Matthias developed empirical rules which enabled him, in a few years, to construct over 800 new superconductors containing some remarkable substances indeed. Most of these substances are alloys or intermetallic compounds, rather than pure elements.

Matthias produced such curiosities as superconducting alloys composed of non-superconducting pure substances, alloys with much higher transition temperatures than any pure substance, alloys in which a small percentage of one substance drastically lowered the superconducting transition temperature, and other alloys in which the same admixture drastically raised the transition temperature. In the hands of Matthias, superconductivity has become a black art.[6]

For a large number, but not for all, superconducting substances, the transition temperature T_c seems to be a function of:

(1) the average volume per atom, V,

(2) the mass M of the average atom, and

(3) the average number n of valence electrons per atom.

[6] Just as in other black arts, however, the achievement most desired by the wizard has been denied to him. According to the empirical rules of Matthias, there were three compounds with predicted transition temperatures well above 20°K (and hence economically useful for laying telephone lines): all three proved chemically unstable!

The dependence is given by

$$T_c = c \frac{V^a}{M} F(n) \tag{3.1}$$

where c is a constant, the exponent a varies between 4 and 5, and $F(n)$ is a purely empirical function reproduced in Fig. 10.1. For transition

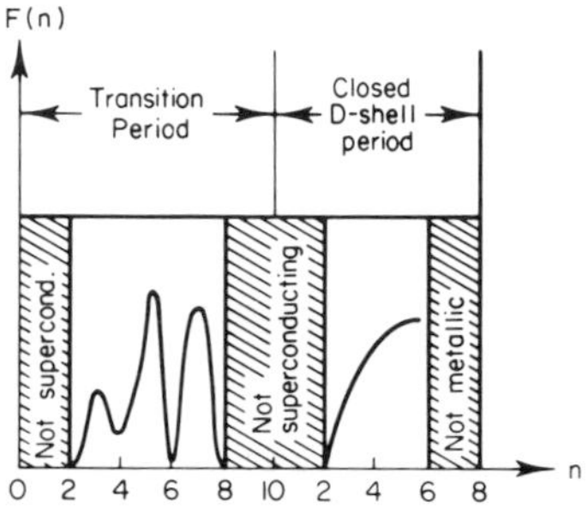

FIG. 10.1. The function $F(n)$ relating the transition temperature of an alloy to the average number n of valence electrons per atom, according to Eq. (3.1). This relationship was suggested by Matthias (57), and the figure is reproduced from that article.

metals, this function shows sharp peaks at 3,5, and 7 valence electrons per atom; it shows minima at 4 and 6 valence electrons per atom, the latter minimum amounting to no observable superconductivity at all; and there is no superconductivity for $n < 2$ or $n > 8$.[7] The transition and non-transition metals behave quite differently, $F(n)$ being smoothly increasing for the latter.

The striking fact is that the "average" V, M, and n may be taken, for alloys and compounds, as simple linear averages of the constituent substances! Thus, Rh with 9 valence electrons per atom is not superconducting (in agreement with Fig. 10.1); neither is sulfur (which is not metallic); but the compound Rh_9S_8, with "average" $n = 7.6$, is a superconductor! So are the compounds $RhSe_{1.75}$, $RhTe_2$, Rh_5P_4, and Rh_5Ge_3.

[7] Since then, iridium, with $n = 9$, has been shown to become superconducting at $T_c = 0.14$°K, by Matthias and collaborators (Hein 62).

By now, the literature on superconducting alloys is enormous, e.g., Banus 62, Berghout 62; Blaugher 61; Bozorth 60; Bucher 61, 61a; Chandrasekhar 58; Compton 61; Corenzwit 59; Darby 62; Gayley 62; Geballe 63; Gendron 62; Giorgi 62; Guggenheim 61; Hake 61; Jansen 61; Kneip 62; Lange 61, 62; Lautz 61, 62; Matthias 58, 58a, 58b, 58c, 58d, 59, 59a, 60, 60a, 61, 61a, 61b, 62, 62a, 63, 63a; Raetz 62; Reed 62; Rose 62; Saur 62; Schröder 57; Wernick 61; Wyman 62.

A complete and systematic review of the field has been given by Matthias (63).

Whereas the simple dependence on atomic volume V and atomic mass M in (3.1) may be within the grasp of theory, the complicated function $F(n)$, with its sharp maxima and minima, is unexplained.[8]

The "averaging" mentioned above is not unrestricted, however. One may "average" between transition metals (Blaugher 61, 62), with considerable success; one may average between non-transition metals (Reeber 60, Coles 62), again with considerable success. But as soon as one tries to apply the rule (3.1) to alloys (or compounds) of transition and non-transition metals, the rule breaks down completely. Indeed, in later work Matthias (62, 63) has presented impressive evidence that the *mechanism* of superconductivity is different in the two regions of the periodic table. The most striking evidence is provided by the isotope effect, which played such a tremendous role in the development of the theory of superconductivity. The isotope effect was observed originally in non-transition elements such as Hg, Sn, Pb, and has since been confirmed in other non-transition elements such as Zn (Geballe 62). But no isotope effect of comparable magnitude (T_c proportional to $M^{-1/2}$) is observed in transition element superconductors such as Ru, Os (Geballe 61, 62; Finnemore 62; Bucher 61c). Some isotope effect is seen in Mo, however. Matthias (63b) interprets this in terms of a superposition of the Fröhlich and the alternative interaction. A similar explanation may hold for the residual isotope effect (T_c proportional to $M^{-0.21}$) in Os (Hein 63a).

Since the isotope effect seems to be a natural consequence of the Fröhlich electron-phonon coupling,[9] this coupling may not be the explanation for superconductivity in transition elements. Further evidence for a non-phonon mechanism is given by Andres (62). Perhaps an exchange interaction between conduction electrons and inner-shell electrons may be the explanation (Herring 58).

From the theoretical point of view, the precise nature of the electron-electron interaction responsible for the electron pairing is not really

[8] Pines (58) has attempted to explain the Matthias rules by a crude application of the BCS criterion for superconductivity; he was not successful. Vonsovskii (58a) fared only slightly better.

The complicated, narrow band structure of some of these alloys (Clogston 61, Blumberg 60) is not encouraging for the theorists.

[9] Swihart (59, 62) has pointed out that the electron-electron Coulomb interaction modifies the exponent in the isotope effect, and the modified exponent has been estimated in some detail by Morel and Anderson (Morel 62). However, this modification gives a wide range of exponent values; the experiments seem to lead to exponents which are either close to $\frac{1}{2}$ or close to 0, again with the exception of Mo and Os (Hein 63a, Matthias 63b).

Attempts to understand the parameters of transition element superconductors on the BCS model are not encouraging; Morin (63) find strongly varying values of the cutoff energy $\hbar\omega_c$.

important. The essential aspect seems to be the existence of a force sufficiently attractive to produce an attractively correlated pair, without leading to complete condensation into an "electron liquid." It may be significant that both the Fröhlich interaction and the postulated exchange interaction in transition elements are indirect, two-step interactions which proceed via some intermediate state.

Of just as much interest is the relation found by Matthias and coworkers (Matthias 58, 58a, 59, 60, 61, 62, 62a; Hein 59; Suhl 58, 59, 62). between superconductivity and ferromagnetism. Not only do magnetic ions, even in small concentrations, have marked effects on the transition temperature, but there are alloy systems in which a continuous change of concentration of one constituent leads from superconducting to ferromagnetic alloys, and others in which one and the same alloy appears to be both superconducting and ferromagnetic! Matthias (63a) summarizes some very strong evidence for a sympathetic relationship between superconductivity and ferromagnetism: a situation favorable to ferromagnetism seems to be favorable to superconductivity as well.

The theoretical situation is highly unclear; the exchange interaction with inner electrons, invoked to explain superconductors without isotope effect, may also play a part in this relationship with ferromagnetism, but the theorists are far from unanimous (Herring 58; Kasuya 58; Vonsovskii 58, 60, 61; Zharkov 58, 60; Anderson 59b; Maradudin 59; Nakamura 59; Suhl 59a, 62a; Garland 62; Peretti 62; Kondo 63).

A theoretical suggestion of great interest has been made by M. J. Buckingham. A Bose-Einstein gas of spinless bosons is a superconductor below the transition point, as we have seen. If the bosons have non-zero spin, however, the condensed Bose gas is a ferromagnet![9a] The single-particle ground state is degenerate in the absence of an applied field; but even the slightest field gives a Zeeman splitting of the ground state levels, and all the particles go into the lowest state of the Zeeman multiplet; a reasonable spin-spin coupling between the bosons produces enough self-consistent field to make the phenomenon spontaneous ferromagnetism.

If there is enough spin-orbit coupling to make the electron pairs, under certain conditions, prefer the spin 1 pair state to the spin 0 pair state, a smooth change-over from superconductivity to ferromagnetism could be understood theoretically. Whether these conjectures have anything to do with the phenomena observed by Matthias is not known.

For alloys consisting of a main substance with only very small admixtures of impurities, Anderson (59) has developed a "theory of dirty superconductors." This theory predicts that the scattering of electrons,

[9a] This point was overlooked by Fisher (60).

due to the impurities, should smooth out any dependence of the energy gap paramater Δ upon crystal directions, and should lead to an initial lowering of the transition temperature. Many experiments (Serin 57; Chanin 59; Muller 59, 60; Miller 60; Richards 61; Quinn 61; Seraphim 61a, 61b; Adkins 62; Masuda 62) support this theory.

Further theoretical work on the effect of admixtures of impurities has been done by Brout (59), Maradudin (59), Nakamura (59, 59a), Suhl (59b), Jones (60a), Kothari (62), Tsuneto (62), and Kondo (63).

Another effect of alloying, of great significance theoretically if it can be confirmed, has been announced by Reif (62): the disappearance of the energy gap when a sufficient percentage of paramagnetic impurities are added. The experiment is performed on "quenched" alloy films formed by evaporation; the basic substance being indium, with up to 1% atomic concentration of iron impurities. The film evaporation technique (Schwidtal 60, Opitz 55) allows great flexibility in the choice of alloy components. The transition temperature T_c (defined as the temperature at which electrical resistance appears) is a decreasing, linear function of impurity concentration; it decreases from 4.2°K for a pure In film, to 2.2°K for a film containing 1% iron. The energy gap is measured in this experiment by the tunneling technique (see Chapter VI, Section 7); initially, Δ decreases sharply with concentration of iron; for concentrations between 0.2 and 0.8%, the energy gap seems to become unsharp, and its mean value keeps decreasing; finally, at concentrations above 0.8%, there is no evidence for any energy gap from these tunneling studies, in spite of the fact that the substance is still superconducting. The same phenomenon is observed with Gd impurity in Pg films, but not with Zn or Tl impurities (which are nonmagnetic ions).

In view of the great importance attributed to the energy gap by some theorists, further study of this phenomenon seems highly desirable. From the point of view of the quasi-chemical equilibrium theory, the absence of an energy gap would be quite unimportant, since the energy gap is considered to be a property of the "normal fluid," and not causally connected with true "superfluid" properties. The present theoretical situation for these alloy films with paramagnetic ions is unsatisfactory; one set of theories (Suhl 59, Baltensperger 59) disagrees with Reif's measurements as well as with other experiments (Müller 60, Parks 62); a calculation by Abrikosov and Gor'kov (Abrikosov 60), based on Green's function techniques (Abrikosov 58, 59, 59a; Gor'kov 58, 59, 59a, 59b, 60; Migdal 58), did predict the disappearance of the energy gap at sufficiently high concentrations of paramagnetic impurities; however, their minimum concentration for this effect is considerably larger than the one observed by Reif, and the physical meaning of the phenomenon is not too clear from the calculation. Later theories seem inconclusive (Phillips 63, Suhl 63).

While more work, both experimental and theoretical, is clearly needed here, it would be most interesting to find an actual experimental substance (in addition to the purely theoretical ideal Bose gas), in which true superconductivity (Meissner effect and persistent currents) exists without any energy gap.

4. Superconducting Thin Films

Since the Pippard coherence length ξ, which measures the "internal size" of the electron pairs in the superconducting state, is of the order of 10^{-4} cm in pure crystals, there is considerable theoretical interest in experiments with films of much smaller thickness. Experiments with films of thicknesses down to about 10^{-6} cm are entirely feasible. Contrary to initial expectations, these films behave rather similarly to the bulk material. The transition temperature alters slightly, but very little; there is a clear Meissner effect, and persistent currents can be set up and maintained indefinitely in ring-shaped thin films.

In fact, superconductivity comes easier to films than to bulk material! The transition temperature of Ga in film form greatly exceeds the bulk value of T_c (Buckel 54, 58); and several materials are superconducting in film form only: Beryllium (Lazarev 58, 60), Bismuth (Hilsch 51, Zavaritskii 52), and, most surprisingly, iron (Mikhailov 60).

If one applies the analogy to an ideal Bose-Einstein gas too directly, this behavior of thin films is hard to understand. One might, after all, be tempted to consider a film of thickness $a \ll \xi$ as equivalent to a two-dimensional Bose gas, and we have seen in Chapter II that the two-dimensional Bose gas does not condense at all. This objection was raised against the Bose-Einstein condensation picture in the early days.

There are two reasons why this objection is invalid: (1) the two-dimensional ideal Bose gas, even though it does not condense in the thermodynamic sense, behaves "almost" like a superconductor as far as Meissner effect and critical field are concerned (May 59b); and (2) the analogy to a two-dimensional Bose gas is misleading anyway. We shall discuss these two points separately.

May (59b) has repeated the Schafroth (55a) calculation for a two-dimensional charged Bose gas. The specific heat curve shows no evidence of any transition; in fact $c_v(T)$ is a perfectly smooth function of temperature, going to zero linearly with T at low temperatures, and approaching the equipartition value $2(\frac{1}{2}N\mathfrak{k})$ at high temperatures. However, the magnetic susceptibility kernel $K(q)$ of Eq. (I, 3.8) shows decided evidence of a "quasi-transition"; that is, there is a very narrow region of temperatures ΔT, around a "quasi-transition temperature" T_q, such that $K(q)$ is "normal" for $T > T_q + \Delta T$, and $K(q)$ leads to nearly complete magnetic field expulsion for $T < T_q - \Delta T$. The magnetic field penetrating the specimen is so small that it would escape detection, and thus the "quasi-Meissner effect" would be indistinguishable from a true Meissner effect. Furthermore, there is also a "quasi-critical field" H_c, such that $B \cong 0$ for $H < H_c$, and $B \cong H - H_c$ for $H > H_c$, just as in the three-dimensional ideal Bose gas (Chapter II).

The fact that these striking magnetic properties are not reflected at all in the specific heat curve shows once more that one cannot reason from purely thermal properties (such as an energy gap with its resulting exponential specific heat curve) directly to electromagnetic properties. May did not consider persistent currents in two-dimensional rings, and the extension of the three-dimensional calculation (Blatt 61, Bloch 62) to the two-dimensional model would be a highly interesting theoretical exercise.

It would be no more than that, however, because the analogy between thin films and a two-dimensional Bose gas with the conventional energy spectrum is highly suspect anyway. In Chapter II, the condition for Bose-Einstein condensation was stated in the form (II, 2.18):

$$\lim_{\varepsilon \to \varepsilon_0} \rho(\epsilon) = 0 \qquad \text{(condition for condensation)} \qquad (4.1)$$

where $\rho(\epsilon)$ is the density of single-particle levels. For the two-dimensional Bose gas in an area L^2 we take $\epsilon_k = \hbar^2 k^2/2M$ and find $\rho(\epsilon) = L^2 M/(2\pi\hbar^2) =$ constant, which violates condition (4.1). However, one does not need anything as strong as a true energy gap to restore (4.1). If the self-consistent calculation for electron pairs leads to any dependence on k weaker than proportionality to k^2, (4.1) is again satisfied. Since the usual three-dimensional case appears to lead to a true energy gap, it is to be expected that the density of pair levels for a thin film is low enough to satisfy (4.1), and thus the pair gas has a true condensation.

The condensation phenomenon for the pair gas in a thin film can be studied by explicit calculation (Blatt 63), provided one restricts oneself to the simplified interaction (IV, 2.5). The basic set of functions in which the pair wave function is to be expanded can no longer be taken as simple plane waves. Ideally, one should determine the set by a variation principle. In practice, it is highly likely that the set of functions suggested by Anderson (59) in his theory of dirty superconductors is applicable to the thin film situation. Anderson suggests using pairs of wave functions which are time-inverses of each other, as the basic functions in the Zumino decomposition of the pair wave function. Letting the slab extend from $x = 0$ to $x = a$, and using periodic boundary conditions in the y- and z-directions, with periodicity constant L, the basic functions have the form

$$\phi_k(x, y, z, \zeta) = u_n(x) \frac{\exp(ik_2 y + ik_3 z)}{L} \alpha(\zeta) \tag{4.2a}$$

and

$$\phi_{-k}(x, y, z, \zeta) = u_n(x) \frac{\exp(-ik_2 y - ik_3 z)}{L} \beta(\zeta) \tag{4.2b}$$

where ζ is the spin coordinate, $\alpha(\zeta)$ denotes a spin function for spin up, $\beta(\zeta)$ a spin function for spin down, k_2 and k_3 are integral multiples of $2\pi/L$, and $u_n(x)$ is a real function which obeys the Schrödinger equation for motion of a particle perpendicular to the slab faces:

$$-\frac{\hbar^2}{2M}\frac{d^2u_n}{dx^2} + V(x)\, u_n(x) = \eta_n u_n(x) \tag{4.3}$$

The formal index k in $\phi_k(x)$ represents the actual indices n, k_2, k_3, and the single-particle energy ϵ_k is related to the energy η_n of the Schrödinger equation (4.3) by

$$\epsilon_k = \eta_n + (\hbar^2/2M)(k_2^2 + k_3^2) \tag{4.4}$$

The potential energy $V(x)$ in (4.3) will be determined later; meanwhile, let us note that the simplified interaction (IV, 2.5) gives rise to the following reduced matrix elements in the base set (4.2):

$$\begin{aligned} \langle k, -k \mid v \mid k', -k' \rangle &= \gamma^2 \int_0^a [u_n(x)\, u_{n'}(x)]^2\, dx \qquad &&\text{for} \quad | \epsilon_k - \mu | < \hbar\omega_c \\ &= 0 &&\text{for} \quad | \epsilon_k - \mu | > \hbar\omega_c \end{aligned} \tag{4.5}$$

The quantity L_k, (IV, 1.15), vanishes identically, and the energy gap function Δ_k assumes the highly "anisotropic" form

$$\begin{aligned} \Delta_k &= \Delta_n \quad \text{(independent of } k_2, k_3) \qquad &&\text{for} \quad | \epsilon_k - \mu | < \hbar\omega_c \\ &= 0 &&\text{for} \quad | \epsilon_k - \mu | > \hbar\omega_c \end{aligned} \tag{4.6}$$

With the forms (4.5) and (4.6), direct numerical solution of the integral equation (IV, 1.26) is possible, since only a finite number of n-values contribute in a finite width slab. The number in question is determined by the condition $\eta_n - \mu < \hbar\omega_c$ where η_n are the eigenvalues in (4.3); if this condition is violated, $\epsilon_k - \mu > \hbar\omega_c$ for all possible values of k_2 and k_3, leading to a vanishing interaction matrix element (4.5).

The number density of electrons is not uniform across the slab; thus, even if we assume a background uniform positive charge density (the jellium model), there exists a net charge density $\rho(x)$ which changes sign in such a way that the total charge is zero. This charge density gives rise to an electrostatic potential $\Phi(x)$; the potential energy $V(x)$ in (4.3) is the energy of an electron in that electrostatic potential. The resulting set of equations must be solved self-consistently, and this can be done only on an electronic computer.

However, the calculation shows that the electrostatic potential is nearly constant throughout the interior of the slab; for reasonable values of the parameters, the "boundary region" near the edges of the slab is only some 2 Angstroms thick. It is therefore a reasonable first approximation to set $V(x)$ in (4.3) equal to a constant, V_0. The solution of (4.3) is then an ordinary sine wave, and the entire calculation can be done analytically (Thompson 63a), with results in excellent agreement with the computer calculation.

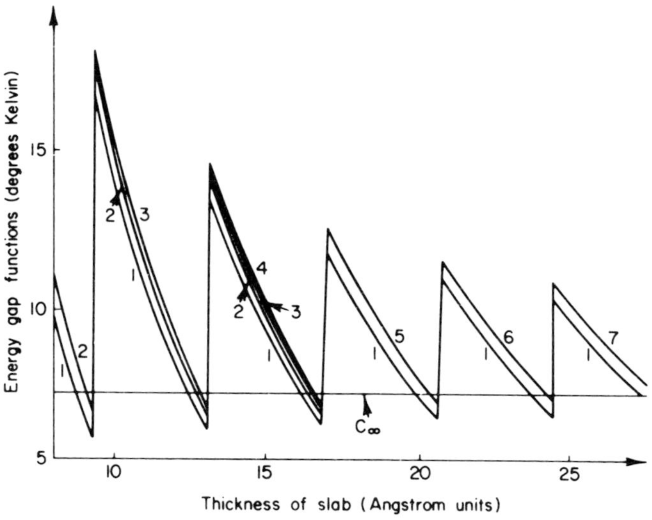

FIG. 10.2. The energy gap parameters Δ_n in a thin film, as functions of the film thickness a. The "shape resonances" occur whenever a new value of n starts to contribute. The distance between these resonances equals one-half of a de Broglie wavelength of an electron right at the Fermi surface. c_∞ is the bulk value of the gap.

The most interesting feature of the results is the existence of very sharp and pronounced "shape resonances" in the energy gap parameters Δ_n, Eq. (4.6), as functions of the width a of the slab. As the width of the slab increases, successively larger values of n begin to give non-zero Δ_n. Whenever a new value of n makes its first appearance, all the Δ_n increase rapidly and significantly above the bulk value of Δ, Eq. (IV, 2.9) Thereafter a smooth decrease takes place, until a width is reached at which the next value of n begins to contribute. These shape resonances [the name is taken from the cloudy crystal ball model of nuclear reactions, see Feshbach (54)] are shown for one case in Fig. 10.2.

The precise positions of the resonances depend on the boundary conditions imposed on the functions $u_n(x)$ at the slab faces; but the distance between resonances is independent of boundary conditions, and is equal to one-half of a de Broglie wavelength of an electron right at the Fermi surface. These resonances arise from the component single-particle wave functions out of which the electron pair wave function is constructed. The resonances are therefore characteristic of the electron pair theory, and cannot arise in the Landau-Ginzburg approximation, in which the pair wave function is replaced by the wave function of a single "superelectron."

The bulk value of Δ is approached for thick slabs through the resonances becoming progressively lower in height as the slab thickness increases; the height of the resonances is roughly inversely proportional to the slab thickness at which they occur.

By solving the same set of equations with zero reduced interaction [i.e., $\gamma^2 = 0$ in (4.5)], one obtains the corresponding Hartree-Fock solution without electron pairing. As expected, the pairing energy makes the superconducting state lower than the normal state, thereby showing explicitly that the finite thickness of the slab does not inhibit the Schafroth condensation. On the contrary, the pairing energy in a slab is, on the average, higher than in the bulk medium; that is, the resonances of Fig. 10.2 lead to gap values which are almost always higher than the bulk value of the gap.

Another property of thin films which can be understood theoretically is their *much higher critical field*, compared to the bulk critical field H_c. In Chapter I, Section 4, we discussed the thermodynamic relation between the Meissner field expulsion and the critical field [see Eqs. (I, 4.3), (I, 4.4), (I, 4.5)]. In that discussion, we noted that the need to expel the magnetic field gives a term in the free energy $G(H, T)$ which favors the normal state compared to the superconducting state. For sufficiently high fields, the normal state has the lower free energy, and is thus stable thermodynamically.

In thin films, the Meissner field expulsion is far from complete. For film thicknesses small compared to the London penetration depth, the magnetic field penetrates almost completely, even in the superconducting state of the film. As a result, it takes much higher applied fields before the free energy associated with the field expulsion is large enough to balance out the basic free energy difference between the normal and superconducting phases.

This effect was first noticed by H. London (35), and is discussed in F. London's book (London 50).[9b] More recently, Toxen (62) and Hauser (62a) have recalculated the critical field, allowing in an approximate way for the different field penetration law associated with the finite pair size (the Pippard correction); this is important since the films are thin compared to the Pippard length. These calculations are in reasonable agreement with some experimental data (Douglass 62a).

The calculations of Toxen (62) and Hauser (62a) involve simplifications of the Landau-Ginzburg non-linear theory; of much more interest is a direct application of that theory, which turns out to lead to amazingly accurate predictions concerning the transition of thin films.[9c] Since the Landau-Ginzburg theory is non-linear, it is not possible to reason from the Meissner field expulsion in weak fields, directly to the thermodynamics of the transition at the critical field. Rather, the non-linear equations must be solved numerically, for different film thicknesses, and different external conditions.

The most interesting consequence of these solutions is that the transition becomes *second-order* (no latent heat) for sufficiently thin films, namely films of thickness $d < 5^{1/2}\lambda_L$. For these films, the Landau-

---

[9b] See also Lininger (63).

[9c] One direct verification is through measurement of the magnetic field penetration through thin films, including the dependence of the penetration depth on the temperature and on the applied magnetic field (Prozorova 58; Khukhareva 58, 61; Schawlow 58; Erlbach 60; Jaggi 60).

The magnetic moment of a tin film, measured by Sevastyanov (61, 62), agrees well with the theoretical calculation of Zharkov (62).

Further confirmation comes from measurements of the critical field as a function of film thickness (Behrndt 60, Toxen 61), and of the critical current (Alekseevski 57, 60; Bremer 58, 59; Rhoderick 62, 62a).

Particularly close agreement with the Ginzburg-Landau theory is obtained from a study of hysteresis in the current transition (Baldwin 63). For reviews of the application of the Ginzburg-Landau theory to these phenomena, see Marcus (62), Douglass (62b), Bardeen (62b), Tong (63).

The preparation of "clean" films is quite an art (Marchand 59; Brandt 60, 61; Kahan 60; Frerichs 62; Lucas 62; Schwidtal 62). Residual gases (Caswell 61) and surface adsorbed gases (Ruhl 59) cause trouble. So do mechanical strains (Toxen 61a; Baumann 56; Blumberg 62, 62a), and perhaps surface electrical charges (Glover 60; Bonfiglioni 62).

Ginzburg "wave function of the superelectrons" approaches zero as the applied field approaches the critical field of the film; since this wave function is proportional to the energy gap parameter Δ, it also determines the thermodynamics of the transition (Douglass 61).[9d] This prediction has been verified beautifully by tunneling experiments which measure the energy gap directly (Douglass 62b). We reproduce Douglass' results on Al films in Fig. 10.3. Further corroboration is obtained by measurements of the thermal conductivity of thin films (Morris 61, Tinkham 62).

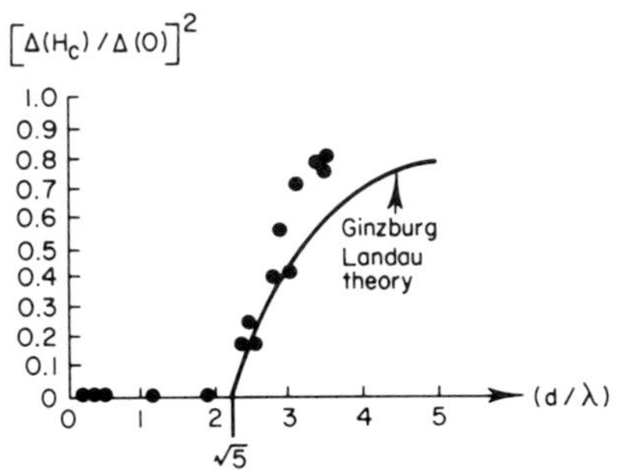

FIG. 10.3. Energy gap in aluminum films, as function of the film thickness, at the critical field. The figure is taken from Douglass (62b), who determined the energy gap values by electron tunneling measurements. As predicted by the Landau-Ginzburg theory, the energy gap goes to zero as the critical field is approached, in sufficiently thin films. Note that the film thicknesses used here are much larger than the ones of Fig. 10.2; thus no resonances are visible.

This behavior of a thin film subjected to a magnetic field depends markedly upon the direction of that field; so far, we have been discussing the case of a field parallel to the surface of the film. For *fields normal to the film*, the flux line picture of Abrikosov is likely to be of great importance (Tinkham 63). The field penetrates the film by way of individual flux lines, spaced some distance apart; as the value of the field is increased, more of these quantized flux lines appear, and their distance apart correspondingly decreases. Eventually, an "upper critical field" is reached at which the flux lines "touch," and the film makes the transition to the normal state. Tinkham (63) has made a calculation along these lines, based upon the Landau-Ginzburg-Abrikosov picture, and he obtains good agreement with experimental data.

[9d] Besides the work of Douglass (61, 61b, 62b), there is a conflicting calculation (based on the microscopic theory) by Gupta and Mathur (Gupta 61, Mathur 62), as well as a Green's function calculation by Nambu and Tuan (Nambu 62). The various theories of the field dependence of the energy gap are compared and discussed by Tong (63).

The Ginzburg-Landau theory also predicts that the change to a second-order transition is inhibited if the film carries a supercurrent, i.e., if different magnetic fields are applied to the two sides of the film. Douglass (62b) gives the theoretical results for the case of zero field on one side of the film, non-zero field on the other side. There is then much less variation of the energy gap parameter as a function of the applied field, all the way up to the critical field, and the energy-gap parameter does *not* vanish when the applied field reaches critical value. This prediction has not yet been verified experimentally.

Application of the Landau-Ginzburg theory to thin films depends upon the fact, noted in Section 2, that their "wave function of the super-electron" has been identified by Gor'kov (58, 59, 59a) with $\Delta(r, r')$ according to (2.2), and $\Delta(r, r')$ has a "range" equal to that of the electron-electron interaction, much less than the pair size. Thus, even though the film may be thin compared to the pair size (the Pippard coherence length), it is usually thick enough to allow us to replace $\Delta(r, r')$ by a function $\Delta(r)$ of only one variable. For certain purposes, it is then possible to treat this function by means of the Landau-Ginzburg theory. However, some effects, notably the "shape resonances" discussed above, are not obtained this way. Another limitation of the Gor'kov derivation, namely, $\Delta \ll \mathrm{t}T$, seems to be borne out experimentally (Douglass 62b, Tinkham 62); at low temperatures, the thin film data deviate appreciably from the Landau-Ginzburg-Gor'kov predictions.

In the region where the Landau-Ginzburg theory is inapplicable, the theoretical situation is less satisfactory (Ittner 60, Toxen 63; for a review, see Bardeen 62b).

There is considerable practical interest in films made of alloys between superconducting and normal metals (Hulm 61, 61a) as well as "sandwich" films with normal and superconducting layers (Parmenter 60; P. Hilsch 61, 62; R. Hilsch 62; Smith 61; Rose-Innes 61; Meissner 58, 59, 60, 62; Simmons 62). A largely phenomenological discussion by Douglass (62) gives much better agreement with experiment than an *ad hoc* extension of the BCS theory due to Cooper (61, 62a).[10] It will be necessary, though, to find a basic theoretical justification for Douglass' phenomenology (DeGennes 63).

Another area in which theory is behind experiment is the transition *time* required to "switch" a film from the superconducting to the normal

[10] According to Cooper, the normal material adjacent to the superconducting layer acts only to "dilute the volume." Thus, films containing a superconducting layer should always become superconducting, which is contrary to experiment (Misener 35; Meissner 58, 59, 60, 62; Shapiro 61), and the nature of the normal metal should make no difference, whereas in fact is does (Douglass 62).

state by application of a magnetic field in excess of the critical field. This problem has been discussed by Nethercot (61), whose qualitative estimates agree with present experimental data, but much more theoretical work is necessary.

The considerable interest in this switching time for thin films (Broom 59, 60, 62; Oliver 60; Feucht 61; Näbauer 61; Kolchin 61) is related to the possibility of using superconducting materials in the construction of computer memories. The original suggestion by M. J. Buckingham (58) has given rise to very intensive investigation, e.g., Newhouse (59), Young (59, 61), Broom (60), Ittner (60a, 61), Kontarev (61), Litting (61), Lock (61, 62), Minnigerode (61), Parkinson (61), Marchand (62a), Volger (62).

5. Conclusion

In concluding this book, we think it is desirable to emphasize the positive successes of the basic theory of superconductivity in understanding and correlating the true "super"-properties (Meissner effect and persistent currents), rather than the limits to this understanding, which have been exhibited so prominently in this last chapter. To appreciate the present situation, we must see it in historical perspective. As late as 1950, when London published his book, there were only vague speculations concerning the basic cause of superconducting behavior in metals. At present, almost all of the basic phenomena are understood, and theorists have been turning their attention to finer details, perhaps somewhat prematurely.

The two outstanding fundamental problems of the theory are: (1) the lifetime of persistent currents in superconducting rings; (2) the regularities found by Matthias in superconducting alloys and compounds, especially the relationship between superconductivity and ferromagnetism.

Appendix A

Some Concepts from Statistical Mechanics

In this Appendix, we present in condensed form those parts of statistical mechanics which are used extensively in the body of the book. We hope that this summary presentation will aid readers who have an incomplete acquaintance with this subject. Such readers should however also consult full texts on the subject, for example the excellent book by Landau and Lifshitz (Landau 58).

1. Ensembles and Entropy

In statistical mechanics, we deal not with one system at a time, but with whole sets of similar systems, called *ensembles*; the systems comprising such an ensemble must be sufficiently similar so that averaging over the ensemble makes sense, but the systems need not be identical in the quantum mechanical sense. For example, we shall be interested in ensembles the systems of which have varying numbers of particles. Contrary to one's first impression, this kind of ensemble is actually closer to normal laboratory practice than ensembles in which all component systems have the same number of particles: One kilogram of water specifies a macroscopic system to sufficient accuracy, but it would be rare indeed to obtain two such macroscopic systems with exactly equal numbers of water molecules. The properties attributed experimentally to "one kilogram of water" are ensemble average properties over an ensemble of systems with similar, but not identical, numbers of particles.

For a start, we make the simplifying assumption that each component system of the ensemble is in some one definite quantum state Ψ_α at a given time, say $t = 0$. In general, different systems of the ensemble are in different quantum states. Let the possible quantum states Ψ_α be mutually orthogonal and normalized (we shall remove this restriction later on), and let us denote by P_α the fraction of systems found to be in quantum state Ψ_α; P_α is then also the probability of finding any one system of the ensemble in state Ψ_α upon carrying out a complete measurement on a system of the ensemble, selected at random.

It is highly convenient to define a *statistical operator*, or *density matrix*, called $\mathcal{U}$, by the statement that $\mathcal{U}$ has eigenfunctions Ψ_α with associated eigenvalues P_α. Since the P_α are probabilities, they are real numbers which satisfy the conditions

$$0 \leqslant P_\alpha \leqslant 1 \qquad \text{for all } \alpha \tag{1.1a}$$

$$\sum_\alpha P_\alpha = 1 \tag{1.1b}$$

Since the P_α are by definition the eigenvalues of the operator $\mathcal{U}$, this operator has the properties

$$0 \leqslant (\Phi, \mathcal{U}\Phi) \leqslant 1 \qquad \text{for all wave functions } \Phi \tag{1.2a}$$

$$\text{Trace}\, \mathcal{U} = 1 \tag{1.2b}$$

The first inequality (1.2a) means that $\mathcal{U}$ is a non-negative Hermitean operator, whereas the second inequality means that $\mathcal{U}$ is bounded from above.

Let Q be a quantum mechanical operator corresponding to some measurable physical property of each component system of the ensemble. The quantum mechanical expectation value of Q for a particular system in state Ψ_α is given by the matrix element $(\Psi_\alpha, Q\Psi_\alpha)$; we obtain the *ensemble average* value of Q by putting

$$\langle Q\rangle_{\text{av}} = \sum_\alpha P_\alpha(\Psi_\alpha, Q\Psi_\alpha) = \text{Trace}\,(Q\mathcal{U}) \tag{1.3}$$

A special case of ensembles is an ensemble consisting of identical systems, all in one and the same quantum state Ψ_α. This special case is included in the general density matrix formalism; it corresponds to a density matrix $\mathcal{U}$ with one particular eigenvalue P_α equal to unity, all other $P_\beta\, (\beta \neq \alpha)$ equal to zero. Clearly, a measurement on only one system of the ensemble cannot determine whether the ensemble is in a pure quantum state, or whether there exists a mixture of states. However, in statistical mechanics it is understood that measurements must be made on a large number of representative systems of the ensemble. When this is done, the distinction between a pure state and a mixture is immediate: In one case, all measured systems of the ensemble are in one and the same quantum state, in the other case, different systems are in different quantum states. The distinction in question is thus entirely objective and clearcut.

The square of the operator $\mathcal{U}$ is an operator with the same eigenfunctions Ψ_α, but associated eigenvalues $(P_\alpha)^2$. If and only if the ensemble corresponds to a pure state, we have the equality $(P_\alpha)^2 = P_\alpha$, for all α; thus the special case of a pure state corresponds to the operator identity

$$\mathcal{U}^2 = \mathcal{U} \qquad \text{for a pure state} \tag{1.4}$$

We may now remove the restriction that all member systems of the ensemble must occupy a set Ψ_α of mutually orthogonal quantum states. Instead, we may start from arbitrary density matrices $\mathcal{U}$ subject to the conditions (1.2) and (1.3), each such density matrix defining a possible ensemble of systems. Although this procedure looks more general, it reduces to the previous one, since non-negative Hermitean operators $\mathcal{U}$ which are bounded from above can always be brought to diagonal form [see von Neumann (32) for a full discussion of density matrices].

The property (1.4) distinguishes between an ensemble describing a pure state, and all other ensembles. Since the case of a pure state is never realized in practical macroscopic systems, we would like to find a less restrictive classification of ensembles according to their degree of "order" and "disorder." For example, if only two quantum states Ψ_1 and Ψ_2 are realized by member systems of the ensemble, the ensemble with probabilities $P_1 = P_2 = \frac{1}{2}$ is more "disordered" than an ensemble with probabilities $P_1 = 0.99$, $P_2 = 0.01$. The latter ensemble is in some reasonable sense "close to" a pure state 1.

Let us compute the expectation value of the particular operator $\ln \mathcal{U}$. According to the general definition (1.3), we have

$$\langle \ln \mathcal{U} \rangle_{\text{av}} = \sum_\alpha P_\alpha(\Psi_\alpha, \ln \mathcal{U}\, \Psi_\alpha) = \sum_\alpha P_\alpha \ln P_\alpha = \text{Trace}\,(\mathcal{U} \ln \mathcal{U}) \tag{1.5}$$

For the two ensembles mentioned above, the expectation values (1.5) are

$$\tfrac{1}{2} \ln \tfrac{1}{2} + \tfrac{1}{2} \ln \tfrac{1}{2} = -\ln 2 = -0.693 \qquad \text{for} \quad P_1 = P_2 = \tfrac{1}{2}$$

and

$$0.99 \ln 0.99 + 0.01 \ln 0.01 = -0.056 \qquad \text{for} \quad P_1 = 0.99,\ P_2 = 0.01$$

respectively. We see that these expectation values are negative [this follows immediately from (1.1a)], and that the more "disordered" ensemble corresponds to a more negative value of $\langle \ln \mathcal{U} \rangle_{av}$. Furthermore, by going to the limit in which P_1 approaches 1, and P_2 approaches 0, we obtain the value zero for (1.5).

For a number of reasons (von Neumann 32) the average (1.5) turns out to give a convenient measure of the degree of disorder of the ensemble associated with the density matrix $\mathcal{U}$. In statistical mechanics, it is conventional to change the sign (so that disordered systems correspond to positive values of the quantity to be defined, the more positive, the more disordered), and to multiply by Boltzmann's constant $\mathfrak{k}$; the resulting quantity is called the *entropy* of the ensemble:

$$S = -\mathfrak{k} \sum_{\alpha} P_{\alpha} \ln P_{\alpha} = -\mathfrak{k} \operatorname{Trace}(\mathcal{U} \ln \mathcal{U}) \tag{1.6}$$

We observe that the entropy vanishes if and only if the ensemble corresponds to a pure quantum state, i.e., if and only if $\mathcal{U}$ satisfies (1.4); otherwise S is positive, the more so, the more "disordered" the ensemble is.[1]

The concept of ensembles and the associated concept of entropy are characteristic of statistical mechanics as opposed to ordinary (quantum) mechanics. It should be noted that the statistical concepts introduced here are *not* restricted to large, macroscopic systems, even though they are used most often in connection with such systems. The initial beam of silver atoms in the Stern-Gerlach experiment is an incoherent mixture, i.e., an ensemble, of atoms with spins up and with spins down. This provides an example of the use of ensembles in the microscopic domain; there are many other examples.[2]

---

[1] In the statistical theory of transmission of information, one considers ensembles of "messages" α with corresponding probabilities P_{α}. The larger the range of possible messages, the more "information" is transmitted over the channel in question. Thus (1.6), without the factor $\mathfrak{k}$, can serve as a mesure of the "information" transmitted by a communications channel (Shannon 48).

Alternatively, we may (but we need not!) think of our ensemble of systems in terms of the information available about any one member system of that ensemble, *before* a measurement is carried out on that particular system. In that case, the "pure state" ensemble (1.4) gives us complete (in the quantum mechanical sense) information, while other ensembles give us only limited information. We would then interpret the entropy (1.6) as a measure of the amount of "neg-information" about any particular system of the ensemble (Brillouin 51). It is important to realize, however, that this rather subjective approach to the entropy concept is by no means necessary; given any ensemble, the probabilities P_{α} and thence the entropy S can be determined, in principle, by a perfectly objective series of measurements on a large representative sample of member systems of the ensemble.

[2] The beam in qucstion can *not* be described by one spin function with equal components for spin up and spin down. Such a spin function is an cigenfunction of σ_x, and therefore describes a polarized beam, with all spins pointing in the $+x$ direction. An unpolarized beam *must* be described by a density matrix.

We also note that the concepts of ensemble and of entropy are not restricted to systems in equilibrium, or even to stationary systems (systems which have properties independent of time). On the contrary, the states Ψ_α which enter the definition of the statistical operator $\mathcal{U}$ need not be eigenstates of the Hamiltonian H describing the time-evolution of member systems of the ensemble.[3] If Ψ_α is a general function, its time-evolution is given by

$$\frac{\partial \Psi_\alpha}{\partial t} = (i\hbar)^{-1} H\Psi_\alpha \tag{1.7}$$

This is a first-order differential equation, which determines $\Psi_\alpha(t)$ for all times t, given the initial function $\Psi_\alpha = \Psi_\alpha(0)$ at time $t = 0$. Thus, all those members systems of our ensemble which are in state $\Psi_\alpha(0)$ at time $t = 0$ will be found in state $\Psi_\alpha(t)$ at time t. The fraction of those systems, P_α, does not change in time. Thus we have obtained the following result: *If the time-development of member systems of the ensemble is given by* (1.7) *with some unique Hamiltonian operator* H, *then the time-development of the density matrix* $\mathcal{U} = \mathcal{U}(t)$ *is as follows: The eigenvalues* P_α *of* $\mathcal{U}$ *are independent of time, whereas the eigenfunctions of* $\mathcal{U}$ *are time-dependent functions* $\Psi_\alpha(t)$ *developing according to Eq.* (1.7).

It can be shown (von Neumann 32, ter Haar 54) that this time-development can be described by the operator equation[4]

$$\frac{d\mathcal{U}}{dt} = (i\hbar)^{-1}(H\mathcal{U} - \mathcal{U}H) \tag{1.8}$$

Either from (1.8), or from the verbal statement in the preceding paragraph, it follows immediately that *the density matrix* $\mathcal{U}$, *and with it all ensemble properties, are independent of time if and only if* $\mathcal{U}$ *commutes with the Hamiltonian* H *of the member systems of the ensemble.* Such ensembles are called *stationary.* The density matrix $\mathcal{U}$ of a stationary ensemble need *not* be a function of the Hamiltonian operator $\mathcal{U} = \mathcal{U}(H)$, since $\mathcal{U}$ may distinguish between states Ψ_α degenerate in H, i.e., between states with the same energy.

---

[3] We shall assume, for the moment, that such a Hamiltonian H exists, i.e., that each member system of our ensemble can be described by ordinary quantum mechanics. We shall return to a discussion of this point in Section 2; see also Chapter IX, Section 10.

[4] Note that the sign in (1.8) is opposite to the usual sign for the time-derivative of an operator in quantum mechanics.

2. Equilibrium Ensembles. (1) The Microcanonical Ensemble

In practice, we are frequently interested in describing systems in thermodynamic equilibrium. Intuitively, we say that a system is in equilibrium if it is not only stationary when undisturbed, but this stationary state is also stable against small external perturbations.[5] It follows that density operators $\mathscr{U}$ describing equilibrium ensembles must be functions of the Hamiltonian H, i.e., cannot distinguish between states with equal energy; for the states in question are mixed with each other by the slightest external perturbation, and such a distinction could not persist.[5a] Thus all equilibrium ensembles have the common property that the probabilities P_α are functions of the energies E_α only.

The simplest equilibrium ensemble is the *microcanonical ensemble*, which is designed to describe experimental conditions in which the systems of the ensemble are isolated and have known energy E. Since the energy of a macroscopic system is never measured to an accuracy corresponding to the quantum mechanical separation of adjacent energy levels, we must specify some finite range of energies ΔE as well as the mean energy E. Since the probabilities P_α are functions of the energies E_α only, the simplest permissible assumption is

$$P_\alpha = P \quad \text{(independent of } \alpha) \qquad \text{for} \quad E - \tfrac{1}{2}\Delta E \leqslant E_\alpha \leqslant E + \tfrac{1}{2}\Delta E \tag{2.1a}$$

$$P_\alpha = 0 \qquad \text{otherwise} \tag{2.1b}$$

This ensemble is always stationary; it is in equilibrium with respect to perturbations provided that the perturbations in question do not add or subtract energy from the member systems of the ensemble. Physically, this means that we are considering *thermally isolated* systems.

Let $\rho(E)\Delta E$ be the number of quantum states Ψ_α in the energy range (2.1a). To satisfy (1.1b), the common value P of the probabilities P_α must then equal

$$P = \frac{1}{\rho(E)\,\Delta E} \tag{2.2}$$

and the entropy (1.6) is given by

$$S = -\mathfrak{k}\,\langle \ln P_\alpha \rangle_{\text{av}} = \mathfrak{k} \ln\,[\rho(E)\,\Delta E] \tag{2.3}$$

[5] In ordinary mechanics, this condition is called "stable equilibrium," as opposed to "unstable" and "neutral" equilibrium conditions. In statistical mechanics, the term equilibrium refers almost always to a stable equilibrium under given external conditions.

[5a] This assumes that energy is the only constant of motion.

Although this entropy value depends explicitly upon the energy range ΔE selected at the beginning, this dependence turns out to be unimportant practically, for ensembles of macroscopic systems. For macroscopic systems, the level density $\rho(E)$ is so enormous that $\ln \Delta E$ is immeasurably small compared to $\ln \rho(E)$. To see this heuristically, let us solve (2.3) for the level density $\rho(E)$:

$$\rho(E) = \frac{1}{\Delta E} \exp\left[\frac{S(E)}{\mathfrak{k}}\right] \tag{2.4}$$

For macroscopic systems, the entropy $S(E)$ is an extensive quantity, i.e., it is proportional to the number of particles or the volume of the system, if the size of the system is increased in such a way that the number density N/V stays constant. Thus the *exponent* in (2.4) is proportional to the volume, so that the exponential factor has superastronomical magnitude.[6]

This consideration also shows several other things: (1) It is practically impossible to determine the quantum state of a macroscopic system, since adjacent quantum states are separated in energy by amounts of order $1/\rho(E)$. (2) The probability P_α of finding a member system of the ensemble in any particular quantum state is unimaginably small, see (2.2) and (2.4). The special case of a "pure state" is of no practical interest in statistical mechanics.

Let us see what happens when we allow our thermally isolated system to expand. For the sake of definiteness, suppose that the system is a gas enclosed in a cylinder with a removable piston, and that the piston is pushed out a small distance by the gas pressure, so that the volume available to the gas is changed from V to $V + dV$. The energy levels E_α of the gas[7] and the associated wave functions Ψ_α depend on the boundary conditions, i.e., they depend on the volume V. If the expansion of the volume proceeds sufficiently slowly, each member system of the ensemble remains in a particular quantum state α, with the wave function Ψ_α and the energy E_α changing "adiabatically" during the process. This process is called "adiabatic" in quantum mechanical terminology as well as in the language of thermodynamics. In thermodynamics, the term "adiabatic" means that no heat is exchanged between a system and its surroundings. Tentatively, we may therefore interpret the absence of heat exchange as no change in the set of probabilities P_α describing

[6] We have not proved that $S(E)$ is indeed extensive; this can be shown for a number of simple systems by explicit calculation.

[7] These are energy levels of the gas as a whole, i.e., of an N-particle system, not energy levels of a single gas molecule.

the ensemble; we note that constancy of all the P_α implies constancy of the entropy S, (1.6).

Although there is no exchange of *heat* under the conditions envisaged here, there is most definitely an exchange of *energy* between the system and its surroundings: By pushing the piston outwards, the gas does an amount of work $p\,dV$ on the piston, where p is the gas pressure; this indeed *defines* the pressure p. This work is done at the expense of internal energy contained in the gas. The internal energy is the ensemble average of the energies E_α of member systems of the ensemble, i.e.,

$$E = \sum_\alpha P_\alpha E_\alpha = \text{Trace}\,(H\mathscr{U}) \qquad \text{Internal energy} \tag{2.5}$$

Thus we have the equality

$$\begin{aligned} p\,dV &= E(V) - E(V + dV) \\ &= \sum_\alpha P_\alpha\,[E_\alpha(V) - E_\alpha(V + dV)] \\ &= -\,dV \sum_\alpha P_\alpha \frac{\partial E_\alpha}{\partial V} \end{aligned}$$

which leads to the following definition of pressure:

$$p = -\left\langle \frac{\partial E_\alpha}{\partial V} \right\rangle_{\text{av}} \tag{2.6}$$

We note that we have used the concept of an "adiabatic" change in this derivation, through the fact that the probabilities P_α have been assumed to be independent of the volume during the expansion. Unchanged P_α implies unchanged entropy S. *If* specification of the volume V and the entropy S suffice to determine the macroscopic state of the system (more precisely, suffice to specify the ensemble from which the systems are taken), *then* Eq. (2.6) can be written in terms of the internal energy E, (2.5), as

$$p = -\left(\frac{\partial E}{\partial V}\right)_S \tag{2.7}$$

where the subscript S implies that the entropy is kept constant during the differentiation. If other quantities must be specified also (for example: a magnetic field as in magnetostriction), (2.7) is insufficient and one must go back to the more fundamental (2.6).

3. Equilibrium Ensembles. (2) The Canonical Ensemble

When the typical system of the ensemble is not thermally isolated, but rather is placed in contact with a heat bath, there is a random exchange of energy between the system and the heat bath, and thus the probability assumption (2.1) is inappropriate. In order to obtain a more appropriate distribution, we pose the following question: Given the ensemble average of the energy (the thermodynamic "internal energy") E, Eq. (2.5), what density matrix (probability distribution) leads to maximum degree of disorder, i.e., to maximum entropy (1.6)?

In its full generality, this is quite a difficult problem, since one cannot assume to start with that the density matrix $\mathscr{U}$ is a function of the Hamiltonian H only. However, for equilibrium ensembles, $\mathscr{U} = \mathscr{U}(H)$ is a necessary condition, and with this assumption the calculation becomes easy: Let the eigenvalues of H be E_α, the eigenvalues of $\mathscr{U}$ (for the same eigenfunctions) be P_α; we then want

$$S/\mathfrak{k} = -\sum_\alpha P_\alpha \ln P_\alpha = \max \tag{3.1a}$$

subject to

$$E = \sum_\alpha P_\alpha E_\alpha = \text{given} \tag{3.1b}$$

$$\sum_\alpha P_\alpha = 1 \tag{3.1c}$$

$$0 \leqslant P_\alpha \leqslant 1 \qquad \text{for all } \alpha \tag{3.1d}$$

This is a conventional minimum problem which can be handled by the method of Lagrange multipliers, except for condition (3.1d), which fortunately turns out to be obeyed automatically by the eventual solution. We multiply (3.1b) by the Lagrange multiplier $-\beta$, and (3.1c) by the Lagrange multiplier $+\beta F^*$, add these to (3.1a), and differentiate with respect to P_α. The result is

$$-\ln P_\alpha - 1 - \beta E_\alpha + \beta F^* = 0$$

Introducing the quantity F by

$$F = F^* - 1/\beta$$

we obtain the solution

$$P_\alpha = \exp\,[\beta(F - E_\alpha)] \tag{3.2}$$

This distribution function defines the *canonical ensemble*. The two constants β and F must be determined by substitution of (3.2) into the simultaneous equations (3.1b) and (3.1c). We shall show, below, that the physical interpretation of F is the thermodynamic free energy of the ensemble, and that $\beta = 1/\mathfrak{k}T$ where $\mathfrak{k}$ is Boltzmann's constant and T the absolute temperature. Substitution of (3.2) into (3.1c) gives the usual definition of the free energy F:

$$\exp(-\beta F) = \sum_{\alpha} \exp(-\beta E_{\alpha}) = \text{Trace}\,[\exp(-\beta H)] \tag{3.3}$$

If the spectrum of H is not bounded from above, i.e., if arbitrarily large values of the energy E_{α} are possible, then the sum in (3.3) converges only for positive values of the parameter β. This is the usual case; but exceptional systems exist in which the spectrum of H is bounded from both sides; for those systems, β may take negative values (negative temperatures).

A first indication that the parameter β is related to temperature can be obtained by considering a composite system, consisting of two independent subsystems (1) and (2). The quantum states α can be labeled by pairs of indices: $\alpha = (\alpha_1, \alpha_2)$, and the total energy E_{α} is a sum of two separate energies:

$$E_{\alpha} = E^{(1)}_{\alpha_1} + E^{(2)}_{\alpha_2} \tag{3.4}$$

Under what conditions, we now ask, is the joint probability distribution for the two separate systems equal to the probability distribution (3.2) for the combined system? The joint probability distribution for the two independent systems is

$$P_{\alpha} = P_{\alpha_1\alpha_2} = P^{(1)}_{\alpha_1} P^{(2)}_{\alpha_2} = \exp[\beta_1(F_1 - E^{(1)}_{\alpha_1})] \exp[\beta_2(F_2 - E^{(2)}_{\alpha_2})] \tag{3.5}$$

This reduces to (3.2) if and only if $\beta_1 = \beta_2$; furthermore, in that case, we have

$$F = F_1 + F_2 \qquad \text{for } \beta_1 = \beta_2$$

It is reasonable to suppose that the process of putting the two independent systems into thermal contact with each other will result in an increase, rather than an decrease, of the total amount of disorder; we now see that the probability distribution of the combined system (and hence also the total amount of disorder) remains the same if and only if the separate systems have equal values of β to start with. In macroscopic terms, two systems which do not exchange heat upon being placed in thermal contact are said to have the same temperature. Identifying (once more) heat exchange with a change in the basic probability distribution, we arrive at the conclusion that equal values of β mean equal temperatures, i.e., the Lagrange parameter β is a function of the absolute temperature T as defined in thermodynamics.

Some very useful equations can be obtained by writing the normalization condition (3.1c) explicitly, and differentiating it with respect to parameters of the system. Substitution of (3.2) into (3.1c) yields

$$\sum_{\alpha} \exp\,[\beta(F - E_\alpha)] = 1 \tag{3.6}$$

Let us differentiate both sides with respect to the parameter β. This yields

$$\sum_{\alpha} \left(\beta \frac{\partial F}{\partial \beta} + F - E_\alpha\right) \exp\,[\beta(F - E_\alpha)] = 0$$

We now use (3.1b) and (3.1c) to write this in the form

$$\beta \frac{\partial F}{\partial \beta} + F - E = 0$$

or, solving for the internal energy E,

$$E = F + \beta \left(\frac{\partial F}{\partial \beta}\right)_V = F - T\left(\frac{\partial F}{\partial T}\right)_V \tag{3.7}$$

where we have anticipated the relation $\beta = 1/\mathfrak{k}T$ in the last step. The subscript V on the partial derivative indicates that the volume of the system (as well as other external parameters) is kept constant during the differentiation.

As a byproduct of this derivation of a standard thermodynamic relationship, we have also obtained the result

$$\left(\frac{\partial P_\alpha}{\partial \beta}\right)_V = (E - E_\alpha)\, P_\alpha \tag{3.8}$$

This equation depends, of course, on the initial assumption (3.2); it does not apply to the microcanonical distribution (2.1).

Next, let us evaluate the entropy S, Eq. (1.6), for the canonical distribution. Straightforward substitution of (3.2) into (1.6) yields

$$S = -\mathfrak{k} \sum_{\alpha} P_\alpha \ln P_\alpha = -\mathfrak{k}\beta \sum_{\alpha} (F - E_\alpha)\, P_\alpha = -\mathfrak{k}\beta(F - E) \tag{3.9}$$

This agrees with the thermodynamic relationship

$$F = E - TS \tag{3.10}$$

if and only if $\mathfrak{k}\beta = 1/T$, i.e., if and only if

$$\beta = 1/\mathfrak{k}T \tag{3.11}$$

Our earlier argument, see Eq. (3.5), established that β is related uniquely to (is a function of) the thermodynamic temperature T; the present calculation has now decided the precise functional relationship between β and T necessary for consistency with our identifications of S as entropy and F as free energy. To complete the argument, we must establish these identifications directly, by comparison with the first and second laws of thermodynamics; this we shall do at the end of this section.

It is highly plausible to assume that the "disorder" of an ensemble of systems in contact with a heat bath increases with time, until a maximum value is reached; and consequently the ensemble with maximum disorder, i.e., the canonical ensemble (2.2), represents the state of thermodynamic equilibrium. Although intuitively acceptable, the assumption that external influences tend to destroy ordering and increase the disorder *is* an assumption, and requires discussion. In particular, the law of time-development (1.8) applicable to an ensemble of *isolated* systems does *not* lead to any increase in disorder with time, i.e., the probability distribution P_α remains independent of time, and with it the entropy S.

From one possible point of view, this is a result of the over-idealization to a completely isolated system; actual systems are never that completely isolated, and the random influences of the remainder of the world may be expected to increase the disorder. This is then an assumption about that part of the real world not included as part of the "system" under consideration, and the forward sense of time is that direction of the time coordinate which leads to the outside influences being (at least approximately) uncorrelated with each other. The negative sense of time, if chosen by mistake, would disclose itself by exceedingly strong systematic correlations between "outside influences" on our typical system of the ensemble.

An alternative point of view, which is much more generally accepted, asserts that the time-development (1.8) should be retained (i.e., no essential feature of the situation is lost by assuming complete isolation of the system), but the definition (1.6) of entropy must be discarded in favor of a "coarse-grained" entropy: Large numbers of quantum states α are placed together into groups γ, say, and we do not distinguish between different states α within the same group γ; in particular, the "fine-grained" probabilities P_α should be replaced by "coarse-grained" probabilities $\hat{P}_\alpha$ which are simply the *averages* of the P_α contained within the appropriate group γ. The coarse-grained entropy is then

$$\hat{S} = -\mathfrak{k} \sum_\gamma N_\gamma \hat{P}_\gamma \ln \hat{P}_\gamma \tag{3.12}$$

where N_γ is the number of quantum states α within the group γ. This concept of coarse-graining is discussed in considerable detail in Tolman (38); the reader is referred to Tolman (38) and Blatt (59a) for further discussion and references.

Let us write (3.1b) in the form

$$\sum_\alpha (E - E_\alpha) P_\alpha = 0$$

and let us differentiate this equation with respect to β. This gives

$$\sum_\alpha \left[\frac{\partial E}{\partial \beta} P_\alpha + (E - E_\alpha) \frac{\partial P_\alpha}{\partial \beta}\right] = 0$$

We now use (3.8) to obtain

$$\frac{\partial E}{\partial \beta} + \langle (E - E_\alpha)^2 \rangle_{\text{av}} = 0$$

and finally, using (3.11) together with the definition of specific heat:

$$c_V = \left(\frac{\partial E}{\partial T}\right)_V \tag{3.13}$$

we obtain Einstein's result for the *equilibrium fluctuations in the energy*:

$$\langle (E - E_\alpha)^2 \rangle_{\text{av}} = \mathfrak{k} T^2 c_V \tag{3.14}$$

Two results follow immediately from (3.14): First, the specific heat c_V cannot be negative; second, the equilibrium fluctuations are exceedingly small in a macroscopic system: The specific heat c_V is proportional to the number of particles N, and thus the relative fluctuation in energy has the N-dependence

$$\frac{(\langle (E - E_\alpha)^2 \rangle_{\text{av}})^{1/2}}{E} \propto \frac{1}{N^{1/2}} \tag{3.15}$$

For N of order 10^{22}, the relative fluctuation in energy is of the order of 1 part in 10^{11}, which is sufficiently small for all practical purposes. Thus, even though the canonical ensemble contains all quantum states α with non-zero probabilities P_α, the main contribution to all ensemble averages arises from an exceedingly narrow region of energies. This is due to the fact that the exponential distribution (3.2) is a sharply decreasing function of energy E_α, whereas the density of states (2.4) is a sharply increasing function of energy; multiplication of these two functions gives rise to a narrow peak in the integrand, at the particular energy E given by (3.1b); the relative width of the peak is determined by (3.14).

Since the contributing region of energies is so narrow, we conclude that *we may identify averages over the canonical and microcanonical ensembles to sufficient accuracy for practical purposes*, in macroscopic systems.

In particular, since the entropy S is a multiple of $\langle \ln P_\alpha \rangle_{\text{av}}$, *we can identify the microcanonical entropy* (2.3) *with the canonical entropy* (3.9) *to an adequate approximation*. This statement can be confirmed by explicit calculation on a number of simple systems; it can also be derived formally by replacing the sum over α in (3.9) by an integral over energies E_α, with weighting factor $\rho(E_\alpha)$, and evaluating the integral by a saddle point approximation. This procedure permits an estimate of the error, which turns out to be of the expected order of magnitude.

Next, let us differentiate the identity (3.6) with respect to the volume V, keeping β constant. This yields

$$\left(\frac{\partial F}{\partial V}\right)_\beta = \left\langle \frac{\partial E_\alpha}{\partial V} \right\rangle_{\text{av}} \tag{3.16}$$

The average on the right-hand side of (3.16) is over the canonical ensemble. We identify this with the average over the microcanonical ensemble and use (2.6) for the latter average; we thus obtain the thermodynamic result

$$p = -\left(\frac{\partial F}{\partial V}\right)_T \tag{3.17}$$

We are now in a position to make the final identification of our statistical quantities with the corresponding thermodynamic quantities, by direct comparison with the first and second laws of thermodynamics. The first law asserts that there exists an "internal energy" E which is a function of the present state of the system (i.e., does not depend on the previous history of the system), and that for slow enough changes in external conditions the internal energy E changes according to the law

$$dE = dQ - p\, dV \tag{3.18}$$

where dQ is the quantity of heat supplied to the system during the change. The second law asserts that there exists another function of the present state of the system, called the entropy S, such that the change in S is related to the heat supplied by

$$dS = \frac{dQ}{T} \tag{3.19}$$

T being the absolute temperature. Whereas the total amount of heat supplied to the system in going from state 1 to state 2 may, and frequently does, depend on the particular path chosen to go from 1 to 2, the change in entropy $S_2 - S_1$ depends only on the initial and final states. We combine (3.18) and (3.19) to obtain

$$dE = T\,dS - p\,dV \tag{3.20}$$

According to the second law of thermodynamics, S is a function of the present state of the system only, and may therefore be taken as one of the variables used to define the state of the system. The volume V is another such variable, and we may therefore consider the internal energy E as a function of S, V, and whatever other variables are required to define the thermodynamic (macroscopic) state of the system uniquely. For simplicity, we restrict ourselves here to systems in which two variables of state suffice. Differentiation of the function $E = E(S, V)$ yields

$$dE = \left(\frac{\partial E}{\partial S}\right)_V dS + \left(\frac{\partial E}{\partial V}\right)_S dV \tag{3.21}$$

Comparison with (3.20) produces the necessary relations:

$$\left(\frac{\partial E}{\partial S}\right)_V = T \qquad \left(\frac{\partial E}{\partial V}\right)_S = -p \tag{3.22}$$

We emphasize that the relations (3.22) embody both the first and second laws of thermodynamics; if we can show that they are satisfied for the quantities which we have identified as E, S, T, etc. from statistical mechanics, then the proof is complete.

The second relationship (3.22) is identical with (2.7), and is established for the canonical ensemble by the fact that ensemble averages over the microcanonical ensemble (from which (2.7) was derived) can be identified with ensemble averages over the canonical ensemble, to sufficient accuracy.

In order to establish the crucial first relation (3.22), we rewrite it in the form

$$\left(\frac{\partial S}{\partial E}\right)_V = \frac{1}{T} \tag{3.23}$$

and we differentiate the definition (1.6) of the quantity which we wish to identify with the thermodynamic entropy. This yields

$$\left(\frac{\partial S}{\partial E}\right)_V = -\mathfrak{k} \sum_\alpha (1 + \ln P_\alpha) \left(\frac{\partial P_\alpha}{\partial E}\right)_V = -\mathfrak{k} \sum_\alpha (1 + \beta F - \beta E_\alpha) \left(\frac{\partial P_\alpha}{\partial E}\right)_V \tag{3.24}$$

where we have used the explicit expression (3.2) in the second step. We now simplify (3.24) as follows: Differentiation of (3.1c) with respect to E yields

$$\sum_\alpha \left(\frac{\partial P_\alpha}{\partial E}\right)_V = 0 \tag{3.25}$$

and differentiation of (3.1b) with respect to E gives

$$\sum_\alpha E_\alpha \left(\frac{\partial P_\alpha}{\partial E}\right)_V = 1 \tag{3.26}$$

Substitution of (3.25) and (3.26) into (3.24) leads us to

$$\left(\frac{\partial S}{\partial E}\right)_V = \mathfrak{k}\beta \tag{3.27}$$

This equation is derived entirely from statistical mechanics, the fundamental definition being that of entropy, (1.6); β appears as a Lagrange multiplier in the extremum problem (3.1).

Now, however, we see that (3.27) has exactly the right form for the thermodynamic relation (3.23), if we accept the further identification $\beta = 1/\mathfrak{k}T$, Eq. (3.11).

This completes the proof that our identification of statistical quantities with thermodynamic quantities is consistent with the basic laws of thermodynamics. These laws apply to equilibrium states, and to processes whose time-dependence is so slow that they may be considered a succession of equilibrium states. When it comes to truly irreversible processes, further problems arise [see the discussion adjoining Eq. (3.12)].

4. Equilibrium Ensembles. (3) The Grand Canonical Ensemble

So far, we have considered ensembles composed of identical systems. It is instructive and useful to consider an ensemble containing systems with different numbers N of particles. Such ensembles are necessary in a number of practical situations; for example, when two metals are placed in contact, electrons can transfer from one to the other, so that the number of electrons in each separate metal is not known *a priori*; a similar situation arises in systems in which chemical reactions are possible; e.g., in the presence of the chemical reaction $H + H = H_2$

the number of hydrogen molecules is not given *a priori*, but must be deduced from the known quantities (the volume V, the temperature T, and the total number of protons).

Let N be the number of particles of a typical system of our ensemble and let H_N be the Hamiltonian of that system; the eigenfunctions $\Psi_{N\alpha}$ and eigenvalues $E_{N\alpha}$ of the Hamiltonian H_N satisfy

$$H_N \Psi_{N\alpha} = E_{N\alpha} \Psi_{N\alpha} \tag{4.1}$$

We assume that the Hamiltonian H_N commutes with N, i.e., the law of conservation of particles holds.[8] Although we shall perform averages over an ensemble of systems with different numbers N of particles, the particle number N in any one system remains constant in time.

The density matrix $\mathscr{U}$ of our ensemble must have the same eigenfunctions $\Psi_{N\alpha}$ as the Hamiltonian, in order to describe a stationary ensemble; furthermore, in order to have an equilibrium ensemble, the eigenvalues $P_{N\alpha}$ of $\mathscr{U}$ must be functions of the energies $E_{N\alpha}$ only. The amount of disorder (entropy) of the ensemble is still given by (1.6), if we think of the index "α" as a combination (N, α) of two indices. More explicitly,

$$S = -\mathfrak{k} \operatorname{Trace} (\mathscr{U} \ln \mathscr{U}) = -\mathfrak{k} \sum_N \sum_\alpha P_{N\alpha} \ln P_{N\alpha} \tag{4.2}$$

In the preceding section, we determined an appropriate ensemble by maximizing the entropy, under certain conditions. We shall do the same here; the conditions are

$$\sum_N \sum_\alpha E_{N\alpha} P_{N\alpha} = E = \text{given} \tag{4.3a}$$

$$\sum_N \sum_\alpha P_{N\alpha} = 1 \tag{4.3b}$$

$$0 \leqslant P_{N\alpha} \leqslant 1 \qquad \text{for all } N \text{ and all } \alpha \tag{4.3c}$$

$$\sum_N \sum_\alpha N P_{N\alpha} = \bar{N} = \text{given} \tag{4.3d}$$

The first three conditions are identical with (3.1b) to (3.1d), respectively; the last condition limits us to ensembles with given average number

[8] This assumption is not necessary, and is sometimes unduly restrictive. For example, one may wish to know the equilibrium number of photons within some given frequency range, contained in a black box kept at temperature T. If the temperature T is sufficiently high (in the interior of a star) one may also wish to know the equilibrium number of electron-positron pairs. However, in systems violating the law of conservation of particles, such as these, it is not possible to write an equation of type (4.1), and the discussion becomes correspondingly more complex. The formalism of creation and destruction operators (Chapter V, Section 1) is then necessary.

of particles $\bar{N}$. We hope, and shall indeed find, that the systems which contribute most to the ensemble average quantities have particle numbers N close to $\bar{N}$.

The maximization of (4.2) subject to (4.3) is done by the method of Lagrange multipliers in a way completely analogous to the derivation of (3.2) from (3.1). The result is the *grand canonical distribution function*

$$P_{N\alpha} = \exp\left[\beta(\Omega + \mu N - E_{N\alpha})\right] \tag{4.4}$$

The quantities β, Ω, and μ arise from Lagrange multipliers; their actual values are determined from the given values of E and $\bar{N}$ by substituting (4.4) into the three simultaneous equations (4.3a), (4.3b), and (4.3d). It will turn out that $\beta = 1/\mathfrak{k}T$ as before, that Ω plays the role of a thermodynamic free energy, and that μ can be identified with the quantity called "chemical potential" in thermodynamics.

An interesting result can be derived by projecting (4.4) onto the sub-ensemble of systems with some given number N. Let F_N be the ordinary free energy for that case, so that the canonical probability distribution is

$$P_{\alpha}^{(\text{can})} = \exp\left[\beta(F_N - E_{N\alpha})\right] \qquad \text{for given } N \tag{4.5}$$

Direct substitution yields

$$P_{N\alpha} = \exp\left[\beta(\Omega + \mu N - F_N)\right] P_{\alpha}^{(\text{can})} \tag{4.6}$$

In words: *The projection of the grand canonical ensemble onto the sub-ensemble with given particle number N is identical with the canonical ensemble, except for the normalization constant in front.* This fact is very useful since the calculation of ensemble averages over the grand canonical ensemble is frequently much easier than over the canonical ensemble; the normalization constant cancels out in the calculation of an ensemble average. This result also establishes the identification (3.11) of β for the grand canonical ensemble.

Substitution of (4.4) into (4.2) gives the following result for the entropy

$$S = -\mathfrak{k} \sum_N \sum_\alpha \beta(\Omega + \mu N - E_{N\alpha})\, P_{N\alpha} = -\mathfrak{k}\beta(\Omega + \mu\bar{N} - E) \tag{4.7}$$

where we have used (4.3) in the last step. Using the identification $\beta = 1/\mathfrak{k}T$, (4.7) can be solved for Ω to give the well-known (to those who know it well) thermodynamic relationship

$$\Omega = E - TS - \mu\bar{N} \tag{4.8}$$

The similarity with the free energy F, Eq. (3.10), indicates that Ω also plays the role of a free energy; we shall refer to Ω as the *grand canonical free energy* or *grand canonical potential.* Ω is a function of the thermodynamic state of the system; the most convenient variables defining the state turn out to be: the volume V, the temperature T, and the chemical potential μ; that is, we shall consider $\Omega = \Omega(V, T, \mu)$ to be a function of these particular three variables. Other quantities, such as the pressure p, the entropy S, the average particle number $\bar{N}$, etc., can be deduced from Ω by differentiation, as we shall see.

Let us determine the probability P_N of finding a system of the ensemble with exactly N particles, irrespective of the quantum state. Using (4.6) and (3.6), we obtain

$$P_N = \sum_\alpha P_{N\alpha} = \exp\left[\beta(\Omega + \mu N - F_N)\right] \tag{4.9}$$

The sum of all the P_N must equal unity; solving for $\exp(-\beta\Omega)$ gives

$$\exp(-\beta\Omega) = \sum_N \exp\left[\beta(\mu N - F_N)\right] \tag{4.10}$$

This allows us to deduce the grand canonical free energy Ω from the ordinary free energies F_N. We may introduce the "chemical activity" parameter z by

$$z = \exp(\beta\mu) \tag{4.11}$$

and write (4.10) in the equivalent form

$$\exp(-\beta\Omega) = \sum_N z^N \exp(-\beta F_N) \tag{4.12}$$

This shows that the function $\exp(-\beta\Omega)$ is a generating function for the set of numbers $\exp(-\beta F_N)$. If we multiply (4.12) on both sides by z^{-k} and integrate around a circle in the complex z-plane, enclosing the origin, only one value of N contributes and we obtain

$$\exp(-\beta F_N) = \frac{1}{2\pi i} \oint \frac{\exp(-\beta\Omega)}{z^{N+1}}\, dz \tag{4.13}$$

For macroscopic systems, this integral can be evaluated by the saddle-point method. The result is, however, obtained more easily from (4.8), by solving that equation for $F = E - TS$.

Substitution of (4.4) into (4.3b) yields

$$\sum_N \sum_\alpha \exp\left[\beta(\Omega + \mu N - E_{N\alpha})\right] = 1 \tag{4.14}$$

We obtain useful equations by differentiating both sides of (4.14). For example, differentiation with respect to μ yields

$$\sum_N \sum_\alpha \left(\frac{\partial \Omega}{\partial \mu} + N \right) P_{N\alpha} = 0$$

or, using (4.3d), and indicating explicitly which variables are kept constant,

$$\bar{N} = -\left[\frac{\partial \Omega}{\partial \mu} \right]_{V,T} \tag{4.15}$$

As a byproduct, we obtain the relation, valid for the grand canonical ensemble only,

$$\frac{\partial P_{N\alpha}}{\partial \mu} = \beta (N - \bar{N}) P_{N\alpha} \tag{4.16}$$

We now differentiate with respect to μ the equation

$$\sum_N \sum_\alpha (N - \bar{N}) P_{N\alpha} = 0$$

and use (4.16). The result is

$$\langle (N - \bar{N})^2 \rangle_{\text{av}} = \mathfrak{k}T \left(\frac{\partial \bar{N}}{\partial \mu} \right)_{V,T} = -\mathfrak{k}T \left(\frac{\partial^2 \Omega}{\partial \mu^2} \right)_{V,T} \tag{4.17}$$

For macroscopic systems, Ω is an extensive quantity (proportional to the volume V if $\bar{N}/V$ is maintained constant), whereas μ is independent of volume. Thus the right-hand side of (4.17) is proportional to $\bar{N}$, and the relative fluctuation in N for the grand canonical ensemble is of order

$$\frac{\Delta N}{\bar{N}} \sim (\bar{N})^{-1/2} \tag{4.18}$$

If $\bar{N}$ is of the order of 10^{22}, this relative fluctuation is sufficiently small for practical purposes. This result together with (4.6) allows us to *identify averages over the canonical and grand canonical ensemble to sufficient accuracy, for macroscopic systems.*

As the result of the smallness of the fluctuation in N, sums over N such as the one occurring in (4.10) get their main contribution from a very narrow range of values of N. In particular, then, the *average* value of N [i.e., $\bar{N}$, Eq. (4.3d)] differs only insignificantly from the value N_0 for which P_N, Eq. (4.9), has its maximum. If we imagine that F_N is a smooth function of N, the maximum value of (4.9) occurs when $\partial(\mu N - F_N)/\partial N = 0$, i.e., when

$$\mu = \left(\frac{\partial F_N}{\partial N} \right)_{V,T} \tag{4.19}$$

This, however, is the standard definition of the *chemical potential* in thermodynamics, and we have therefore proved that the Lagrange multiplier μ of statistical mechanics must be identified with the thermodynamic chemical potential.[9] Equation (4.19) also shows that μ is an intensive quantity.

Next, let us differentiate (4.14) with respect to the volume V, keeping β and μ constant. This yields

$$\left(\frac{\partial \Omega}{\partial V}\right)_{T,\mu} = \left\langle \frac{\partial E_{N\alpha}}{\partial V} \right\rangle_{\text{av}}$$

The average here is over the grand canonical ensemble. However, we may identify that average with one over the canonical ensemble, and the latter with an average over the microcanonical ensemble, because of the smallness of all fluctuations. Using (2.5), we therefore obtain an equation for the pressure

$$p = -\left(\frac{\partial \Omega}{\partial V}\right)_{T,\mu} \tag{4.20}$$

If we consider Ω as a function of V, T, and μ, then only one of these variables of state is extensive, namely the volume V itself. T and μ are intensive variables. Thus, for macroscopic systems, Ω is proportional to the volume V, and the derivative in (4.20) is equal to the ratio Ω/V, to sufficient accuracy. We therefore have the identity

$$\Omega(V, T, \mu) = -pV \tag{4.21}$$

This allows us to deduce the equation of state (pressure as a function of volume, temperature, and particle number) directly, by eliminating the chemical potential μ between Eqs. (4.15) and (4.21).[10]

Having differentiated (4.14) with respect to V and with respect to μ, let us now differentiate it with respect to β. The result is

$$\sum_N \sum_\alpha \left(\Omega + \beta\left(\frac{\partial \Omega}{\partial \beta}\right)_{V,\mu} + \mu N - E_{N\alpha}\right) P_{N\alpha} = 0$$

---

[9] Systems exist, for example superconductors, in which F_N is not quite that smooth as a function of N. As a result of the electron pairing in superconductors, there is a significant difference between even and odd values of N (a pairing energy). In such cases, the derivative (4.19) must be replaced by a difference quotient $(F_N - F_{N'})/(N - N')$, with $1 \ll N - N' \ll N$.

[10] Since Ω is directly proportional to the pressure, this thermodynamic potential is sometimes simply called "pressure." This is somewhat misleading, however. Ω is a proper thermodynamic potential, i.e., a quantity from which all other thermodynamic quantities can be derived by differentiation, and which is a minimum in thermodynamic equilibrium, only if Ω is expressed as a function of the particular three variables of state V, T, and μ. If any other variables are used, for example, V, T, and N, the product pV does not play the role of a thermodynamic potential.

or, using (4.3) and solving for the internal energy E,

$$E = \Omega + \beta \left(\frac{\partial \Omega}{\partial \beta}\right)_{V,\mu} - \mu \left(\frac{\partial \Omega}{\partial \mu}\right)_{V,\beta}$$

$$= \Omega - T \left(\frac{\partial \Omega}{\partial T}\right)_{V,\mu} - \mu \left(\frac{\partial \Omega}{\partial \mu}\right)_{V,T} \tag{4.22}$$

This allows us to find the internal energy E, as a function of V, T, and μ, by straightforward differentiations. However, normally we want E as a function of V, T, and $\bar{N}$, and to obtain this we must eliminate μ between (4.15) and (4.22).

A particularly simple result is obtained for the entropy S by substituting (4.22) into (4.7), and using (4.15); we get

$$S = -\left(\frac{\partial \Omega}{\partial T}\right)_{V,\mu} \tag{4.23}$$

In order to find the specific heat c_V, it is best to start from $d\underset{\sim}{Q} = TdS$ (the second law of thermodynamics) together with the observation that $d\underset{\sim}{Q} = c_V dT$ for heating at constant volume and constant particle number. Thus we obtain

$$c_V = T \left(\frac{\partial S}{\partial T}\right)_{V,N} \tag{4.24}$$

The partial derivative here requires careful evaluation, since N rather than μ is to be kept constant. We write

$$\left(\frac{\partial S}{\partial T}\right)_{V,N} = \left(\frac{\partial S}{\partial T}\right)_{V,\mu} + \left(\frac{\partial S}{\partial \mu}\right)_{V,T} \left(\frac{\partial \mu}{\partial T}\right)_{V,N}$$

$$= -\left(\frac{\partial^2 \Omega}{\partial T^2}\right)_{V,\mu} - \left(\frac{\partial^2 \Omega}{\partial T\, \partial \mu}\right)_{V} \left(\frac{\partial \mu}{\partial T}\right)_{V,N} \tag{4.25}$$

In order to find $(\partial \mu/\partial T)_{V,N}$ we differentiate (4.15) with respect to T, keeping V and N constant. This gives

$$0 = -\left(\frac{\partial^2 \Omega}{\partial T\, \partial \mu}\right)_{V} - \left(\frac{\partial^2 \Omega}{\partial \mu^2}\right)_{V\,T} \left(\frac{\partial \mu}{\partial T}\right)_{V\,N}$$

We solve this equation for $(\partial \mu/\partial T)_{V,N}$, substitute it into (4.25), and the latter into (4.24), to obtain the final result

$$c_V = T \left\{ -\left(\frac{\partial^2 \Omega}{\partial T^2}\right)_{V,\mu} + \left[\left(\frac{\partial^2 \Omega}{\partial T\, \partial \mu}\right)_{V}\right]^2 \Big/ \left(\frac{\partial^2 \Omega}{\partial \mu^2}\right)_{V,T} \right\} \tag{4.26}$$

This completes the derivation of the important thermodynamic quantities from the grand canonical potential $\Omega(V, T, \mu)$. The comparative complication of working with μ rather than N as a variable of state is more than made up by the simplicity with which Ω can be evaluated from statistical mechanics, compared to the evaluation of F_N. Furthermore, in many important cases (e.g., chemical reactions) N is simply not a suitable variable of state.

5. The Fermi-Dirac and Bose-Einstein Distributions

In a system of non-interacting particles, called an ideal gas, the quantum states α of the system as a whole can be described by specifying occupation numbers n_k of single-particle states k. The set of all occupation numbers is called a "configuration"

$$\alpha = \{n_0, n_1, n_2, \cdots, n_k, \cdots\} \tag{5.1}$$

The total number of particles is given by

$$N = \sum_k n_k \tag{5.2}$$

and the energy of the N-particle system in state α is related to the single-particle energies ϵ_k through

$$E_\alpha = \sum_k \epsilon_k n_k^{(\alpha)} \tag{5.3}$$

Here we have added the superscript α to the occupation numbers n_k for configuration α.

The quantum statistics of the particles affects the possible values of the occupation numbers: For Bose-Einstein statistics, all non-negative integral values are permitted; for Fermi-Dirac statistics, the possible values of each n_k are 0 and 1, only.

The evaluation of the free energy F from Eq. (3.3) would suffice to define all the thermodynamic properties of such a gas; but the condition of given N, (5.2), makes this evaluation exceedingly complicated. We shall therefore follow an alternative route, by evaluating the grand canonical potential Ω from Eq. (4.14), which we rewrite in the form

$$\exp(-\beta\Omega) = \sum_N \sum_\alpha \exp[\beta(\mu N - E_{N\alpha})] \tag{5.4}$$

When we use (5.2) for N and (5.3) for $E_{N\alpha}$, the explicit dependence on N

disappears; summation over all configurations α for given N, followed by summation over all N, is equivalent to an independent summation over all occupation numbers n_k. We thus obtain

$$\exp(-\beta\Omega) = \sum_{n_0}\sum_{n_1}\sum_{n_2}\cdots\sum_{n_k}\cdots \exp\left[\beta\sum_k(\mu-\epsilon_k)\,n_k\right] \tag{5.5}$$

We define $\mathfrak{u}_k$ by

$$\mathfrak{u}_k = \exp[\beta(\mu-\epsilon_k)] \tag{5.6}$$

and note that the summand in (5.5) is the product of factors $(\mathfrak{u}_k)^{n_k}$. Let us introduce the single sums S_k by the definition

$$S_k = \sum_{n_k}\exp[\beta(\mu-\epsilon_k)\,n_k] = \sum_{n_k}(\mathfrak{u}_k)^{n_k} \tag{5.7}$$

It follows from the independence[11] of the separate sums in (5.5) that the multiple sum of the product is a product of single sums S_k, i.e.,

$$\exp(-\beta\Omega) = \prod_k S_k \tag{5.8}$$

This expression applies to both Fermi-Dirac and Bose-Einstein statistics; the two forms of quantum statistics differ only in the range of summation in (5.7). For Fermi-Dirac statistics, we have

$$S_k = \sum_{n_k=0}^{1}(\mathfrak{u}_k)^{n_k} = 1+\mathfrak{u}_k \qquad \text{(F. D.)} \tag{5.9}$$

and thus, from (5.8) and (5.6),

$$\Omega(V,T,\mu) = -\mathfrak{k}T\sum_k \ln\{1+\exp[\beta(\mu-\epsilon_k)]\} \qquad \text{(F.D.)} \tag{5.10}$$

The detailed probability $P_{N\alpha}$ of a definite quantum state (configuration) α of exactly N particles is given by (4.4); using (5.2), (5.3), and (5.10), this becomes

$$P_{N\alpha} = \exp(\beta\Omega)\prod_k(\mathfrak{u}_k)^{n_k} = \prod_k\frac{(\mathfrak{u}_k)^{n_k}}{1+\mathfrak{u}_k} \qquad \text{(F.D.)} \tag{5.11}$$

(5.11) together with the definition (5.6) defines the full *Fermi-Dirac distribution law.*

In practice, we are interested only in averages over this distribution;

[11] This is the point at which the use of the grand canonical ensemble is essential. In the canonical ensemble, N must be kept fixed, and this requirement introduces a coupling between the sums over the individual occupation numbers n_k.

of particular concern is the average occupation number $\bar{n}_k$ defined by

$$\bar{n}_k = \sum_N \sum_\alpha n_k^{(\alpha)} P_{N\alpha} = \exp(\beta\Omega) \sum_{n_k} n_k(\mathrm{u}_k)^{n_k} \prod_{l \neq k} S_l \tag{5.12}$$

For Fermi-Dirac statistics, we have

$$\sum_{n_k=0}^{1} n_k(\mathrm{u}_k)^{n_k} = 0 + \mathrm{u}_k = \mathrm{u}_k$$

and the factors S_l for $l \neq k$ cancel against corresponding factors in $\exp(\beta\Omega)$; thus we obtain the average numbers

$$\bar{n}_k = \frac{\mathrm{u}_k}{1 + \mathrm{u}_k} = \frac{1}{\exp[\beta(\epsilon_k - \mu)] + 1} \qquad \text{(F. D.)} \tag{5.13}$$

This set of average occupation numbers is often called the Fermi-Dirac distribution function; but, strictly speaking, the full probability distribution of a Fermi-Dirac gas is the projection, onto the subspace of given N, of (5.11).

As a check on (5.13), we may evaluate the average total number of particles $\bar{N}$ from the thermodynamic expression (4.15); carrying out the differentiation on (5.10), and comparing with (5.13), we obtain the expected result

$$\bar{N} = \sum_k \bar{n}_k \tag{5.14}$$

Similarly, formula (4.22) for the thermodynamic internal energy reduces to the expected result

$$E = \sum_k \epsilon_k \bar{n}_k \tag{5.15}$$

In Appendix B, Section 1, we show that a similar formula can be used for any one-particle operator J with matrix elements $\langle k \mid j \mid k' \rangle$ between single-particle states k and k'; only the diagonal matrix elements enter into the ensemble average result (B, 1.12).

The corresponding calculation for the Bose-Einstein gas is equally simple; the sum (5.7) now goes over all non-negative integers, and yields

$$S_k = \sum_{n_k=0}^{\infty} (\mathrm{u}_k)^{n_k} = \frac{1}{1 - \mathrm{u}_k} \qquad \text{(B.E.)} \tag{5.16}$$

and thus, from (5.8) and (5.6),

$$\Omega(V, T, \mu) = \mathfrak{t}T \sum_k \ln\{1 - \exp[\beta(\mu - \epsilon_k)]\} \qquad \text{(B.E.)} \tag{5.17}$$

The detailed probability distribution in the Bose-Einstein case is

$$P_{N\alpha} = \exp(\beta\Omega) \prod_k (\mathrm{u}_k)^{n_k} = \prod_k (1 - \mathrm{u}_k)(\mathrm{u}_k)^{n_k} \qquad \text{(B.E.)} \qquad (5.18)$$

Formula (5.12) for the average occupation number still applies, but the range of summation now extends to infinity, with the result

$$\sum_{n_k=0}^{\infty} n_k(\mathrm{u}_k)^{n_k} = 0 + \mathrm{u}_k + 2(\mathrm{u}_k)^2 + \cdots = \frac{\mathrm{u}_k}{(1 - \mathrm{u}_k)^2}$$

We therefore obtain

$$\bar{n}_k = \frac{\mathrm{u}_k}{1 - \mathrm{u}_k} = \frac{1}{\exp[\beta(\epsilon_k - \mu)] - 1} \qquad \text{(B.E.)} \qquad (5.19)$$

Equations (5.14) and (5.15) still hold, and can be derived in the same way by performing differentiations on (5.17).

6. Minimum Principles in Statistical Mechanics

In Sections 3 and 4, we derived the canonical and grand canonical distribution laws, respectively, by maximizing the disorder (entropy) of an equilibrium ensemble $\mathcal{U} = \mathcal{U}(H)$ under specified external conditions.

We now ask whether these distribution laws maximize the entropy for *all* ensembles (all density matrices $\mathcal{U}$) under the specified external conditions; i.e., we now drop the restriction that $\mathcal{U}$ must be a function of the Hamiltonian H only. We replace the restricted extremum problem (3.1) by the more general problem: Find a Hermitean operator $\mathcal{U}$ which maximizes

$$S = -\mathfrak{k}\, \text{Trace}\, (\mathcal{U} \ln \mathcal{U}) = \max \qquad (6.1a)$$

subject to the conditions

$$\text{Trace}\, (H\mathcal{U}) = E = \text{given} \qquad (6.1b)$$

$$\text{Trace}\, (\mathcal{U}) = 1 \qquad (6.1c)$$

$$0 \leqslant (\Phi, \mathcal{U}\Phi) \leqslant 1 \text{ for all square integrable } \Phi \qquad (6.1d)$$

These are the immediate generalizations of (3.1a)–(3.1d), respectively.

In his book, von Neumann (32) proves[12] that the density matrix $\mathcal{U}$

[12] See pages 202 to 210 of the Dover edition.

which solves (6.1) is a function of H, i.e., that the general problem (6.1) reduces directly to the simpler problem (3.1), and thence to the solution

$$\mathscr{U} = \exp [\beta(F - H)] \tag{6.2}$$

where β and F are c-numbers; (6.2) is the operator form of (3.2), the P_α being the eigenvalues of $\mathscr{U}$.

The derivation of (6.2) from (6.1) is quite complicated, and we shall not reproduce it here. Once this theorem is accepted, however, a number of powerful minimum principles follow easily from it.

Let us plot, in the S-E plane, all pairs of values of S and E which can be attained by means of density matrices $\mathscr{U}$ satisfying (6.1c) and (6.1d). Such a plot is shown schematically in Fig. A.1, for ordinary

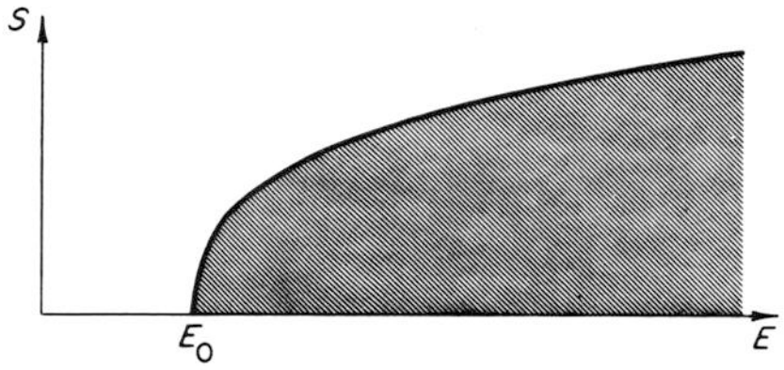

FIG. A.1. Entropy S vs. internal energy E, for a normal system. The shaded region contains possible pairs of values of S and E, obtained from density matrices $\mathscr{U}$ satisfying (6.1c) and (6.1d). The boundary curve corresponds to maximum S for given E, and contains the equilibrium states, with density matrices (6.2). E is bounded from below by the ground state energy E_0, but is not bounded from above. The slope of the boundary curve is related to the absolute temperature T by the thermodynamic relation $(\partial S/\partial E)_V = 1/T$. This slope is positive and is a decreasing function of E. In the limit $E \to E_0$, $\mathscr{U}$ becomes a projection operator onto the ground state, S approaches zero, and the slope becomes infinite (T approaches zero). The figure shows directly that the boundary curve also corresponds to minimum E for given S, i.e., the minimum problem (6.3).

systems. The possible values of S and E lie in the shaded portion of the S-E plane; the boundary of the shaded region represents the maximum values of S for given E, which means the equilibrium ensembles (6.2). The slope of this boundary curve is related to the absolute temperature T by the thermodynamic relation (3.23); in particular, the slope becomes infinite as E approaches the ground state energy E_0 (and S correspondingly approaches 0). For ordinary system, the temperature T is always non-negative, and thus the boundary curve in Fig. A.1 continues sloping upwards.

Let us restrict ourselves to such ordinary systems for the moment, and let us ask for the solution to the following minimum problem:

$$E = \text{Trace}\,(H\mathscr{U}) = \min \tag{6.3a}$$

subject to

$$-\mathfrak{k}\,\mathrm{Trace}\,(\mathscr{U}\ln\mathscr{U}) = S = \text{given} \tag{6.3b}$$

$$\mathrm{Trace}\,(\mathscr{U}) = 1 \tag{6.3c}$$

$$0 \leqslant (\Phi, \mathscr{U}\Phi) \leqslant 1 \text{ for all square integrable } \Phi \tag{6.3d}$$

Looking at Fig. A.1, we are asking for the lowest E for given S. It is clear that the solutions lie on the same boundary curve as before, and hence (6.2) represents the solutions to the minimum problem (6.3), as well as the solutions to the maximum problem (6.1), for ordinary (positive temperature) systems.

We must point out, however, that exceptional systems exist (e.g., systems of spins) in which the Hamiltonian H has a maximum eigenvalue E_m as well as a minimum eigenvalue E_0. For such systems, the possible points in the S-E plane are represented schematically in Fig. A.2.

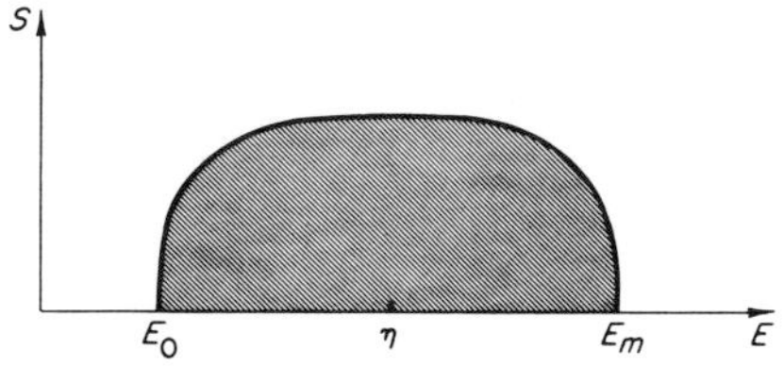

FIG. A.2. Entropy S vs. internal energy E, for an exceptional system in which the eigenvalues of the Hamiltonian are bounded from above (by a maximum energy E_m) as well as from below. The shaded region contains possible pairs of values of S and E, obtained from density matrices $\mathscr{U}$ satisfying (6.1c) and (6.1d). The boundary curve still represents maximum S for given E, and contains the equilibrium states. Now, however, the slope of the boundary curve changes sign at $E = \eta$. For $E > \eta$, the temperature T is negative. As E moves from its minimum value E_0 to its maximum value E_m, the temperature goes from 0 (at $E = E_0$) to $+\infty$ (E just below η), to $-\infty$ (E just above η), and then approaches zero from the negative side as E approaches E_m. Thus negative temperatures are "higher" than all positive temperatures. The parameter $\beta = 1/\mathfrak{k}T$ is actually more appropriate here, going smoothly from $+\infty$ at $E = E_0$, through 0 at $E = \eta$, to $-\infty$ at $E = E_m$. The minimum problem $E = \min$ for given S leads to only the positive temperature side of the boundary curve. To obtain the negative temperature side of the boundary curve, we must ask for maximum E for given S.

Von Neumann's theorem still applies, i.e., the boundary curve of maximum S for given E contains the equilibrium ensembles (6.2); but the slope of this boundary curve is no longer purely positive; rather, for energies $E \geqslant \eta$ the slope is negative, corresponding according to (3.23) to a negative temperature.

For such systems, the minimum problem (6.3) does not give all equilibrium ensembles, but merely those corresponding to the positive temperature (positive slope) part of the boundary curve in Fig. A.2. The negative temperature portion is obtained by requiring (6.3a) to be a maximum rather than a minimum, under the same restraints.

The minimum principle (6.3) in turn can be used to prove a minimum principle for the free energy F, namely

$$F = \text{Trace}\,(H\mathscr{U}) + \mathfrak{k}T\,\text{Trace}\,(\mathscr{U}\ln\mathscr{U}) = \min \tag{6.4a}$$

subject to

$$\text{Trace}\,(\mathscr{U}) = 1 \tag{6.4b}$$

$$0 \leqslant (\Phi, \mathscr{U}\Phi) \leqslant 1 \text{ for all square integrable } \Phi \tag{6.4c}$$

We assert that the solution of (6.4) is the operator (6.2), with $\beta = 1/\mathfrak{k}T$ and F equal to the expectation value (6.4a).

The proof of this assertion proceeds in two steps. In the first step, we restrict ourselves to density matrices $\mathscr{U}$ with a given value of S, (6.3b). In the second step, we allow the given value S to vary.

The first step reduces directly to the minimum problem (6.3), with the known solution

$$\mathscr{U}^* = \exp\,[\beta^*(F^* - H)] \tag{6.5}$$

Equation (6.4b) provides a relation between the parameters β^* and F^*, namely

$$\text{Trace}\exp\,[\beta^*(F^* - H)] = 1 \tag{6.6}$$

We differentiate both sides of (6.6) with respect to β^* to obtain

$$\text{Trace}\,(H\mathscr{U}^*) = \frac{\partial(\beta^*F^*)}{\partial\beta^*} \tag{6.7}$$

We now proceed to the second step of the proof, i.e., we now allow β^* to vary, with F^* determined as a function of β^* by (6.6), and we look for the minimum value of (6.4a). Straightforward substitution yields

$$\begin{aligned} F = \text{Trace}\,(H\mathscr{U}^*) + \mathfrak{k}T\,\text{Trace}\,(\mathscr{U}^*\ln\mathscr{U}^*) &= \text{Trace}\,\{[H + \mathfrak{k}T\beta^*(F^* - H)]\,\mathscr{U}^*\} \\ &= \mathfrak{k}T\beta^*F^* + (1 - \mathfrak{k}T\beta^*)\,\text{Trace}\,(H\mathscr{U}^*) \end{aligned} \tag{6.8}$$

In order to find the minimum of this quantity, we differentiate with respect to β^*; use of the identity (6.7) then gives

$$\frac{\partial F}{\partial\beta^*} = (1 - \mathfrak{k}T\beta^*)\frac{\partial}{\partial\beta^*}\,\text{Trace}\,(H\mathscr{U}^*) = 0$$

and thus the minimum value of F is obtained when the parameter β^* equals $1/\mathfrak{k}T$; for this value of β^*, (6.8) gives $F = F^*$, i.e., F as defined by the minimum problem (6.4) is the ordinary free energy at temperature T, and $\mathscr{U}$ is the canonical density matrix (6.2), as asserted.

The minimum principle (6.4) is applicable only for positive temperatures T. For negative values of T, one has to look for a maximum of (6.4a); solutions of that problem exist only for the exceptional systems mentioned earlier.

(6.4) is a statistical mechanical version of the thermodynamic statement that the free energy F is a minimum for systems in thermal equilibrium, for given volume V and temperature T. In the same way, (6.3) is the analog of the thermodynamic statement that the internal energy E is a minimum for systems in thermal equilibrium, if the systems are kept thermally insulated (constant entropy S).

Equation (6.4) can also be considered to be the generalization to statistical mechanics of the well-known Rayleigh-Ritz minimum principle of quantum mechanics. In fact, (6.4) reduces to the ordinary Rayleigh-Ritz principle in the limit $T \to 0$; the corresponding density matrix $\mathscr{U}$ is a projection operator onto the ground state Ψ_0 of the Hamiltonian H, and F becomes the ground state energy E_0.

In quantum mechanics, the Rayleigh-Ritz principle provides a powerful approximation procedure; the ground state wave function Ψ_0 is replaced by a trial function Ψ_t depending upon a set of parameters c_1, c_2, $\cdots$, and one determines an upper bound on the ground state energy E_0 by minimizing the expectation value $(\Psi_t, H\Psi_t)$ with respect to the parameters c_i.

Similarly, the minimum principle (6.4) serves as the basis of an approximation procedure in statistical mechanics. One replaces the true density matrix $\mathscr{U}$, (6.2), by a trial density matrix $\mathscr{U}_t$ satisfying (6.4b) and (6.4c), and one evaluates

$$F_t = \text{Trace}\,(H\mathscr{U}_t) + \mathfrak{k}T\,\text{Trace}\,(\mathscr{U}_t \ln \mathscr{U}_t) \tag{6.9}$$

The minimum principle (6.4) implies that

$$F_t \geqslant F \tag{6.10}$$

Thus, if $\mathscr{U}_t$ depends on certain parameters c_1, c_2, $\cdots$, one may minimize the trace (6.9) with respect to these parameters, to obtain the best approximation to the correct F and the correct $\mathscr{U}$.

As a useful illustration of this procedure, we use it now to derive a minimum principle of Peierls (38). The practical difficulty in using (6.9) lies in the evaluation of Trace $(\mathscr{U}_t \ln \mathscr{U}_t)$; we circumvent this difficulty by constructing a trial operator $\mathscr{U}_t$ in diagonal form, as follows: Let Φ_1, Φ_2, Φ_3, $\cdots$ be an orthonormal complete set chosen at will.

Let $\mathscr{U}_t$ be diagonal with respect to this set, with eigenvalues P_n given by

$$P_n = \exp\left(\frac{F^* - H_{nn}}{\mathfrak{k}T}\right) \tag{6.11}$$

where H_{nn} is the expectation value of H over the wave function Φ_n

$$H_{nn} = (\Phi_n, H\Phi_n) \tag{6.12}$$

and where F^* is determined from the condition (6.4b), which we write in the form

$$\exp(-F^*/\mathfrak{k}T) = \sum_n \exp(-H_{nn}/\mathfrak{k}T) \tag{6.13}$$

The operator $\mathscr{U}_t$ constructed in this fashion clearly satisfies (6.4b) and (6.4c). Let us now evaluate F_t, (6.9), for this choice of $\mathscr{U}_t$. Since $\mathscr{U}_t$ is by construction diagonal in the set Φ_n, the traces in (6.9) can be written as single sums over n, and yield

$$F_t = \sum_n H_{nn}P_n + \mathfrak{k}T \sum_n P_n \ln P_n = F^* \sum_n P_n = F^* \tag{6.14}$$

Thus, (6.10) can be applied to deduce Peierls' theorem: *The trial free energy F^* defined by* (6.13) *with an arbitrary orthonormal set Φ_n is greater than or equal to the true free energy F.*

Finally, we derive a minimum principle for the grand canonical free energy Ω. The principle in question reads[13]:

$$\Omega = \text{Trace}\,(H\mathscr{U}) + \mathfrak{k}T\,\text{Trace}\,(\mathscr{U}\ln\mathscr{U}) - \mu\,\text{Trace}\,(N_{\text{Op}}\mathscr{U}) = \min \tag{6.15a}$$

subject to

$$\text{Trace}\,(\mathscr{U}) = 1 \tag{6.15b}$$

$$0 \leqslant (\Phi, \mathscr{U}\Phi) \leqslant 1 \qquad \text{for all square integrable } \Phi \tag{6.15c}$$

In (6.15a), T and μ are given quantities, and the theorem asserts that the density matrix $\mathscr{U}$ which minimizes (6.15a) is the grand canonical density matrix

$$\mathscr{U} = \exp\left(\frac{\Omega + \mu N_{\text{Op}} - H}{\mathfrak{k}T}\right) \tag{6.16}$$

with $\Omega = \Omega(V, T, \mu)$ equal to (6.15a) and identical with the grand canonical potential defined in Section 4.

---

[13] N_{Op} is the operator for the number of particles.

The proof of (6.15) is trivial, since all we need to do is introduce the operator

$$H^* = H - \mu N_{\text{Op}} \tag{6.17}$$

whereupon (6.15) reduces to (6.4) with H replaced by H^*. Since we did not use any special properties of the operator H in the proof of (6.4), the solution to (6.16) can be read off from the solution (6.2) of (6.4), by replacing H by H^*. With an obvious change of notation, this solution becomes identical with (6.16), as asserted. The proof that (6.15a) with (6.16) for $\mathscr{U}$ reduces to the ordinary grand canonical potential is equally trivial.

The minimum principle (6.15) can again be used as the basis of an approximation procedure. We replace the true density matrix $\mathscr{U}$, (6.16), by a trial density matrix $\mathscr{U}_t$ satisfying (6.15b) and (6.15c), and we evaluate the trial grand canonical free energy Ω_t from

$$\Omega_t = \text{Trace}\,(H\mathscr{U}_t) + \mathfrak{k}T\,\text{Trace}\,(\mathscr{U}_t \ln \mathscr{U}_t) - \mu\,\text{Trace}\,(N_{\text{Op}}\mathscr{U}_t) \tag{6.18}$$

This trial value of Ω is greater than or equal to the true value; if $\mathscr{U}_t$ contains adjustable parameters, we may minimize Ω_t with respect to these parameters. One application of this procedure is given in Section 1 of Appendix B.

Appendix B

Mathematical Formulation of the Quasi-Chemical Equilibrium Theory

1. The Husimi Theory

In Chapter VII, we wrote down the trial density matrix of the quasichemical equilibrium theory, Eqs. (VII, 2.6), (VII, 2.11), and (VII, 2.12). We now write equivalent expressions, using creation and destruction operators.

We start with the independent-particle approximation of Husimi, i.e., we ignore all pairing for the moment. The unsymmetrized density matrix is then a simple Kronecker product of one-particle operators $\langle k \mid \hat{U} \mid k' \rangle$ [see (VII, 2.2)]. Even then, the final expression is complicated because of the antisymmetrization, Eq. (VII, 2.11), and the sum over all numbers of particles N, Eq. (VII, 2.12).

Without loss of generality, we may assume that the states $\mid k \rangle$ are eigenstates of the operator $\hat{U}$, i.e.,

$$\langle k \mid \hat{U} \mid k' \rangle = \delta_{kk'} \hat{\mathrm{u}}_k$$

Although the formulas can be written down in terms of arbitrary states, the formulas simplify greatly for these particular basis states. For example, the Husimi trial density matrix, applied to any configuration

$$\gamma = \{n_1, n_2, \cdots, n_k, \cdots\} \tag{1.1}$$

has the effect of multiplying the wave function by a factor (i.e., the configurations are eigenfunctions of $\mathscr{U}$); the factor is made up of a product of simple factors, namely:

(i) the factor $\exp(N\beta\mu) = \exp(\beta\mu \sum_k n_k) = \prod_k \exp(\beta\mu n_k)$

(ii) a factor $\hat{u}_k$ for each state appearing in the configuration, i.e., for each k with $n_k = 1$.

We can combine these factors by introducing the one-particle operator U through

$$\langle k \mid U \mid k' \rangle = \exp(\beta\mu) \langle k \mid \hat{U} \mid k' \rangle \tag{1.2}$$

with eigenvalues

$$u_k = \exp(\beta\mu)\, \hat{u}_k \tag{1.3}$$

Let $\mathscr{V}$ stand for the unnormalized trial density matrix of the Husimi theory; we then have

$$\mathscr{V} \mid \{n_1, n_2, \cdots, n_k, \cdots\}\rangle = \prod_k (u_k)^{n_k} \mid \{n_1, n_2, \cdots, n_k, \cdots\}\rangle \tag{1.4}$$

Using the fact that n_k is the eigenvalue of the operator $a_k^+ a_k$ (see Chapter V, Section 1), we have the operator form

$$\mathscr{V} = \prod_k (u_k)^{a_k^+ a_k} \tag{1.5}$$

In order to normalize the trial density matrix properly, we require the trace of this operator. The trace is the sum of all the eigenvalues (1.4), i.e.,

$$\begin{aligned} \text{Trace } \mathscr{V} &= \sum_\gamma \left\{ \prod_k (u_k)^{n_k} \right\} \\ &= \prod_k \left\{ \sum_{n_k=0,1} (u_k)^{n_k} \right\} = \prod_k (1 + u_k) \end{aligned} \tag{1.6}$$

Thus, the normalized trial density matrix of Husimi is

$$\mathscr{U}_H = \frac{\mathscr{V}}{\text{Trace } \mathscr{V}} = \prod_k \frac{(u_k)^{n_k}}{1 + u_k} \tag{1.7}$$

Next, we evaluate the expectation value of the operator $n_k = a_k^+ a_k$. For all $l \neq k$, the calculation of Trace $(n_k \mathscr{V})$ is the same as for the Trace $\mathscr{V}$. For $l = k$ we get a factor

$$\sum_{n_k=0}^{1} n_k (u_k)^{n_k} = u_k$$

and so we obtain

$$\text{Trace } (n_k \mathscr{V}) = u_k \prod_{l \neq k} (1 + u_l) \tag{1.8}$$

whence

$$\bar{n}_k = \frac{\text{Trace}\,(n_k \mathscr{V})}{\text{Trace}\,\mathscr{V}} = \frac{\mathrm{u}_k}{1 + \mathrm{u}_k} \tag{1.9}$$

and, conversely,

$$\mathrm{u}_k = \frac{\bar{n}_k}{1 - \bar{n}_k} \tag{1.10}$$

The most general one-particle operator J is given by (V, 1.22). The expectation value of this operator is known if we know the trace:

$$\text{Trace}\,(a_k^+ a_l \mathscr{V})$$

If $k = l$, the result is (1.8); on the other hand, if $k \neq l$, the result is zero, since the operator $\mathscr{V}$ is diagonal in the occupation numbers n_k, whereas the operator $a_k^+ a_l$ increases n_k by 1, and decreases n_l by 1. Thus, we have the result

$$\frac{\text{Trace}\,(a_k^+ a_l \mathscr{V})}{\text{Trace}\,\mathscr{V}} = \delta_{kl} \bar{n}_k \tag{1.11}$$

whence

$$\frac{\text{Trace}\,(J\mathscr{V})}{\text{Trace}\,\mathscr{V}} = \sum_k \langle k \mid j \mid k \rangle\, \bar{n}_k \tag{1.12}$$

The most general two-particle operator is given by (V, 1.26). We thus require the trace

$$\text{Trace}\,(a_k^+ a_l^+ a_{l'} a_{k'} \mathscr{V})$$

Since $\mathscr{V}$ is diagonal in the occupation numbers, the above trace vanishes unless either $k = k'$ and $l = l'$, or else $k = l'$ and $l = k'$; by (V, 1.8), it vanishes for $k = l$, and by (V, 1.12) it vanishes for $k' = l'$. The first case of interest is then $k = k' \neq l = l'$. Using the commutation rules (V, 1.7), (V, 1.13), and (V, 1.18), we get

$$\text{Trace}\,(a_k^+ a_l^+ a_l a_k \mathscr{V}) = \text{Trace}\,(a_k^+ a_k a_l^+ a_l \mathscr{V}) = \bar{n}_k \bar{n}_l\, \text{Trace}\,\mathscr{V}$$

where the final result is obtained in a way similar to the derivation of (1.8) and (1.9). The other possibility, $k = l' \neq l = k'$ gives the same result except for a minus sign. Thus, finally,

$$\frac{\text{Trace}\,(a_k^+ a_l^+ a_{l'} a_{k'} \mathscr{V})}{\text{Trace}\,\mathscr{V}} = (\delta_{kl}\, \delta_{k'l'} - \delta_{kl'}\, \delta_{lk'})\, \bar{n}_k \bar{n}_l \tag{1.13}$$

whence

$$\frac{\text{Trace}\,(T\mathscr{V})}{\text{Trace}\,\mathscr{V}} = \tfrac{1}{2} \sum_{k,l} (\langle kl \mid t \mid kl \rangle - \langle kl \mid t \mid lk \rangle)\, \bar{n}_k \bar{n}_l \tag{1.14}$$

These expectation values of the Husimi theory go over directly into Hartree-Fock expectation values if we restrict ourselves to one unique configuration; for then, $\bar{n}_k = n_k$ is zero for an unoccupied state, and unity for an occupied state. In the Husimi theory, on the other hand, we have a statistical average over many configurations, so that $\bar{n}_k$ in general is neither zero nor one, but somewhere in between.

If we write the Hamiltonian of the system in the form

$$H = T + V = \sum_k \epsilon_k a_k^+ a_k + \tfrac{1}{2} \sum_{k,l,k',l'} \langle kl \mid v \mid k'l' \rangle \, a_k^+ a_l^+ a_{l'} a_{k'} \tag{1.15}$$

then its statistical expectation value, the internal energy $\bar{E}$, is

$$\bar{E} = \frac{\text{Trace}\,(H\mathscr{V})}{\text{Trace}\,\mathscr{V}} = \sum_k \epsilon_k \bar{n}_k + \tfrac{1}{2} \sum_{k,l} (\langle kl \mid v \mid kl \rangle - \langle kl \mid v \mid lk \rangle)\, \bar{n}_k \bar{n}_l \tag{1.16}$$

This value is exact if H is an independent-particle Hamiltonian, i.e., if the interaction matrix elements all vanish. Otherwise, (1.16) is an approximation.

In order to get the best possible approximation in the general case, we use the variation principle (VII, 1.6). Since the trial density matrix, (1.7), is a product over states k, its logarithm is the sum

$$\ln \mathscr{U}_H = \sum_k [n_k \ln (\mathfrak{u}_k) - \ln (1 + \mathfrak{u}_k)] \tag{1.17}$$

and the Husimi approximation to the entropy reads

$$S = -\mathfrak{k}\, \text{Trace}\,(\mathscr{U}_H \ln \mathscr{U}_H) = -\mathfrak{k} \sum_k [\bar{n}_k \ln (\mathfrak{u}_k) - \ln (1 + \mathfrak{u}_k)] \tag{1.18}$$

The quantity to be minimized is

$$\begin{aligned} \Omega &= \bar{E} - \mu \bar{N} - TS \\ &= \sum_k [(\epsilon_k - \mu + \mathfrak{k}T \ln \mathfrak{u}_k)\, \bar{n}_k - \mathfrak{k}T \ln (1 + \mathfrak{u}_k)] \\ &\quad + \tfrac{1}{2} \sum_{k,l} (\langle kl \mid v \mid kl \rangle - \langle kl \mid v \mid lk \rangle)\, \bar{n}_k \bar{n}_l \end{aligned} \tag{1.19}$$

The independent parameters for the variation are the numbers $\mathfrak{u}_k$; $\bar{n}_k$ is a function of $\mathfrak{u}_k$, given by (1.9). The calculation is straightforward, and the result can be written most simply as follows: we introduce the quantity L_k by

$$L_k = \tfrac{1}{2} \sum_l (\langle kl \mid v \mid kl \rangle - \langle kl \mid v \mid lk \rangle)\, \bar{n}_l \tag{1.20}$$

so that $2L_k$ is the average interaction felt by a particle in state k, due to all the other particles. We introduce a "renormalized energy" $\hat{\epsilon}_k$ through

$$\hat{\epsilon}_k = \epsilon_k + 2L_k \tag{1.21}$$

We obtain the minimum value for (1.19) by setting

$$\mathrm{u}_k = \exp\left[\beta(\mu - \hat{\epsilon}_k)\right] = \exp\left[\beta(\mu - \epsilon_k - 2L_k)\right] \tag{1.22}$$

(1.9) then yields

$$\bar{n}_k = \frac{1}{\exp\left[\beta(\hat{\epsilon}_k - \mu)\right] + 1} \tag{1.23}$$

In spite of its innocuous appearance, (1.23) is a non-linear integral equation; the energies $\hat{\epsilon}_k$, (1.21), themselves depend on $\bar{n}_k$ through (1.20)!

If the system is an independent-particle system, L_k vanishes, $\hat{\epsilon}_k = \epsilon_k$, and (1.23) is just the ordinary Fermi distribution law.

In the general case, a good low-temperature approximation is obtained by using (VII, 2.4) for $\hat{\epsilon}_k$, instead of the exact (1.21). This amounts to ignoring the temperature dependence of L_k, (1.20), which is a very reasonable approximation indeed.

Substituting the expressions (1.20)–(1.23) into (1.16), the internal energy in the Husimi approximation is

$$\bar{E} = \sum_k (\epsilon_k + L_k)\, \bar{n}_k \tag{1.24}$$

We note that, unlike (1.21), there is no factor 2 multiplying L_k in (1.24). This difference of a factor 2 is characteristic of the Hartree-Fock theory: without it, each particle-particle interaction would contribute twice to the energy expectation value (1.24).

If we use (1.10), we can put the Husimi entropy (1.18) into a neater form, namely

$$\begin{aligned} S &= -\mathfrak{k}\, \mathrm{Trace}\,(\mathscr{U}_H \ln \mathscr{U}_H) \\ &= -\mathfrak{k} \sum_k \left[\bar{n}_k \ln \bar{n}_k + (1 - \bar{n}_k) \ln (1 - \bar{n}_k)\right] \end{aligned} \tag{1.25}$$

We note that this form follows directly from the trial density matrix (1.7), no matter what values we put in for the parameters u_k, or the corresponding $\bar{n}_k$.

In the limit of zero temperature, $\beta \to \infty$, Eq. (1.23) leads to $\bar{n}_k = 0$ for $\hat{\epsilon}_k > \mu$, and $\bar{n}_k = 1$ for $\hat{\epsilon}_k < \mu$. The entropy (1.25) vanishes, as it must, and the internal energy (1.24) reduces to the ordinary Hartree-Fock value.

2. Trial Density Matrix with Pair Correlations

In order to introduce pair correlations into the trial density matrix, we make use of the decomposition (VII, 4.1) of the basic pair matrix $\hat{W}$:

$$\langle kl \mid \hat{W} \mid k'l' \rangle = \sum_{\alpha} \delta_\alpha \hat{w}_\alpha(k, l)\, \hat{w}_\alpha^*(k', l') \tag{2.1}$$

where $\delta_\alpha = \pm 1$, only. If all the δ_α are $+1$, the operator $\hat{W}$ is non-negative. We shall assume this henceforth (see Chapter VII, Section 3, for a discussion of that assumption).

In order to allow conveniently for the factors $\exp(N\beta\mu)$ in (VII, 2.12), we introduce the functions $w_\alpha(k, l)$ by

$$w_\alpha(k, l) = \exp(\beta\mu)\, \hat{w}_\alpha(k, l) \tag{2.2}$$

We use these functions to construct second-quantized operators W_α through

$$W_\alpha = \frac{1}{\sqrt{2}} \sum_{k,l} w_\alpha^*(k, l)\, a_k a_l \tag{2.3}$$

The second-quantized pair correlation operator for the case of just two particles is then given by

$$\tfrac{1}{2} \exp(2\beta\mu) \sum_{k,l,k',l'} \langle kl \mid \hat{W} \mid k'l' \rangle\, a_k^+ a_l^+ a_{l'} a_{k'} = \sum_{\alpha} \delta_\alpha W_\alpha^+ W_\alpha \tag{2.4}$$

If all the $\delta_\alpha = +1$, the operator on the right-hand side of (2.4) is clearly non-negative; we note, also, that there are no cross-terms of type $W_\alpha^+ W_\beta$, $\beta \neq \alpha$, on the right of (2.4).

The idea of the subsequent construction of a trial density matrix operator $\mathscr{U}$ is as follows: acting on some configuration γ specified by occupation numbers n_k, the operator $\mathscr{U}$ first "destroys" pairs of particles, by use of the operators (2.3); whatever particles remain, are considered unpaired and are acted on by the Husimi operator $\mathscr{V}$, Eq. (1.5); last, we "re-create" the missing pairs of particles by use of the Hermitean conjugate operators W_β^+. Special precautions are needed to ensure, first, that the total particle number is not altered in the end, and, second, that the various W_α and W_β^+ can eventually be combined into sums of type (2.4).

These precautions are formulated most easily by means of "labeling operators" A_α (Blatt 58) which are purely formal operators obeying Bose-Einstein commutation rules:

$$[A_\alpha, A_\beta^+]_- = \delta_{\alpha\beta}, \qquad [A_\alpha, A_\beta]_- = [A_\alpha^+, A_\beta^+]_- = 0 \tag{2.5}$$

The operators $A_\alpha^+ A_\alpha$ have eigenvalues 0, 1, 2, $\cdots$; we shall think of these as numbers of "pairs of type α." We introduce the projection operator

$$\omega = \text{projection onto the formal state with } A_\alpha^+ A_\alpha = 0, \qquad \text{all } \alpha \tag{2.6}$$

We employ these labeling operators to construct the operator

$$Q = \sum_\alpha W_\alpha A_\alpha^+ \tag{2.7}$$

If we think of W_α as the operator which destroys a "physical pair of type α," and A_α^+ as the operator which creates a "formal pair of type α," then Q creates a formal pair of type α for every physical pair of type α which it destroys. We shall use the numbers of formal pairs to keep track of the numbers of physical pairs.

It is shown in (Blatt 58) that the quasi-chemical equilibrium density matrix (VII, 2.6), (VII, 2.11), (VII, 2.12) can be written in the form

$$\mathscr{U} = C\omega \exp(Q^+) \mathscr{V} \exp(Q)\, \omega \tag{2.8}$$

where C is the same c-number constant which appears in (VII, 2.12). The form (2.8) applies if all the δ_α in (2.1) are positive.[1]

We do not go through the proof of the equivalence of (2.8) and (VII, 2.12) in full detail, but only sketch it roughly.

It follows immediately from the commutation rules that the terms $W_\alpha A_\alpha^+$ in (2.7) all commute with each other. We therefore may expand the exponential $\exp(Q)$ as follows:

$$\exp(Q) = \prod_\alpha \exp(W_\alpha A_\alpha^+) = \prod_\alpha \left\{ \sum_{n_\alpha=0}^{\infty} \frac{(W_\alpha A_\alpha^+)^{n_\alpha}}{n_\alpha!} \right\}$$

$$= \sum_{\{n_\alpha\}} \prod_\alpha \frac{(W_\alpha A_\alpha^+)^{n_\alpha}}{n_\alpha!} \tag{2.9}$$

[1] Otherwise, we must also introduce the operator

$$Q' = \sum_\alpha \delta_\alpha W_\alpha A_\alpha^+$$

and we must replace (2.8) by

$$\mathscr{U} = C\omega \exp(Q^+) \mathscr{V} \exp(Q')\, \omega \tag{2.8$'$}$$

This expression turns out to be equivalent to (VII,2.12) with an arbitrary pair correlation matrix $\hat{W}$. However, the operator $\mathscr{U}$ defined by (2.8′) with arbitrary W_α is not necessarily non-negative and therefore must not be used as a trial operator in the variation principle (VII,1.6).

where $\{n_\alpha\}$ denotes a configuration of pair occupation numbers

$$\{n_\alpha\} = \{n_0, n_1, \cdots, n_\alpha, \cdots\} \tag{2.10}$$

With each such configuration, there is associated a total number of pairs N_2 given by

$$N_2 = \sum_\alpha n_\alpha \tag{2.11}$$

When we substitute the expansion (2.9), and its Hermitean conjugate, into (2.8), we obtain

$$\mathscr{U} = C \sum_{\{n_\alpha\}} \sum_{\{n'_\beta\}} \omega \prod_\alpha \frac{(W_\alpha^+ A_\alpha)^{n_\alpha}}{n_\alpha!} \mathscr{V} \prod_\beta \frac{(W_\beta A_\beta^+)^{n'_\beta}}{n'_\beta!} \omega \tag{2.12}$$

We now make use of the following operator identity, which is a direct consequence of the commutation rules (2.5):

$$\omega \prod_\alpha (A_\alpha)^{n_\alpha} \prod_\beta (A_\beta^+)^{m_\beta} \omega = \omega \prod_\alpha n_\alpha! \, \delta_{n_\alpha m_\alpha} \tag{2.13}$$

to obtain

$$\mathscr{U} = C\omega \sum_{\{n_\alpha\}} \left\{ \prod_\alpha \frac{(W_\alpha^+)^{n_\alpha}}{n_\alpha!} \right\} \mathscr{V} \left\{ \prod_\alpha (W_\alpha)^{n_\alpha} \right\} \tag{2.14}$$

At this stage, the projection operator ω onto the "labeling vacuum" serves no further function, and may be omitted.

Rather than establishing the equivalence of (2.14) and (VII, 2.12) generally, let us restrict ourselves to two-particle states, i.e.,

$$N = \sum_k n_k = 2$$

Since each operator W_α, (2.3), annihilates two particles, there can be at most one such operator active for two-particle states. In terms of the pair configurations (2.10) and (2.11), this means that N_2 is either zero or one; in the latter case, exactly one $n_\alpha = 1$, all the others vanish.

Letting $\mathscr{U}_2$ stand for the projection of the full operator $\mathscr{U}$, (2.8) or (2.14), onto the two-particle space, we therefore obtain the result

$$\mathscr{U}_2 = C\left(\mathscr{V} + \sum_\alpha \frac{1}{1!} W_\alpha^+ \mathscr{V} W_\alpha\right) \tag{2.15}$$

However, since W_α annihilates two particles, the operator $\mathscr{V}$ in $W_\alpha^+ \mathscr{V} W_\alpha$ has no particles at all to operate on, and therefore gives unity; we thus have

$$\mathscr{U}_2 = C\left(\mathscr{V} + \sum_\alpha W_\alpha^+ W_\alpha\right) \tag{2.16}$$

According to (2.4), with all $\delta_\alpha = +1$ by assumption, the sum over α is just the second-quantized expression for the pair correlation operator in the two-particle space. The operator $\mathscr{V}$, in any subspace, is the Husimi operator for independent particles. Thus (2.16) is substantially equivalent to (VII, 2.6a), the only difference being that (2.16) is automatically antisymmetrized according to (VII, 2.11).

The proof of the equivalence for larger numbers of particles is more involved, but follows the same kind of reasoning. As a result of the identity (2.13), all surviving terms contain exactly as many operator factors W_α as W_α^+, separately for each α. These can then always be combined into sums like the one occurring in (2.16), i.e., the final results can be written in terms of the pair correlation operator (2.4) and powers thereof. The factorials in (2.13) provide necessary combinatorial factors.

It is possible (Blatt 58) to extend the Ansatz (2.8) to include correlations of higher order (triplet correlations, quadruplet correlations, etc.) It is necessary to introduce separate labelling operators $A_\alpha^{(k)}$ for correlations of order k. It turns out that these labeling operators obey Bose-Einstein commutation rules for even values of k, and Fermi-Dirac anticommutation rules for odd values of k. This is a formal expression of the statistics of composite particles.

3. The Quenching Identity and Expectation Values

Let J be some operator, for example the single-particle operator (V, 1.22). We wish to find the statistical expectation value of this operator, given by

$$\langle J \rangle = \text{Trace}\,(J\mathscr{U}) \tag{3.1}$$

We transform this by using the property Trace (AB) = Trace (BA)

$$C\,\text{Trace}\,[J\omega \exp(Q^+)\,\mathscr{V} \exp(Q)\,\omega] = C\,\text{Trace}\,[\mathscr{V} \exp(Q)\,\omega J \omega \exp(Q^+)] \tag{3.2}$$

At this stage, it turns out to be convenient to introduce the operator

$$P = \sum_\alpha W_\alpha A_\alpha \tag{3.3}$$

This differs from Q, Eq. (2.7), by destroying (rather than creating) a formal pair of type α for every physical pair of type α which it destroys. By expanding exponentials in power series, and using (2.13), it can be shown that (3.2) can be replaced by

$$\text{Trace}\,(J\mathscr{U}) = C\,\text{Trace}\,[\mathscr{V}\omega \exp(P)\,J \exp(P^+)\,\omega] \tag{3.4}$$

In general, J contains creation and destruction operators. By using the anticommutation rules, it is always possible to write J as a sum of operators in "normal form," this form being defined by the requirement that all creation operators appear to the left of all destruction operators. The "normal form" is commonly used in quantum field theory.

For our purposes at the moment, however, we shall use the anticommutation rules in the opposite direction, so as to write J as a sum of operators in "anti-normal form," in which all creation operators appear to be *right* of all destruction operators. Since P contains only destruction operators, and P^+ contains only creation operators, (3.4) then becomes a sum of terms of type[2]

$$\text{Trace}\,[\mathscr{V} F(a_k)\, G(a_k^+)]$$

where F is a function of destruction operators a_k only, and G is a function of creation operators a_k^+ only.

As an example, let us consider $F(a) = a_k$ and $G(a^+) = a_l^+$, so that the Trace becomes $\text{Trace}\,(\mathscr{V} a_k a_l^+)$. Since the operator $\mathscr{V}$ is diagonal in the occupation numbers, the trace vanishes unless $k = l$. In that case, however, we have

$$\text{Trace}\,(\mathscr{V}\, a_k a_k^+) = \text{Trace}\,[\mathscr{V}(1 - a_k^+ a_k)] = (1 - \bar{n}_k)\,\text{Trace}\,\mathscr{V}$$

We now show that the factor multiplying Trace $\mathscr{V}$ can be written as a vacuum expectation value, as follows: we introduce "quenched" creation and destruction operators $\tilde{a}_k^+$ and $\tilde{a}_k$ by the definitions

$$\tilde{a}_k = (1 - \bar{n}_k)^{1/2}\, a_k = (1 + \text{u}_k)^{-1/2}\, a_k \tag{3.5a}$$

$$\tilde{a}_k^+ = (1 - \bar{n}_k)^{1/2}\, a_k^+ = (1 + \text{u}_k)^{-1/2}\, a_k^+ \tag{3.5b}$$

We then assert the following equality:

$$\text{Trace}\,(\mathscr{V}\, a_k a_l^+) = \langle 0 \mid \tilde{a}_k \tilde{a}_l^+ \mid 0 \rangle\, \text{Trace}\,\mathscr{V} \tag{3.6}$$

The vacuum expectation value vanishes if $k \neq l$, and for $k = l$ we recover the earlier result.

By generalizing the same argument, we may establish the general *quenching identity*

$$\text{Trace}\,[\mathscr{V} F(a)\, G(a^+)] = \langle 0 \mid F(\tilde{a})\, G(\tilde{a}^+) \mid 0 \rangle\, \text{Trace}\,\mathscr{V} \tag{3.7}$$

[2] We are not concerned with the labeling operators in this part of the discussion; the labeling operators commute with all the a_k and a_k^+, and act like c-numbers as far as the quenching identity goes.

We now apply this result to the general expectation value (3.4). Since the projection operator ω gives unity when applied to the vacuum state, we need not write it explicitly for a vacuum expectation value. Thus, *for all operators J in anti-normal form,*

$$\text{Trace}\,(J\mathscr{U}) = C\,\langle 0\,|\,\exp(\tilde{P})\,\tilde{J}\exp(\tilde{P}^{+})\,|\,0\rangle\ \text{Trace}\,\mathscr{V} \tag{3.8}$$

where $\tilde{P}$, $\tilde{J}$, and $\tilde{P}^{+}$ are obtained from P, J, and P^{+}, *respectively, by means of the quenching operation which replaces all a_k by $\tilde{a}_k$, all a_k^{+} by $\tilde{a}_k^{+}$, according to* (3.5).

As a first application of this quenching identity, let us consider the normalization integral Trace $\mathscr{U}$, i.e., we let J be the unit operator. We then get an equation for the constant C in (2.8), namely,

$$\text{Trace}\,\mathscr{U} = 1 = C\,\langle 0\,|\,\exp(\tilde{P})\exp(\tilde{P}^{+})\,|\,0\rangle\ \text{Trace}\,\mathscr{V} \tag{3.9}$$

According to (1.6) and (III, 8.7), the Trace $\mathscr{V}$ is related directly to the grand canonical free energy $\Omega^{0}(V, T, \mu)$ of an ideal Fermi gas by

$$\text{Trace}\,\mathscr{V} = \exp(-\beta\Omega^{0}) \tag{3.10}$$

It is natural, therefore, to define a grand canonical free energy for the pairs, $\Omega'(V, T, \mu)$, by using the other factor in (3.9), i.e., we define

$$\exp(-\beta\Omega') = \langle 0\,|\,\exp(\tilde{P})\exp(\tilde{P}^{+})\,|\,0\rangle \tag{3.11}$$

so that we may write (3.9) in the form

$$\text{Trace}\,\mathscr{U} = 1 = C\exp[-\beta(\Omega^{0} + \Omega')] \tag{3.12}$$

The proof that the vacuum expectation value (3.11) is equal to the complicated sum (III, 8.11) will not be given here.

Since the operator $\tilde{P}$ is clearly important in this theory, let us have a more detailed look at it. Starting from (2.1) and (2.2), let us define the quenched functions $\tilde{w}_{\alpha}(k, l)$ by

$$\begin{aligned}\tilde{w}_{\alpha}(k, l) &= (1 - \bar{n}_k)^{1/2}(1 - \bar{n}_l)^{1/2}\,w_{\alpha}(k, l)\\ &= \exp(\beta\mu)\,(1 - \bar{n}_k)^{1/2}(1 - \bar{n}_l)^{1/2}\,\hat{w}_{\alpha}(k, l)\end{aligned} \tag{3.13}$$

Using (2.3), (3.3), (3.5), and (3.13), we obtain the explicit result

$$\tilde{P} = 2^{-1/2}\sum_{\alpha}\sum_{k,l}\tilde{w}_{\alpha}^{*}(k, l)\,a_k a_l A_{\alpha} \tag{3.14}$$

The original decomposition (2.1) is not unique; the number of positive terms and the number of negative terms cannot be altered, but the functions $\hat{w}_{\alpha}(k, l)$ have considerable freedom. We could make (2.1) unique by requiring the $\hat{w}_\alpha(k, l)$ to be mutually orthogonal. However, this would *not* be a convenient choice, since the operator of main interest is $\tilde{P}$, (3.14), which contains the quenched functions $\tilde{w}_\alpha(k, l)$. Rather, we now impose the condition that the *quenched* functions $\tilde{w}_\alpha(k, l)$ are *mutually orthogonal*. They can then be written as multiples of a normalized orthogonal set $\varphi_\alpha(k, l)$:

$$\tilde{w}_\alpha(k, l) = \mathfrak{v}_\alpha^{1/2}\, \varphi_\alpha(k, l) \tag{3.15}$$

and

$$(\varphi_\alpha, \varphi_\beta) = \sum_{k,l} \varphi_\alpha^*(k, l)\, \varphi_\beta(k, l) = \delta_{\alpha\beta} \tag{3.16}$$

The $\mathfrak{v}_\alpha$ are the eigenvalues (assumed all positive), and the $\varphi_\alpha(k, l)$ are the normalized eigenfunctions, of the "quenched correlation matrix"

$$\begin{aligned} \langle kl \mid \tilde{W} \mid k'l' \rangle &= (1 - \bar{n}_k)^{1/2} (1 - \bar{n}_l)^{1/2} \langle kl \mid \hat{W} \mid k'l' \rangle (1 - \bar{n}_{k'})^{1/2} (1 - \bar{n}_{l'})^{1/2} \exp(2\beta\mu) \\ &= \sum_\alpha \mathfrak{v}_\alpha \varphi_\alpha(k, l)\, \varphi_\alpha^*(k', l') \end{aligned} \tag{3.17}$$

Except for the c-number factor $\exp(2\beta\mu)$, this is the same matrix as (III, 8.8) and (III, 8.13).

The eigenfunctions $\varphi_\alpha(k, l)$ of the quenched pair correlation matrix play the part of "molecular eigenfunctions" in this theory. We can see this formally by introducing the "pair creation operator" b_α^+ according to (V, 1.9):

$$b_\alpha^+ = 2^{-1/2} \sum_{k,l} \varphi_\alpha(k, l)\, a_l^+ a_k^+ \tag{3.18}$$

It was shown in Chapter V, Section 1, that this operator, when acting on the vacuum state, produces a two-particle state with the two particles described by the pair wave function $\varphi_\alpha(k, l)$. The corresponding pair destruction operator is the Hermitean conjugate of (3.18), namely,

$$b_\alpha = 2^{-1/2} \sum_{k,l} \varphi_\alpha^*(k, l)\, a_k a_l \tag{3.19}$$

and comparison with (3.14) yields the neat form

$$\tilde{P} = \sum_\alpha \mathfrak{v}_\alpha^{1/2} b_\alpha A_\alpha \tag{3.20}$$

If we think of the "pairs" as "quasi-bosons," then the operators (3.18) and (3.19) are the creation and destruction operators for the "quasi-bosons." The complex structure of these particles shows itself in the commutation rules, which differ from the elementary Bose commutation rules (V, 1.28). Direct application of the anticommutation rules (V, 1.7), (V, 1.13), and (V, 1.18) to the operators (3.18) and (3.19) leads to the following results:

$$[b_\alpha, b_\beta]_- = [b_\alpha^+, b_\beta^+]_- = 0 \tag{3.21a}$$

$$[b_\alpha, b_\beta^+]_- = \delta_{\alpha\beta} + \sum_{k,k'} \left\{ \sum_l \varphi_\beta(k, l)\, \varphi_\alpha^*(l, k') \right\} a_k^+ a_{k'} \tag{3.21b}$$

Equation (3.21a) is the same as for elementary bosons, but (3.21b) contains an extra term, which depends on the "crossed density matrices"[3]

$$\langle k \mid q_\alpha^\beta \mid k' \rangle = \sum_l \varphi_\beta(k, l)\, \varphi_\alpha^*(k', l) \tag{3.22}$$

The eigenfunctions $\varphi_\alpha(k, l)$ of interest are all antisymmetric; symmetric contributions are annihilated automatically in (3.18) and (3.19), since the expressions $a_l^+ a_k^+$ and $a_k a_l$ are antisymmetric in k and l.

We may think of (3.22) as a one-particle density matrix of mixed type; for $\beta = \alpha$, (3.22) is the conventional one-particle density matrix for one of the two particles contained in the pair α. For $\beta \neq \alpha$, on the other hand, (3.22) is a kind of overlap integral, which vanishes if the wave functions φ_α and φ_β are concentrated in different regions.

If the pairs are well separated, in some sense, then it may be a reasonable approximation to ignore the overlap integrals (3.22). We now proceed to show that in this approximation, (3.11) is indeed equal to the ideal Bose gas result (III, 8.18). The proof is simple; if we ignore the overlap term in (3.21b), the different contributions α are independent of each other, and thus combination of (3.11) and (3.20) gives approximately

$$\exp(-\beta\Omega') \simeq \prod_\alpha \langle 0 \mid \exp(\mathfrak{v}_\alpha^{1/2} b_\alpha A_\alpha) \exp(\mathfrak{v}_\alpha^{1/2} b_\alpha^+ A_\alpha^+) \mid 0 \rangle$$

We now expand both exponentials in power series, and use the identity (2.13) to get

$$\exp(-\beta\Omega') \simeq \prod_\alpha \left\{ \sum_{n=0}^{\infty} \frac{(\mathfrak{v}_\alpha)^n}{n!} \langle 0 \mid (b_\alpha)^n (b_\alpha^+)^n \mid 0 \rangle \right\}$$

[3] This definition differs from the one in Blatt (60a) by a minus sign. As defined here, q_β^α with $\beta = \alpha$ is a non-negative operator in k-space, and differs from the operator ρ in (VIII,2.3) only by the c-number factor $2x$.

In this approximation, the vacuum expectation values on the right are equal to $n!$, and we obtain the result

$$\exp(-\beta\Omega') \cong \prod_{\alpha} (1 - \mathfrak{v}_\alpha)^{-1} \qquad \text{(neglecting overlap)} \tag{3.23}$$

This is precisely the approximation (III, 8.18).

Unfortunately, this approximation is completely useless in the theory of superconductivity, since the pairs are large compared to the average distance between electrons. The pair size is the Pippard length, of order 10^{-4} cm, whereas electron-electron distances are of the order of 10^{-8} cm. Thus the overlap integrals, far from being negligible, are in fact huge; the second term in (3.21b) cannot be ignored, and the theory is correspondingly much more complicated.

Before proceeding with the treatment of the commutation rules (3.21), we apply the quenching identity (3.7) to obtain reduced forms for the expectation values of single-particle and two-particle operators (Blatt 60a). As a first example, we take the number operator N_{Op}, which we write in "antinormal" form

$$N_{\text{Op}} = \sum_k (1 - a_k a_k^+) \tag{3.24}$$

Use of (3.8) yields:

$$\text{Trace}\,(N_{\text{Op}}\mathscr{U}) = C\langle 0 \mid \exp(\tilde{P}) \sum_k (1 - \tilde{a}_k \tilde{a}_k^+) \exp(\tilde{P}^+) \mid 0\rangle \,\text{Trace}\,\mathscr{V} \tag{3.25}$$

We now use (3.5) to write

$$1 - \tilde{a}_k \tilde{a}_k^+ = 1 - (1 - \bar{n}_k)\, a_k a_k^+ = \bar{n}_k + (1 - \bar{n}_k)\, a_k^+ a_k \tag{3.26}$$

We substitute (3.26) into (3.25), and use Eq. (3.9) to eliminate the constant C. This procedure leads to

$$\text{Trace}\,(N_{\text{Op}}\mathscr{U}) = \sum_k \bar{n}_k + \frac{\langle 0 \mid \exp(\tilde{P})\, \Sigma_k (1 - \bar{n}_k)\, a_k^+ a_k \exp(\tilde{P}^+) \mid 0\rangle}{\langle 0 \mid \exp(\tilde{P}) \exp(\tilde{P}^+) \mid 0\rangle} \tag{3.27}$$

The two contributions on the right of (3.27) have simple interpretations: the first is the number of unpaired particles, the second is the number of particles bound up in the correlated pairs. The quenching factor $(1 - \bar{n}_k)$ ensures that fully occupied states k, with $\bar{n}_k = 1$, cannot be used for pair correlations.

Next, we apply the same technique to a general single-particle operator J, Eq. (V, 1.22). We introduce the quenched matrix elements $\langle k \mid \tilde{j} \mid k' \rangle$ and the quenched operator $\tilde{J}$ by[4]

$$\langle k \mid \tilde{j} \mid k' \rangle = (1 - \bar{n}_k)^{1/2} \langle k \mid j \mid k' \rangle (1 - \bar{n}_{k'})^{1/2} \tag{3.28a}$$

$$\tilde{J} = \sum_{k,k'} \langle k \mid \tilde{j} \mid k' \rangle a_k^+ a_{k'} = \sum_{k,k'} \langle k \mid j \mid k' \rangle \tilde{a}_k^+ \tilde{a}_{k'} \tag{3.28b}$$

In terms of this operator, the statistical expectation value assumes the form (Blatt 60a)

$$\text{Trace}\,(J\mathscr{U}) = \sum_k \bar{n}_k \langle k \mid j \mid k \rangle + \frac{\langle 0 \mid \exp(\tilde{P})\, \tilde{J} \exp(\tilde{P}^+) \mid 0 \rangle}{\langle 0 \mid \exp(\tilde{P}) \exp(\tilde{P}^+) \mid 0 \rangle} \tag{3.29}$$

The first term on the right-hand side of (3.29) is the contribution of the unpaired particles, which is equal to (1.12) of the Husimi theory; the second term gives the contribution of the particles in correlated pairs. (3.27) is a special case of (3.29).

Finally, we consider two-particle operators T, Eq. (V, 1.26). The calculation is entirely analogous (Blatt 60a). Qualitatively, we expect three distinct contributions: (i) from two unpaired particles, (ii) one unpaired particle, the other particle paired, (iii) two particles both within pairs (not necessarily within a pair of the same type, however). This is indeed what we find. It is convenient to define the operator $\tilde{T}^{(1)}$ by

$$\tilde{T}^{(1)} = \sum_{k,k'} \langle k \mid \tilde{t}^{(1)} \mid k' \rangle a_k^+ a_{k'} \tag{3.30a}$$

and

$$\langle k \mid \tilde{t}^{(1)} \mid k' \rangle = \sum_l \bar{n}_l (1 - \bar{n}_k)^{1/2} (1 - \bar{n}_{k'})^{1/2} [\langle kl \mid t \mid k'l \rangle + \langle lk \mid t \mid lk' \rangle - \langle kl \mid t \mid lk' \rangle - \langle lk \mid t \mid k'l \rangle] \tag{3.30b}$$

If we think of $\langle kl \mid t \mid k'l' \rangle$ as the matrix element of an interaction, then (3.30) is the operator for the interaction between two particles, one unpaired in state l, the other a member of a pair, quenched in the usual way and averaged over the entire distribution of unpaired particles.

We also introduce the fully quenched pair operator $\tilde{T}$ through

$$\tilde{T} = \tfrac{1}{2} \sum_{k,l,k',l'} \langle kl \mid \tilde{t} \mid k'l' \rangle a_k^+ a_l^+ a_{l'} a_{k'} \tag{3.31a}$$

$$\langle kl \mid \tilde{t} \mid k'l' \rangle = (1 - \bar{n}_k)^{1/2} (1 - \bar{n}_l)^{1/2} \langle kl \mid t \mid k'l' \rangle (1 - \bar{n}_{k'})^{1/2} (1 - \bar{n}_{l'})^{1/2} \tag{3.31b}$$

[4] Note that this operator is written in normal form, not in anti-normal form. We prefer to write final results in terms of operators in normal form.

The expression for the expectation value of the operator T, Eq. (V, 1.26), is then

$$\begin{aligned}\mathrm{Tr}\,(T\mathscr{U}) &= \tfrac{1}{2}\sum_{k,l} (\langle kl \mid t \mid kl\rangle - \langle kl \mid t \mid lk\rangle)\,\bar{n}_k\bar{n}_l \\ &+ \tfrac{1}{2}\frac{\langle 0 \mid \exp(\tilde{P})\,\tilde{T}^{(1)}\exp(\tilde{P}^+) \mid 0\rangle}{\langle 0 \mid \exp(\tilde{P})\exp(\tilde{P}^+) \mid 0\rangle} \\ &+ \tfrac{1}{2}\frac{\langle 0 \mid \exp(\tilde{P})\,\tilde{T}\exp(\tilde{P}^+) \mid 0\rangle}{\langle 0 \mid \exp(\tilde{P})\exp(\tilde{P}^+) \mid 0\rangle}\end{aligned} \tag{3.32}$$

The three contributions to (3.32) are the expected ones, outlined before; the single-particle contribution is identical with the Husimi theory result (1.14), as expected.

4. The Dyson Transformation

In order to handle the awkward commutation rule (3.21b), we make use of a technique developed by Dyson (56) for the theory of spin-waves.[4a] In the Dyson formalism, we establish a correspondence between the "physical pair" operators b_α^+, (3.18), and purely "ideal" pair creation operators B_α^+, which satisfy ordinary Bose-Einstein commutation rules by assumption. We then consider a modified problem, in which the operators all obey "ideal" commutation rules, but are correspondingly more complicated functions of the basic operators B_α and B_α^+.

The Dyson correspondence is defined as follows. We start with "physical boson configurations" $\mid \{n_\alpha\} >$ defined by

$$\mid \{n_\alpha\}\rangle = \prod_\alpha \frac{(b_\alpha^+)^{n_\alpha}}{(n_\alpha!)^{1/2}} \mid 0\rangle \tag{4.1}$$

[4a] *Note added in proof.* Dr. M. Girardeau called to my attention some fundamental difficulties with the use of the Dyson transformation. The set of states (4.1) is not merely complete, but overcomplete. Thus, the coefficients in (4.7) are not defined uniquely, and the same lack of uniqueness plagues (4.9).

Fortunately, Matsubara and Thompson (Matsubara 63, Thompson 63) have succeeded in deriving the final results (4.24) and (4.29) by a completely different method, without use of the Dyson transformation.

This provides an *a posteriori* justification for the recipe (Blatt 62b) of carrying out all closure reductions, based on (4.21), just as soon as they become possible, and to continue until no further reduction can be made; the recipe was tested originally by comparison with a more straightforward technique of calculation for the special case of the Schafroth-condensed state (Blatt 60a).

This section may be skipped in favor of reading Thompson (63).

The states (4.1) are a complete set of states, but they are not orthogonal to each other. As an example, consider the overlap integral between the configurations $|\{1_\alpha 1_\beta\}>$ and $|\{1_\gamma 1_\delta\}>$; this integral is

$$\langle 1_\alpha 1_\beta \mid 1_\gamma 1_\delta \rangle = \langle 0 \mid b_\alpha b_\beta b_\gamma^+ b_\delta^+ \mid 0 \rangle = \delta_{\alpha\gamma}\,\delta_{\beta\delta} + \delta_{\alpha\delta}\,\delta_{\beta\gamma} - C_{\gamma\delta}^{\alpha\beta} \tag{4.2}$$

where

$$C_{\gamma\delta}^{\alpha\beta} = 4 \sum_{k_1 k_2 k_3 k_4} \varphi_\alpha^*(k_1 k_2)\, \varphi_\beta^*(k_3 k_4)\, \varphi_\gamma(k_2 k_3)\, \varphi_\delta(k_4 k_1)$$

$$= 4\,\mathrm{tr}_1\,(q_\alpha^\gamma q_\beta^\delta) \tag{4.3}$$

If the overlap integral $C_{\gamma\delta}^{\alpha\beta}$ can be ignored, (4.2) is the desired orthogonality integral; but of course, (4.3) is by no means small, actually.

Although the physical boson configurations (4.1) are neither orthogonal nor normalized in general, there is an important exception to this rule: the vacuum state $|\,0\rangle$ is not only normalized, but is also orthogonal to all other states (4.1).

We now follow Dyson by establishing a correspondence between the physical boson states (4.1) and operators J acting on them, and "ideal" boson states and "Dyson image" operators J_D acting on them. The ideal boson states are defined in terms of an ideal boson vacuum state $|\,0)$ and ideal boson operators B_α, B_β^+ with commutation rules

$$[B_\alpha\,, B_\beta^+]_- = \delta_{\alpha\beta}, \qquad [B_\alpha\,, B_\beta]_- = [B_\alpha^+, B_\beta^+]_- = 0 \tag{4.4}$$

The ideal boson configuration $|\,\{n_\alpha\})$ is defined by

$$|\,\{n_\alpha\}) = \prod_\alpha \frac{(B_\alpha^+)^{n_\alpha}}{(n_\alpha!)^{1/2}}\,|\,0) \tag{4.5}$$

The states (4.5) are mutually orthogonal and normalized:

$$(\{n_\alpha\} \mid \{n'_\alpha\}) = \prod_\alpha \delta_{n_\alpha n'_\alpha} \tag{4.6}$$

We now set up a linear mapping $J \to J_D$ of operators on the two spaces. Let coefficients $J(\{n_\alpha\}, \{n'_\alpha\})$ be *defined* by

$$J\,|\,\{n'_\alpha\}\rangle = \sum_{\{n_\alpha\}} J(\{n_\alpha\}, \{n'_\alpha\})\,|\,\{n_\alpha\}\rangle \tag{4.7}$$

Since the states $|\,\{n_\alpha\}\rangle$ are not an orthonormal set, the coefficients in (4.7) are not the usual matrix elements, i.e.,

$$J(\{n_\alpha\}, \{n'_\alpha\}) \neq \langle \{n_\alpha\} \mid J \mid \{n'_\alpha\}\rangle$$

However, since the vacuum state $|0\rangle$ is normalized and orthogonal to all the other states (4.1), we do have the useful identity

$$J(0, 0) = \langle 0 | J | 0 \rangle \tag{4.8}$$

We use the coefficients $J(\{n_\alpha\}, \{n'_\alpha\})$ to *define* the Dyson image operator J_D by means of

$$J_D \,|\, \{n'_\alpha\}) = \prod_{\{n_\alpha\}} J(\{n_\alpha\}, \{n'_\alpha\}) \,|\, \{n_\alpha\}) \tag{4.9}$$

The orthonormality of the ideal states $|\{n_\alpha\})$ gives then the expected relation between the coefficients in (4.9) and the ideal matrix elements:

$$J(\{n_\alpha\}, \{n'_\alpha\}) = (\{n_\alpha\} \,|\, J_D \,|\, \{n'_\alpha\}) \tag{4.10}$$

The mapping $J \rightarrow J_D$ is affine and linear[4a]; i.e., the Dyson image of the sum of two operators is the sum of their Dyson images, and the same is true for a product. The unit operator in the physical space maps into the unit operator in the ideal space. However, the mapping does *not* preserve scalar products, and therefore the Dyson image of the adjoint operator J^+ is *not* the adjoint of the Dyson image J_D of J. A useful method for generating Dyson images is described in Blatt (58). Let us quote the results here, for some relevant operators:

$$(b_\alpha^+)_D = B_\alpha^+ \tag{4.11a}$$

$$(b_\alpha)_D = B_\alpha - \tfrac{1}{2} \sum_{\beta\gamma\delta} C_{\gamma\delta}^{\alpha\beta} B_\beta^+ B_\gamma B_\delta \tag{4.11b}$$

$$(\tilde{P}^+)_D = \sum_\alpha \mathfrak{v}_\alpha^{1/2} A_\alpha^+ B_\alpha^+ \equiv R^+ \tag{4.11c}$$

$$(\tilde{P})_D = \sum_\alpha \mathfrak{v}_\alpha^{1/2} A_\alpha B_\alpha - \tfrac{1}{2} \sum_{\alpha\beta\gamma\delta} \mathfrak{v}_\alpha^{1/2} C_{\gamma\delta}^{\alpha\beta} A_\alpha B_\beta^+ B_\gamma B_\delta \equiv R + S \tag{4.11d}$$

where the last two equations also serve as definitions of the operators R, R^+, and S, all of which act in the ideal space.

Combination of (4.8) and (4.10) leads to the general relation

$$\langle 0 | J | 0 \rangle = (0 | J_D | 0) \tag{4.12}$$

i.e., vacuum expectation values in the two spaces are equal to each other.

Thus, combination of (3.11) and (4.11) gives the following expression for the grand canonical free energy of the pairs Ω':

$$\exp(-\beta\Omega') = (0 | \exp(R + S) \exp(R^+) | 0) \tag{4.13}$$

The advantage of (4.13) over (3.11) is the fact that all operators appearing in (4.13) obey ordinary Bose-Einstein commutation rules. By means of the Dyson transformation, we have succeeded in eliminating Fermi-Dirac operators completely.

The exponential function $\exp(R + S)$ is rather complicated, because R and S fail to commute with each other. The subsequent development of the theory depends crucially upon a simplification of $\exp(R + S)$. We shall quote the result here; the proof is contained in Matsubara (60). We define the operator M by[5]

$$\langle k \mid M \mid k' \rangle = 2 \sum_{\alpha,\beta} \langle k \mid q_\alpha^\beta \mid k' \rangle \, \mathfrak{v}_\alpha^{1/2} A_\alpha B_\beta \tag{4.14}$$

We note that the operators A_α and B_β all commute mutually. Next, we define the operator L by

$$L = \tfrac{1}{2}\,\mathrm{tr}_1 \ln(1 + M) = \tfrac{1}{2} \sum_k \langle k \mid \ln(1 + M) \mid k \rangle \tag{4.15}$$

Then the following operator identity holds:

$$\exp(R + S) = \exp(S)\exp(L) \tag{4.16}$$

When we substitute (4.16) into (4.13), the result can be further simplified. The operator S defined by (4.11d) has a B_β^+ as its left-most operator in the ideal space. Since $(0 \mid B_\beta^+$ vanishes for every β, we have the identity

$$(0 \mid \exp(S) = (0 \mid \left[1 + \frac{S}{1!} + \frac{S^2}{2!} + \cdots\right] = (0 \mid \tag{4.17}$$

whence

$$\exp(-\beta\Omega') = (0 \mid \exp(L)\exp(R^+) \mid 0) \tag{4.18}$$

In spite of the complicated appearance of the operator L, (4.14) and (4.15), Eq. (4.18) is a very considerable simplification over (3.11) or (4.13). Unlike (3.11), all operators appearing in (4.18) obey ordinary Bose-Einstein commutation rules; and $\exp(L)\exp(R^+)$ is an operator in "anti-normal" form, which is not true of (4.13).

Having reduced the normalization integral (3.11), let us now do a similar reduction for the pair contribution to the expectation value of a single-particle operator, i.e., the second term on the right-hand side of (3.29). We first require the Dyson image of the operator $\tilde{J}$ defined by (3.28). It is shown in Blatt (62b) that

$$(\tilde{J})_D = 2 \sum_{\alpha,\beta} \mathrm{tr}_1(\tilde{\jmath} q_\alpha^\beta)\, B_\alpha^+ B_\beta \tag{4.19a}$$

---

[5] This is the negative of the operator M in Matsubara (60), their equation (2.19), since we have defined q_α^β with opposite sign.

where

$$\mathrm{tr}_1(\tilde{j}q_\alpha^\beta) = \sum_{k,k'} \langle k \mid \tilde{j} \mid k' \rangle \langle k' \mid q_\alpha^\beta \mid k \rangle \tag{4.19b}$$

We thus obtain the identity

$$\langle 0 \mid \exp(\tilde{P})\, \tilde{J} \exp(\tilde{P}^+) \mid 0 \rangle = 2 \sum_{\alpha,\beta} \mathrm{tr}_1(\tilde{j}q_\alpha^\beta)\, (0 \mid \exp(L)\, B_\alpha^+ B_\beta \exp(R^+) \mid 0) \tag{4.20}$$

It is possible to reduce this expression further, by making use of the closure relation for the pair wave functions:

$$\sum_\alpha \varphi_\alpha^*(k, k')\, \varphi_\alpha(l, l') = \delta_{kl}\, \delta_{k'l'} - \delta_{kl'}\, \delta_{k'l} \tag{4.21}$$

The calculation (Blatt 62b) is somewhat complicated, but the final result is simple. We define the operator g by

$$\langle k \mid g \mid k' \rangle = \left\langle k \left| \frac{M}{1 + M} \right| k' \right\rangle \tag{4.22}$$

and we define

$$\mathrm{tr}_1(\tilde{j}g) = \sum_{k,k'} \langle k \mid \tilde{j} \mid k' \rangle \langle k' \mid g \mid k \rangle \tag{4.23}$$

Then (4.20) can be reduced to

$$\langle 0 \mid \exp(\tilde{P})\, \tilde{J} \exp(\tilde{P}^+) \mid 0 \rangle = (0 \mid \exp(L)\, \mathrm{tr}_1(\tilde{j}g) \exp(R^+) \mid 0) \tag{4.24}$$

We note that the operator appearing on the right-hand side of (4.24) is in anti-normal form; this is not true for the right-hand side of (4.20).

The reduction of expectation values of two-particle operators, (3.32), is somewhat more lengthy (Blatt 62b), and we shall restrict ourselves to quoting results. The first term on the right-hand side of (3.32) requires no reduction; the second term involves a one-particle operator, $\tilde{T}^{(1)}$, and can therefore be reduced according to (4.24); in order to write the final result for the last term of (3.32), we require certain definitions, namely,

$$\langle kl \mid p \mid k'l' \rangle = \langle k \mid g \mid k' \rangle \langle l \mid g \mid l' \rangle - \langle k \mid g \mid l' \rangle \langle l \mid g \mid k' \rangle \tag{4.25}$$

$$\mathrm{tr}_2(\tilde{t}p) = \sum_{k,l,k',l'} \langle kl \mid \tilde{t} \mid k'l' \rangle \langle k'l' \mid p \mid kl \rangle \tag{4.26}$$

$$\psi_\alpha(k, l) = 2^{-1/2} \sum_{k'} \left[\left\langle k \left| \frac{1}{1+M} \right| k' \right\rangle \varphi_\alpha(k', l) - \left\langle l \left| \frac{1}{1+M} \right| k' \right\rangle \varphi_\alpha(k', k) \right] \tag{4.27}$$

$$(\psi_\alpha, \tilde{t}\psi_\beta) = \sum_{k,l,k',l'} \psi_\alpha^*(k, l) \langle kl \mid \tilde{t} \mid k'l' \rangle \psi_\beta(k', l') \tag{4.28}$$

Then

$$\langle 0 \mid \exp(\tilde{P})\, \tilde{T} \exp(\tilde{P}^+) \mid 0\rangle$$
$$= \tfrac{1}{2}(0 \mid \exp(L)\, [\mathrm{tr}_2\,(\tilde{t}p) + \sum_{\alpha,\beta} (\psi_\alpha\,,\, \tilde{t}\psi_\beta)\, \mathfrak{v}_\alpha^{1/2} A_\alpha B_\beta] \exp(R^+) \mid 0) \qquad (4.29)$$

5. Specialization to One Pair State Only

If there is Schafroth-condensation of pairs, one pair state is immeasurably more important than all the other pair states. This is the assumption underlying the entire theory of superconductivity; the only place where non-condensed pairs *must* be allowed for, is in the proof of the Schafroth-condensation, which was discussed at the end of Chapter III. The actual calculation (Matsubara 60) is based on the general formula (4.18), but is exceedingly complex and will not be given here.

For all the other work, if suffices to include only one pair state. That is, we put

$$\mathfrak{v}_0 \neq 0, \qquad \mathfrak{v}_1 = \mathfrak{v}_2 = \mathfrak{v}_3 = \cdots = 0 \qquad (5.1)$$

Since there are factors $\mathfrak{v}_\alpha^{1/2}$ in all the relevant operators, it follows that in all sums over α, β, etc., only the term $\alpha = \beta = \cdots = 0$ survives. The operators A_α and B_β enter only in the combination $A_0 B_0$ and $A_0^+ B_0^+$; we shall drop the subscript 0 throughout, henceforth.

The operator L, (4.15), becomes

$$L = \tfrac{1}{2}\,\mathrm{tr}_1 \ln (1 + 2q\mathfrak{v}^{1/2} AB) \qquad (5.2)$$

and the pair contribution to the grand canonical free energy is given by (4.18) as

$$\exp(-\beta\Omega') = (0 \mid \exp(L) \exp(\mathfrak{v}^{1/2} A^+ B^+) \mid 0) \qquad (5.3)$$

Let $f(x)$ be an analytic function of x. We prove below the identity

$$(0 \mid f(AB) \exp(\mathfrak{v}^{1/2} A^+ B^+) \mid 0) = \int_0^\infty dt \exp(-t)\, f(\mathfrak{v}^{1/2} t) \qquad (5.4)$$

Proof. We start by expanding $f(x)$ in a power series:

$$f(x) = \sum_{n=0}^{\infty} c_n x^n \qquad (5.5)$$

and we expand both factors in (5.4) to obtain

$$(0 \mid f(AB) \exp(\mathfrak{v}^{1/2}A^+B^+) \mid 0) = \sum_{n,m} \frac{c_n(\mathfrak{v})^{m/2}}{m!} (0 \mid (AB)^n (A^+B^+)^m \mid 0) \tag{5.6}$$

As a result of the Bose commutation rules, the expectation values on the right-hand side of (5.6) are equal to

$$(0 \mid (AB)^n(A^+B^+)^m \mid 0) = \delta_{nm}(n!)^2 \tag{5.7}$$

and thus the right-hand side of (5.6) equals

$$I = \sum_{n=0}^{\infty} n!\, c_n(\mathfrak{v}^{1/2})^n \tag{5.8}$$

This is not quite in the right form to use (5.5); however, we can get it into that form by using the identity

$$n! = \int_0^\infty dt\, t^n \exp(-t) \tag{5.9}$$

When we substitute (5.9) into (5.8), interchange the order of summation and integration, and use (5.5) for the sum over n, we obtain (5.4).

We now apply the identity (5.4) to (5.3). We define the function $L(t)$ by

$$L(t) = \tfrac{1}{2}\, tr_1 \ln(1 + 2\mathfrak{v}tq) = \tfrac{1}{2}\sum_k \langle k \mid \ln(1 + 2\mathfrak{v}tq) \mid k\rangle \tag{5.10}$$

Application of (5.4) to (5.3) gives:

$$\exp(-\beta\Omega') = \int_0^\infty dt \exp[-t + L(t)] \tag{5.11}$$

In order to study the behavior of this integral, it is convenient to assume simple pairing, and expand the logarithm in (5.10) in a power series. This expansion reads (assuming simple pairing for the moment)

$$\begin{aligned} L(t) &= \tfrac{1}{2}\sum_k \ln(1 + 2\mathfrak{v}t \mid \varphi_k \mid^2) \\ &= \mathfrak{v}t - (\mathfrak{v}t)^2 \sum_k \mid \varphi_k \mid^4 + \cdots \end{aligned}$$

where we have used the normalization integral of the pair wave function to simplify the leading term. The quantities $\mid \varphi_k \mid^2$ are of order $1/V$, $\mid\varphi_k\mid^4$ of order $(1/V)^2$, and thus the last sum over k is of order $1/V$. The n'th term in the expansion of $L(t)$ is of order t^n/V^{n-1}.

When we substitute this expansion into (5.11), we find completely different behavior, depending upon whether $\mathfrak{v} = \mathfrak{v}_0$ is smaller than, or larger than, unity. For $\mathfrak{v} < 1$, the dominant contribution to the integral comes from small values of t, so that we may ignore all but the leading term of the expansion of $L(t)$. The result is

$$\exp(-\beta\Omega') = \int_0^\infty dt \exp(-t + \mathfrak{v}t) = (1 - \mathfrak{v})^{-1} \qquad \text{for } \mathfrak{v} < 1$$

which is equivalent to (III, 9.9).

On the other hand, for $\mathfrak{v} > 1$, the dominant region is large t, $t \sim V$, so that the power series expansion of the logarithm is much less useful.[6] Rather, there is now a sharp maximum of the integrand at a value of t given by (III, 9.16). A conventional saddle point integration applied to (5.11) then yields (III, 9.17), except for terms of order $1/V$.

Next, we consider expectation values of single-particle operators, as given by (3.29) and (4.24). For the case of one pair state only, the infinite sum in the definition (4.14) of the operator M reduces to just one term[7]

$$\langle k \mid M \mid k' \rangle = 2 \langle k \mid q_0^0 \mid k' \rangle \, \mathfrak{v}_0^{1/2} A_0 B_0 \tag{5.12}$$

Since g, Eq. (4.22), is a function of M, we conclude that the operator $\exp(L)\,\mathrm{tr}_1(\tilde{\jmath}g)$ on the right-hand side of (4.24) is a function $f(A_0B_0) = f(AB)$ of AB, and we can therefore employ the identity (5.4). We define the operators $\rho(x)$ and $g(x)$ by (we drop subscripts zero now)

$$\langle k \mid \rho(x) \mid k' \rangle = 2x \langle k \mid q \mid k' \rangle \tag{5.13}$$

$$\langle k \mid g(x) \mid k' \rangle = \left\langle k \left| \frac{\rho}{1+\rho} \right| k' \right\rangle \tag{5.14}$$

With these definitions, application of (5.4) to (4.24) yields

$$\langle 0 \mid \exp(\tilde{P})\, \tilde{J} \exp(\tilde{P}^+) \mid 0 \rangle = \int_0^\infty dt \exp[-t + L(t)]\, \mathrm{tr}_1 [\tilde{\jmath}\, g(\mathfrak{v}t)] \tag{5.15}$$

---

[6] Stopping with the first *two* terms of the power series, i.e., the terms written explicitly above, is permitted if the pairs are well separated in space. This, however, is not true of the electron pairs in superconductors; for those, the power series expansion of $L(t)$ fails to converge, in the interesting region of t.

[7] Since there is a factor $v_\alpha^{1/2}$ but no factor $v_\beta^{1/2}$ in (4.14), it is not obvious that the sum over β can be ignored; it is nevertheless true, for the following reason: the operator $\exp(R^+)$ in (4.24), when expanded in power series, contains A_α^+ and B_α^+ only in the combination $(A_\alpha^+ B_\alpha^+)^{n_\alpha}$, i.e., to the same power. Thus, if A_α $(\alpha \neq 0)$ does not occur in the expansion of M, then the only non-vanishing contribution to the vacuum expectation value comes from the term $n_\alpha = 0$ in the expansion of $\exp(R^+)$, and this in turn implies that non-zero powers of B_α in the series for M leads to zero vacuum expectation values.

This is a general result, true for all values of $\mathfrak{v}$. In the limit of small $\mathfrak{v}$ one may expand everything in powers of $\mathfrak{v}$, and verify the first few terms directly. We are interested, however, in values of $\mathfrak{v} > 1$, for which the integrand of (5.15) has a saddle point at a value of t proportional to the volume V, see (III, 9.16) and (III, 9.22).

Compared to the exponential factor, the factor $\text{tr}_1\,[\tilde{\jmath}g(\mathfrak{v}t)]$ is a slowly varying function of t. Thus, in a saddle point integration, we may replace t by its saddle point value, and take this factor outside the integral sign. Comparison with (4.18) and (5.11) then gives

$$\frac{\langle 0 \mid \exp(\tilde{P})\,\tilde{J}\exp(\tilde{P}^+) \mid 0\rangle}{\langle 0 \mid \exp(\tilde{P})\exp(\tilde{P}^+) \mid 0\rangle} = tr_1\,[\tilde{\jmath}g(\mathfrak{v}t)]\,\{1 + O(1/V)\} \tag{5.16}$$

Thus, (3.29) becomes, except for terms of order $1/V$,

$$\text{Trace}\,(J\mathscr{U}) = \sum_k \bar{n}_k \langle k \mid j \mid k\rangle + \text{tr}_1\,[\tilde{\jmath}\,g(\mathfrak{v}t)] \tag{5.17}$$

If there are no unpaired particles, we have $\bar{n}_k = 0$, all k, and $\tilde{\jmath} = j$. Equation (5.17) then reduces to (VIII, 2.6). If, furthermore, there is simple pairing, the operators $\rho(x)$, (5.13), and $g(x)$, (5.14), are diagonal in k-space, with diagonal values given by (VIII, 2.4) and (IV, 1.7), respectively, and (VIII, 2.6) reduces to (IV, 1.8).

Another special case of interest is simple pairing, but not all particles bound in pairs, i.e., $\bar{n}_k \neq 0$. The operator g is then diagonal, with eigenvalues given by (IV, 1.7). We define the quantities ν_k by

$$\nu_k = \bar{n}_k + (1 - \bar{n}_k)\,g_k(\mathfrak{v}t) \tag{5.18}$$

Since g is diagonal now, only the diagonal elements of $\tilde{\jmath}$ contribute to (5.17), and (5.17) reduces to

$$\text{Trace}\,(J\mathscr{U}) = \sum_k \nu_k \langle k \mid j \mid k\rangle \qquad \text{(simple pairing)} \tag{5.19}$$

In particular, for the number operator $J = N_{\text{op}}$, $\langle k \mid j \mid k\rangle = 1$, and we obtain Eq. (VII, 4.5).

For two-particle operators T, we can carry out completely analogous reductions. The operators p, Eq. (4.25), and ψ_α, Eq. (4.27), become functions of AB only, so that we may use the identity (5.4) to reduce (4.29). Replacing AB by $x = \mathfrak{v}t$, (4.25) reduces to (VIII, 2.11), and (4.27) for $\alpha = 0$ reduces to (VIII, 2.8) except for a factor $(\mathfrak{v}t)^{1/2}$. Defining ψ by (VIII, 2.8), i.e., with the extra factor $(\mathfrak{v}t)^{1/2}$ in the numerator, (4.29) reduces to

$$\langle 0 \mid \exp(\tilde{P})\,\tilde{T}\exp(\tilde{P}^+) \mid 0\rangle = \tfrac{1}{2}\int_0^\infty dt\,\exp[-t + L(t)]\,\{tr_2\,(\tilde{t}p) + (\psi, \tilde{t}\psi)\} \tag{5.20}$$

Just as before, this integral can be approximated by the saddle point method. Under those conditions, combination of (3.32), (5.16), and (5.20) gives

$$\text{Trace}\,(T\mathscr{U}) = \tfrac{1}{2}\sum_{k,l} (\langle kl \mid t \mid kl\rangle - \langle kl \mid t \mid lk\rangle)\, \bar{n}_k \bar{n}_l$$

$$+ \tfrac{1}{2}\,\text{tr}_1\,(\tilde{t}^{(1)} g) + \tfrac{1}{2}\,\text{tr}_2\,(\tilde{t}p) + \tfrac{1}{2}\,(\psi, \tilde{t}\psi) \qquad (5.21)$$

This general result allows specialization in various ways:

(i) If there is simple pairing, but $\bar{n}_k \neq 0$, then g and p are diagonal, and $\psi(k, k')$ has the simple pairing property. Equation (5.21) then reduces directly to the last two sums in (VII, 4.7). The first sum on the right of (VII, 4.7) is the single-particle operator contribution, obtained from (5.19).

(ii) For general pairing, but all particles paired, we have $\bar{n}_k = 0$, all k, and hence $\tilde{t}^{(1)} = 0$ from (3.30b). Only the last two terms of (5.21) survive, $\tilde{t} = t$, and we thus get (VIII, 2.15).

(iii) If all particles are paired *and* the pairing is simple, both (VII, 4.7) and (VIII, 2.15) reduce to (IV, 1.11).

In this Appendix, we shall not go any further; most of the results used in the book have now been derived. The outstanding exception is the result (VII, 4.9) for the entropy (Thompson 63); the proof is too lengthy for inclusion here.

Although all results in the theory of superconductivity have been obtained by the specialization to a single pair state (i.e., an extreme Schafroth condensation), the formalism of Section 4 is not restricted to that case. In particular, a perturbation expansion to allow for the presence of non-condensed pairs is quite possible (Thompson 63b). The special results for only one pair state can be obtained without the use of the Dyson formalism (Blatt 60a), but the special method used for one pair state does not allow extension to the general case.

BIBLIOGRAPHY and AUTHOR INDEX

The Roman numeral after the reference entry denotes the chapter number the reference is cited in; the Arabic number denotes the section within the chapter.

Abraham, M.

49 and R. BECKER. "The Classical Theory of Electricity and Magnetism," 2nd English Edition, Hafner, New York, 1949. I,4, 6.

Abrikosov, A. A.

52 *Dokl. Akad. Nauk SSSR* **86,** 489 (1952). I, 5; X, 2.

57 *J. Exptl. Theoret. Phys. (USSR)* **32,** 1442 (1957) [English translation: *Soviet Phys.—JETP* **5,** 1174 (1957)]. 1,5;X,2.

57a *J. Phys. Chem. Solids* **2,** 199 (1957). I,5; X,2.

58 and L. P. GOR'KOV. *J. Exptl. Theoret. Phys. (USSR)* **35,** 1558 (1958). [English translation: *Soviet Phys.—JETP* **8,** 1090 (1959)]. VIII, 7; X,2, 3.

58a L. P. GOR'KOV, and I. M. KHALATNIKOV. *J. Exptl. Theoret. Phys. (USSR)* **35,** 265 (1958) [English translation: *Soviet Phys.—JETP* **8,** 182 (1959)]. VI,7.

59 and L. P. GOR'KOV. *J. Exptl. Theoret. Phys. (USSR)* **36,** 319 (1959) [English translation: *Soviet Phys.—JETP* **9,** 220 (1959)]. X,2, 3.

59a L. P. GOR'KOV and I. E. DZYALOSHINSKII. *J. Exptl. Theoret. Phys. (USSR),* **36,** 900 (1959) [English translation: *Soviet Phys.—JETP* **9,** 636 (1959)]. X,2, 3.

59b L. P. GOR'KOV, and I. M. KHALATNIKOV. *J. Exptl. Theoret. Phys. (USSR)* **37,** 187 (1959) [English translation: *Soviet Phys.—JETP* **10,** 132 (1960)]. VI,7.

60 and L. P. GOR'KOV. *J. Exptl. Theoret. Phys. (USSR)* **39,** 1781 (1960) [English translation: *Soviet Phys.—JETP* **12,** 1243 (1961)]. X,2, 3.

61 and L. P. GOR'KOV. *J. Exptl. Theoret. Phys. (USSR)* **39,** 480 (1960). [English translation: *Soviet Phys.—JETP* **12,** 337 (1961)]. VIII,7.

62 and L. P. GOR'KOV. *J. Exptl. Theoret. Phys. (USSR)* **42,** 1088 (1962) [English translation: *Soviet Phys.—JETP* **15,** 752 (1962)]. VIII,7.

[**See also** KHALATNIKOV, I. M., 59]

Adkins, C. J.

62 *Proc. Roy. Soc.* **A268,** 276 (1962). VI,7.

Adler, J. G.

63 and J. S. ROGERS. *Phys. Rev. Letters* **10,** 217 (1963). VI,7.

Aharanov, Y.

59 and D. BOHM. *Phys. Rev.* **115,** 485 (1959). I,7; IX,4,

61 and D. BOHM. *Phys. Rev.* **123,** 1511 (1961). IX,4.

Albertoni, S.

60 P. BOCCHIERI, and A. LOINGER. *J. Math. Phys.* **1,** 244 (1960). IX, 10.

Alekseevski, N. E.

57 and M. N. MIKHEEVA. *Soviet Phys.—JETP* **4,** 810 (1957). X,4.

60 and M. N. MIKHEEVA. *J. Exptl. Theoret. Phys. (USSR)* **38,** 292 (1960) [English translation: *Soviet Phys.—JETP* **11,** 211 (1960)]. X,4.

61 and N. N. MIKHAILOV. *J. Exptl. Theoret. Phys.* (*USSR*) **41**, 1809 (1961) [English translation: *Soviet Phys. JETP* **14**, 1287 (1962)]. X, 2.

Alers, G. A.
60 *Phys. Rev.* **119**, 1532 (1960). VI, 5.
61 and D. L. WALDORF. *Phys. Rev. Letters* **6**, 677 (1961). VI,5.
62 and D. L. WALDORF. *IBM J. Res. Develop.* **6**, 89 (1962). VI,5.

Alers, P. B.
57 *Phys. Rev.* **105**, 104 (1957). X,2.
59 *Phys. Rev.* **116**, 1483 (1959). X,2.

Ambegaokar, V.
61 and L. P. KADANOFF. *Nuovo Cimento* **22**, 914 (1961). VIII, 6.

Anderson, A. C.
61 G. Z. SALINGER, W. A. STEYERT, and J. C. WHEATLEY. *Phys. Rev. Letters* **6**, 331 (1961); **7**, 295 (1961). V,6.

Anderson, P. W.
58 *Phys. Rev.* **110**, 827 (1958). V,6; VI,1; VIII,6.
58a *Phys. Rev.* **112**, 1900 (1958). V,6; VI,1; VIII,6.
59 *Phys. Chem. Solids* **11**, 26 (1959). VI,7; VIII,7; X,3.
59a *Phys. Rev. Letters* **3**, 325 (1959). VIII,7.
59b and H. SUHL. *Phys. Rev.* **116**, 898 (1959). X,3.
61 and P. MOREL. *Phys. Rev.* **123**, 1911 (1961). V,6.
62 *Phys. Rev. Letters* **9**, 309 (1962). IX,10; X,2.
63 and J. M. ROWELL. *Phys. Rev. Letters* **10**, 230 (1963). VI,7.
[**See also** BRUECKNER, K. A., 60; MOREL, P., 62; ROWELL, J. M., 63]

Andres, K.
62 J. L. OLSEN, and H. ROHRER. *IBM J. Res. Develop.* **6**, 84 (1962). VI,5; X,3.

Andrew, E. R.
48 *Proc. Roy. Soc.* **A194**, 98 (1948). X,2.

Androes, G. M.
59 and W. D. KNIGHT. *Phys. Rev. Letters* **2**, 386 (1959). VIII,7.
61 and W. D. KNIGHT. *Phys. Rev.* **121**, 779 (1961). VIII, 7.
[**See also** KNIGHT, W. D., 56]

Andronikashvili, E. L.
46 *J. Phys.* (*USSR*) **10**, 201 (1946). II,7.

Aron, P. R.
62 and H. C. HITCHCOCK. *J. Appl. Phys.* **33**, 2242 (1962). X,2.

Arp, V. D.
61 R. H. KROPSCHOT, J. H. WILSON, W. F. LOVE, and R. PHELAN. *Phys. Rev. Letters* **6**, 452 (1961). X,2.
[**See also** KROPSCHOT, R. H., 61]

Arvieu, R.
60 and M. VENERONI. *Compt. Rend.* **250**, 992, 2155 (1960). III,9.

Autler, S. H.
60 *Rev. Sci. Instr.* **31**, 369 (1960). X,2.
62 E. S. ROSENBLUM, and K. H. GOOEN. *Phys. Rev. Letters* **9**, 489 (1962). X,2.

Azbel, M. I.
58 and I. M. LIFSHITZ. *Soviet Phys.—JETP* **6**, 609 (1958). VI,7.

Baird, D. C.

59 *Can. J. Phys.* **37**, 129 (1959). X,2.

Balashova, B. M.

57 and Iu. Sharvin. *Soviet Phys.—JETP* **4**, 54 (1957). X,2.

Baldwin, J. P.

63 *Phys. Letters (Neth.)* **3**, 223 (1963). X, 4.

Balian, B.

60 and C. DeDominicis. *Nucl. Phys.* **16**, 502 (1960). VII,1.

60a and C. DeDominicis. *Compt. Rend.* **250**, 3285, 4111 (1960). VII,1.

Balian, R.

61 C. Bloch, and C. DeDominicis. *Nucl. Phys.* **25**, 529 (1961). V,6; VII,1.

61a C. Bloch, and C. DeDominicis. *Nucl. Phys.* **27**, 294 (1961). V,6; VII,1.

62 L. H. Nosanow, and N. R. Werthamer. *Phys. Rev. Letters* **8**, 372 (1962); Erratum, *Phys. Rev. Letters* **8**, 466 (1962). V,6.

63 *Phys. Rev.* **131**, 1553 (1963). V,6; VIII,7.

Ballmoos, F. von

60 *Z. Angew. Phys.* **12**, 1 (1960). X,2.

Baltensperger, W.

59 *Helv. Phys. Acta* **32**, 197 (1959). X,3.
[**See also** Daunt, J. G., 62]

Banus, M. D.

62 T. B. Reed, H. C. Gatos, M. C. Lavine, and J. A. Kafalas. *Phys. Chem. Solids* **23**, 971 (1962). X,3.

Baranger, M.

60 *Phys. Rev.* **120**, 957 (1960). III,9.

61 *Phys. Rev.* **122**, 992 (1961). III,9; VIII,1.

62 A Reformulation of the Theory of Pairing Correlations. ONR contract Nonr. 760(15), Tech. Rept. No. 10, Carnegie Inst. of Technology, Pittsburgh, Pennsylvania (November 1962). *Phys. Rev.* **130**, 1244 (1963). VIII,1,6.

63 "Lecture Notes from the Theoretical Physics Summer School at Cargese 1962." Benjamin, New York, 1963. VIII,1.

Bardasis, A.

61 and J. R. Schrieffer. *Phys. Rev.* **121**, 1050 (1961). VI,1; VII,5.

Bardeen, J.

41 *Phys. Rev.* **59**, 928 (1941). I,9.

50 *Phys. Rev.* **79**, 167 (1950). 1,10.

50a *Phys. Rev.* **80**, 567 (1950). I,10.

51 *Phys. Rev.* **81**, 829 (1951). I,9.

51a *Phys. Rev.* **82**, 978 (1951). I,9.

51b *Rev. Mod. Phys.* **23**, 261 (1951). I,10.

55 *Phys. Rev.* **87**, 192 (1952). I,8.

55a and D. Pines. *Phys. Rev.* **99**, 1140 (1955). 1,9; V,3,6.

56 "Handbuch der Physik" (S. Flügge, ed.), 2nd ed., Vol. 15, p. 274. Springer, Berlin, 1956. I,10.

57 L. N. Cooper, and J. R. Schrieffer. *Phys. Rev.* **108**, 1175 (1957). III,1; IV,1, 2; V,1, 2, 3, 6; VI,3, 5, 7; VII,1.5; VIII,6.

57a *Nuovo Cimento* **5**, 1764 (1957). VIII,6.

58 *Physica, Suppl.* **24**, 27 (1958). V,2.

59 *Phys. Rev. Letters* **1**, 399 (1959). IX, 3.
59a G. RICKAYZEN and L. TEWORDT. *Phys. Rev.* **113**, 982 (1959). VI,7.
60 and G. RICKAYZEN. *Phys. Rev.* **118**, 936 (1960). V,2.
61 and J. R. SCHRIEFFER. Recent Developments in Superconductivity, *Progr. Low Temp. Phys.* **3**, Chapter VI (1961). I,6, 10; III,1; V,1, 2, 3; VI,6; VII,3.
61a *Phys. Rev. Letters* **7**, 162 (1961). IX,8.
61b *Phys. Rev. Letters* **6**, 57 (1961). VI, 7.
62 *Phys. Rev. Letters* **9**, 147 (1962). VI,7.
62a *IBM J. Res. Develop.* **6**, 3 (1962). V,2.
62b *Rev. Mod. Phys.* **34**, 667 (1962). IX,36,10; X,4.
63 *Phys. Today*, **16**, 19 (1963). I,10; V, 2.
[**See also** MATTIS, D. C., 58.)

Barnes, L. J.
63 and J. R. DILLINGER. *Phys. Rev. Letters* **10**, 287 (1963). VI,7.

Batrakov, G. F.
62 O. R. MIS'KEVICH, and E. TROINAR. *J. Exptl. Theoret. Phys.* (*USSR*) **42**, 1171 (1962). X,2.

Baumann, F.
56 *Nachr. Akad. Wiss. Göttingen, Math.-Physik Kl.* **IIa**, 285 (1956). X,4.

Bayman, B. F.
60 *Nucl. Phys.* **15**, 33 (1960). III, 9; V,2.

Bean, C. P.
62 M. V. DOYLE, and A.G. PINCUS. *Phys. Rev. Letters* **9**, 93 (1962). X,2.
62a *Phys. Rev. Letters* **8**, 250 (1962). X,2.

Beck, F.
55 *Phys. Rev.* **98**, 852 (1955). I,5.

Becker, R.
33. "Theorie der Elektrizität, Band II, Elektronen-theorie." Teubner, Leipzig, 1933. I,2; II,5.
[**See also** ABRAHAM, M., 49]

Behrndt, M. E.
60 R. H. BLUMBERG, and G. R. GIEDD. *IBM J. Res. Develop.* **4**, 184 (1960). X,4.

Bell, J. S.
61 and E. J. SQUIRES. *Advan. Phys.* **10**, 211 (1961). III,7; VII,3; VIII,6.
63 *Phys. Rev.* **129**, 1896 (1963). VII,1.

Belyaev, S. T.
59 *Kgl. Danske Videnskab. Selskab., Mat. Fys. Medd.* **31**, No. 11 (1959). III,9.
60 *Physica, Suppl.* **26**, 181 (1960). VI,1.

Berghout, C. W.
62 *Phys. Letters* (*Neth.*) **1**, 292 (1962). X,3.

Bergmann, P. G.
55 and J. L. LEBOWITZ, *Phys. Rev.* **99**, 578 (1955). IX, 10.
[**See also** LEBOWITZ, J. L., 57; WILLIS, C. R., 62a]

Berlincourt, T. G.
59 *Phys. Rev.* **114**, 969 (1959). X,2.
61 R. R. HAKE, and D. H. LESLIE. *Phys. Rev. Letters* **6**, 671 (1961). X,2.

62 and R. R. HAKE, *Phys. Rev. Letters* **9**, 293 (1962). X,2.
63 and R. R. HAKE. *Phys. Rev.* **131**, 140 (1963). X,2.
[**See also** HAKE, R. R., 62, 62a]

Betbeder-Matibet, O.
61 and P. NOZIÈRES. *Compt. Rend.* **252**, 3943 (1961). V,6.

Betterton, J. O., Jr.
61 R. W. BOOM, G. D. KNEIP, R. E. WORSHAM, and C. E. ROOS. *Phys. Rev. Letters* **6**, 532 (1961). X,2.
[**See also** KNEIP, G. D., Jr., 62]

Bezuglyi, P. A.
58 and A. A. GALKIN. *Soviet Phys.—JETP* **7**, 164 (1958). VI,7.
59 A. A. GALKIN, and A. P. KOROLYUK. *J. Exptl. Theoret. Phys.* (*USSR*) **36**, 1951 (1959) [English translation: *Soviet Phys.—JETP* **9**, 1388 (1959)]. VI,7.
61 A. A. GALKIN, and A. P. KOROLYUK. *Soviet Phys.—JETP* **12**, 4 (1961). VI, 7

Biondi, M. A.
58 A. T. FORRESTER, M. P. GARFUNKEL, and C. B. SATTERTHWAITE. *Rev. Mod. Phys.* **30**, 1109 (1958). I,8; V,2.
59 and M. P. GARFUNKEL. *Phys. Rev. Letters* **2**, 143 (1959); *Phys. Rev.* **116**, 853, 862 (1959). I,8; VI,7.

Blatt, J. M.
52 and V. F. WEISSKOPF. "Theoretical Nuclear Physics." Wiley, New York, 1952. VI,7.
54 and S. T. BUTLER. *Phys. Rev.* **96**, 1141 (1954). II,7.
55 S. T. BUTLER, and M. R. Schafroth. *Phys. Rev.* **100**, 481 (1955). I,3.
55a and S. T. BUTLER. *Phys. Rev.* **100**, 476 (1955). II,7.
56 *Nuovo Cimento* **4**, 430 (1956). III,5.
56a *Nuovo Cimento* **4**, 465 (1956). III,5.
58 and T. MATSUBARA. *Progr. Theoret. Phys. (Kyoto)* **20**, 553 (1958). III,8, 9; VII,4; B,4.
58a and T. MATSUBARA. *Progr. Theoret. Phys. (Kyoto)* **20**, 781 (1958). VIII, 6.
59 T. MATSUBARA, and R. M. MAY. *Progr. Theoret. Phys. (Kyoto)* **21**, 745 (1959). VIII,6.
59a *Progr. Theoret. Phys. (Kyoto)* **22**, 745 (1959). IX,10; A,3.
60 *Progr. Theoret. Phys. (Kyoto)* **23**, 447 (1960). III,1; V,2.
60a *J. Australian Math. Soc.* **1**, 465 (1960). IV,1; V,2; VII,4; VIII,2; B,4, 5.
60b *Progr. Theoret. Phys. (Kyoto)* **24**, 851 (1960). VIII,3, 4.
61 *Progr. Theoret. Phys. (Kyoto)* **26**, 761 (1961). II,7; IX,1, 5, 6, 8, 9, 10; X,4.
61a *Phys. Rev. Letters.* **7**, 82 (1961). II,7; IX,5.
62 *Progr. Theoret. Phys. (Kyoto)* **27**, 1137 (1962). III,9.
62a K. W. BÖER, and W. BRANDT. *Phys. Rev.* **126**, 1691 (1962). III,1.
62b *J. Australian Math. Soc.* **2**, 464 (1962). B,4.
63 and C. J. THOMPSON. *Phys. Rev. Letters* **10**, 332 (1963). X,4.
[**See also** KATZ, A., 59; MCKENNA, J., 62; MATSUBARA, T., 60; SCHAFROTH, M. R., 56, 57; THOMPSON, C. J., 63a, 63b]

Blaugher, R. D.
61 and J. K. HULM. *Phys. Chem. Solids* **19**, 134 (1961), X,3.
61a B. S. CHANDRASEKHAR, J. K. HULM, E. CORENZWIT, and B. T. MATTHIAS. *Phys. Chem. Solids* **21**, 252 (1961). X,3.
62 A. TAYLOR, and J. K. HULM. *IBM J. Res. Develop.* **6**, 116 (1962). X, 3.

62a and J. K. Hulm, *Phys. Rev.* **125**, 474 (1962). X,2.
[**See also** Hein, R. A., 63; Hulm, J. K., 61, 61a]

Bloch, C.

59 and C. DeDominicis. *Nucl. Phys.* **10**, 181 (1959). III,7; V,6.
59a and C. DeDominicis. *Nucl. Phys.* **10**, 509 (1959). III,7; V,6.
60 *Physica* **26**, 562 (1960). VII,1.
62 and A. Messiah. *Nucl. Phys.* **39**, 95 (1962). V,2; VIII,2.
[**See also** Balian, R., 61, 61a]

Bloch, F.

62 and H. E. Rorschach. *Phys. Rev.* **128**, 1697 (1962). II,7; IX,1,6,9; X,4.

Blum, L.

63 *Nuovo Cimento* **28**, 25 (1963). V,6.

Blumberg, R. H.

62 and D. P. Seraphim. *J. Appl. Phys.* **33**, 163 (1962). X,4.
62a *J. Appl. Phys.* **33**, 1822 (1962). X,3.
[**See also** Behrndt, M. E., 60, Douglas, D. H., Jr., 62a]

Blumberg, W. E.

60 J. Eisinger, V. Jaccarino, and B. T. Matthias. *Phys. Rev. Letters* **5**, 149 (1960). IX,3.

Bogoliubov, N. N.

46 *J. Phys.* (*USSR*) **10**, 256, 265 (1946). [Full English translation: "Studies in Statistical Mechanics," Vol. I, p. 5. North-Holland, Amsterdam, 1962]. VI,3.
47 *J. Phys. (USSR)* **9**, 23 (1947). II,6; IV,1.
47a and K. P. Gurov. *J. Exptl. Theoret. Phys.* (*USSR*) **17**, 614 (1947). VI,3; IX,2.
57 D. N. Zubarev, and Yu. A. Tserkovnikov. *Dokl. Akad. Nauk SSSR* **117**, 778 (1957) [English translation: *Soviet Phys. Doklady* **2**, 535 (1957)].
58 V. V. Tolmachov, and D. V. Shirkov. "A New Method in the Theory of Superconductivity." Dubna, 1958 (in English).
Fortschr. Physik **6**, 605 (1958) (in German).
Consultants Bureau, Inc., New York (1959) (in English) I,10; III,1; IV,1, 2. V,1, 2, 3, 5, 6; VII,3, 5; VIII,6.
58a *Nuovo Cimento* **7**, 6, 794 (1958). III,1; IV,1; V,3.
58b *J. Exptl. Theoret. Phys.* (*USSR*) **34**, 58, 73 (1958) [English translation: *Soviet Phys.—JETP* **7**, 41, 51 (1958)]. III,1.
58c *Dokl. Akad. Nauk SSSR* **119**, 52 (1958) [English translation: *Soviet Phys. Doklady* **3**, 279 (1958)]. III,9.
58d *Dokl. Akad. Nauk SSSR* **119**, 2 (1958) [English translation: *Soviet Phys. Doklady* **3**, 292 (1959)]. V,4; VIII,1.
59 *Usp. Fiz. Nauk* **67**, 549 (1959) [English translation: *Soviet Phys.—Uspekhi* **2**, 236 (1959)]. VIII,1,6.
60 *Physica, Suppl.* **26**, 1 (1960). V,2.
60a D. N. Zubarev, and Yu. A. Tserkovnikov, *J. Exptl. Theoret. Phys. (USSR)* **39**, 120 (1960) [English translation: *Soviet Phys.—JETP* **12**, 88 (1961)]. V,2; VI,3; VII,1.

Bohm, D.

51 and D. Pines. *Phys. Rev.* **82**, 625 (1951). I,9.
53 and D. Pines. *Phys. Rev.* **92**, 609 (1953). I,9; V,6.
[**See also** Aharanov, 59, 61 Pines, D., 52]

Bohr, A.
58 B. R. Mottelson, and D. Pines. *Phys. Rev.* **110**, 936 (1958). III,9.
62 and B. R. Mottelson. *Phys. Rev.* **125**, 495 (1962). IX,8.

Bömmel, H. E.
54 *Phys. Rev.* **96**, 220 (1954). VI,7.

Bonfiglioli, G.
62 R. Malvano, and B. B. Goodman. *J. Appl. Phys.* **33**, 2564 (1962). X,4.

Bon Mardion, G.
62 B. B. Goodman, and A. Lacaze. *Phys. Letters* **2**, 321 (1962). X,2.

Boorse, H. A.
59 *Phys. Rev. Letters* **2**, 391 (1959). VI,5.
60 A. T. Hirschfeld, and H. Leupold. *Phys. Rev. Letters* **5**, 246 (1960). VI,5.
[**See also** Cook, D. B., 50; Hirshfeld, A. T., 62; Kaplan, R., 59]

Born, M.
48 and K. C. Cheng. *J. Phys. Radium* **9**, 249 (1948). I,9.
48a and K. C. Cheng. *Nature* **161**, 968 (1948). I,9.
48b and K. C. Cheng. *Nature* **161**, 1017 (1948). I,9.

Bowen, D. H.
60 and G. O. Jones. *Proc. Roy. Soc.* **A254**, 522 (1960). VI,5.

Bozorth, R. M.
60 D. D. Davis and A. J. Williams. *Phys. Rev.* **119**, 1570 (1960). X,3.
60a A. J. Williams, and D. D. Davis. *Phys. Rev. Letters* **5**, 148 (1960). X,2.

Brandt, N. B.
60 and N. I. Ginzburg. *J. Exptl. Theoret. Phys.* (*USSR*) **39**, 1554 (1960) [English translation: *Soviet Phys.—JETP* **12**, 1082 (1961)]. X,4.
61 and N. I. Ginzburg. *Fiz. Tverd. Tela* **3**, 3461 (1961) [English translation: *Soviet Phys.—Solid State* **3**, 2510 (1962)]. X,4.

Bremer, J. W.
58 and V. L. Newhouse. *Phys. Rev. Letters* **1**, 282 (1958). X,4.
59 and V. L. Newhouse. *Phys. Rev.* **116**, 309 (1959). X,4.
[**See also** Newhouse, V. L., 59]

Brémond, B.
63 and J. G. Valatin, *Nucl. Phys.* **41**, 640 (1963). III,9.

Brenig, W.
61 *Phys. Rev. Letters* **7**, 337 (1961). IX,5, 7.

Brillouin, L.
51 *J. Appl. Phys.* **22**, 334 (1951). A,1.
58 *J. Phys. Radium* **19**, 112, 184 (1958). IX,1.

Broom, R. F.
59 and E. H. Rhoderick. *Phys. Rev.* **116**, 344 (1959). X,4.
60 and E. H. Rhoderick. *Brit. J. Appl. Phys.* **11**, 292 (1960). X,4.
61 *Nature* **190**, 992 (1961). IX, 10.
62 and E. H. Rhoderick. *Proc. Phys. Soc.* **79**, 586 (1962). X,4.

Brout, R.
57 *Phys. Rev.* **108**, 515 (1957). I,9.
59 and H. Suhl. *Phys. Rev. Letters* **2**, 387 (1959). X,3.
[**See also** Sawada, K., 57a]

Brovetto, P.
62 and V. Canuto. *Phys. Letters (Neth.)* **3**, 124 (1962). III,9.

Brueckner, K. A.
58 "The Many-Body Problem," Les Houches p. 47. Wiley, New York, 1958. III,7; VII,3; VIII,6.
60 T. Soda, P. W. Anderson, and P. Morel. *Phys. Rev.* **118**, 1442 (1960). V,6.
[**See also** Gell-Mann, M., 57; Sawada, K., 57a]

Bryant, C. A.
60 and P. H. Keesom. *Phys. Rev. Letters* **4**, 460 (1960).
Phys. Rev. **123**, 491 (1961). VI,5.
61 and P. H. Keesom. *J. Chem. Phys.* **35**, 1149 (1961). VI,5.

Bucher, E.
61 F. Heiniger, and J. Müller. *Helv. Phys. Acta* **34**, 843 (1961). X,3.
61a F. Heiniger, and J. Müller. *Helv. Phys. Acta* **34**, 413 (1961). X,3.
61b and J. Müller. *Helv. Phys. Acta* **34**, 410 (1961). X,3.
61c D. Gross, and J. L. Olsen. *Helv. Phys. Acta* **34**, 775 (1961). X,3.

Buckel, W.
54 and R. Hilsch. *Z. Physik* **138**, 109 (1954). X,4.
58 *Proc. 5th Intern. Conf. Low Temp. Phys. Chem.* p. 326. (1958) X,4.

Buckingham, M. J.
54 and M. R. Schafroth. *Proc. Phys. Soc.* **A67**, 828 (1954). I,10.
57 *Nuovo Cimento* **5**, 1763 (1957). VIII, 6.
58 *Bull. Inst. Intern. Froid* **1**, 237 (1958). X,4.
61 and W. M. Fairbank. *Progr. Low Temp. Phys.* **3**, Chapter III (1961). I,1; VI,3.

Budnick, J. I.
56 E. A. Lynton, and B. Serin. *Phys. Rev.* **103**, 286 (1956). X,2.
60 *Phys. Rev.* **119**, 1578 (1960). I,1.
[**See also** Quinn, D. J., III., 61; Seraphim, D. P., 61a]

Buikov, M. V.
60 *Soviet Phys. Doklady* **5**, 544 (1960). V,6.

Burstein, E.
61 D. N. Langenberg, and B. N. Taylor. *Phys. Rev. Letters* **6**, 92 (1961). VI,7.
[**See also** Taylor, B. N., 63]

Burton, E. F.
34 "The Phenomenon of Superconductivity." Univ. of Toronto Press, Toronto, Canada, 1934. I,10.

Byers, N.
61 and C. N. Yang. *Phys. Rev. Letters* **7**, 46 (1961). IX,5.

Cabibbo, N.
63 and S. Doniach. *Phys. Letters (Neth.)* **4**, 29 (1963). IX,10.

Calverley, A.
60 and A. C. Rose-Innes. *Proc. Roy. Soc.* **A255**, 267 (1960), X,2.

Casimir, H. B. G.
34 and C. J. Gorter. *Z. Physik* **35**, 963 (1934). VI,5.
45 *Rev. Mod. Phys.* **17**, 343 (1945). IX,2.

55 *in* "N. Bohr and the Development of Physics" (W. Pauli, ed.), p. 118. Pergamon, London, 1955. I,6; III,1, 3.
[**See also** GORTER, C. J., 34]

Caswell, H. L.
61 *J. Appl. Phys.* **32**, 2641 (1961). X,4.

Challis, L. J.
61 *Proc. 7th Intern. Conf. Low Temp. Phys., Toronto, Ont., 1960* p. 466 (1961). VI,7.
62 *Proc. Phys. Soc.* **80**, 759 (1962). VI,7.

Chandrasekhar, B. S.
58 and J. K. HULM, *Phys. Chem. Solids* **7**, 259 (1958). X,3.
61 and I. A. RAYNE, *Phys. Rev. Letters* **6**, 3 (1961). VI,5.
63 J. K. HULM, C. K. JONES, and A. PATTERSON. *Bull. Am. Phys. Soc.* **8**, 294 (1963). X,2.
[**See also** BLAUGHER, R. D., 61a]

Chang, G. K.
62 R. E. JONES, and A. M. TOXEN. *IBM J. Res. Develop.* **6**, 112 (1962). VI,7.
[**See also** TOXEN, A. M., 62a]

Chanin, G.
59 E. A. LYNTON, and B. SERIN, *Phys. Rev.* **114**, 719 (1959). X,3.

Chaudhuri, K. D.
59 *Z. Physik* **155**, 290 (1959). VI,7.
60 K. MENDELSSOHN, and M. W. THOMPSON. *Cryogenics* **1**, 47 (1960). X,2.
62 *Phys. Chem. Solids* **23**, 1341 (1962). VI,7.

Cheng, K. C.
49 *Nature* **163**, 247 (1949). I,9.
[**See also** BORN, M., 48, 48a, 48b]

Chester, G. V.
61 *Advan. Phys.* **10**, 357 (1961). V,3, 6.

Chester, M.
62 *J. Appl. Phys.* **33**, 643 (1962). VI,5.

Chiou, C.
63 R. A. CONNELL, and D. P. SERAPHIM. *Phys. Rev.* **129**, 1070 (1963). X,2.
[**See also** SERAPHIM, D. P., 61b]

Chopra, K. L.
59 and T. S. HUTCHISON. *Can. J. Phys.* **37**, 614 (1959). VI,7.
59a and T. S. HUTCHISON. *Can. J. Phys.* **37**, 1100 (1959). VI,7.

Chotkevich, V. I.
50 and V. R. GOLIK. *J. Exptl. Theoret. Phys. (USSR)* **20**, 427 (1950). X,2.

Chun-Sian, Chen
59 and Chow SHIH-HSUN. *Soviet Phys.—JETP* **9**, 885 (1959). V,6.

Chynoweth, A. G.
62 R. A. LOGAN, and D. E. THOMAS, *Phys. Rev.* **125**, 877 (1962). VI,7.
[**See also** ROWELL, J. M., 62]

Cioffi, P. P.
62 *J. Appl. Phys.* **33**, 875 (1962). I,2.

Claiborne, L. T.
63 and N. G. EINSPRUCH. *Phys. Rev. Letters* **10**, 49 (1963). X,2.

Clogston, A. M.
61 and V. JACCARINO. *Phys. Rev.* **121**, 1357 (1961). VI,5; X,3.
62 A. C. GOSSARD, V. JACCARINO, and Y. YAFET. *Phys. Rev. Letters* **9**, 262 (1962). VIII,7.
62a *Phys. Rev. Letters* **9**, 266 (1962). X,2.
[**See also** MATTHIAS, B. T., 61]

Cochran, J. F.
60 and P. COCHRAN, *Proc. 7th, Intern. Conf. Low Temp. Phys. Toronto, Ont., 1960* p. 288 (1961). I,1.
61 *Phys. Rev.* **121**, 1688 (1961). I,1.
62 *Ann. Phys. (NY)* **19**, 177 (1962). I,1.
[**See also** LEWIS, H. R., 60; SHIFFMAN, C. A., 63]

Cody, G. D.
58 *Phys. Rev.* **111**, 1078 (1958). VI,5.

Cohen, E.
62 "Fundamental Problems in Statistical Mechanics." North Holland, Amsterdam 1962. IX,10.

Cohen, H.
61 and W. L. MIRANKER. *J. Math. Phys.* **2**, 575 (1961). VI,5.

Cohen, M. H.
62 L. M. FALICOV, and J. C. PHILLIPS. *Phys. Rev. Letters* **8**, 316 (1962). VI,7.

Coles, B. R.
62 *IBM J. Res. Develop.* **6**, 68 (1962). X,3.

Collins
57 as quoted by J. W. CROWE, *IBM J. Res. Develop.* **1**, 294 (1957). IX,10.

Compton, V. B.
61 E. CORENZWIT, J. P. MAITA, B. T. MATTHIAS, and F. J. MORIN. *Phys. Rev.* **123**, 1567 (1961). X,3.
[**See also** MATTHIAS, B. T., 59, 61a, 62a, 63]

Connell, R. A.
63 *Phys. Rev.* **129**, 1952 (1963). X,2.
[**See also** CHIOU, C., 63; SERAPHIM, D. P., 59]

Connolly, A.
62 and K. MENDELSSOHN. *Proc. Roy. Soc.* **A366**, 429 (1962). VI,7.

Cook, D. B.
50 M. W. ZEMANSKY, and H. A. BOORSE, *Phys. Rev.* **80**, 737 (1950). X,2.

Cooper, L. N.
56 *Phys. Rev.* **104**, 1189 (1956). III,1; V,2.
59 Lectures at the 1959 Summer Institute of Brandeis University, Lecture Notes by Z. Fried. V,2.
59a *Phys. Rev. Letters* **3**, 17 (1959). VI,5.
60 *Am. J. Phys.* **28**, 91 (1960). V,2.
61 *Phys. Rev. Letters* **6**, 689 (1961). X,4.
62 *Phys. Rev. Letters* **8**, 367 (1962). IV,4; VIII,7.
62a *IBM J. Res. Develop.* **6**, 75 (1962). X,4.
[**See also** BARDEEN, J., 57]

Corak, W. S.
56 and C. B. SATTERTHWAITE. *Phys. Rev.* **102**, 662 (1956). I,1.

Corenzwit, E.
59 *Phys. Chem. Solids* **9**, 93 (1959). X,3.
[**See also** BLAUGHER, R. D., 61a; COMPTON, V. B., 61; DEVLIN. G. E., 60; GEBALLE, T. H., 61, 62, 63; HEIN, R. A., 59, 62; MATTHIAS, B. T. 58, 58a 58b, 59, 60a, 61, 61a, 62a, 63b; SUHL, H., 58]

Cornish, F. H. J.
53 and J. L. OLSEN. *Helv. Phys. Acta* **26**, 369 (1953). X,2.

Cottrell, A. H.
53 "Dislocations and Plastic Flow in Crystals" Oxford Univ. Press (Clarendon), London and New York, 1953. X,1.

Craig, R. A.
63 *Nuovo Cimento* **27**, 257 (1963). VI,7.

Culler, G. J.
62 B. D. FRIED, R. W. HUFF, and J. R. SCHRIEFFER. *Phys. Rev. Letters* **8**, 399 (1962). V,6.

Dang, Giu Do
63 and A. KLEIN. *Phys. Rev.* **130**, 2572 (1963). I,9.

Darby, J. B., Jr.
62 and S. T. ZEGLER. *Phys. Chem. Solids* **23**, 1825 (1962). X,3.

Daunt, J. G.
46 and K. MENDELSSOHN. *Proc. Roy. Soc.* **A185**, 225 (1946). I,8.
49 and C. V. HEER, *Phys. Rev.* **76**, 1324 (1949). X,2.
61 and J. L. OLSEN. *Phys. Rev. Letters* **6**, 267 (1961). VI,5.
62 M. KREITMAN, W. BALTENSPERGER, and J. L. OLSEN. *Cryogenics* **2**, 212 (1962). X,2.
[**See also** SMITH, T. S., 52]

Davies, E. A.
60 *Proc. Roy. Soc.* **A255**, 407 (1960). X,2.

Dayem, A. H.
62 and R. J. MARTIN. *Phys. Rev. Letters* **8**, 246 (1962). VI,7.

Deaver, B. S.
61 and W. M. FAIRBANK. *Phys. Rev. Letters* **7**, 43 (1961). I,7; III,1; IX,1, 5.

Debye, P.
12 *Ann. Physik* **39** (4), 789 (1912). I,10.
23 and E. HÜCKEL. *Physik. Z.* **24**, 185 (1923). I,9.

Decker, D. L.
58 D. E. MAPOTHER, and R. W. SHAW. *Phys. Rev.* **112**, 1888 (1958). V,6.

DeDominicis, C.
62 *J. Math. Phys.* **3**, 983 (1962). VII,1.
63 *J. Math. Phys.* **4**, 255 (1963). VII,1.
[**See also** BALIAN, B., 60, 60a; BALIAN, R., 61, 61a; BLOCH, C., 59, 59a]

DeFeo, P.
62 and G. SACERDOTI. *Phys. Letters* **2**, 264 (1962). IX,10.

DeGennes, P. G.
63 and E. GUYON. *Phys. Letters* **3**, 168 (1963). X,4.

De Groot, S. R.

51 "Thermodynamics of Irreversible Processes," Interscience, New York and North Holland, Amsterdam, 1951. IX,2,10.

59 *Rendi Scuola Intern. Fis. "Enrico Fermi"* **10**, 131 (Varenna, 1959). IX,2,10.

62 and P. Mazur, "Non-Equilibrium Thermodynamics." North-Holland, Amsterdam, 1962. IX,2,10.

63 *J. Math. Phys.* **4**, 147 (1963). IX,2,10.

DeSorbo, W.

60 *Phys. Rev. Letters* **4**, 406 (1960). X,2.

60a *Proc. 7th, Intern. Conf. Low Temp. Phys. Toronto, Ont., 1960* p. 367 (1961). X,2.

62 and V. L. Newhouse. *J. Appl. Phys.* **33**, 1004 (1962). X,2.

Detwiler, D. P.

52 and H. A. Fairbank. *Phys. Rev.* **88**, 1049 (1952). X,2.

Devlin, G. E.

60 and E. Corenzwit. *Phys. Rev.* **120**, 1964 (1960). VI,5.

[**See also** Schawlow, A. L., 58, 59, 59a]

Devons, S.

62 *Sci. Progr. (London)* **50**, 246 (1962). X,2.

DeWit, R.

60 *Solid State Phys.* **10**, 249 (1960). X,1.

Dietrich, I.

62 *Z. Naturforsch.* **17a**, 94 (1962). VI,7.

Doll, R.

58 *Z. Physik* **153**, 207 (1958). I,3.

61 and M. Näbauer, *Phys. Rev. Letters* **7**, 51 (1961); *Z. Physik* **169**, 526 (1962). I,7; III,1; IX,1,5.

[**See also** Meissner, W., 55, 59]

Domb, C.

60 *Advan. Phys.* **9**, 149, 245 (1960). VI,3.

61 and M. F. Sykes. *J. Math. Phys.* **2**, 63 (1961). VI,3.

Douglass, D. H., Jr.

61 *Phys. Rev. Letters* **6**, 346 (1961). VIII,7; X,4.

61a *Phys. Rev. Letters* **7**, 14 (1961). VI,5; VIII,7.

61b *Phys. Rev.* **124**, 735 (1961). VI,5; VIII,7; X,4.

62 *Phys. Rev. Letters* **9**, 155 (1962). X,4.

62a and R. H. Blumberg. *Phys. Rev.* **127**, 2038 (1962). X,4.

62b *IBM J. Res. Develop.* **6**, 44 (1962). X,4.

[**See also** Simmons, W. A., 62]

Doulat, J.

59 *Compt. Rend.* **249**, 2017 (1959). X,2.

Drangeid, K. E.

62 and R. Sommerhalder, *Phys. Rev. Letters* **8**, 467 (1962). VIII,7.

Dresden, M.

62 "A Study of Models in Non-Equilibrium Statistical Mechanics". "Studies in Statistical Mechanics," Vol. I, p. 299. North-Holland, Amsterdam, 1962. IX,10.

Dresselhaus, G.
60 and M. S. DRESSELHAUS. *Phys. Rev.* **120**, 1971 (1960). VI, 7.
60a and M. S. DRESSELHAUS. *Phys. Rev.* **118**, 77 (1960). VI,7.
62 and M. S. DRESSELHAUS, *Phys. Rev. Letters* **125**, 1212 (1962). VI,7.
[**See also** DRESSELHAUS, M. S., 60b]

Dresselhaus, M. S.
60b and G. DRESSELHAUS. *Phys. Rev. Letters* **3**, 401 (1960). VI,7.
[**See also** DRESSELHAUS, G., 60, 60a, 62]

Dubeck, L.
63 P. LINDENFELD, E. A. LYNTON, and H. ROHRER. *Phys. Rev. Letters* **10**, 98 (1963). VI,7; X,2.

Du Bois, D. F.
59 *Ann. Phys. (NY)* **7**, 174 (1959); **8**, 24 (1959). I,9.

Duijvestijn, A. J. W.
59 *IBM J. Res. Develop.* **3**, 132 (1959). VI,5.

Dyson, F. J.
56 *Phys. Rev.* **102**, 1217 (1956). B,4.

Edwards, S. F.
58 *Phil. Mag.* **3**, 1020 (1958). VIII,7.

Ehrenfest, P.
33 *Commun. Phys. Lab. Univ. Leiden Suppl.* **75b** (1933). I,1.

Einstein, A.
24 *Ber. Berlin. Akad.* 261 (1924). II,2.
25 *Ber. Berlin. Akad.* 3 (1925). II,2.

Eisenstein, J.
54 *Rev. Mod. Phys.* **26**, 277 (1954). I,10.

Eliashberg, G. M.
60 *J. Exptl. Theoret. Phys. (USSR)* **38**, 966 (1960); **39**, 1437 (1960) [English translation: *Soviet Phys. JETP* **11**, 696 (1960); **12**, 1000 (1961)]. V,6.
62 *J. Exptl. Theoret. Phys. (USSR)* **43**, 1105 (1962) [English translation: *Soviet Phys.—JETP* **16**, 780 (1963)]. VI,6.

Emery, V. J.
60 *Nucl. Phys.* **19**, 154 (1960). V,6.
60a and A. M. SESSLER. *Phys. Rev.* **119**, 43, 248 (1960). V,6.

Engelsberg, S.
62 *Phys. Rev.* **126**, 1251 (1962). VIII,7.
63 and J. R. SCHRIEFFER. *Phys. Rev.* **131**, 993 (1963). VI,7.

Englman, R.
63 *Phys. Rev.* **129**, 551 (1963). I,10.

Erlbach, E.
60 R. L. GARWIN, and M. P. SARACHIK. *IBM J. Res. Develop.* **4**, 107 (1960). X,4.
[**See also** SARACHIK, M. P., 60]

Eshelby, J. D.
56 *Solid State Phys.* **3**, 79 (1956). X,1.

Faber, T. E.
52 *Proc. Roy. Soc.* **A214**, 392 (1952). VI,5.
53 *Proc. Roy. Soc.* **A219**, 75 (1953). VI,5.
54 *Proc. Roy. Soc.* **A223**, 174 (1954). VI,5.
55 and A. B. PIPPARD. *Proc. Roy. Soc.* **A231**, 336 (1955). I, 3; VI,7.
55a *Proc. Roy. Soc.* **A231**, 353 (1955). VI,5,7.
55b and A. B. PIPPARD. *Progr. Low Temp. Phys.* **1**, Chapter 9 (1955). I,10; VI,7.
58 *Proc. Roy. Soc.* **A248**, 460 (1958). X,2.

Ferrell, R. A.
59 *Phys. Rev. Letters* **3**, 262 (1959). VIII,7.
61 *Phys. Rev. Letters* **6**, 541 (1961). VI,5.
[**See also** TINKHAM, M., 59]

Feshbach H.
54 C. E. PORTER, and V. F. WEISSKOPF. *Phys. Rev.* **96**, 448 (1954). X,4.

Feucht, D. L.
61 and J. B. WOODFORD, Jr. *J. Appl. Phys.* **32**, 1882 (1961), X,4.

Feynman, R. P.
53 *Proc. Intern. Conf. Theoret. Phys. Kyoto Tokyo, Japan, 1953* (1953). II,1.

Fierz, M.
51 *Helv. Phys. Acta* **24**, 357 (1951). III,6.

File, J.
63 and R. G. MILLS. *Phys. Rev. Letters* **10**, 93 (1963). IX,10; X,2.

Fink, H. J.
59 *Can. J. Phys.* **37**, 474 (1959). IX,10.
59a *Can. J. Phys.* **37**, 485 (1959). VI,5.

Finnemore, D. K.
62 and D. E. MAPOTHER. *Phys. Rev. Letters* **9**, 288 (1962). X,3.

Fisher, J. C.
60 *Australian J. Phys.* **13**, 446 (1960). VIII,7; X,3.
61 and I. GIAEVER. *J. Appl. Phys.* **32**, 172 (1961). VI, 7.

Fisher, M. E.
59 *Phys. Rev.* **113**, 969 (1959). VI,3.
63 *J. Math. Phys.* **4**, 278 (1963). VI,3.

Fleischer, R. L.
62 *Phys. Letters* **3**, 111 (1962). X,2.

Foldy, L. L.
61 *Phys. Rev.* **124**, 649 (1961). II,6; VI,1.

Franck, J. P.
61 and D. L. MARTIN. *Can. J. Phys.* **39**, 1320 (1961). I,1.

Franz, W.
61 *Z. Naturforsch.* **16a**, 436 (1961). VI,7.

Frerichs, R.
62 *J. Appl. Phys.* **33**, 1898 (1962). X,4.

Fröhlich, H.
50 *Phys. Rev.* **79**, 845 (1950). I,10; V,2, 3, 6.
50a *Proc. Phys. Soc.* **A63**, 778 (1950). I,10; IV,2; V,3.

52 *Proc. Roy. Soc.* **A215**, 291 (1952). I,10.
54 *Proc. Roy. Soc.* **A223**, 296 (1954). IV,2.
54a *Advan. Phys.* **3**, 325 (1954). I,10.
61 *Rept. Progr. Phys.* **24**, 1 (1961). I,10.

Fukuda, N.
59 Y. WADA, and S. OTAKE. *Progr. Theoret. Phys. (Kyoto)* **21**, 343 (1959). VIII,6
63 K. FUKUSHIMA, and T. INAGAKI. *Progr. Theoret. Phys. (Kyoto)* **29**, 206 (1963) IX,3.
[**See also** FUKUSHIMA, K., 62; SAWADA, K., 57a; WADA, Y., 58, 59]

Fukushima, K.
62 and N. FUKUDA. *Progr. Theoret. Phys. (Kyoto)* **28**, 809 (1962). V,6.
[**See also** FUKUDA, N., 63]

Furth, H. P.
62 and M. A. LEVINE. *J. Appl. Phys.* **33**, 747 (1962). X,2.

Galasiewicz, Z.
60 *Progr. Theoret. Phys. (Kyoto)* **23**, 197 (1960). VIII,7.
61 *Bull. Acad. Polon. Sci. Ser. Sci. Math. Astron. Phys.* **9**, 605 (1961). V,2.

Galitskii, V. M.
58 *J. Exptl. Theoret. Phys. (USSR)* **34**, 151 (1958) [English translation: *Soviet Phys.—JETP* **7**, 104 (1958)]. V,6.
58a and A. B. MIGDAL. *J. Exptl. Theoret. Phys. (USSR)* **34**, 139 (1958) [English translation: *Soviet Phys.—JETP* **7**, 96 (1958)]. V,6.
60 *Physica, Suppl.* **26**, 174 (1960). VI,1.
[**See also** GOR'KOV, L. P., 61; MIGDAL, A. B., 58, VAKS, V. G., 1]

Gallagher, C. J.
62 and V. G. SOLOVIEV. *Kgl. Danske Videnskab. Selskab, Math.-Fys. Skrifter* **2**, 2 (1962). III,9.

Garfinkel, M.
61 and D. E. MAPOTHER. *Phys. Rev.* **122**, 459 (1961). VI,5.

Garland, J. W.
62 *Proc. 8th Conf. Low Temp. Phys., London*, 1962. X,3.

Gayley, R. I., Jr.
62 E. A. LYNTON, and B. SERIN. *Phys. Rev.* **126**, 43 (1962). X,3.

Geballe, T. H.
61 B. T. MATTHIAS, G. W. HULL, and E. CORENZWIT, *Phys. Rev. Letters* **6**, 275 (1961). X,3.
62 B. T. MATTHIAS, E. CORENZWIT, and G. W. HULL, JR. *Phys. Rev. Letters* **8**, 313 (1962). IV,3; X,3.
63 B. T. MATTHIAS, V. B. COMPTON, E. CORENZWIT, and G. W. HULL, JR. *Phys. Rev.* **129**, 182 (1963). X,3.
[**See also** HEIN, R. A., 62; HULM, J. K., 61; MATTHIAS, B. T., 62a, 63, 63b]

Geilikman, B. T.
58 *J. Exptl. Theoret. Phys. (USSR)* **34**, 1042 (1958) [English translation: *Soviet Phys.—JETP* **7**, 721 (1958)]. VI,7.
58a and V. Z. KRESIN. *Dokl. Akad. Nauk SSSR* **123**, 259 (1958) [English translation: *Soviet Phys. Doklady* **3**, 1161 (1958)]. VI,7.

61 and V. Z. KRESIN. *J. Exptl. Theoret. Phys.* (*USSR*) **41**, 1142 (1961) [English translation: *Soviet Phys.—JETP* **14**, 816 (1962)]. VI,7.
61a and V. Z. KRESIN. *Soviet Phys.—JETP* **12**, 352 (1961). VI,7.
61b and V. Z. KRESIN. *J. Exptl. Theoret. Phys.* (*USSR*) **40**, 970 (1961) [English translation: *Soviet Phys.—JETP* **13**, 677 (1961)]. VI,5.

Gell-Mann, M.
57 and K. A. BRUECKNER. *Phys. Rev.* **106**, 364 (1957). I,9; VIII,6.

Gendron, M. F.
62 and R. E. JONES. *Phys. Chem. Solids* **23**, 405 (1962). X,3.

Giaever, I.
60 *Phys. Rev. Letters* **5**, 147 (1960). VI,7.
60a *Phys. Rev. Letters* **5**, 464 (1960). VI,7.
61 and K. MEGERLE. *Phys. Rev.* **122**, 1101 (1961). VI,7.
61a *Electronics and Communications* **9**, 25 (1961). VI,7.
62 H. R. HART, JR., and K. MEGERLE. *Phys. Rev.* **126**, 941 (1962). VI,7.
[**See also** FISHER, J. C., 61]

Gibbons, D. F.
59 and C. A. RENTON. *Phys. Rev.* **114**, 1257 (1959). VI,5.

Ginsberg, D. M.
59 P. L. RICHARDS, and M. TINKHAM. *Phys. Rev. Letters* **3**, 337 (1959). VI,7.
60 and M. TINKHAM. *Phys. Rev.* **118**, 990 (1960). V,2; VI,7.
62 *Phys. Rev. Letters* **8**, 204 (1962). VI,7.
62a and J. D. LESLIE. *IBM J. Res. Develop.* **6**, 55 (1962). VI,7.
62b *Am. J. Phys.* **30**, 433 (1962). V,2.

Ginzburg, V. L.
45 *J. Phys. (USSR)* **9**, 305 (1945). I,5.
46 "Superconductivity." Moscow, 1946. (in Russian). I,10.
50 and L. LANDAU. *J. Exptl. Theoret. Phys.* (*USSR*) **20**, 1064 (1950). I,5; X,2.
50a *Nuovo Cimento* **2**, 1234 (1950). I,5; X,2.
50b *Usp. Fiz. Nauk* **42**, 169, 333 (1950). I,10.
52 *J. Exptl. Theoret. Phys.* (*USSR*) **23**, 236 (1952). I,5; X,2.
52a *Usp. Fiz. Nauk* **48**, 25 (1952). I,10; II,1; X,2.
52b *Dokl. Akad. Nauk SSSR* **83**, 385 (1952). I,5; X,2.
53 *Fortschr. Physik* **1**, 99 (1953). I,5, 10; II,1; X,2.
55 *Nuovo Cimento* **2**, 1234 (1955). I,5; X,2.
56 *J. Exptl. Theoret. Phys.* (*USSR*) **30**, 593 (1956) [English translation: *Soviet Phys.—JETP* **3**, 621 (1956)]. I,5; X,2.
56a *Dokl. Akad. Nauk. SSSR* **110**, 358 (1956). I,5; X,2.
57 *Soviet Phys.—JETP* **4**, 594 (1957). I,5; X,2.
58 *J. Exptl. Theoret. Phys.* (*USSR*) **34**, 113 (1958) [English translation: *Soviet Phys.—JETP* **7**, 78 (1958)]. I,5; X,2.
58a *Dokl. Akad. Nauk SSSR* **118**, 464 (1958) [English translation: *Soviet Phys.—Doklady* **3**, 102 (1958)]. I,5; X,2.
59 *J. Exptl. Theoret. Phys. (USSR)* **36**, 1930 (1959) [English translation: *Soviet Phys.—JETP* **9**, 1372 (1959)]. I,5; VIII,7; X,2.
59a and A. I. SHALNIKOV. *J. Exptl. Theoret. Phys.* (*USSR*) **37**, 399 (1959) [English translation: *Soviet Phys.—JETP* **10**, 285 (1960)]. IX,10.
61 *J. Exptl. Theoret. Phys. (USSR)* **41**, 828 (1961) [English translation: *Soviet Phys.—JETP* **14**, 594 (1962)]. VIII,7.

62 *J. Exptl. Theoret. Phys.* (*USSR*) **42**, 299 (1962) [English translation: *Soviet Phys.—JETP* **15**, 207 (1962)]. IX,8.

Giorgi, A. L.

62 E. G. Szklarz, E. K. Storms, A. L. Bowman, and B. T. Matthias. *Phys. Rev.* **125**, 837 (1962). X,3.

Girardeau, M.

62 *J. Math. Phys.* **3**, 131 (1962); Erratum **3**, 1058 (1962). VII,1.

63 Formulation of the Many-Body Problem for Composite Particles. Preprint EFINS-63-10, Univ. of Chicago (March 1963); *J. Math. Phys.* **4**, 1096 (1963). B,4.

Gittleman, J. I.

61 *Bull. Am. Phys. Soc.* **6**, 268 (1961). VI, 7.

62 and S. Bozowski. *Phys. Rev.* **128**, 646 (1962). VI,7.

[**See also** Serin, B., 59]

Glauber, R. J.

63 *J. Math. Phys.* **4**, 294 (1963). VI,3.

Glover, R. E., III

56 and M. Tinkham. *Phys. Rev.* **104**, 844 (1956). V,2; VI,7.

57 and M. Tinkham. *Phys. Rev.* **108**, 243 (1957). V,2; VI,7.

60 and M. D. Sherrill. *Phys. Rev. Letters* **5**, 248 (1960). X,4.

63 *Bull. Am. Phys. Soc.* **8**, 295 (1963). X,2.

Goedemoed, S. H.

63 A. Van der Giessen, D. DeKlerk, and C. J. Gorter. *Phys. Letters* **3**, 250 (1963). X,2.

Goldberger, M. L.

52 and E. N. Adams, II. *J. Chem. Phys.* **20**, 240 (1952). VI,3.

Goodman, B. B.

61 *Phys. Rev. Letters* **6**, 597 (1961). X,2, 3.

62 *IBM J. Res. Develop.* **6**, 63 (1962). X,2.

62a *Phys. Letters* (*Neth.*) **1**, 215 (1962). X,2.

[**See also** Bonfiglioli, G., 62; BonMardian, G., 62; Hulm, J. K., 57]

Gor'kov, L. P.

58 *J. Exptl. Theoret. Phys.* (*USSR*) **34**, 735 (1958) [English translation: *Soviet Phys.—JETP* **7**, 505 (1958)]. V,6; VIII,1; X,2, 3, 4.

59 *J. Exptl. Theoret. Phys.* (*USSR*) **36**, 1918 (1959) [English translation: *Soviet Phys.—JETP* **9**, 1364 (1959)]. VIII,7; X,2, 3, 4.

59a *J. Exptl. Theoret. Phys.* (*USSR*) **37**, 833 (1959) [English translation: *Soviet Phys.—JETP* **10**, 593 (1960)]. VIII,7; X,2, 3, 4.

59b *J. Exptl. Theoret. Phys.* (*USSR*) **37**, 1407 (1959) [English translation: *Soviet Phys.—JETP* **10**, 998 (1960)]. X,2, 3.

60 *Proc. 7th Intern. Conf. Low Temp. Phys., Toronto, Ont., 1960* p. 315 (1961). X,2, 3.

61 and V. M. Galitskii. *J. Exptl. Theoret. Phys.* (*USSR*) **40**, 1124 (1961) [English translation: *Soviet Phys.—JETP* **13**, 792 (1961)]. V,6.

[**See also** Abrikosov, A. A., 58, 58a, 59, 59a, 59b, 60, 61, 62]

Gorter, C. J.

33 *Arch. Teyler* **7**, 387 (1933). I,4.

34 and H.B.G. Casimir. *Physica* **1**, 305 (1934). I,4; X,2.

35 *Physica* **2**, 449 (1935). X,2.
49 *Physica* **15**, 55 (1949). I,10.
55 *Progr. Low Temp. Phys.* **1**, 1 (1955). I,10.
61 *Ned. Tijdschr. Natuurk.* **26**, 269 (1961). I,10.
62 *Phys. Letters (Neth.)* **1**, 69 (1962). X,2.
62a *Phys. Letters (Neth.)* **2**, 26 (1962). X,2.
[**See also** CASIMIR, H. B. G., 34; GOEDEMOED, S. H., 63]

Gottfried, K.
60 and L. PICMAN. *Kgl. Danske Videnskab. Selskab, Mat.-Fys. Medd.* **32**, No. 13 (1960). VI,1.

Graham, G. M.
58 *Proc. Roy. Soc.* **A248**, 522 (1958). VI,7.

Grassman, V. P.
36 *Physik. Z.* **37**, 569 (1936). IX, 10.

Green, H. S.
51 *J. Chem. Phys.* **19**, 955 (1951). VI,3.
52 *Proc. Phys. Soc.* **A45**, 1022 (1952). III,5.
52a *J. Chem. Phys.* **20**, 1274 (1952). VI,3.
52b "The Molecular Theory of Fluids." Wiley (Interscience), New York, 1952. IX,2.
62 *Z. Physik* **170**, 129 (1962). VI,3.
63 and C. A. HURST. "Order-Disorder Phenomena." Wiley (Interscience), New York, 1963. VI,3.
[**See also** HURST, C. A., 60]

Green, M. S.
55 U.S.A. National Bureau of Standards Rept. 3327, August 1955. IX,2.
58 *Phys. Rev. Letters* **1**, 409 (1958). VII,1.
58a *Physica* **24**, 393 (1958). IX,2.
60 *J. Chem. Phys.* **33**, 1403 (1960). VII,1.

Greenwood, D. A.
62 *Proc. Phys. Soc.* **80**, 226 (1962). VI,7.

Grenier, C.
55 *Compt. Rend.* **240**, 2302 (1955). X,2.

Griffin, J. J.
60 and M. RICH. *Phys. Rev.* **118**, 850 (1960). III,9.

Grin, Yu. T.
61 *J. Exptl. Theoret. Phys. (USSR)* **41**, 410 (1961) [English translation: *Soviet Phys.—JETP* **14**, 320 (1962)]. I,3.

Gropper, L.
36 *Phys. Rev.* **50**, 963 (1936). III,5.
37 *Phys. Rev.* **51**, 1108 (1937). III,5.
39 *Phys. Rev.* **55**, 1095 (1939). III,5.

Gross, D.
60 and J. L. OLSEN. *Cryogenics* **1**, 91 (1960). VI,5.

Gross, E. P.
56 and J. L. LEBOWITZ. *Phys. Rev.* **104**, 1528 (1956). IX, 10.
[**See also** BUCHER, E., 61c]
63 *J. Math. Physics* **4**, 195 (1963). IX,3.

Guénault, A. M.
61 *Proc. Roy. Soc.* **A262**, 420 (1961). VI,7.
Guggenheim, J.
61 F. Hulliger, and J. Müller. *Helv. Phys. Acta* **34**, 408 (1961). X,3.
Gupta, K. K.
59 and V. S. Mathur. *Phys. Rev.* **115**, 75 (1959). VIII,6.
61 and V. S. Mathur. *Phys. Rev.* **121**, 107 (1961). VIII,6,7; X,4.

Haag, R.
62 *Nuovo Cimento* **25**, 287 (1962). V,2.
Hagenow, K. U. von
62 and H. Koppe. *IBM J. Res. Develop.* **6**, 12 (1962). VIII,7.
Hahn, E. L.
50 *Phys. Rev.* **80**, 580 (1950). IX,10.
53 *Phys. Today* **6**, 4 (1953). IX,10.
Hake, R. R.
61 *Phys. Rev.* **123**, 1986 (1961). X,3.
62 D. H. Leslie, and T. G. Berlincourt. *Phys. Rev.* **127**, 170 (1962). X,3.
62a T. G. Berlincourt, and D. H. Leslie. *IBM J. Res. Develop.* **6**, 119 (1962). X,2.
[**See also** Berlincourt, T. G., 61, 62, 63]
Hammond, R. H.
60 and W. D. Knight. *Phys. Rev.* **120**, 762 (1960). VI,7.
[**See also** Knight, W. D., 56]
Harper, P. G.
61 *Proc. Phys. Soc.* **77**, 299 (1961). VI,7.
Harrison, W. A.
61 *Phys. Rev.* **123**, 85 (1961). VI,7.
Haug, A.
56 *Z. Physik* **146**, 75 (1956). I,10.
Hauser, J. J.
62 and E. Buehler. *Phys. Rev.* **125**, 142 (1962). X,2.
62a and E. Helfand. *Phys. Rev.* **127**, 386 (1962). X,2, 4.
62b *J. Appl. Phys.* **33**, 3074 (1962). X,2.
62c *Phys. Rev. Letters* **9**, 423 (1962). X,2.
63 and H. C. Theuerer. *Phys. Rev.* **129**, 103 (1963). X,2.
Hebel, L. C.
57 and C. P. Slichter. *Phys. Rev.* **107**, 401 (1957). VI,7.
59 and C. P. Slichter. *Phys. Rev.* **113**, 1504 (1959). VI,7.
59a Phys. Rev. **116**, 79 (1959). VI,7.
Hein, R. A.
57 W. E. Henry, and N. M. Wolcott. *Phys. Rev.* **107**, 1517 (1957). X,2.
59 R. L. Falge, Jr., B. T. Matthias, and C. Corenzwit. *Phys. Rev. Letters* **2**, 500 (1959). X,3.
61 and R. L. Falge, Jr. *Phys. Rev.* **123**, 407 (1961). X,2.
62 J. W. Gibson, B. T. Matthias, T. H. Geballe, and E. Corenzwit. *Phys. Rev. Letters* **8**, 408 (1962). IV,3; X,2.

63 J. W. Gibson, M. R. Pablo, and R. D. Blaugher. *Phys. Rev.* **129**, 136 (1963). VI,5.

63a and J. W. Gibson. *Phys. Rev.* **131**, 1105 (1963). X,3.
[**See also** Steele, M. C., 53]

Heine, V.

58 and A. B. Pippard. *Phil. Mag.* **3**, 1046 (1958). VIII,7

Heisenberg, W.

48 Two Lectures, Cambridge Univ. (1948). I,10.

Hempstead, C. F.

62 and Y. B. Kim. *Bull. Am. Phys. Soc.* **7**, 309 (1962). IX,10.

63 and Y. B. Kim. *Bull. Am. Phys. Soc.* **8**, 79 (1963). X,2.
[**See also** Kim, Y. B., 62, 63, 63a]

Henin, F.

61 P. Resibois, and F. Andrews, *J. Math. Physics* **2**, 68 (1961). IX,2.
[**See also** Prigogine, J., 60]

Herring, C.

58 and H. Suhl. *Physica, Suppl.* **24**, 184 (1958). X,3.

Hildebrandt, A. F.

62 D. D. Elleman, F. C. Whitmore, and R. Simpkins. *J. Appl. Phys.* **33**, 2375 (1962). I,2.

62a H. Wahlquist, and D. D. Elleman. *J. Appl. Phys.* **33**, 1798 (1962). I,2.

Hilsch, P.

61 and R. Hilsch. *Naturwissenschaften* **48**, 549 (1961). X,4.

62 *Z. Physik* **167**, 511 (1962). X,4.

Hilsch, R.

51 Superconductive Properties of Metal Layers. *Proc. Intern. Conf. Low Temp. Phys. Oxford*, 1951 p. 119. X,4.

62 D. Korn, and G. von Minnigerode. *Z. Physik* **167**, 501 (1962). X,4.
[**See also** Buckel, W., 54; Hilsch, P., 61]

Hinrichs, C. H.

61 and C. A. Swenson. *Phys. Rev.* **123**, 1106 (1961). VI,5.

Hirshfeld, A. T.

60 and H. Leupold. *Phys. Rev. Letters* **5**, 246 (1960). VI,5.

62 H. A. Leupold, and H. A. Boorse. *Phys. Rev.* **127**, 1501 (1962). VI,5.
[**See also** Boorse, H. A., 60]

Hone, D.

62 *Phys. Rev. Letters* **8**, 370 (1962). V,6.

Horwitz, N. H.

62 and H. V. Bohm. *Phys. Rev. Letters* **9**, 313 (1962). VI,5, 7.

Houtappel, R. M. F.

50 *Physica* **16**, 425 (1950). VI,3.

Huang, K.

51 *Proc. Phys. Soc.* **A64**, 867 (1951). I,10.

57 and C. N. Yang, *Phys. Rev.* **105**, 767 (1957). II,6.

57a C. N. Yang, and J. M. Luttinger. *Phys. Rev.* **105**, 776 (1957). II,6.

60 *Phys. Rev.* **119**, 1129 (1960). II,6.
[**See also** LEE, T. D., 57]

Hubbard, J.
57 *Proc. Roy. Soc.* **A243**, 336 (1957). I,9.

Hulm, J. K.
50 *Proc. Roy. Soc.* **A204**, 98 (1950). VI,7.
53 *Phys. Rev.* **90**, 1116 (1953). X,2.
57 and B. B. GOODMAN. *Phys. Rev.* **106**, 659 (1957). X,2.
61 R. D. BLAUGHER, T. H. GEBALLE, and B. T. MATTHIAS. *Phys. Rev. Letters* **7**, 302 (1961). X,4.
61a and R. D. BLAUGHER. *Phys. Rev.* **123**, 1569 (1961). X,4.
[**See also** BLAUGHER, R. D., 61, 61a, 62, 62a; CHANDRASEKHAR, B. S., 58, 63; SCHAWLOW, A. L., 59a]

Hurst, C. A.
60 and H. S. GREEN. *J. Chem. Phys.* **33**, 1059 (1960). VI,3.
63 Solution of Plane Ising Lattices by the Pfaffian Method. Preprint, Univ. of Adelaide (1963). VI,3.
[**See also** GREEN, H. S., 63]

Husimi, K.
40 *Proc. Phys. Math. Soc. Japan* **22**, 264 (1940). III, 7; VI,3; VII, 2,5.
50 and I. SYOZI, *Progr. Theoret. Phys. (Kyoto)* **5**, 177, 341 (1950). VI,3.

Itterbeck, A. van
45 *Soc. Roy. Belge Ingrs. Industriels Rev. Mém.* p. 47 (1945). I,10.

Ittner, W. B., III
58 *Phys. Rev.* **111**, 1483 (1958). VI,5.
60 *Phys. Rev.* **119**, 1591 (1960). X,4.
60a *Solid State J.* **1**, 44 (1960). X,4.
61 and C. J. KRAUS. *Sci. Am.* **205**, 124 (1961). X,4.
[**See also** QUINN, D. J., III,62]

Iwamoto, F.
60 *Progr. Theoret. Phys. (Kyoto)* **23**, 871 (1960). V,6.

Jaccarino, V.
62 and M. PETER, *Phys. Rev. Letters* **9**, 290 (1962). X,2.
[**See also** BLUMBERG, W. E., 60; CLOGSTON, A. M., 61, 62]

Jaggi, R.
58 J. MÜLLER, and R. SOMMERHALDER. *Helv. Phys. Acta* **31**, 637 (1958). X,2.
59 and R. SOMMERHALDER. *Helv. Phys. Acta* **32**, 167 (1959). VI,7.
60 and R. SOMMERHALDER. *Helv. Phys. Acta* **33**, 1 (1960). X,4.

Janner, A.
62 L. VAN HOVE, and E. VERBOVEN. *Physica* **28**, 1341 (1962). IX,10.

Jansen, H. G.
61 *Z. Physik* **162**, 275 (1961). X,3.

Jennings, L. D.
58 and C. A. SWENSON. *Phys. Rev.* **112**, 31 (1958). VI,5.

Jones, R. E., Jr.
60 and A. M. Toxen. *Phys. Rev.* **120**, 1167 (1960). VI,7.
60a *IBM J. Res. Develop.* **4**, 23 (1960). X,3.
[**See also** Chang, G.K., 62; Gendron, M. F., 62; Toxen, A. M., 62a]

Josephson, B. D.
62 *Phys. Letters (Neth.)* **1**, 251 (1962). VI,7.

Jurisson, J.
62 and R. J. Oakes. *Phys. Letters (Neth.)* **2**, 187 (1962). X,2.

Justi, E.
46 *Naturwissenschaften* **33**, 292, 329 (1946). I,10.

Kac, M.
52 and J. C. Ward. *Phys. Rev.* **88**, 1332 (1952). VI.3.

Kadanoff, L. P.
61 and P. C. Martin. *Phys. Rev.* **124**, 670 (1961). V,6; VI,7; VIII,1, 6, 7.
[**See also** Ambegaokar, 61; Martin, P. C., 59a]

Kahan, G. J.
60 R. B. Delano, Jr., A. E. Brennemann, and R. T. C. Tsui. *IBM J. Res. Develop.* **4**, 173 (1960). X,4.

Kahn, B.
38 and G. E. Uhlenbeck. *Physica* **5**, 399 (1938). III,7.

Kamigaki, K.
61 *J. Phys. Soc. Japan* **16**, 1141 (1961). VI,7.

Kamper, R. A.
62 *Phys. Letters* **2**, 290 (1962). X, 2.

Kan, L. S.
61 B. G. Lazarev, and V. I. Makarov. *J. Exptl. Theoret. Phys. (USSR)* **40**, 457 (1961) [English translation: *Soviet Phys.—JETP* **13**, 317 (1961). VI,5.

Kano, K.
53 and S. Naya. *Progr. Theoret. Phys. (Kyoto)* **10**, 158 (1953). VI,3.

Kapitza, P. L.
41 *J. Exptl. Theoret. Phys. (USSR)* **11**, 1 (1941). VI,7.

Kaplan, R.
59 A. H. Nethercot, Jr., and H. A. Boorse. *Phys. Rev.* **116**, 270 (1959). VI,7.

Kasteleyn, P. W.
63 *J. Math. Physics* **4**, 287 (1963). VI,3.

Kasuya, T.
58 *Progr. Theoret. Phys. (Kyoto)* **20**, 980 (1958). X,3.

Katz, A.
59 and J. M. Blatt. *Nucl. Phys.* **23**, 612 (1959). III,9.
60 A. de Shalit and I. Talmi. *Nuovo Cimento* **16**, 485 (1960). V,2.
63 *Nuclear Physics* **42**, 394, 416 (1963). IV, 2; V,6.

Kaufman, B.
49 *Phys. Rev.* **76**, 123 (1949). VI,3.
49a and L. Onsager. *Phys. Rev.* **76**, 135 (1949). VI,3.

Kawasaki, K.
62 and H. Mori. *Progr. Theoret. Phys. (Kyoto)* **28**, 784 (1962). VI, 7.

Keesom, P. H.
63 and B. J. C. VAN DER HOEVEN, JR. *Phys. Letters* **3**, 360 (1963). VI,5,7.
[**See also** BRYANT, C. A., 60, 61]

Keesom, W. H.
24 *Rappt. Congr. Solvay* 289 (1924). I,4.

Keller, J. B.
58 *Phys. Rev.* **111**, 1497 (1958). VI,5.
61 and B. ZUMINO. *Phys. Rev. Letters* **7**, 164 (1961). IX,8.

Kenworthy, D. J.
61 and D. TER HAAR. *Phys. Rev.* **123**, 1181 (1961). VI,5.
61a and D. TER HAAR. *Physica* **27**, 1189 (1961). V,6.
62 M. J. ZUCKERMANN, D. M. BRINK, and D. TER HAAR. *Phys. Letters (Neth.)* **1**, 35 (1962). VI,5.

Kerman, A. K.
61 R. D. LAWSON, and M. H. MACFARLANE. *Phys. Rev.* **124**, 162 (1961). III,9.

Khaikin, M. S.
58 *Soviet Phys.—JETP* **7**, 961 (1958). VI,7.

Khalatnikov, I. M.
59 and A. A. ABRIKOSOV. *Advan. Phys.* **8**, 45 (1959). VIII,7.
59a *Soviet Phys.—JETP* **9**, 1296 (1959). VI,7.
[**See also** ABRIKOSOV, 58a, 59b; LANDAU, L. D., 47, 47a]

Khanna, K. M.
62 *Progr. Theoret. Phys. (Kyoto)* **28**, 205 (1962). III,9.

Khukhareva, I. S.
58 *Soviet Phys.—JETP* **6**, 234 (1958). X,4.
61 *J. Exptl. Theoret. Phys. (USSR)* **41**, 728 (1961) [English translation: *Soviet Phys.—JETP* **14**, 526 (1962)]. X,4.

Kim, Y. B.
62 C. F. HEMPSTEAD, and A. R. STRNAD. *Phys. Rev. Letters* **9**, 306 (1962). IX,10; X,2.
63 C. F. HEMPSTEAD, and A. R. STRNAD. *Phys. Rev.* **129**, 528 (1963). X,2.
63a R. D. DUNLAP, and C. F. HEMPSTEAD. *Bull. Am. Phys. Soc.* **8**, 308 (1963). X,2.

Kinsel, T.
62 E. A. LYNTON, and B. SERIN. *Phys. Letters (Neth.)* **3**, 30 (1962). I,5; X,2.
63 E. A. LYNTON, and B. SERIN. *Bull. Am. Phys. Soc.* **8**, 294 (1963). X,2.

Kirkwood, J. G.
33 *Phys. Rev.* **44**, 31 (1933). VI,3.
34 *Phys. Rev.* **45**, 116 (1934). VI,3.

Kisslinger, L. S.
60 and R. A. SORENSEN. *Kgl. Danske Videnskab. Selskab., Mat.-Fys. Medd.* **32**, No. 9 (1960). III,9.
62 *Nucl. Phys.* **35**, 114 (1962). III,9.

Kitano, Y.
53 and H. NAKANO. *Progr. Theoret. Phys. (Kyoto)* **9**, 370 (1953). I,10.

Kittel, C.
53 "Introduction to Solid State Physics," Wiley, New York, 1953. X,1.

Klein, A.
58 and R. PRANGE. *Phys. Rev.* **112**, 994, 1008 (1958). V,6.
62 *Nuovo Cimento* **24**, 788 (1962). V,6.
62a *Phys. Letters (Neth.)* **1**, 311 (1962). V,6.
62b *Nuovo Cimento* **23**, 919 (1962). V,6.
[**See also**, DANG, GIU, DO, 63]

Klein, R.
61 and L. B. LEDER. *Phys. Rev.* **124**, 1050 (1961). VI,7.

Kneip, G. D., Jr.
62 J. O. BETTERTON, JR. D. S. EASTON, and J. O. SCARBROUGH. *J. Appl. Phys.* **33**, 754 (1962). X,3.
[**See also** BETTERTON, J. O., JR., 61]

Knight, W. D.
55 *Solid State Phys.* **2**, 93 (1955). VIII,7.
56 G. M. ANDROES, and R. H. HAMMOND. *Phys. Rev.* **104**, 852 (1956). VIII,7.
[**See also** ANDROES, G. M., 59, 61; HAMMOND, R. H., 60; NOER, R. J., 61]

Kohn, W.
51 and VACHASPATI. *Phys. Rev.* **83**, 462 (1951). I,10; V,5, 6.
57 *Phys. Rev.* **105**, 509 (1957). III,7.
57a and J. M. LUTTINGER. *Phys. Rev.* **108**, 590 (1957). IX,1, 2.
[**See also** LUTTINGER, J. M., 58]

Kolchin, A. M.
61 YU G. MIKHAILOV, N. M. REINOV, A. V. RUMYANTSEVA, A. P. SMIRNOV, and V. N. TOTUBALIN. *J. Exptl. Theoret. Phys. (USSR)* **40**, 1543 (1961) [English translation: *Soviet Phys.—JETP* **13**, 1083 (1961)]. X,4.

Kondo, J.
63 *Progr. Theoret. Phys. (Kyoto)* **29**, 1 (1963). V,6; VI, 5; X,3.

Kontarev, V. Ya.
61 *Zh. Tekhn. Fiz.* **31**, 854 (1961). X,4.

Koppe, H.
50 *Ergeb. Exakt. Naturw.* **23**, 283, (1950). I,9, 10.
54 *Fortschr. Physik* **1**, 420 (1954). I,**9**.
58 and B. MÜHLSCHLEGEL. *Z. Physik* **151**, 613 (1958). VII, 1.
[**See also** HAGENOW, K. U. VON, 62]

Kothari, D. S.
62 V. C. MATHUR, and N. PANCHAPAKESAN. *Phys. Letters (Neth.)* **2**, 235 (1962). X,3.

Kramers, H. A.
41 and G. H. WANNIER. *Phys. Rev.* **60**, 252 (1941). VI,3.
41a and G. H. WANNIER. *Phys. Rev.* **60**, 263 (1941). VI,3.

Kresin, V. Z.
59 *J. Exptl. Theoret. Phys.* **36**, 1947 (1959) [English translation: *Soviet Phys.—JETP* **9**, 1385 (1959)]. VI,7.
[**See also** GEILIKMAN, B. T., 61, 61a, 61b]

Kromminga, A. J.
62 and M. BOLSTERLI. *Phys. Rev.* **128**, 2887 (1962). II,6.

Kröner, E.
58 "Kontinuumstheorie der Versetzungen and Eigenspannungen." Springer, Berlin, 1958. X,1.

Kropschot, R. H.
61 and V. Arp. *Cryogenics* **2**, 1 (1961). X,2.
[**See also** Arp, V. D., 61]

Kubo, R.
56 and Y. Obata. *J. Phys. Soc. Japan* **11**, 547 (1956). VIII,7.
56a *Can. J. Phys.* **34**, 1274 (1956). IX,2.
57 *J. Phys. Soc. Japan* **12**, 570 (1957). IX,1, 2.
59 "Lectures in Theoretical Physics," (W. E. Brittin and L. G. Dunham), Interscience, New York, 1959. IX,10.
62 *J. Phys. Soc. Japan* **17**, 1100 (1962). VIII,6.
63 *J. Math. Physics* **4**, 174 (1963). IX,10.

Kulik, I. O.
62 *J. Exptl. Theoret. Physics (USSR)* **43**, 1489 (1962), English translation: *Soviet Physics—JETP* **16**, 1052 (1963)]. VI,5.

Kunzler, J. E.
61 *Rev. Mod. Phys.* **33**, 501 (1961). X,2.
61a E. Buehler, F. S. L. Hsu, and J. H. Wernick. *Phys. Rev. Letters* **6**, 89 (1961). X,2.
61b E. Buehler, F. S. L. Hsu, B. T. Matthias, and C. Wahl. *J. Appl. Phys.* **32**, 325 (1961). X,2.
[**See also** Morin, F. J., 62]

Kuper, C. G.
59 *Advan. Phys.* **8**, 1 (1959). I,10, VIII,6.

Kvasnikov, I. A.
58 *Soviet Phys. Doklady* **3**, 318 (1958). VII,5.
58a *Soviet Phys. Doklady* **3**, 329 (1958). VII, 5.
58b and V. V. Tolmachev. *Dokl. Akad. Nauk SSSR* **120**, 273 (1958) [English translation: *Soviet Phys. Doklady* **3**, 553 (1959)]. VII,5.

Lalevic, B.
62 *Phys. Rev.* **128**, 1070 (1962). X,2.

Landau, L. D.
37 *Physik. Z. Sowjetunion* **11**, 129 (1937). X,2.
38 *Nature* **141**, 688 (1938). I,5; X,2.
41 *J. Phys. (USSR)* **5**, 71 (1941). II,3, 7; IX,3.
43 *J. Phys. (USSR)* **7**, 99 (1943). I,5; X,2.
44 *J. Phys. (USSR)* **8**, 1 (1944). II,3.
47 and I. M. Khalatnikov. *J. Exptl. Theoret. Phys. (USSR)* **19**, 637 (1947). II,3.
47a and I. M. Khalatnikov. *J. Exptl. Theoret. Phys. (USSR)* **19**, 709 (1947). II,3.
47b *J. Phys. (USSR)* **11**, 91 (1947). X,2.
49 *Phys. Rev.* **75**, 884 (1949). II,3.
58 and E. M. Lifshitz. "Statistical Physics" (E. and R. F. Peierls, translators). Pergamon, London and New York, 1958. III,3; A,1.
59 and E. M. Lifshitz. "Quantum Mechanics." Pergamon, London and New York, 1959. V,1.

Lang, D. W.
63 *Nuclear Physics* **42**, 353 (1963). III,9.

Lange, F.
61 *Monatsber. Deut. Akad. Wiss. Berlin* **3**, 376 (1961). X,3.

62 *J. Exptl. Theoret. Phys.* (*USSR*) **42**, 42 (1962) [English translation: *Soviet Phys.—JETP* **15**, 29 (1962)]. X,3.

Langer, J. S.

62 *Phys. Rev.* **127**, 5 (1962). VIII,7.
[**See also** SUHL, H., 62]

Laredo, S. J.

55 *Proc. Roy. Soc.* **A229**, 473 (1955). VI,7.
55a and A. B. PIPPARD. *Proc. Cambridge Phil. Soc.* **51**, 369 (1955). X,2.

Laue, M. von

52 "Theory of Superconductivity." Academic Press, New York, 1952. I,2,10.

Lautz, G.

61 and D. SCHNEIDER. *Z. Naturforsch.* **16a**, 1368 (1961). X,3.
62 and D. SCHNEIDER. *Z. Naturforch.* **17a**, 54 (1962). X,3.

Lazarev, B. G.

44 and A. GALKIN. *J. Phys. (USSR)* **8**, 371 (1944). X,2.
58 A. I. SUDOVTSEV, and A. P. SMIRNOV. *Soviet Phys.—JETP* **6**, 816 (1958). X, 4.
60 A. I. SUDOVTSEV, and E. E. SEMENENKO. *Soviet Phys.—JETP* **10**, 1035 (1960). X,4.
[**See also** KAN, L. S., 61]

LeBlanc, M. A. R.

61 and W. A. LITTLE. *Proc. 7th, Intern. Conf. Low Temp. Phys. Toronto, Ont., 1960* p. 362 (1961). X,2.
61a *Phys. Rev.* **124**, 1423 (1961). X,2.
62 *IBM J. Res. Develop.* **6**, 122 (1962). X,2.

Lebowitz, J. L.

57 and P. G. BERGMANN. *Ann. Phys. (NY)* **1**, 1 (1957). IX,10.
57a and H. L. FRISCH. *Phys. Rev.* **107**, 917 (1957). IX,10.
62 and A. SHIMONY. *Phys. Rev.* **128**, 1945 (1962). IX,10.
[**See also** BERGMANN, P. G., 55; GROSS, E.P., 56.]

Lee, T. D.

57 K. HUANG, and C. N. YANG. *Phys. Rev.* **106**, 1135 (1957). II, 6.
57a and C. N. YANG. *Phys. Rev.* **105**, 1119 (1957). II,6.
58 and C. N. YANG. *Phys. Rev.* **109**, 1755 (1958). II,6.
59 and C. N. YANG. *Phys. Rev.* **113**, 1406 (1959). II, 6.
59a and C. N. YANG. *Phys. Rev.* **113**, 1165 (1959). II,6; VII,1.
59b and C. N. YANG. *Phys. Rev.* **116**, 25 (1959). II, 6.
60 and C. N. YANG. *Phys. Rev.* **117**, 12 (1960). II,6; VI,1.
60a and C. N. YANG. *Phys. Rev.* **117**, 22 (1960). II,6.
60b *Phys. Rev.* **117**, 897 (1960). II,6.

Lewis, H. R.

60 J. F. COCHRAN, H. FRAUENFELDER, D. E. MAPOTHER, and R. N. PEACOCK. *Z. Physik* **158**, 26 (1960). VIII,7.

Lieb, E. H.

63 *Bull. Am. Phys. Soc.* **8**, 346 (1963). II,6.
[**See also** MATTIS, D. C., 61]

Lifshitz, E. M.

50 *J. Exptl. Theoret. Phys.* (*USSR*) **20**, 834 (1950). VI,5.
51 and IU. SHARWIN. *Dokl. Akad. Nauk SSSR* **79**, 783 (1951). X,2.
[**See also** AZBEL, M. I., 58; LANDAU, L. D., 58, 59]

Lininger, W.
63 and F. ODEH. *Phys. Rev. Letters* **10**, 47 (1963). X,4.

Lipkin, H. J.
62 M. PESHKIN, and L. J. TASSIE. *Phys. Rev.* **126**, 116 (1962). IX,8.

Litting, C. N. W.
61 *Brit. J. Appl. Phys.* **12**, 207 (1961). X,4.

Little, W. A.
59 *Can. J. Phys.* **37**, 334 (1959). VI,7.
61 *Phys. Rev.* **123**, 435 (1961). VI,7.
62 and R. D. PARKS. *Phys. Rev. Letters* **9**, 9 (1962). IX,6.
62a *IBM J. Res. Develop.* **6**, 31 (1962). VI,7.
[**See also** LEBLANC, M. A. R., 61]

Liu, S. H.
62 *Phys. Rev.* **125**, 1244 (1962). V,6.

Livingston, J. D.
63 *Phys. Rev.* **129**, 1943 (1963). X,2.

Lock, J. M.
61 *Cryogenics* **2**, 65 (1961). X,4.
62 *Rept. Progr. Phys.* **25**, 38 (1962). X,4.

London, F.
35 and H. LONDON. *Physica* **2**, 341 (1935). I,2, 6.
36 *Physica* **3**, 450 (1936). X,2.
36a *Nature* **137**, 991 (1936). X,2.
37 "Une conception nouvelle de la supraconductibilite." *Actualites Sci. et Ind.* No. 458 (1937). I,2.
38 *Phys. Rev.* **54**, 947 (1938). II,1, 2, 3.
50 "Superfluids. Vol. I: Macroscopic Theory of Superconductivity." Wiley, New York, 1950. I,2, 6, 7, 10; IX,1, 5, 8; X,2, 4.

London, H.
35 *Proc. Roy. Soc.* **A152**, 650 (1935). X,2, 4.
[**See also** LONDON, F., 35]

Luban, M.
62 *Phys. Rev.* **128**, 965 (1962). II,6.

Lucas, M. S. P.
62 and D. T. MEYER. *Nature* **193**, 766 (1962). X,4.

Lüders, G.
62 *Z. Naturforsch.* **17a**, 181 (1962). IX,8.
62a *Z. Naturforsch.* **17**, 47 (1962). V,2.

Ludwig, G.
63 *Z. Physik* **171**, 476 (1963). IX,10.
63a *Z. Physik* **173**, 232, (1963). IX,10.

Luttinger, J. M.
50 *Phys. Rev.* **80**, 727 (1950). I,9.
58 and W. KOHN. *Phys. Rev.* **109**, 1892 (1958). IX,1, 2.
60 and J. C. WARD. *Phys. Rev.* **118**, 1417 (1960). V,6; VII,1.

Lynton, E. A.
58 and B. SERIN. *Phys. Rev.* **112**, 70 (1958). X,2.

62 and D. McLachlan. *Phys. Rev.* **126**, 40 (1962). VI,5.
[**See also** Budnick, J. I., 56; Chanin, G., 59; Dubeck, E., 63; Gayley, R. I., Jr., 62; Kinsel, T., 62, 63; Serin, B., 57, 59]

Machlup, S.
53 and L. Onsager, *Phys. Rev.* **91**, 1512 (1953). IX,2.

McKenna, J.
62 and J. M. Blatt. *Progr. Theoret. Phys. (Kyoto)* **27**, 511 (1962). V,3, 6; VIII,3 7.

Mackinnon, L.
55 *Phys. Rev.* **98**, 1181 (1955). VI,7.
57 *Phys. Rev.* **106**, 70 (1957). VI,7.
59 and A. Myers. *Proc. Phys. Soc.* **73**, 291 (1959). VI,7.

Mackintosh, A. R.
61 *Proc. 7th, Intern. Conf. Low Temp. Phys. Toronto, Ont., 1960* p. 240 (1961). VI,7.

McLean, W. L.
62 *Proc. Phys. Soc.* **79**, 572 (1962). VIII,7.

Makei, B. V.
58 *J. Exptl. Theoret. Phys. (USSR)* **34**, 312 (1958) [English translation: *Soviet Phys.—JETP* **7**, 217 (1958)]. I,5; X,2.

Maki, K.
62 and T. Tsuneto. *Progr. Theoret. Phys. (Kyoto)* **28**, 163 (1962). VI,7.
62a and T. Tsuneto. *Progr. Theoret Phys. (Kyoto)* **27**, 228 (1962). IX,8.
63 *Progr. Theoret. Phys. (Kyoto)* **29**, 10 (1963). IX,6.
63a *Progr. Theoret. Phys. (Kyoto)* **29**, 333 (1963). IX,6.

Maleev, S. V.
62 *J. Exptl. Theoret. Phys. (USSR)* **43**, 1044 (1962) [English translation: *Soviet Phys.—JETP* **16**, 738 (1963)]. V,6.

Mapother, D. E.
62 *IBM J. Res. Develop.* **6**, 77 (1962); *Phys. Rev.* **126**, 2021 (1962). VI,5.
[**See also** Decker, D. L.. 58; Finnemore, Dok, 62; Garfinkel, M., 61; Lewis, H. R., 60; Mould, R. E., 62; Otter, F. A., 62; Shaw, R. W., 60, 60a]

Maradudin, A.
59 and J. Peretti. *Compt. Rend.* **248**, 2856 (1959). X,3.

Marchand, J. F.
59 and A. Venema. *Phillips Res. Rept.* **14**, 427 (1959). X,4.
62 and J. Volger. *Phys. Letters (Neth.)* **2**, 118 (1962). X,2.
62a *Phys. Letters (Neth.)* **2**, 57 (1962), X,4.

Marcus, P. M.
62 Critical Conditions for Thin Superconducting Films in Ginzburg-Landau Theory. *Proc. Intern. Conf. Low Temp. Phys. London* 1962. X,4.
[**See also** Seraphim, D. P., 61, 62]

Martin, D. L.
61 *Proc. Phys. Soc.* **78**, 1482 (1961). VI,5.
61a *Proc. Phys. Soc.* **78**, 1489 (1961). VI,5.
[**See also** Franck, J. P., 61]

Martin, P. C.
59 and J. Schwinger. *Phys. Rev.* **115**, 1342 (1959). V,6; VII,1.

59a and L. P. KADANOFF. *Phys. Rev. Letters* **3**, 322 (1959). VIII,7.
63 *J. Math. Phys.* **4**, 208 (1963). IX,3.
[**See also** KADANOFF, L. P., 61]

Marumori, T.
60 *Progr. Theoret. Phys.* (*Kyoto*) **24**, 331 (1960). III,9.

Maschke, E. K.
63 and W. WILD. *Z. Physik* **172**, 314 (1963). V,6; VI,1.

Masuda, Y.
57 *J. Phys. Soc. Japan* **12**, 523 (1957). VIII,7.
62 and A. G. REDFIELD. *Phys. Rev.* **125**, 159 (1962). VI,7.
62a *Phys. Rev.* **126**, 1271 (1962). VI,7.
62b *IBM J. Res. Develop.* **6**, 24 (1962). VI,7.

Mathur, V. S.
62 N. PANCHAPAKESAN, and R. P. SAXENA. *Phys. Rev. Letters* **9**, 374 (1962). VIII,7; X,4.
[**See also** GUPTA, K. K., 59, 61; KOTHARI, D. S., 62]

Matsubara, T.
55 *Progr. Theoret. Phys.* (*Kyoto*) **14**, 351 (1955). V,6; VI,3.
60 and J. M. BLATT. *Progr. Theoret. Phys.* (*Kyoto*) **23**, 451 (1960). III,8; B,4.
63 and C. J. THOMPSON. *J. Australian Math. Soc.* to be published (1963). VII,4.
[**See also** BLATT, J. M., 58, 58a, 59; THOMPSON, C. J., 63]

Matthias, B. T.
57 *Progr. Low Temp. Phys.* **2**, 138. I,10; X,3.
58 H. SUHL and E. CORENZWIT. *Phys. Rev. Letters* **1**, 92 (1958). X,3.
58a H. SUHL and E. CORENZWIT. *Phys. Rev. Letters* **1**, 449 (1958). X,3.
58b E. CORENZWIT, and W. H. ZACHARIASEN. *Phys. Rev.* **112**, 89 (1958) X,3.
58c and S. GELLER. *Phys. Chem. Solids* **4**, 318 (1958). X,3.
58d and W. H. ZACHARIASEN. *Phys. Chem. Solids* **7**, 98 (1958). X,3.
59 V. B. COMPTON, H. SUHL, and E. CORENZWIT. *Phys. Rev.* **115**, 1597 (1959). X,3.
59a *Phys. Chem. Solids* **10**, 342 (1959). X,3.
60 and H. SUHL. *Phys. Rev. Letters* **4**, 51 (1960). X,3.
60a H. SUHL, and E. CORENZWIT. *Phys. Chem. Solids* **13**, 156 (1960). X,3.
61 A. M. CLOGSTON, H. J. WILLIAMS, E. CORENZWIT, and R. C. SHERWOOD. *Phys. Rev. Letters* **7**, 7 (1961). X,3.
61a V. B. COMPTON, and E. CORENZWIT. *Phys. Chem. Solids* **19**, 130 (1961). X,3.
61b *Rev. Mod. Phys.* **33**, 499 (1961). X,3.
62 *IBM J. Res. Develop.* **6**, 250 (1962). X,3.
62a T. H. GEBALLE, V. B. COMPTON, E. CORENZWIT, and G. W. HULL, JR. *Phys. Rev.* **128**, 588 (1962). X,3.
63 T. H. GEBALLE, and V. B. COMPTON. *Rev. Mod. Phys.* **35**, 1 (1963). I,10; V,6; X,3.
63a *Phys. Today* **16**, 21 (1963). V,6; X,3.
63b T. H. GEBALLE, E. CORENZWIT, and G. W. HULL, JR. *Phys. Rev.* **129**, 1025 (1963). X,3.
[**See also** BLAUGHER, R. D., 61a; BLUMBERG, W. E., 60; COMPTON, V. B., 61; GEBALLE, T. H., 61, 62, 63; GIORGI, A. L., 62; HEIN, R. A., 59, 62; HULM, J. K., 61; KUNZLER, J. E., 61b; SUHL, H., 58, 59, 59a, 59b, 62; WERNICK, J. H., 61]

Mattis, D. C.
58 and J. BARDEEN. *Phys. Rev.* **111**, 412 (1958). VI,7; VIII,7.
61 and E. LIEB. *J. Math. Phys.* **2**, 602 (1961). V,2.
62 *IBM J. Res. Develop.* **6**, 258 (1962). VIII,7.

Maxwell, E.
50 *Phys. Rev.* **78**, 477 (1950). I,10.

May, R. M.
59 and M. R. SCHAFROTH. *Phys. Rev.* **115**, 1446 (1959). VIII,6.
59a and M. R. SCHAFROTH. *Proc. Phys. Soc.* **74**, 153 (1959). X,2.
59b *Phys. Rev.* **115**, 254 (1959). X,4.
59c *Progr. Theoret. Phys. (Kyoto)* **22**, 12 (1959). III,8.
63 and M. STRONGIN. *Phys. Rev. Letters* **10**, 212 (1963). X,2.

Mazur, P.
59 *Rendi. Scuola Intern. Fis. "Enrico Fermi"* **10**, 167 (Varenna, 1959). IX,2,10.
60 and E. MONTROLL. *J. Math. Phys.* **1**, 70 (1960). IX, 10.
61 *Proc. Intern. Summer Course on Fundamental Problems in Statistical Mechanics, Nijenrode, Netherlands*, 1961. IX,2,10.
[**See also** DE GROOT, S. R., 62]

Meeron, E.
58 *Phys. Fluids* **1**, 246 (1958). VII,1.
60 *J. Math. Phys.* **1**, 192 (1960). VII,1.

Meijer, P. H. E.
62 T. TANAKA, and J. BARRY. *J. Math. Phys.* **3**, 793 (1962). VI,3.

Meissner, H.
55 *Phys. Rev.* **97**, 1627 (1955). X,2.
56 *Phys. Rev.* **101**, 31 (1956). X,2.
56a *Phys. Rev.* **103**, 39 (1956). X,2.
58 *Phys. Rev.* **109**, 686 (1958). X,4.
58a *Phys. Rev.* **109**, 668 (1958). X,2.
58b and R. ZDANIS. *Phys. Rev.* **109**, 681 (1958). X,2.
58c *Phys. Rev.* **109**, 1479 (1958). X,2.
59 *Phys. Rev. Letters* **2**, 458 (1959). X,4.
59a *Phys. Rev.* **113**, 1183 (1959). VI,5.
60 *Phys. Rev.* **117**, 672 (1960). X,4.
62 *IBM J. Res. Develop.* **6**, 71 (1962). X,4.
[**See also** MEISSNER, W., 51, 51a, 52, 53]

Meissner, W.
33 and R. OCHSENFELD. *Naturwiss.* **21**, 787 (1933). I,2.
48 and G. U. SCHUBERT. *Phys. and Chem. Solids* **2**, 143 (1948). I,10.
51 F. SCHMEISSNER, and H. MEISSNER. *Z. Physik* **130**, 521 (1951). X,2.
51a F. SCHMEISSNER, and H. MEISSNER. *Z. Physik* **130**, 529 (1951). X,2.
52 F. SCHMEISSNER, and H. MEISSNER. *Z. Physik* **132**, 529 (1952). X,2.
53 F. SCHMEISSNER, and H. MEISSNER. *Phys. Rev.* **90**, 709 (1953). X,2.
55 and R. DOLL. *Z. Physik* **140**, 340 (1955). X,2.
59 and R. DOLL. *Z. Physik* **154**, 524 (1959); **156**, 488 (1959), VI,5.

Meixner, J.
41 *Ann. Physik* **39**, 333 (1941). IX,2.
42 *Ann. Physik* **41**, 409 (1942). IX,2.

43 *Ann. Physik* **43**, 244 (1943). IX,2.
43a *Z. Physik. Chem. (Frankfort)* **B53**, 235 (1943). IX,2.
59 and H. G. REIK, "Encyclopaedia of Physics," Vol. III, Part 2. Julius Springer, Berlin, 1959. IX,2,10.
63 *J. Math. Phys.* **4**, 154 (1963). IX,2.

Mendelssohn, K.
35 and J. R. MOORE. *Proc. Roy. Soc.* **A151**, 334 (1935); **A152**, 34 (1935). X,2.
48 *Rept. Progr. Phys.* **12**, 270 (1948). I,10.
50 and J. L. OLSEN. *Proc. Phys. Soc.* **A63**, 2 (1950). X,2.
50a and J. L. OLSEN. *Phys. Rev.* **80**, 859 (1950). X,2.
52 C. F. C. SQUIRE, and T. S. TEASDALE. *Phys. Rev.* **87**, 589 (1952). X,2.
55 *Progr. Low Temp. Phys.* **1**, Chapter X (1955). I,10.
58 *Physica, Suppl.* **24**, 53 (1958). VI,7.
60 "Cryophysics," Chapter 6. Wiley (Interscience), New York, 1960. I,10.
60a and C. A. SHIFFMAN. *Proc. Roy. Soc.* **A255**, 199 (1960). VI,7; X,2.
62 *IBM J. Res. Develop.* **6**, 27 (1962). "Exp. Work on Supercon." at Oxford. VI,7.
[**See also** CHAUDHURI, K. D., 60; CONNOLLY, A., 62; DAUNT, J. G., 46]

Mercereau, J. E.
62 and T. K. HUNT. *Phys. Rev. Letters* **8**, 243 (1962). IX,6, 10.

Mermin, N. D.
63 *Ann. Phys. (NY)* **21**, 99 (1963). VII,1.

Meshkovsky, A.
47 and A. SHALNIKOV. *J. Phys. (USSR)* **11**, 1 (1947). I,5; X,2.
47a and A. SHALNIKOV. *J. Exptl. Theoret. Phys. (USSR)* **17**, 851 (1947). I,5; X,2.

Migdal, A. B.
58 and V. M. GALITSKII. *J. Exptl. Theoret. Phys. (USSR)* **34**, 139 (1958) [English translation: *Soviet Phys.—JETP* **7**, 96 (1958)]. V,6.
[**See also** GALITSKII, V. M., 58a]
58a *J. Exptl. Theoret. Phys. (USSR)* **34**, 1438 (1958). [English translation: *Soviet Phys.—JETP* **7**, 996 (1958)]. VI, 7.
59 *Nucl. Phys.* **13**, 655 (1959). III,9; V,6.

Mikhailov, I. N.
63 Partially Projected Functions in a Superfluid Nuclear Model. Preprint P-1187 Dubna (1963). III,9.

Mikhailov, Yu. G.
60 E. R. MIKULIN, N. M., REINOV, and A. P. SMIRNOV. *Soviet Phys.—Tech. Phys.* **4**, 844 (1960). X,4.
[**See also** KOLCHIN, A. M., 61]

Miller, P. B.
59 *Phys. Rev.* **113**, 1209 (1959). VIII,7.
60 *Phys. Rev.* **118**, 928 (1960). VI,7; VIII,7; X,3.
61 *Phys. Rev.* **121**, 435 (1961). VI,7.

Milne, J. G. C.
61 *Phys. Rev.* **122**, 387 (1961). I,1.

Minnigerode, G. von
59 *Z. Physik* **154**, 442 (1959). X,2.
61 *Naturwissenschaften* **48**, 636 (1961). VI,7; X,4.

Misener, A. D.
35 and J. O. Wilhelm. *Trans. Roy. Soc. Can. Sect. III* **29**, 5 (1935). X,4.

Mittelstaedt, P.
61 *Nucl. Phys.* **25**, 522 (1961). V,2.
62 *Z. Physik* **167**, 439 (1962). V,2; VII,1.

Mitter, H.
61 and K. Yamazaki. *Nucl. Phys.* **30**, 683 (1961). V,6.

Montroll, E. W.
63 R. B. Potts, and J. C. Ward, *J. Math. Physics* **4**, 308 (1963). VI,3.
[**See also** Mazur, P., 60]

Morel, P.
58 *Phys. Rev. Letters* **1**, 244 (1958). V,6.
59 *Phys. Chem. Solids* **10**, 277 (1959). VI,5.
62 and P. W. Anderson. *Phys. Rev.* **125**, 1263 (1962). IV,3, 4; V,6; VI,4; VII,3; X,3.
[**See also** Anderson, P. W., 61; Brueckner, K. A., 60]

Mori, H.
62 *Progr. Theoret. Phys.* (*Kyoto*) **28**, 763 (1962). VIII,6.
62a I. Oppenheim, and J. Ross. "Studies in Statistical Mechanics," Vol. I, p. 213. North-Holland, Amsterdam, 1962. IX,2.
[**See also** Kawasaki, K., 62]

Morin, F. J.
62 J. P. Maita, H. J. Williams, R. C. Sherwood, J. H. Wernick, and J. E. Kunzler. *Phys. Rev. Letters* **8**, 275 (1962). X,2.
63 and J. P. Maita. *Phys. Rev.* **129**, 1115 (1963). X,3.
[**See also** Compton, V. B., 61]

Morita, T.
60 and K. Hiroike. *Progr. Theoret. Phys.* (*Kyoto*) **23**, 1003 (1960). VII,1.
61 and K. Hiroike. *Progr. Theoret. Phys.* (*Kyoto*) **25**, 537 (1961). VII,1.

Morris, D. E.
61 and M. Tinkham. *Phys. Rev. Letters* **6**, 600 (1961). VI,7; X,4.

Morse, R. W.
57 and H. V. Bohm. *Phys. Rev.* **108**, 1094 (1957). VI,7.
59 *Progr. Cryog.* **1**, 220 (1959). VI,7.
59a T. Olsen, and J. D. Gavenda. *Phys. Rev. Letters* **3**, 15 (1959); Erratum, *ibid.* **4**, 193 (1959). VI,7.
59b and H. V. Bohm. *J. Acoust. Soc. Am.* **31**, 1523 (1959). VI,7.
62 *IBM J. Res. Develop.* **6**, 58 (1962). VI,7.

Moskalenko, V. A.
58 *Soviet Phys. Doklady* **3**, 1171 (1958). V,5; VI,4; VII,5.
59 *Fiz. Metal. i Metalloved.* **8**, 503 (1959). V,6.

Mottelson, B. R.
60 and J. G. Valatin. *Phys. Rev. Letters* **5**, 511 (1960). III,9.
[**See also** Bohr, A., 58, 62]

Mould, R. E.
62 and D. E. Mapother. *Phys. Rev.* **125**, 33 (1962). X,2.

Mühlschlegel, B.
59 *Z. Physik* **155**, 313 (1959). VI,4, 5.
59a *Z. Physik* **156**, 235 (1959). V,2.
60 *Sitzber. Math.-Naturw. Kl. Bayer. Akad. Wiss. Muenchen* **10**, Sect. 3 (1960). VII,1.
62 *J. Math. Phys.* **3**, 522 (1962). VI,3.
[**See also** KOPPE, H., 58]

Müller, C.
62 and E. SAUR. *Z. Physik* **170**, 154 (1962). X,2.

Müller, J.
59 *Helv. Phys. Acta* **32**, 141 (1959). X,3.
60 and M. RISI. *Helv. Phys. Acta* **33**, 459 (1960). X,3.
[**See also** BUCHER, E., 61, 61a, 61b; GUGGENHEIM, J., 61; JAGGI, R., 58]

Näbauer, M.
58 *Z. Physik* **152**, 328 (1958). VI,5.
61 *Z. Physik* **163**, 119 (1961). VI,5; X,4.
[**See also** DOLL, R., 61]

Nagasaki, M.
63 Critical Temperature of Nuclear Matter. Preprint Rikkyo Univ. Tokyo (December 1962), III,9.

Nakajima, S.
59 *Progr. Theoret. Phys. (Kyoto)* **21**, 357 (1959). VIII,6.
59a *Progr. Theoret. Phys. (Kyoto)* **22**, 430 (1959). VIII,6.
59b *Progr. Theoret. Phys. (Kyoto)* **21**, 358 (1959). IX,9.
63 and M. WATABE. *Progr. Theoret. Phys. (Kyoto)* **29**, 341 (1963). I,10.

Nakamura, K.
59 *Progr. Theoret. Phys. (Kyoto)* **22**, 156 (1959). X,3.
59a *Progr. Theoret. Phys. (Kyoto)* **21**, 435 (1959). X,3.

Nambu, Y.
60 *Phys. Rev.* **117**, 648 (1960). V,6; VIII,1, 6, 7.
62 and S. F. TUAN. *Phys. Rev.* **128**, 2622 (1962). VIII,7; X,4.

Nesbet, R. K.
62 *Phys. Rev.* **126**, 2014 (1962). I,10.
62a *Phys. Rev.* **128**, 139 (1962). I,10.

Nethercot, A. H., Jr.
61 *Phys. Rev. Letters* **7**, 226 (1961). X,4.

Neumann, J. von
32 "Mathematische Grundlagen der Quantenmechanik." Springer, Berlin, 1932. A,1,6.

Newell, G. F.
50 *Phys. Rev.* **79**, 876 (1950). VI,3.

Newhouse, V. L.
59 and J. W. BREMER, *J. Appl. Phys.* **30**, 1458 (1959). X,4.
[**See also** BREMER, A. J., 58, 59; DESORBO, W., 62]

Nicol, J.
60 S. SHAPIRO, and P. H. SMITH. *Phys. Rev. Letters* **5**, 461 (1960). VI,7.
[**See also** SHAPIRO, S., 61, 62; SMITH, P. H., 61]

Nigam, B. P.
61 *Progr. Theoret. Phys. (Kyoto)* **25**, 436 (1961). V,6.

Nilsson, S. G.
61 and O. PRIOR. *Kgl. Danske Videnskab. Selskab, Mat. - Fys. Medd.* **32**, No. 16, (1961). III,9.

Ninham, B.
63 *Phys. Letters (Neth.)* **4**, 278 (1963). II,6.

Nishiyama, T.
62 Some Properties of Dielectric Constant and Phonon Frequency in a Superconductor. Tech. Rept No. 259, Univ. of Maryland, College Park, Maryland (June 1962). VIII,6.

Noer, R. J.
61 and W. D. KNIGHT. *Bull. Am. Phys. Soc.* **6**, 122 (1961). VIII,7.

Ogg, R. A.
46 *Phys. Rev.* **69**, 243, 544 (1946). III,1; IV,4.

Oliphant, T. A.
60 and W. TOBOCMAN. *Phys. Rev.* **119**, 502 (1960). VIII,6.

Oliver, D. J.
60 M. J. RAYNER, and E. H. RODERICK. *Nature* **187**, 492 (1960). X,4.

Olsen, J. L.
52 and C. A. RENTON. *Phil. Mag.* **43**, 946 (1952). X,2.
53 and H. M. ROSENBERG. *Advan. Phys.* **2**, 28 (1953). I,10.
58 *Rev. Sci. Instr.* **29**, 537 (1958). X, 4.
[**See also** ANDRES, K., 62; BUCHER, E., 61c; CORNISH, F. H. J., 53; DAUNT, J. G., 49, 62; GROSS, D., 60; MENDELSSOHN, K., 50, 50a; MORSE, R. W., 59a]

Onnes, H. K.
11 *Commun. Phys. Lab. Univ. Leiden* **12**, 119, 120, 122b, 124c (1911). I,6; IX,1.
14 *Leiden Comm.* 140b, 141b (1914). IX,10.

Onsager, L.
31 *Phys. Rev.* **37**, 405 (1931). VI,7.
32 *Phys. Rev.* **38**, 2265 (1932). VI,7.
32a and R. M. FUOSS. *J. Phys. Chem.* **36**, 2689 (1932). VI,7.
44 *Phys. Rev.* **65**, 117 (1944). VI,3.
53 and S. MACHLUP, *Phys. Rev.* **91**, 1505 (1953). IX,2.
61 *Phys. Rev. Letters* **7**, 50 (1961). I,7; III,1, 8; IX,5.
[**See also** KAUFMAN, B., 49a; PENROSE, O., 56; MACHLUP, S., 53]

Opitz, W.
55 *Z. Physik* **141**, 263 (1955). X,3.

Oppenheim, I.
57 and J. ROSS. *Phys. Rev.* **107**, 28 (1957). VI,3.
[**See also** MORI, H., 62a]

Osaka, Y.
62 *J. Phys. Soc. Japan* **17**, 547 (1962). I,9.

Otter, F. A.
62 and D. E. Mapother. *Phys. Rev.* **125**, 1171 (1962). X,2.

Pais, A.
63 *Proc. Natl. Acad. Sci. U. S. A.* **49**, 34 (1963). VI,3.

Pal, M. K.
63 and D. Mitra, *Nuclear Physics* **42**, 221 (1963). III,9.

Parikh, J. C.
62 *Phys. Rev.* **128**, 1530 (1962). IX,6.

Parkinson, D. H.
61 *Brit. J. Appl. Phys.* **12**, 353 (1961). X,4.

Parks, R. D.
62 *Bull. Am. Phys. Soc.* **7**, 322 (1962). X,3.
[**See also** Little, W. A., 62]

Parmenter, R. H.
59 *Phys. Rev.* **116**, 1390 (1959). X,2.
60 *Phys. Rev.* **118**, 1173 (1960). X,4.
61 *Phys. Rev. Letters* **7**, 274 (1961). VI,7.
62 *RCA Rev.* **23**, 323 (1962). X,4.

Parry, W. E.
62 and R. E. Turner. *Phys. Rev.* **128**, 929 (1962). II,6.

Pavlikovski, A.
62 and V. Rybarska. *J. Exptl. Theoret. Phys. (USSR)* **43**, 543 (1962) [English translation: *Soviet Phys.—JETP* **16**, 388 (1963). III,9.
[**See also** Volkov, M. K., 62]

Peierls, R. E.
36 *Proc. Roy. Soc.* **A155**, 613 (1936). X,2.
38 *Phys. Rev.* **54**, 918 (1938). A,6.
55 "Quantum Theory of Solids." Oxford Univ. Press, London and New York, 1955. I,9; II,5; IV,3; IX,1, 2.

Penrose, O.
51 *Phil. Mag.* **42**, 1373 (1951). II,6.
56 and L. Onsager. *Phys. Rev.* **104**, 576 (1956). II,6.

Peretti, J.
62 *Phys. Letters* **2**, 275 (1962). X,3.
[**See also** Maradudin, A., 59]

Peshkin, M.
61 I. Talmi, and L. J. Tassie, *Ann. Phys. (NY)* **12**, 426 (1961). IX,4, 5.
[**See also** Lipkin, H. J., 62; Tassie, L. J., 61]
62 and W. Tobocman. *Phys. Rev.* **127**, 1865 (1962). IX,5.

Peter, M.
58 *Phys. Rev.* **109**, 1857 (1958). I,3; VIII,7.
[**See also** Jaccarino, V., 62]

Pfennig, H.
59 *Z. Physik* **155**, 332 (1959). IX,2.

Phillips, J. C.

63 *Phys. Rev. Letters* **10**, 96 (1963). X,3.
[**See also** COHEN, M. H., 62; ROWELL, J. M., 62]

Phillips, N. E.

58 *Phys. Rev. Letters* **1**, 363 (1958). VI,5.
59 *Phys. Rev.* **114**, 676 (1959). VI,2, 4, 5.

Pines, D.

52 and D. BOHM. *Phys. Rev.* **85**, 338 (1952). I,9.
53 *Phys. Rev.* **92**, 626 (1953). I,9.
58 *Phys. Rev.* **109**, 280 (1958). IV,3; V,6; VII,3; X,3.
58a and J. R. SCHRIEFFER. *Nuovo Cimento* **10**, 496 (1958). VIII,6.
58b and J. R. SCHRIEFFER. *Phys. Rev. Letters* **1**, 407 (1958). VIII,6.
[**See also** BARDEEN, J., 55a; BOHM, D., 51, 53; BOHR, A., 58]

Pippard, A. B.

50 *Proc. Roy. Soc.* **A203**, 210 (1950). I,5.
50a *Phil. Mag.* **41**, 243 (1950). VI,5.
53 *Proc. Roy. Soc.* **A216**, 547 (1953). I,3, 5; II,5; VI,7.
53a *Nature* **172**, 896 (1953). I,3, 5.
54 *Advan. Electron. Electron Phys.* **6**, 1 (1954). I,10; VI,7.
58 *Physica, Suppl.* **24**, 48 (1958). V,2.
[**See also** FABER, T. E., 56, 55b; HEINE, V., 58; LAREDO, S. J., 55a]

Pitzer, K. S.

56 *Proc. Natl. Acad. Sci. U. S. A* **42**, 665 (1956). III,1.

Pokrovskii, V. L.

61 *J. Exptl. Theoret. Phys. (USSR)* **40**, 641 (1961) [English translation: *Soviet Phys.—JETP* **13**, 447 (1961)]. VI,5.
61a *J. Exptl. Theoret. Phys. (USSR)* **40**, 143 (1961) [English translation: *Soviet Phys.—JETP* **13**, 100 (1961)]. VI,7.
61b *J. Exptl. Theoret. Phys. (USSR)* **40**, 898 (1961) [English translation: *Soviet Phys.—JETP* **13**, 628 (1961)]. VI,7.
61c and V. A. TOPONOGOV. *J. Exptl. Theoret. Phys. (USSR)* **40**, 1112 (1961) [English translation: *Soviet Phys.—JETP* **13**, 785 (1961)]. VI,7.
61d and M. S. RYVKIN. *J. Exptl. Theoret. Phys. (USSR)* **40**, 1859 (1961) [English translation: *Soviet Phys.—JETP* **13**, 1306 (1961)]. VI,7.
62 and M. S. RYVKIN. *J. Exptl. Theoret. Phys. (USSR)* **43**, 92 (1962) [English translation: *Soviet Phys.—JETP* **16**, 67 (1963)]. VI,5.
62a and M. S. RYVKIN. *J. Exptl. Theoret. Phys. (USSR)* **43**, 900 (1962) [English translation: *Soviet Phys.—JETP* **16**, 639 (1963)]. VI,7.
62b and S. K. SAVVINYKH. *J. Exptl. Theoret. Phys. (USSR)* **43**, 564 (1962) [English translation: *Soviet Phys.—JETP* **16**, 404 (1963)]. VI,7.

Potts, R. B.

55 *Proc. Phys. Soc.* **A68**, 145 (1955). VI,3.
[**See also** MONTROLL, E. W., 63]

Prange, R. E.

61 *Nucl. Phys.* **28**, 369 (1961). I,3.
61a *Nucl. Phys.* **28**, 376 (1961). I,3.
62 Tunneling from a Many-Particle Point of View. Tech. Rept. No. 284, Univ. of Maryland, College Park, Maryland (December 1962). VI,7.

62a Dielectric Constant of a Superconductor. Tech. Rept. No. 266, Univ. of Maryland, College Park, Maryland (October 1962); *Phys. Rev.* **129**, 2495 (1963). VI,5; VIII,6.
[**See also** KLEIN, A., 58]

Prigogine, I.
47 "Étude thermodynamique des phénomènes irréversibles." Dunod, Paris and Desoer, Liège, Belgium, 1947. IX,2.
49 *Physica* **15**, 272 (1949). IX,2.
60 and F. HENIN. *J. Math. Phys.* **1**, 349 (1960). IX,10.
62 "Non-Equilibrium Statistical Mechanics." Interscience, New York, 1962. IX,2,10.

Privorotskii, I. A.
62 *J. Exptl. Theoret. Phys. (USSR)* **42**, 450 (1962) [English translation: *Soviet Phys.—JETP* **15**, 315 (1962)]. VI,7.
62a *J. Exptl. Theoret. Phys. (USSR)* **43**, 1331 (1962), [English translation: *Soviet Physics—JETP* **16**, 945 (1963)]. VI,7.

Prosperi, G. M.
60 and A. SCOTTI. *J. Math. Phys.* **1**, 218 (1960). IX,10.

Prozorova, L. A.
58 *J. Exptl. Theoret. Phys. (USSR)* **34**, 14 (1958) [English translation: *Soviet Phys.—JETP* **7**, 9 (1958)]. VI,7; X,2,4.

Quinn, D. J., III
61 and J. I. BUDNICK. *Phys. Rev.* **123**, 466 (1961). X,3.
62 and W. B. ITTNER, III. *J. Appl. Phys.* **33**, 748 (1962). IX,10.
[**See also** SERAPHIM, D. P., 61b]

Raetz, K.
62 and E. SAUR. *Z. Physik* **169**, 315 (1962). X,3.

Ranninger, J.
63 and W. THIRRING. *Z. Physik* **171**, 312 (1963). IX,2.

Rapoport, L. P.
62 and A. G. KRYLOVETSKII. *Dokl. Akad. Nauk SSSR* **145**, 771 (1962) [English translation: *Soviet Phys. Doklady* **7**, 703 (1963)]. X,2.

Read, W. T.
53 "Dislocations in Crystals." McGraw-Hill, New York, 1953. X,1.

Redfield, A. G.
59 *Phys. Rev. Letters* **3**, 85 (1959). VI,7.
[**See also** MASUDA, Y., 62]

Reeber, M. D.
60 *Phys. Rev.* **117**, 1476 (1960). X,3.

Reed, T. B.
62 and H. C. GATOS. *J. Appl. Phys.* **33**, 2657 (1962). X,3.
[**See also** BANUS, M. D., 62]

Reif, F.
57 *Phys. Rev.* **106**, 208 (1957). VIII,7.
62 and M. A. Woolf. *Phys. Rev. Letters* **9**, 315 (1962). X,3.

Renton, C. A.
55 *Phil. Mag.* **46**, 47 (1955). X,2.
[**See also** Olsen, J. L., 52]

Résibois, P.
59 *Physica* **25**, 725 (1959). IX,2.
63 *J. Math. Physics* **4**, 166 (1963). IX,2.

Reynolds, C. A.
50 B. Serin, W. H. Wright, and L. B. Neskitt. *Phys. Rev.* **78**, 487 (1950). I,10.

Rhoderick, E. H.
62 *Proc. Roy. Soc.* **A267**, 231 (1962). X,4.
62a and E. M. Wilson. *Nature* **194**, 1167 (1962). X,4.
[**See also** Broom, R. F., 59, 60, 62]

Rice, S.
60 and H. Frisch, *Annual Rev. Phys. Chem.* **11**, 187 (1960). IX,2.
61 J. G. Kirkwood, and R. Harris, *Physica* **26**, 717 (1961). IX,2.

Richards, P. L.
58 and M. Tinkham. *Phys. Rev. Letters* **1**, 318 (1958). I,8.
60 and M. Tinkham. *Phys. Rev.* **119**, 575 (1960). VI,7.
61 *Phys. Rev. Letters* **7**, 412 (1961). VI,7; X,3.
62 *Phys. Rev.* **126**, 912 (1962). I,5; VI,7.
[**See also** Ginsberg, D. M., 59]

Richardson, R. W.
63 *Phys. Letters* **3**, 277 (1963). V,2.

Rickayzen, G.
58 *Phys. Rev.* **111**, 817 (1958). VIII,6.
59 *Phys. Rev.* **115**, 795 (1959). V,6; VI,1; VIII,6.
59a *Phys. Rev. Letters* **2**, 90 (1959). VIII,6.
[**See also** Bardeen, J., 59a, 60]

Rogers, K. T.
60 Thesis. Univ. of Illinois, Urbana, Illinois (1960). IX,6.

Rohrer, H.
59 *Phil. Mag.* **4**, 1207 (1959). VI,5.
[**See also** Andres, K., 62; Dubeck, L., 63]

Roos, O. von
61 and J. S. Smuidzinas. *Phys. Rev.* **121**, 941 (1961). I,9.
62 *Phys. Rev.* **128**, 911 (1962). I,9.

Rose, R. M.
62 and J. Wulff. *J. Appl. Phys.* **33**, 2394 (1962). X,3.

Rose-Innes, A. C.
61 and B. Serin. *Phys. Rev. Letters* **7**, 278 (1961). X,4.
61a and G. M. Taylor. *Am. J. Phys.* **29**, 268 (1961). I,6.
[**See also** Calverley, A., 60]

Rothwarf, A.
63 and M. Cohen. *Phys. Rev.* **130**, 1401 (1963). VI,7.

Rowell, J. M.

62 A. G. Chynoweth, and J. C. Phillips. *Phys. Rev. Letters* **9**, 59 (1962). VI,7.

63 P. W. Anderson, and D. E. Thomas. *Phys. Rev. Letters* **10**, 334 (1963). V,6; VI,7.

[**See also** Anderson, A. C., 63]

Rowell, P. M.

60 *Proc. Roy. Soc.* **A254**, 542 (1960). VI,7.

Ruhl, W.

59 *Z. Physik* **157**, 247 (1959). X,4.

Rumer, Y. B.

60 *Soviet Phys.—JETP* **10**, 409 (1960). III,1.

Rushbrooke, G. S.

62 and H. I. Scoins. *J. Math. Phys.* **3**, 176 (1962). VI,3.

62*a* and J. Eve. *J. Math. Phys.* **3**, 185 (1962). VI,3.

Rutgers, A. J.

33 *Commun. Phys. Lab. Univ. Leiden Suppl.* **20**, No. 75b (1933). I,4; VI,5.

36 *Physica* **3**, 999 (1936). I,4.

Saenz, A. W.

55 and R. C. O'Rourke. *Rev. Mod. Phys.* **27**, 381 (1955). VI,3.

Sano, M.

63 and S. Yamasaki. *Progr. Theoret. Phys. (Kyoto)* **29**, 397 (1963). III,9.

Sarachik, M. P.

60 G. S. Garwin, and E. Erlbach. *Phys. Rev. Letters* **4**, 52 (1960). VIII,7.

[**See also** Erlbach, E., 60]

Satterthwaite, C. B.

62 *Phys. Rev.* **125**, 873 (1962). VI,7.

[**See also** Biondi, M. A., 58; Corak, W. S., 56]

Saur, E.

62 and P. Schult. *Z. Physik* **167**, 170 (1962). X,2.

62a and J. Wurm. *Naturwissenschaften* **49**, 127 (1962). X,2.

[**See also** Müller, C., 62; Raetz, K., 62]

Sawada, K.

57 *Phys. Rev.* **106**, 372 (1957). I,9.

57a K. A. Brueckner, N. Fukuda, and R. Brout. *Phys. Rev.* **108**, 507 (1957). I,9.

60 *Phys. Rev.* **119**, 2090 (1960). I,9.

Sawicki, J.

61 *Ann. Phys. (NY)* **13**, 237 (1961). V,6.

Schachenmeier, R.

51 *Z. Physik* **129**, 1 (1951). I,9.

51a *Z. Physik* **130**, 243 (1951). I,9.

Schafroth, M. R.

51 *Helv. Phys. Acta* **24**, 645 (1951). I,10; VI,3.

54 *Phys. Rev.* **96**, 1442 (1954). III,1, 3; IV,4; IX,5.

54a *Phys. Rev.* **96**, 1149 (1954). II,1.

55 *Phys. Rev.* **100**, 502 (1955). I,3.

55a *Phys. Rev.* **100**, 463 (1955). II,1, 4, 5; VIII,7; IX,9; X,4.

56 and J. M. BLATT. *Nuovo Cimento* **4**, 786 (1955). I,2.
57 S. T. BUTLER, and J. M. BLATT. *Helv. Phys. Acta* **30**, 93 (1957). III,1, 6, 7, 9.
58 *Phys. Rev.* **111**, 72 (1958). VIII,3.
60 *Solid State Phys.* **10**, 295 (1960). I,2, 10; IV,2; X,2.
[**See also** BLATT, J. M., 55; BUCKINGHAM, M. J., 54; MAY, R. M., 59,59a]

Schawlow, A. L.
56 *Phys. Rev.* **101**, 573 (1956). X,2.
58 and G. E. DEVLIN. *Phys. Rev.* **110**, 1011 (1958). X,2.
58a *Phys. Rev.* **109**, 1856 (1958). I,3; VIII,7; X,4.
59 and G. E. DEVLIN. *Phys. Rev.* **113**, 120 (1959). VIII,7.
59a G. E. DEVLIN, and J. K. HULM. *Phys. Rev.* **116**, 626 (1959). X,2.

Schiff, L. I.
49 "Quantum Mechanics." McGraw-Hill, New York, 1949. I,5; II,1 4; III,7; V,1, 3; IX,8.

Schindler, A. I.
63 and D. J. GILLESPIE. *Phys. Rev.* **130**, 953 (1963). VI,7.

Schirber, J. E.
61 and C. A. SWENSON. *Phys. Rev.* **123**, 1115 (1961). VI,5.
62 and C. A. SWENSON. *Phys. Rev.* **127**, 72 (1962). VI,5.

Schmitt, R. W.
61 *Phys. Today* **14**, 12, 38 (1961). VI,6.

Schrieffer, J. R.
59 *Phys. Rev. Letters* **3**, 323 (1959). VIII,7.
60 *Physica, Suppl.* **26**, 124 (1960). VI,1.
62 and D. M. GINSBERG. *Phys. Rev. Letters* **8**, 207 (1962). VI,7.
63 D. J. SCALAPINO, and J. W. WILKINS. *Phys. Rev. Letters* **10**, 336 (1963). V,6; VI,7.
63a and J. W. WILKINS. *Phys. Rev. Letters* **10**, 17 (1963). VI,7.
[**See also** BARDASIS, A., 61; BARDEEN, J., 57, 61; EISENSTEIN, J., 63; PINES, D., 58a, 58b]

Schröder, E.
57 *Z. Naturforsch.* **12a**, 247 (1957). X,3.

Schubnikow, L. W.
36 W. J. CHOTKEWITSCH, J. D. SCHEPELEW, and J. N. RJABININ. *Physik Z. Sowjetunion* **10**, 165 (1936). X,2.

Schwabl, F.
62 and W. THIRRING. *Nuovo Cimento* **25**, 175 (1962). II,4; IX,8.

Schwidtal, K.
60 *Z. Physik* **158**, 563 (1960). X,3.
62 *Z. Physik* **169**, 564 (1962). X,4.

Seraphim, D. P.
59 and R. A. CONNELL. *Phys. Rev.* **116**, 606 (1959). VI,5; X,2.
61 and P. M. MARCUS. *Phys. Rev. Letters* **6**, 680 (1961). VI,5.
61a D. T. NOVICK, and J. I. BUDNICK. *Acta Met.* **9**, 446 (1961). X,3.
61b C. CHIOU, and D. J. QUINN. *Acta Met.* **9**, 861 (1961). X, 3.
62 and P. M. MARCUS. *IBM J. Res. Develop.* **6**, 94 (1962). VI,5.
[**See also** BLUMBERG, R. H., 62; CHIOU, C., 63]

Serin, B.

55 *Progr. Low Temp. Phys.* **1**, Chapter 7 (1955). I,10; II,5.

56 "Handbuch der Physik" (S. Flügge, ed.), Vol. 15, p. 210 Springer, Berlin, 1956. I,6,10.

57 E. A. LYNTON, and M. ZUCKER. *Phys. Chem. Solids* **3**, 165 (1957). X,3.

59 E. A. LYNTON, and J. GITTLEMAN. *Helv. Phys. Acta* **32**, 138 (1959). X,2.

[**See also** BUDNICK, J. I., 56; CHANIN, G., 59; GAYLEY, R. I. JR., 62; KINSEL, T., 62, 63; LYNTON, E. A., 58; REYNOLDS, C. A., 50; ROSE-INNES, A. C., 61]

Sevast'yanov, B. K.

61 *J. Exptl. Theoret. Phys.* (*USSR*) **40**, 52 (1961) [English translation: *Soviet Phys.—JETP* **13**, 35 (1961)]. X,4.

62 *J. Exptl. Theoret. Phys.* (*USSR*) **42**, 1212 (1962) [English translation: *Soviet Phys.—JETP* **15**, 840 (1962)]. X,4.

Sewell, G. L.

59 *Proc. Phys. Soc.* **74**, 340 (1959). VIII,6.

Shalnikov, A. I.

45 *J. Phys. (USSR)* **9**, 202 (1945). I,5; X,2.

[**See also** MESHKOVSKY, A., 47, 47a]

Shannon, C. E.

48 *Bell System Tech. J.* **27**, 379 (1948). A,1.

Shapiro, S.

61 P. H. SMITH, J. L. MILES, and J. NICOL, *Phys. Rev. Letters* **6**, 686 (1961). X,4.

62 P. H. SMITH, J. NICOL, J. L. MILES, and P. F. STRONG. *IBM J. Res. Develop.* **6**, 34 (1962). VI,7.

[**See also** NICOL, J., 60; SMITH, P. H., 61]

Shapoval, E. A.

61 *J. Exptl. Theoret. Phys. (USSR)* **41**, 877 (1961) [English translation: *Soviet Phys.—JETP* **14**, 628 (1962)]. X,2.

Sharvin, Y. V.

51 *Zh. Eksptl. Teor. Fiz.* **21**, 658 (1951). I,5.

59 *Soviet Phys.—JETP* **11**, 216 (1960). X,2.

60 and V. F. GANTMAKHER. *Soviet Phys.—JETP* **11**, 1052 (1960). X,2.

Shaw, R. W.

60 and D. E. MAPOTHER. *Phys. Rev.* **118**, 1474 (1960). X,2.

60a D. E. MAPOTHER, and D. C. HOPKINS. *Phys. Rev.* **120**, 88 (1960). VI,5.

[**See also** DECKER, D. L., 58]

Sherrill, M. F.

61 and H. H. EDWARDS. *Phys. Rev. Letters* **6**, 460 (1961). VI,7.

[**See also** GLOVER, R. E., III, 60]

Shibuya, Y.

55 and S. TANUMA. *Phys. Rev.* **98**, 938 (1955). X,2.

58 *J. Phys. Soc. Japan* **13**, 684 (1958). X,2.

58a *J. Phys. Soc. Japan* **13**, 695 (1958). X,2.

Shiffman, C. A.

63 J. F. COCHRAN, and M. GARBER. *Bull. Am. Phys. Soc.* **8**, 307 (1963). VI,5.

[**See also** MENDELSSOHN, K., 60a]

Shirkov, D. V.
59 *Soviet Phys.—JETP* **9**, 421 (1959). V,6.
60 *Soviet Phys.—JETP* **10**, 127 (1960). V,6.
[**See also** BOGOLIUBOV, D. V., 58]

Shoenberg, D.
52 "Superconductivity." Cambridge Univ. Press, London and New York, 1952. I,10.
53 *Nuovo Cimento* [10], *Suppl.* **4**, 459 (1953). I,10.
60 "Superconductivity." Cambridge Univ. Press, London and New York, 1960. I,10; X,2.

Siegert, A. J. F.
52 *J. Chem. Phys.* **20**, 572 (1952). VI,3.

Silverberg, L.
63 and A. WINTHER. *Phys. Letters* **3**, 158 (1963). III,9.

Simmons, W. A.
62 and D. H. DOUGLAS, JR. *Phys. Rev. Letters* **9**, 153 (1962). X,4.

Slater, J. C.
37 *Phys. Rev.* **51**, 195 (1937). I,9.
37a Phys. Rev. **52**, 214 (1937). I,9.

Smith, P. H.
61 S. SHAPIRO, J. MILES, and J. NICOL, *Phys. Rev. Letters* **6**, 686 (1961). X,4.
[**See also** NICOL, J., 60; SHAPIRO, S., 61, 62]

Smith, T. S.
52 and J. G. DAUNT. *Phys. Rev.* **88**, 1172 (1952). X,2.

Soloviev, V. G.
58 *Soviet Phys. Doklady* **3**, 1176, 1179 (1958). III,9.
60 *Dokl. Akad. Nauk SSSR* **133**, 325 (1960) [English translation: *Soviet Phys. Doklady* **5**, 778 (1961)]. III,9.
61 *J. Exptl. Theoret. Phys.* (*USSR*) **40**, 654 (1961) [English translation: *Soviet Phys.—JETP* **13**, 456 (1961)]. III,9.
61a *Kgl. Danske Videnskab. Selskab, Mat.-Fys. Skrifter* **1**, 11 (1961). III,9.
61b *Izv. Akad. Nauk SSSR Ser. Fiz.* **25**, 1198 (1961). III,9.
62 *Phys. Letters* (*Neth.*) **1**, 202 (1962). III,9.
62a *Dokl. Akad. Nauk SSSR* **144**, 1281 (1962) [English translation: *Soviet Phys. Doklady* **7**, 548 (1962)]. III,9.
62b *J. Exptl. Theoret. Phys.* (*USSR*) **43**, 246 (1962) [English translation: *Soviet Phys.—JETP* **16**, 176 (1963)]. III,9.
[**See also** GALLAGHER, C. J., 62; VOLKOV, M. K., 62; VOROS, T., 62]

Sommerhalder, R.
61 and H. THOMAS. *Helv. Phys. Acta* **34**, 29 (1961). VIII,7.
61a and H. THOMAS. *Helv. Phys. Acta* **34**, 265 (1961). VIII,7.
[**See also** DRANGEID, K. E., 62; JAGGI, R., 58, 59, 60]

Spiewak, M.
59 *Phys. Rev.* **113**, 1479 (1959). VI,7.

Squire, C. F.
55 *Progr. Low-Temp. Phys.* **1**, Chapter 8 (1955). I,10; VI,5.
[**See also** MENDELSSOHN, K., 52; THOMPSON, J. C., 54]

Steele, M. C.
52 *Phys. Rev.* **87**, 1137 (1952). X,2.
53 and R. A. HEIN. *Phys. Rev.* **92**, 243 (1953). X,2.

Steiner, K.
37 and H. SCHOENECK. *Physik. Z.* **38**, 887 (1937). X,2.
49 *Z. Naturforsch.* **4a**, 271 (1949). X,2.

Stephen, M. J.
62 *Proc. Phys. Soc.* **79**, 994 (1962). VIII,7.

Strässler, S.
63 and P. WYDER. *Phys. Rev. Letters* **10**, 225 (1963). X,2.

Stromberg, T. F.
62 and C. A. SWENSON. *Phys. Rev. Letters* **9**, 370 (1962). I,5; X,4.

Strongin, M.
62 G. O. ZIMMERMAN, and H. A. FAIRBANK. *Phys. Rev.* **128**, 1983 (1962). V,6.
[**See also** MAY, R. M., 63]

Stumpf, H.
56 *Z. Naturforsch.* **11a**, 259 (1956). I,10.

Sturge, M. D.
58 *Proc. Roy. Soc.* **A246**, 570 (1958). I,3.

Suffczynski, M.
62 *Phys. Rev.* **128**, 1538 (1962). V,6.

Suhl, H.
58 B. T. MATTHIAS, E. CORENZWIT, and W. H. ZACHARIASEN, *Phys. Rev.* **112**, 89 (1958). X,3.
59 and B. T. MATTHIAS. *Phys. Rev.* **114**, 977 (1959). X,3.
59a B. T. MATTHIAS, and L. R. WALKER. *Phys. Rev. Letters* **3**, 552, (1959). V,6; X,3.
59b and B. T. MATTHIAS. *Phys. Rev. Letters* **2**, 5 (1959). X,3.
61 and N. R. WERTHAMER. *Phys. Rev.* **122**, 359 (1961). I,9.
62 D. R. FREDKIN, J. S. LANGER, and B. T. MATTHIAS, *Phys. Rev. Letters* **9**, 63 (1962). X,3.
62a *J. Phys. Soc. Japan* **17**, Suppl. B-1, 106 (1962). X,3.
63 and D. R. FREDKIN. *Phys. Rev. Letters* **10**, 131 (1963). X,3.
[**See also** ANDERSON, P. W., 59b; BROUT, R., 59; HERRING, C., 58; MATTHIAS, B. T., 58, 58a, 59, 60, 60a; WERTHAMER, N. R., 62]

Suzuki, R.
60 and M. AKANO. *Nuovo Cimento* **16**, 570 (1960). IX,6.
61 and M. AKANO. *Nuovo Cimento* **21**, 559 (1961). VIII,7.

Swartz, P. S.
62 and C. H. ROSNER. *J. Appl. Phys.* **33**, 2292 (1962). X,2.
62a *Phys. Rev. Letters* **9**, 448 (1962). X,2.

Swenson, C. A.
62 *IBM* J. *Res. Develop.* **6**, 82 (1962). VI,5.
[**See also** HINRICHS, C. H., 61; JENNINGS, L. D., 58; SCHIRBER, J. E., 61, 62; STROMBERG, T. F., 62]

Swihart, J. C.
59 *Phys. Rev.* **116**, 45 (1959). V,6; VI, 4; X,3.
59a *Phys. Rev.* **116**, 346 (1959). VI,4.

62 *IBM J. Res. Develop.* **6**, 14 (1962). V,6; X,3.
62a Private communication (1962). V,6.
62b Theory of the Phase Transition in a Cylindrical Superconductor. Research Paper RC-718, IBM Watson Research Center, Yorktown Heights, New York, (July 1962). VI,5.

Sykes, M. F.
61 *J. Math. Phys.* **2**, 52 (1961). VI,3.
[**See also** DOMB, C., 61]

Syozi, I.
51 *Progr. Theoret. Phys.* (*Kyoto*) **6**, 306 (1951). VI,3.
[**See also** HUSIMI, K., 50]

Tamura, T.
61 and T. UDAGAWA. *Progr. Theoret. Phys.* (*Kyoto*) **26**, 947 (1961). III,9.

Tassie, L. J.
61 and M. PESHKIN. *Ann. Phys.* (*NY*) **16**, 177 (1961). IX,4, 5.
[**See also** LIPKIN, H. J., 62; PESHKIN, M., 61]

Taylor, A. W. B.
61 *Proc. Phys. Soc.* **78**, 1372 (1962). VIII,6.

Taylor, B. N.
63 and E. BURSTEIN. *Phys. Rev. Letters* **10**, 14 (1963). VI,7.
[**See also** BURSTEIN, E., 61]

Teasdale, T. S.
53 and H. E. RORSCHACH, JR. *Phys. Rev.* **90**, 709 (1953). X,2.
[**See also** MENDELSSOHN, K., 52]

Terasawa, T.
62 *Nucl. Phys.* **39**, 563 (1962). III,9.

Ter Haar, D.
54 "Elements of Statistical Mechanics." Rinehart, New York, 1954. III,4; A,1.
[**See also** KENWORTHY, D. J., 61, 61a, 62]

Tewordt L.
60 *Z. Naturforsch.* **15a**, 490 (1960). V,6.
62 *Phys. Rev.* **127**, 371 (1962). VI,7.
62a *Phys. Rev.* **128**, 12 (1962). VI,7.
63 *Phys. Rev.* **129**, 657 (1963). VI,7.
[**See also** BARDEEN, J., 59a]

Thompson, C. J.
63 and T. MATSUBARA. *Progr. Theoret. Phys.* (*Kyoto*) **29**, 494 (1963). VII,4; B,4,5.
[**See also** MATSUBARA, T., 63]
63a and J. M. BLATT. *Phys. Letters* **5**, 6 (1963). X,4.
63b and J. M. BLATT. *J. Australian Math. Soc.* in press (1963) B,5.
[**See also** BLATT, J. M., 63]

Thompson, J. C.
54 and C. F. SQUIRE. *Phys. Rev.* **96**, 287 (1954). X,2.

Thouless, D. J.
60 *Ann. Phys.* (*NY*) **10**, 553 (1960). V,6.

60a *Phys. Rev.* **117**, 1256 (1960). V,2.
60b *Nucl. Phys.* **22**, 78 (1960). III,9.
61 and D. R. TILLEY. *Proc. Phys. Soc.* **77**, 1175 (1961); Erratum, *ibid.* **80**, 320 (1962). I,6.
61a "The Quantum Mechanics of Many-Body Systems." Academic Press, New York, 1961. I,10.

Tien, P. K.
63 and J. P. GORDON. *Phys. Rev.* **129**, 647 (1963). VI,7.

Tinkham, M.
56 *Phys. Rev.* **104**, 845 (1956). V,2.
58 *Phys. Rev.* **110**, 26 (1958). V,2.
58a *Physica, Suppl.* **24**, 35 (1958). I,8; VI,7.
58b and R. E. GLOVER, III. *Phys. Rev.* **110**, 778 (1958). I,8; VI,7.
59 and R. A. FERRELL. *Phys. Rev. Letters* **2**, 331 (1959). VI,7.
62 *IBM J. Res. Develop.* **6**, 49 (1962). X,4.
63 *Phys. Rev.* **129**, 2413 (1963). X,4.
[**See also** GINSBERG, D. M., 59, 60; GLOVER, R. E., III,56, 57; MORRIS, D. E., 61; RICHARDS, P. L., 58, 60]

Tisza, L.
38 *Compt. Rend.* **206**, 1035 (1938). II,3.
38a *Compt. Rend.* **206**, 1186 (1938). II,3.
40 *J. Phys. Radium* **1**, 165 (1940). II,3.
40a *J. Phys. Radium* **1**, 350 (1940). II,3.
47 *Phys. Rev.* **72**, 838 (1947). II,3.
49 *Phys. Rev.* **75**, 885 (1949). II,3.
50 *Phys. Rev.* **80**, 717 (1950). I,9.
51 *Phys. Rev.* **84**, 163 (1951). I,9.
52 *Natl. Bur. Std. (U.S.) Circ.* No. 519 (1952). I,9.

Tolmachev, V. V.
58 and S. V. TIABLIKOV. *Soviet Phys.—JETP* **7**, 46 (1958). V,3.
61 *Dokl. Akad. Nauk SSSR* **140**, 563 (1961) [English translation: *Soviet Phys.—Doklady* **6**, 800 (1962)]. V,6.
62 *Dokl. Akad. Nauk (SSSR)* **146**, 1312 (1962) [English translation: *Soviet Physics—Doklady* **7**, 924 (1963)]. V,6.
[**See also** BOGOLIUBOV, N. N., 58; KVASNIKOV, I. A., 58b]

Tolman, R. C.
38 "The Principles of Statistical Mechanics." Oxford Univ. Press, London, 1938. IX,10; A,3.

Tong, B. Y.
63 *Phys. Rev.* **130**, 1322 (1963). X,4.

Tonks, L.
29 and J. LANGMUIR. *Phys. Rev.* **33**, 195 (1929). I,9.

Townsend, P.
61 and J. SUTTON. *Proc. Phys. Soc.* **78**, 309 (1961). VI,7.
62 and J. SUTTON. *Phys. Rev.* **128**, 591 (1962). VI,7.

Toxen, A. M.
61 *Phys. Rev.* **123**, 442 (1961). X,4.
61a *Phys. Rev.* **124**, 1018 (1961). X,4.

62 *Phys. Rev.* **127**, 382 (1962). X,4.
62a G. K. CHANG, and R. E. JONES. *Phys. Rev.* **126**, 919 (1962). VI,7.
63 and M. J. BURNS. *Phys. Rev.* **130**, 1808 (1963). X,4.
[**See also** CHANG, G. K., 62; JONES, R. E., JR., 60]

Trimmer, J. D.
62 Hotel Management in Ergodia. *Phys. Today* **15**, 28 (1962). IX,10.

Tsuneto, T.
60 *Phys. Rev.* **118**, 1029 (1960). VI,7.
61 *Phys. Rev.* **121**, 402 (1961). VI,7.
62 *Progr. Theoret. Phys. (Kyoto)* **28**, 857 (1962). VI,7; VIII,7; X,3.
[**See also** MAKI, K., 62, 62a]

Uhlenbeck, G. E.
36 and E. BETH. *Physica* **3**, 729 (1936). III,5.
37 and E. BETH. *Physica* **4**, 915 (1937). III,5.
62 and G. W. FORD. "Studies in Statistical Mechanics," Vol. I, p. 119. North-Holland, Amsterdam, 1962. VI,3.
[**See also** KAHN, B., 38]

Ursell, H. D.
27 *Proc. Cambridge Phil. Soc.* **23**, 685 (1927). III,4.

Utiyama, T.
51 *Progr. Theoret. Phys. (Kyoto)* **6**, 907 (1951). VI,3.

Vaks, V. G.
61 V. M. GALITSKII, and A. I. LARKIN. *J. Exptl. Theoret. Phys. (USSR)* **41**, 1655 (1961) [English translation: *Soviet Phys.—JETP* **14**, 1177 (1962)]. VI,1

Valatin, J. G.
58 *Nuovo Cimento* **7**, 843 (1958). III,1; IV,1; V,3,4.
58a *Physica*, Suppl. **24**, 136 (1958). VI,1.
61 *Phys. Rev.* **122**, 1012 (1961). VIII,1,6.
[**See also** BRÉMOND, B., 63; MOTTELSON, B. R., 60]

Van Hove, L.
55 *Physica* **21**, 517 (1955). IX,10.
57 *Physica* **23**, 441 (1957). IX,10.
[**See also** JANNER, A., 62]

Van Leuwen, J.
59 J. GROENEVELD, and J. DE BOER. *Physica* **25**, 792 (1959). VII,1.

Velibekov, E. R.
62 *Physica* **28**, 711 (1962). I,6; VI,1.
62a *Dokl. Akad. Nauk SSSR* **142**, 1268 (1962) [English translation: *Soviet Phys. Doklady* **7**, 134 (1962)]. VI,1.
62b *Fiz. Tverd. Tela* **4**, 1752 (1962). VI, 1.

Verboven, E.
63 *J. Math. Physics* **4**, 266 (1963). IX,10.
[**See also** JANNER, A., 62]

Vick, F. A.
49 *Sci. Progr.* **37**, 268 (1949). I,10.

Vlasov, A.
50 "Theory of Many-Particle Systems." Gostekhizdat, 1950. VI,3.

Volger, J.
62 and P. S. Admiraal. *Phys. Letters (Neth.)* **2**, 257 (1962). X,4.
[**See also** Marchand, J. F., 62]

Volkov, M. K.
62 A. Pavlikovski, B. Rybarska, and V. G. Soloviev. Preprint E-1154, Dubna (1962). III,9.

Vonsovskii, S. V.
58 and M. S. Svirskii. *Soviet Phys. Doklady* **3**, 949 (1958). V,6; X,3.
58a and M. S. Svirskii. *Dokl. Akad. Nauk SSSR* **120**, 269 (1958) [English translation: *Soviet Phys. Doklady* **3**, 549 (1959)]. V,6; X,3.
60 and M. S. Svirskii. *Soviet Phys.—JETP* **10**, 1060 (1960). X,3.
61 and M. S. Svirskii. *Soviet Phys.—JETP* **12**, 272 (1961). V,6; X,3.

Voros, T.
62 V. G. Soloviev, and T. Siklos. *Izv. Akad. Nauk SSSR Ser. Fiz.* **26**, (1962). III,9.

Wada, Y.
58 F. Takano, and N. Fukuda. *Progr. Theoret. Phys. (Kyoto)* **19**, 597 (1958). V,2.
59 and N. Fukuda. *Progr. Theoret. Phys. (Kyoto)* **22**, 775 (1959). V,2; VIII,6.
61 *Progr. Theoret. Phys. (Kyoto)* **26**, 321 (1961). III,9.
[**See also** Fukuda, N., 59]

Wannier, G. H.
50 *Phys. Rev.* **79**, 357 (1950). VI,3.
[**See also** Kramers, H. A., 41, 41a]

Watson, G. N.
48 "Theory of Bessel Functions," 2nd ed. Cambridge Univ. Press, London and New York, 1948. IV,4; VI,2.

Webber, R. T.
53 and D. A. Spohr. *Phys. Rev.* **91**, 414 (1953). X,2.

Weiss, P. R.
58 and E. Abrahams. *Phys. Rev.* **111**, 722 (1958). VIII,7.

Weisskopf, V. F.
30 and E. Wigner. *Z. Physik* **63**, 54 (1930); **65**, 18 (1930). IX,2.
[**See also** Blatt, J. M., 52; Feshbach, H., 54]

Welker, H.
38 *Physik. Z.* **39**, 920 (1938). VIII,6.

Weller, W.
61 *Z. Naturforsch.* **16a**, 1240 (1961). VIII,1,6.
62 *Z. Naturforsch.* **17a**, 182 (1962). IX,8.
62a *Phys. Letters (Neth.)* **1**, 222 (1962); *Phys. Status Solidi* **2**, 1342 (1962). IX,8.

Wentzel, G.
51 *Phys. Rev.* **83**, 168 (1951). I,10; V,5, 6.
57 *Phys. Rev.* **108**, 1593 (1957). I,9.
58 *Phys. Rev.* **111**, 1488 (1958). VIII,6.
59 *Phys. Rev. Letters* **2**, 33 (1959). VIII,6.

60 *Helv. Phys. Acta* **33**, 859 (1960). VI,3; VII,5.
61 "Werner Heisenberg und die Physik unserer Zeit," p. 189. Vieweg, Braunschweig, 1961. IX,6.

Wernick, J. H.
61 and B. T. MATTHIAS. *J. Chem. Phys.* **34**, 2194 (1961). X,3.
[**See also** KUNZLER, J. E., 61a; MORIN, F. J., 62]

Werthamer, N. R.
62 and H. SUHL. *Phys. Rev.* **125**, 1402 (1962). I,9.
[**See also** BALIAN, R., 62; SUHL, H., 61]

Weyerer, H.
58 *Z. Naturforsch.* **13a**, 286 (1958). III, 1.

White, G. K.
62 *Phil. Mag.* **7**, 271 (1962). VI,5.

Wigner, E.
32 *Nachr. Ges. Wiss. Gottingen* **31**, 546 (1932). III,9; VI,7; IX,2.
33 and F. SEITZ. *Phys. Rev.* **43**, 804 (1933). I,9; IV,4.
34 and F. SEITZ. *Phys. Rev.* **46**, 509 (1934). I,9.
34a *Phys. Rev.* **46**, 1002 (1934). I,9; VIII,6.
38 *Trans. Faraday Soc.* **34**, 678 (1938). I,9; VIII,6.
[**See also** WEISSKOPF, V. F., 30]

Wild, W.
60 *Z. Physik* **158**, 322 (1960). V,6.
[**See also** MASCHKE, E. K., 63]

Williams, D. L.
62 *Proc. Roy. Soc.* **79**, 594 (1962). VI,7.

Willis, C. R.
62 *Phys. Rev.* **127**, 1405 (1962). IX,10.
62a and P. G. BERGMANN. *Phys. Rev.* **128**, 391 (1962). IX,10.

Wiser, N.
63 *Phys. Rev.* **129**, 62 (1963). V,6.

Wu, T. T.
61 *J. Math. Phys.* **2**, 105 (1961). II,6.

Wyman, L. L.
62 J. R. CUTHILL, G. A. MOORE, J. J. PARK, and H. YOKOWITZ, *J. Res. Natl. Bur. Std. (US)* **66A**, 351 (1962). X,3.

Yamamoto, T.
51 *Progr. Theoret. Phys. (Kyoto)* **6**, 533 (1951). VI,3.

Yamazaki, K.
61 *Nucl. Phys.* **23**, 139 (1961). V,6.
[**See also** MITTER, H., 61]

Yang, C. N.
62 *Rev. Mod. Phys.* **34**, 694 (1962). I,3; II,6; IX,5.
[**See also** LEE, T. D., 57, 57a, 58, 59, 59a, 59b, 60, 60a, 60b; BYERS, N., 61; HUANG, K., 57, 57a]

Yaqub, M.
60 *Cryogenics* **1**, 101 (1960). VI,5.
62 and C. Hohenemser, *Phys. Rev.* **127**, 2028 (1962). VI,5.

Yosida, K.
58 *Phys. Rev.* **111**, 1255 (1958). III,1; V,3, 4, 6; VI,1.
58a *Phys. Rev.* **110**, 769 (1958). VIII,7.
59 *Progr. Theoret. Phys. (Kyoto)* **21**, 731 (1959). VI,1; VIII,6.

Yoshida, S.
61 *Phys. Rev.* **123**, 2122 (1961). III,9.

Young, D. R.
59 *Progr. Cryogenics* **1**, 1 (1959). X,4.
61 *Brit. J. Appl. Phys.* **12**, 359 (1961). X,4.

Yvon, J.
35 *Actualites sci. et ind.* No. **203** (1935). VII,1.

Zavaritskii, N. V.
52 *Dokl. Akad. Nauk SSSR* **86**, 687 (1952). X,4.
58 *Soviet Phys.—JETP* **6**, 837 (1958); **7**, 773 (1958). VI,7.
60 *Soviet Phys.—JETP* **10**, 1069 (1960); **11**, 1207 (1960). VI,7.
61 *J. Exptl. Theoret. Phys. (USSR)* **41**, 657 (1961) [English translation: *Soviet Phys.—JETP* **14**, 470 (1962)]. VI,7.
62 *J. Exptl. Theoret. Phys. (USSR)* **43**, 1123 (1962) [English translation: *Soviet Phys.—JETP* **16**, 793 (1963)]. VI,7.

Zharkov, G. F.
58 *J. Exptl. Theoret. Phys. (USSR)* **34**, 412 (1958) [English translation: *Soviet Phys.—JETP* **7**, 286 (1958)]. X,3.
60 *Soviet Phys.—JETP* **10**, 1259 (1960). X,3.
62 *J. Exptl. Theoret. Phys. (USSR)* **42**, 1397 (1962) [English translation: *Soviet Phys.—JETP* **15**, 968 (1962)]. X,4.

Ziman, J. M.
55 *Proc. Cambridge Phil. Soc.* **51**, 707 (1955). I,10.
62 *Phys. Rev. Letters* **8**, 272 (1962). V,6.

Zubarev, D. N.
58 and I. A. Tserkovnikov, *Soviet Phys. Doklady* **3**, 603 (1958). II,6.
58a and I. A. Tserkovnikov, *Dokl. Akad. Nauk SSSR* **122**, 6 (1958) [English translation: *Soviet Phys. Doklady* **3**, 986 (1959)]. V,5; VI,4.
60 *Soviet Phys. Doklady* **5**, 570 (1960). V,5; VI,4.
[**See also** Bogoliubov, N. N., 57, 60a]

Zuckermann, M. J.
63 and D. M. Brink. *Phys. Letters* **4**, 76 (1963). VI,5.
[**See also** Kenworthy, D. J., 62]

Zumino, B.
62 *J. Math. Phys.* **3**, 1055 (1962). V,2; VIII,2.
[**See also** Keller, J. B., 61]

Subject Index

PURE AND APPLIED PHYSICS

A Series of Monographs and Textbooks

Consulting Editors

H. S. W. Massey
University College, London, England

Keith A. Brueckner
University of California, San Diego
La Jolla, California

1. F. H. Field and J. L. Franklin, Electron Impact Phenomena and the Properties of Gaseous Ions. (Revised edition, 1970.)
2. H. Kopfermann, Nuclear Moments, English Version Prepared from the Second German Edition by E. E. Schneider.
3. Walter E. Thirring, Principles of Quantum Electrodynamics. Translated from the German by J. Bernstein. With Corrections and Additions by Walter E. Thirring.
4. U. Fano and G. Racah, Irreducible Tensorial Sets.
5. E. P. Wigner, Group Theory and Its Application to the Quantum Mechanics of Atomic Spectra. Expanded and Improved Edition. Translated from the German by J. J. Griffin.
6. J. Irving and N. Mullineux, Mathematics in Physics and Engineering.
7. Karl F. Herzfeld and Theodore A. Litovitz, Absorption and Dispersion of Ultrasonic Waves.
8. Leon Brillouin, Wave Propagation and Group Velocity.
9. Fay Ajzenberg-Selove (ed.), Nuclear Spectroscopy. Parts A and B.
10. D. R. Bates (ed.), Quantum Theory. In three volumes.
11. D. J. Thouless, The Quantum Mechanics of Many-Body Systems.
12. W. S. C. Williams, An Introduction to Elementary Particles. (Second edition, 1971.)
13. D. R. Bates (ed.), Atomic and Molecular Processes.
14. Amos de-Shalit and Igal Talmi, Nuclear Shell Theory.
15. Walter H. Barkas. Nuclear Research Emulsions. Part I.
    Nuclear Research Emulsions. Part II. *In preparation*
16. Joseph Callaway, Energy Band Theory.
17. John M. Blatt, Theory of Superconductivity.
18. F. A. Kaempffer, Concepts in Quantum Mechanics.
19. R. E. Burgess (ed.), Fluctuation Phenomena in Solids.
20. J. M. Daniels, Oriented Nuclei: Polarized Targets and Beams.
21. R. H. Huddlestone and S. L. Leonard (eds.), Plasma Diagnostic Techniques.
22. Amnon Katz, Classical Mechanics, Quantum Mechanics, Field Theory.
23. Warren P. Mason, Crystal Physics in Interaction Processes.
24. F. A. Berezin, The Method of Second Quantization.
25. E. H. S. Burhop (ed.), High Energy Physics. In four volumes.

26. L. S. Rodberg and R. M. Thaler, Introduction to the Quantum Theory of Scattering.

27. R. P. Shutt (ed.), Bubble and Spark Chambers. In two volumes.

28. Geoffrey V. Marr, Photoionization Processes in Gases.

29. J. P. Davidson, Collective Models of the Nucleus.

30. Sydney Geltman, Topics in Atomic Collision Theory.

31. Eugene Feenberg, Theory of Quantum Fluids.

32. Robert T. Beyer and Stephen V. Letcher, Physical Ultrasonics.

33. S. Sugano, Y. Tanabe, and H. Kamimura, Multiplets of Transition-Metal Ions in Crystals.

34. Walter T. Grandy, Jr., Introduction to Electrodynamics and Radiation.

35. J. Killingbeck and G. H. A. Cole, Mathematical Techniques and Physical Applications.

In preparation

Herbert Uberall, Electron Scattering from Complex Nuclei. Parts A and B.